Sustainable Downstream Processing of Microalgae for Industrial Application

Sustainable Downstream Processing of Microalgae for Industrial Application

Edited by
Kalyan Gayen, Tridib Kumar Bhowmick
and Sunil K. Maity

CRC Press
Taylor & Francis Group
Boca Raton London New York Leiden

CRC Press is an imprint of the
Taylor & Francis Group, an **informa** business

CRC Press
Taylor & Francis Group
6000 Broken Sound Parkway NW, Suite 300
Boca Raton, FL 33487–2742

First issued in paperback 2023

ISBN 13: 978-0-367-13556-0 (hbk)
ISBN 13: 978-1-032-65369-3 (pbk)
ISBN 13: 978-0-429-02797-0 (ebk)

DOI: 10.1201/9780429027970

Library of Congress Cataloging-in-Publication Data

Names: Sivakumar Babu, G. L., author.
Title: Pavement drainage : theory and practice / G.L. Sivakumar Babu, Prithvi S. Kandhal, Nivedya Mandankara Kottayi, Rajib Basu Mallick, Amirthalingam Veeraragavan.
Description: Boca Raton : Taylor & Francis, a CRC title, part of the Taylor & Francis imprint, a member of the Taylor & Francis Group, the academic division of T&F Informa, plc, 2020. | Includes bibliographical references.
Identifiers: LCCN 2019017875 | ISBN 9780815353607 (hardback : acid-free paper)
Subjects: LCSH: Road drainage. | Runoff.
Classification: LCC TE215 .S485 2020 | DDC 625.7/34—dc23
LC record available at https://lccn.loc.gov/2019017875

Visit the Taylor & Francis Web site at
www.taylorandfrancis.com

and the CRC Press Web site at
www.crcpress.com

Contents

SECTION 1: Biomolecules from Microalgae for Commercial Production

SECTION 2: Sustainable Approach for Industrial-Scale Operations

SECTION 3:   *Optimized Downstream Processing of Microalgae*

Preface

Progressive depletion of fossil fuel reservoirs and adverse environmental impacts due to the use of fossil fuels are posing severe threats to our civilization. Therefore, the demand for the large-scale production of bio-fuels and bio-based products from renewable sources is growing worldwide for the sustainability of our society and maintaining cleanliness of the earth. In this regard, the microalgae (both microalgae and cyanobacteria) are under prime research focus for the production of bio-fuels due to their exorbitant photosynthetic efficiency and fast growth rate with high lipid content compared to territorial plants. However, the large-scale production of algal biomass and downstream processing of harvested algae is facing several challenges, particularly, the economic viability of the microalgal bio-fuels.

Apart from bio-fuels, the microalgae synthesize many bio-molecules such as pigments (e.g., chlorophyll, carotenoid), protein (e.g., lectin, phycobiliprotein), and carbohydrates (e.g., agar, carrageenan, alginate, fucodian), which are available in the various forms of microalgal products. Therefore, developing a strategy for large-scale production of bio-fuels using microalgal biomass with co-production of value-added macromolecules is essential. This strategy will eventually make both bio-based products and bioenergy economically viable.

With these obstacles in mind, without a doubt, there is a substantial scope for the improvement in the economics of microalgal bio-fuels and other value-added co-products. Hence, this book focuses on i) biomolecules from microalgae for commercial production, ii) a sustainable approach for industrial-scale operations, and iii) optimized downstream processing of microalgae. Each of these topics is presented as an individual section and composed of several chapters written by renounced academicians and industry experts from all over the globe. The individual chapters are rich in up-to-date research progress and provide comprehensive information about the past and current research in a particular field with a guideline for the upcoming investigators. It may be noted that in this edition, a significant emphasis has been given to the industry experts who have experience in dealing with the shortcomings of the microalgal products. The industry experts contributed around half of the book chapters.

We hope that scientific knowledge of the distinguished academicians and applied research information by industry experts will enrich the upcoming researchers and attract industries to implement a sustainable integrated biorefinery. This book provides the latest information related to the downstream processing of microalgae for sustainable production of commodity and specialty products, such as pigments, protein-rich food supplements, biopolymers, and bio-fuel. It also provides the roadmap for the sustainable industrialization of the microalgae-based value-added chemicals. Altogether, this book will boost microalgal research; provide economic benefit in the industry; and help students, emerging researchers, and academician update the knowledge base.

Kalyan Gayen

Tridib Kumar Bhowmick

Sunil K. Maity

Editor

Kalyan Gayen is an academician in the field of Chemical Engineering and is working at the National Institute of Technology Agartala, India. He received his PhD degree from the Indian Institute of Technology, Bombay. He did his post-doctoral research from the University of California. His current research interests are i) microalgae and cyanobacteria-based biofuels and bioproducts, ii) conversion of lignocellulosic biomass into liquid biofuels and value-added chemicals, ii) metabolic network analysis, iv) systems biology, and v) fermentation technology. He executed a number of externally funded projects. As a research outcome, he published more than thirty research articles and eight book chapters and presented at several national and international conferences. His recent publications are as follows:

Recent Publications

Publications in Reputed Journals

1. Natalia Garcia-Reyero, Edward J Perkins, **Kalyan Gayen**, Jason E Shoemaker, Philipp Antczak, Lyle Burgoon, Francesco Falciani, Steve Gutsell, Geoff Hodges, Aude Kienzler, Dries Knapen, Mary McBride, Catherine Willett, Francis J Doyle, "Chemical Hazard Prediction and Hypothesis Testing Using Quantitative Adverse Outcome Pathways", Alternatives to animal experimentation (ALTEX), 36 (1), 91–102 (2019).

2. Ashmita Ghosh, Saumyakanti Khanra, Gopinath Halder, Tridib Kumar Bhowmick and **Kalyan Gayen**, "Diverse cyanobacteria resource from north east region of India for valuable biomolecules: Phycobiliprotein, carotenoid, carbohydrate and lipid", *Current Biochemical Engineering*, 5, 1–13 (2018).

3. Dibyajyoti Haldar, **Kalyan Gayen** and Dwaipayan Sen, "Enumeration of monosugars' inhibition characteristics on the kinetics of enzymatic hydrolysis of cellulose", *Process Biochemistry*, 72, 130–136 (2018).

4. Dibyajyoti Haldar, Dwaipayan Sen and **Kalyan Gayen**, "Enzymatic hydrolysis of banana stems (Musa acuminata): Optimization of process parameters and inhibition characterization", *International Journal of Green Energy*, 15 (6), 406–413 (2018).

5. Saumyakanti Khanra, Madhumanti Mondal, Gopinath Halder, O N Tiwar, **Kalyan Gayen** and Tridib Kumar Bhowmick, "Downstream processing of microalgae for pigments, protein and carbohydrate in industrial application: A review", *Food and Bioproducts Processing*, 110, 60–84 (2018).

6. Debika Choudhury, **Kalyan Gayen** and Supreet Saini, "Dynamic control of arabinose and xylose utilization in E. Coli", *Canadian Journal of Chemical Engineering*, 96 (9), 1881–1887 (2018).

7. Madhumanti Mondal, Ashmita Ghosh, **Kalyan Gayen**, Gopinath Halder and O. N. Tiwari, "Carbon dioxide bio-fixation by Chlorella sp. BTA 9031 towards biomass and lipid production: Optimization using Central Composite Design approach", *Journal of Carbon Dioxide Utilization*, 22, 317–329 (2017)

8. Ashmita Ghosh, Saumyakanti Khanra, Madhumanti Mondal, Gopinath Halder, O N Tiwari, Tridib Kumar Bhowmick and **Kalyan Gayen**, "Effect of macronutrient supplements on growth and biochemical compositions in photoautotrophic cultivation of isolated Asterarcys sp. (BTA9034)", *Energy Conversion and Management*, 149, 39–51 (2017)

9. Ashmita Ghosh, Saumyakanti Khanra, Madhumanti Mondal, Thingujam Indrama Devi, Gopinath Halder, O N Tiwari, Tridib Kumar Bhowmick and **Kalyan Gayen**, "Biochemical characterization of microalgae collected from north east region of India advancing towards the algae based commercial production", *Asia-Pacific Journal of Chemical Engineering*, 12 (5), 745–754 (2017).

10. Madhumanti Mondal, S. Goswami, Ashmita Ghosh, G. Oinam, O. N. Tiwari, Papita Das, **Kalyan Gayen**, M. K. Mandal and G. N. Halder, "Production of biodiesel from microalgae through biological carbon capture: A review", *3 Biotech*, 7 (2), 99 (2017).

11. Madhumanti Mondal, Ashmita Ghosh, O. N. Tiwari, **Kalyan Gayen**, Papita Das, M. K. Mandal and G. N. Halder, "Influence of carbon sources and light intensity on biomass and lipid production of Chlorella sorokiniana BTA 9031 isolated from coalfield under various nutritional modes", *Energy Conversion and Management*, 145, 247–254 (2017).

12. Madhumanti Mondal, Ashmita Ghosh, Gunapati Oinam, O. N. Tiwari, **Kalyan Gayen** and G. N. Halder, "Biochemical Responses to Bicarbonate Supplementation on Biomass and Lipid Productivity of Chlorella Sp. BTA9031 Isolated from Coalmine Area", *Environmental Progress & Sustainable Energy*, 36 (5), 1498–1506 (2017).

13. Dibyajyoti Haldar, Dwaipayan Sen and **Kalyan Gayen**, "Development of spectrophotometric method for the analysis of multi-component carbohydrate mixture of different moiety", *Applied Biochemistry and Biotechnology*, 181, 4, 1416–1434 (2017).

14. Madhumanti Mondal, Ashmita Ghosh, Aribam S Sharma, O N Tiwari, **Kalyan Gayen**, Mrinal K Mandal and G. N. Halder, "Mixotrophic cultivation of Chlorella sp. BTA 9031 and Oocystis sp. BTA 9032 isolated from coal field using various carbon sources for biodiesel production", *Energy Conversion and Management*, 124, 297–304 (2016).

15. Madhumanti Mondal, Saumyakanti Khanra, O. N. Tiwari, **Kalyan Gayen** and G. N. Halder, "Role of Carbonic Anhydrase on the Way to Biological Carbon Capture through microalgae—A Mini Review",

Environmental Progress & Sustainable Energy, 35 (6), 1605–1615 (2016).

16. Dibyajyoti Haldar, Dwaipayan Sen and **Kalyan Gayen**, "A review on the production of fermentable sugars from lignocellulosic biomass through conventional and enzymatic route—A comparison", *International Journal of Green Energy*, 13 (12), 1232–1253 (2016).

17. Ashmita Ghosh, Saumyakanti Khanra, Madhumanti Mondal, Gopinath Halder, O N Tiwari, Supreet Saini, Tridib Kumar Bhowmick and **Kalyan Gayen**, "Progress towards isolation of strains and genetically engineered strains of microalgae for production of biofuel and other value added chemicals: A review", *Energy Conversion and Management*, 113, 104–118 (2016).

18. Manish Kumar, Supreet Saini, **Kalyan Gayen**, "Elementary mode analysis reveals that Clostridium acetobutylicum modulates its metabolic strategy under external stress", *Molecular BioSystems*, (Royal Society of Chemistry, UK), 10 (8), 2090–2105 (2014).

19. Manish Kumar, Supreet Saini, **Kalyan Gayen**, "Acetone-Butanol-Ethanol (ABE) fermentation analysis using only high performance liquid chromatography", *Analytical Methods* (Royal Society of Chemistry, UK), 6 (3), 774–781 (2014).

20. Meghna Rajvanshi, **Kalyan Gayen**, K. V. Venkatesh, "Lysine overproducing Corynebacterium glutamicum is characterized by a robust linear combination of two optimal phenotypic states", *Systems and Synthetic Biology*, 7, 51–62 (2013).

21. Manish Kumar, **Kalyan Gayen**, Supreet Saini, "Role of extracellular cues to trigger the metabolic phase shifting from acidogenesis to solventogenesis in Clostridium acetobutylicum", *Bioresource Technology*, 138, 55–62 (2013).

22. Yogesh Goyal, Manish Kumar, **Kalyan Gayen**, "Metabolic engineering for enhanced hydrogen production: A review", *Canadian Journal of Microbiology*, 59, 59–78 (2013).

23. Manish Kumar, Yogesh Goyal, Abhijit Sarkar, **Kalyan Gayen**, "Comparative economic assessment of ABE fermentation based on cellulosic and non-cellulosic feedstocks", *Applied Energy*, 93, 193–204, (2012).

24. Manish Kumar, **Kalyan Gayen**, "Developments in Bio-butanol production: New insights", *Applied Energy*, 88, 1999–2010 (2011).

25. **Kalyan Gayen**, Manish Kumar, Meghna Rajvanshi and K. V. Venkatesh, "Metabolic consequences of anaplerotic reactions of Corynebacterium glutamicum during growth on glucose and lactate through elementary mode", *IJBB*, 7, 1, 115–132 (2011).

26. Jason E Shoemaker, **Kalyan Gayen**, Natàlia Garcia-Reyero, Edward J. Perkins, Daniel L. Villeneuve, Li Liu, Francis J Doyle III, "Fathead Minnow Steroidogenesis: In silico analyses reveals tradeoffs between nominal target efficacy and robustness to cross-talk", *BMC Systems Biology*, 4, 89 (2010).

Book Chapters

1. Dibyajyoti Haldar, Mriganka Sekhar Manna, Dwaipayan Sen, Tridib Kumar Bhowmick, **Kalyan Gayen**, "Microbial fuel cell in wastewater treatment", *Enzymatic and Microbial Fuel Cells*, Editors: Inamuddin and Abdullah M. Asiri, Materials Research Forum (MRF), USA, 2018, In press.

2. Saumyakanti Khanra, **Kalyan Gayen**, Gopinath Halder, Tridib Kumar Bhowmick, Gunapati Oinam and O. N. Tiwari, "Lipid derived products from microalgae: Downstream processing for industrial application", *The Role of Photosynthetic Microbes in Agriculture and Industry*, Editors: Keshawan and Tripathi, Narendra Kumar and Gerard Abraham, Nova Science Publishers, Inc., New York, USA, 2018. ISBN: 978-1-53614-033-0.

3. Manish Kumar, Tridib Kumar Bhowmick, Supreet Saini and **Kalyan Gayen**, "Current status and challenges in biobutanol production", *Bioenergy and Biofuels*, Editor: Ozcan Konur, CRC Press, Taylor and Francis, 2017. ISBN-10: 1138032816. ISBN-13: 978-1138032811.

4. Ankita Mazumder, Sunil Maity, Dwaipayan Sen and **Kalyan Gayen**, "Process development for hydrolysate optimization from lignocellulosic biomass towards biofuel production", *Alcohols and Bio-alcohols: Characteristics, Production and Use*, Editors: Angelo Basile and Francesco Dalena, Nova Science Publishers, Inc., New York, USA, 2014. ISBN: 978-1-63463-187-7.

5. Manish Kumar and **Kalyan Gayen**, "Biobutanol: The future biofuel", *Biomass Conversion: The Interface of Biotechnology, Chemistry and Materials Science*, Editors: C. Baskar, S. Baskar and R. Dhillon, Springer-Verlag, Germany, New York and Japan, 2012. ISBN: 978-3-642-28417-5.

6. Theresa Yuraszeck, Peter Chang, **Kalyan Gayen**, Eric Kwei, Henry Mirsky, and Francis J. Doyle III, "Methods for in silico biology: Model construction and analysis", *Systems Biology in Drug Discovery and Development*, Editors: Daniel L. Young and Seth Michelson, John Wiley & Sons, Inc., 2011. ISBN: 978-1-118-01643-5.

Tridib Kumar Bhowmick is an assistant professor in the Department of Bioengineering at the National Institute of Technology Agartala, Tripura, India. He received his PhD degree from the Department of Chemical Engineering, Indian Institute of Technology, Bombay, India. During his PhD, he worked on Indian traditional medicine. He did his postdoctoral research in the Institute for Bioscience and Biotechnology Research, University of Maryland, College Park, Maryland, from 2008 to 2013. His research is focused on nanomaterials and targeted delivery of therapeutic molecules. He produced several items on nanomedicine. Currently, he is involved in exploring the northeast region of India, representing a biodiversity hotspot with endemic flora and fauna, to identify promising microalgal strains as a resource for alternative renewable energy production. His broad goal of interest is to understand the characteristics of biomaterials at nano-scale level and is looking forward their novel application possibilities. His goal in teaching students is to provide the basic foundations that will allow each student to develop his or her understanding of and capabilities in the area of biotechnology and bioengineering.

Recent Publications

Publications in Reputed Journals

1. **Tridib Bhowmick**, Akkihebbal K Suresh, Shantaram G Kane, Ajit Joshi, Jayesh R Bellare, "Physicochemical characterization of Indian traditional medicine, Jasada Bhasma: Detection of nanoparticles containing non-stoichiometric Zinc Oxide" *Journal of Nanoparticle Research*, 11 (3), 655–664 (2009).

2. **Tridib Bhowmick**, Akkihebbal K Suresh, Shantaram G Kane, Ajit Joshi, Jayesh R Bellare, "Nanoparticles containing non-stoichiometric ZnO in Jasada Bhasma and other Zinc containing compounds prevent free radicals mediated cellular damage in Saccharomyces cerevisiae", *International Journal of Green Nanotechnology: Biomedicine*, 1 (1), B69–B89 (2009).

3. **Tridib Bhowmick**, Diana Yoon, Minal Patel, John Fisher, Sheryl Ehrman, "In vitro toxicity of cisplatin functionalized silica nanoparticles on chondrocytes", *Journal of Nanoparticle Research*, 12 (8), 2757 (2010).

4. **Tridib Bhowmick**, Kishan Kumar, Yang Shen, Yuan Chia Kuo, Carmen Garnacho, Silvia Muro, Janet Hsu, Daniel Serrano, "Enhanced endothelial delivery and biochemical effects of α-Galactosidase by ICAM-1-targeted nanocarriers for Fabry disease", *Journal of Control Release*, 149 (3), 323–331 (2011).

5. **Tridib Bhowmick,** John Leferovich, Bahrat Burmann, Benjamin Pichette, Vladimir Muzykantov, David Eckmann, Silvia Muro, "Optimizing endothelial targeting by modulating the antibody density and particle concentration of Anti-ICAM coated carriers, Andres Calderon", *Journal of Control Release*, 150 (1), 37–44 (2011).

6. **Tridib Bhowmick**, Erik Berk, Xiumin Cui, Vladimir Muzykantov, Silvia Muro, "Effect of flow on endothelial endocytosis of nanocarriers targeted to ICAM-1", *Journal of Controlled Release*, 157 (3), February 10, 2012, 485–492 (2011).

7. Maria Ansar, Daniel Serrano, Iason Papademetriou, **Tridib Bhowmick**, Silvia Muro, "Biological functionalization of drug delivery carriers to bypass size restrictions of receptor-mediated endocytosis independently from receptor targeting", *ACS Nano*, 12 (3), 10597–10611 (2013).

8. Ashmita Ghosh, Saumyakanti Khanra, Madhumanti Mondal, Gopinath Halder, O N Tiwari, Supreet Saini, **Tridib Kumar Bhowmick** and Kalyan Gayen, "Progress towards isolation of strains and genetically engineered strains of microalgae for production of biofuel and other value added chemicals: A review", *Energy Conversion and Management*, 113, 104–118 (2016).

9. Ashmita Ghosh, Saumyakanti Khanra, Madhumanti Mondal, Gopinath Halder, O N Tiwari, **Tridib Kumar Bhowmick** and Kalyan Gayen, "Effect of macronutrient supplements on growth and biochemical compositions in photoautotrophic cultivation of isolated Asterarcys sp. (BTA9034)", *Energy Conversion and Management*, 149, 39–51 (2017).

10. Ashmita Ghosh, Saumyakanti Khanra, Madhumanti Mondal, Thingujam Indrama Devi, Gopinath Halder, O N Tiwari, **Tridib Kumar Bhowmick** and Kalyan Gayen, "Biochemical characterization of microalgae collected

from north east region of India advancing towards the algae based commercial production", *Asia-Pacific Journal of Chemical Engineering*, 12 (5), 745–754 (2017).

11. Saumyakanti Khanra, Madhumanti Mondal, Gopinath Halder, O N Tiwar, Kalyan Gayen and **Tridib Kumar Bhowmick**, "Downstream processing of microalgae for pigments, protein and carbohydrate in industrial application: A review", *Food and Bioproducts Processing*, Online (2018).

12. Ashmita Ghosh, Saumyakanti Khanra, Gopinath Halder, **Tridib Kumar Bhowmick** and Kalyan Gayen, "Diverse cyanobacteria resource from north east region of India for valuable biomolecules: Phycobiliprotein, carotenoid, carbohydrate and lipid", *Current Biochemical Engineering*, Online (2018).

Book Chapters

1. Manish Kumar, **Tridib Kumar Bhowmick**, Supreet Saini and Kalyan Gayen, "On", *Bioenergy and Biofuels*, Editor: Ozcan Konur, CRC Press, Taylor and Francis, 2017. ISBN-10: 1138032816. ISBN-13: 978-1138032811.

2. Saumyakanti Khanra, Kalyan Gayen, Gopinath Halder, **Tridib Kumar Bhowmick**, Gunapati Oinam and O N Tiwari, "Lipid derived products from microalgae: Downstream processing for industrial application", *The Role of Photosynthetic Microbes in Agriculture and Industry*, Editors: Keshawanand Tripathi, Narendra Kumar and Gerard Abraham, Nova Science Publishers, Inc., New York, USA, 2018. ISBN: 978-1-53614-033-0.

3. Dibyajyoti Haldar, Mriganka Sekhar Manna, Dwaipayan Sen, **Tridib Kumar Bhowmick**, Kalyan Gayen, "Microbial fuel cell in wastewater treatment", *Enzymatic and Microbial Fuel Cells*, Editors: Inamuddin and Abdullah M. Asiri, Materials Research Forum (MRF), USA, 2018, In press.

Sunil K. Maity is an academician in the field of Chemical Engineering. He is currently working as a Professor at the Indian Institute of Technology Hyderabad, India. Prior to this, he served about two and half years as an Assistant Professor at the National Institute of Technology Rourkela, India. He received his PhD degree from the Indian Institute of Technology Kharagpur. His current research interests are i) biorefinery for biofuels and chemicals, ii) heterogeneous catalysis and chemical reaction engineering, and iii) process design and techno-economic analysis. He is currently working on the hydrodeoxygenation of vegetable oils, steam and oxidative steam reforming, the production of gasoline, butylene, and aromatics from bio-butanol, the hydroxyalkylation-alkylation reaction, and the production of butyl levulinate from furfuryl alcohol. He executed a number of externally funded projects in the past. As a research outcome, he published more than twenty-five research articles and two book chapters and presented at several national and international conferences. His recent publications relevant to this book are listed below.

Recent Publications

Publications in Reputed Journals

1. S Mailaram, **SK Maity**, "Techno-economic evaluation of two alternative processes for production of green diesel from karanja oil: A pinch analysis approach", *J. Renewable Sustainable Energy*, **2019**, 11, 025906.
2. P Kumar, **SK Maity**, D Shee, "Role of NiMo alloy and Ni species in the performance of NiMo/alumina catalysts for hydrodeoxygenation of stearic acid: A kinetic study", *ACS Omega*, **2019**, 4 (2), 2833–2843.
3. SR Yenumala, **SK Maity**, D Shee, "Reaction mechanism and kinetic modeling for the hydrodeoxygenation of triglycerides over alumina supported nickel catalyst", *Reaction Kinetics, Mechanisms and Catalysis*, **2017**, 120, 109–128.
4. SR Yenumala, **SK Maity**, D Shee, "Hydrodeoxygenation of Karanja oil over supported nickel catalysts: Influence of support and nickel loading", *Catalysis Science & Technology*, **2016**, 6, 3156–3165.
5. VCS Palla, D Shee, **SK Maity**, "Conversion of n-butanol to gasoline range hydrocarbons, butylenes and aromatics", *Applied Catalysis A: General*, **2016**, 526, 28–36.
6. **SK Malty**, "Opportunities, recent trends and challenges of integrated biorefinery: Part I", *Renewable and Sustainable Energy Reviews*, **2015**, 43, 1427–1445.
7. **SK Maity**, "Opportunities, recent trends and challenges of integrated biorefinery: Part II", *Renewable and Sustainable Energy Reviews*, **2015**, 43, 1446–1466.
8. V Dhanala, **SK Maity**, D Shee, "Oxidative steam reforming of isobutanol over Ni/γ-Al$_2$O$_3$ catalysts: A comparison with thermodynamic equilibrium analysis", *Journal of Industrial and Engineering Chemistry*, **2015**, 27, 153–163.
9. V Dhanala, **SK Maity**, D Shee, "Roles of supports (γ-Al$_2$O$_3$, SiO$_2$, ZrO$_2$) and performance of metals (Ni, Co, Mo) in steam reforming of isobutanol", *RSC Advances*, **2015**, 5, 52522–52532.
10. P Kumar, SR Yenumala, **SK Maity**, D Shee, "Kinetics of hydrodeoxygenation of stearic acid using supported nickel catalysts: Effects of supports", *Applied Catalysis A: General*, **2014**, 471, 28–38.
11. VCS Palla, D Shee, **SK Maity**, "Kinetics of hydrodeoxygenation of octanol over supported nickel catalysts: A mechanistic aspect", *RSC Advances*, **2014**, 4, 41612–41621.
12. V Dhanala, **SK Maity**, D Shee, "Steam reforming of isobutanol for production of synthesis gas over Ni/γ-Al$_2$O$_3$ catalysts", *RSC Advances*, **2013**, 3, 24521–24529.
13. **SK Maity**, "Correlation of solubility of single gases/ hydrocarbons in polyethylene using PC-SAFT", *Asia-Pacific Journal of Chemical Engineering*, **2012**, 7, 406–417.

14. SR Yenumala, **SK Maity**, "Thermodynamic evaluation of dry reforming of vegetable oils for production of synthesis gas", *Journal of Renewable and Sustainable Energy*, **2012**, 4, 043120 (18 pages).
15. VP Yadav, **SK Maity**, P Biswas, RK Singh, "Kinetics of esterification of ethylene glycol with acetic acid using cation exchange resin catalyst", *Chemical and Biochemical Engineering Quarterly*, **2011**, 25 (3), 359–366.
16. SR Yenumala, **SK Maity**, "Reforming of vegetable oil for production of hydrogen: A thermodynamic analysis", *International Journal of Hydrogen Energy*, **2011**, 36, 11666–11675.

Book Chapters

1. P Kumar, M Varkolu, S Mailaram, A Kunamalla, **SK Maity**, "Chapter 12: Biorefinery polyutilization systems: Production of green transportation fuels from biomass", *Polygeneration with Polystorage for Chemical and Energy Hubs*, Editor: Kaveh Rajab Khalilpour, Academic Press, Elsevier, Burlington, MA, **2019**, 373–407. ISBN 978-0-12-813306-4.
2. A Mazumder, **SK Maity**, D Sen, K Gayen, "Chapter 3: Process development for hydrolysate optimization from lignocellulosic biomass towards biofuel production", *Alcohols and Bio-alcohols: Characteristics, Production and Use*, Editors: Angelo Basile and Francesco Dalena, Nova Science Publishers, Inc., New York, USA, **2014**, 41–76. ISBN: 978-1-63463-187-7.

Contributors

F. C. Thomas Allnutt
NuLode LLC
Glenelg, Maryland

A. Anandraj
Centre for Algal Biotechnology
Faculty of Natural Sciences
Umlazi, Durban, South Africa

Sharad Bhartiya
Department of Chemical Engineering
Indian Institute of Technology Bombay,
Powai, Mumbai, India

Tridib Kumar Bhowmik
Department of Bioengineering
National Institute of Technology
Agartala, West Tripura, India

J. K. Bwapwa
Faculty of Natural Sciences, Faculty of
Engineering
Centre for Algal Biotechnology
Umlazi, Durban, South Africa

Debasish Das
Department of Biosciences &
Bioengineering
Indian Institute of Technology
Guwahati, India

Santanu Dasgupta
Reliance Technology Group
Reliance Industries Limited Ghansoli,
India

A Chandralekha Devi
Department of Food Engineering
CSIR-Central Food Technological
Research Institute (CSIR-CFTRI)
Mysuru, India

Kalyan Gayen
Department of Chemical Engineering
National Institute of Technology
Agartala, West Tripura, India

Sridharan Govindachary
Reliance Technology Group
Reliance Industries Limited Ghansoli,
India

Ravi Prakash Gupta
DBT-IOC Centre for Advanced
Bio-energy Research
Research & Development Centre Indian
Oil Corporation Limited Sector-13,
Faridabad 121007, India

Soumyajit Sen Gupta
Department of Mechanical and
Aerospace Engineering University of
Florida Gainesville, Florida

Pannaga P. Jutur
Omics of Algae Group, International
Centre for Genetic Engineering and
Biotechnology, New Delhi, India

Chitranshu Kumar
Reliance Technology Group
Reliance Industries Limited Ghansoli,
India

G. Raja Krishna Kumar
Reliance Technology Group
Reliance Industries Limited Ghansoli,
India

Sunil Kumar Maity
Department of Chemical Engineering
Indian Institute of Technology
Hyderabad, Telangana, India

Mriganka Sekhar Manna
Department of Chemical Engineering
National Institute of Technology
Agartala, West Tripura, India

Anshu Shankar Mathur
DBT-IOC Centre for Advanced
Bio-energy Research
Research & Development Centre Indian
Oil Corporation Limited Sector-13,
Faridabad 121007, India

Preeti Mehta
DBT-IOC Centre for Advanced
Bio-energy Research
Research & Development Centre
Indian Oil Corporation Limited
Faridabad, India

Muthusivaramapandian Muthuraj
Department of Bioengineering
National Institute of Technology
Agartala, West Tripura, India

T. Mutanda
Faculty of Natural Sciences
Centre for Algal Biotechnology
Umlazi, Durban, South Africa

Asha A. Nesamma
Omics of Algae Group
International Centre for Genetic
Engineering and Biotechnology
New Delhi, India

Emeka G. Nwoba
Algae R & D Centre
School of Veterinary and
Life Sciences
Murdoch University
Murdoch, Australia

Christiana Ogbonna
Department of Plant Science and
Biotechnology
University of Nigeria
Nsukka, Nigeria

Chetan Paliwal
Omics of Algae Group
International Centre for Genetic
Engineering and Biotechnology
New Delhi, India

Amol Pandharbale
Algae to Oil (A2O)
Reliance Technology Group
Reliance Research &
Development Centre
Reliance Industries Limited
Navi Mumbai, India

Jigar Rameshbhai Patel
Algae to Oil (A2O)
Reliance Technology Group
Reliance Research &
Development Centre
Reliance Industries Limited
Navi Mumbai, India

K. S. M. S. Raghavarao
Department of Food Engineering CSIR-
Central Food Technological Research
Institute (CSIR-CFTRI) Mysuru, India

Meghna Rajvanshi
Reliance Technology Group
Reliance Industries Limited Ghansoli,
India

Uma Shankar Sagaram
Reliance Technology Group
Reliance Industries Limited Ghansoli,
India

Sambit Sarkar
Department of Chemical Engineering
National Institute of Technology
Agartala, West Tripura, India

Kritika Singh
DBT-IOC Centre for Advanced
Bio-energy Research
Research & Development Centre Indian
Oil Corporation Limited Faridabad, India

Avinash Sinha
Algae to Oil (A2O)
Reliance Technology Group
Reliance Research & Development Centre
Reliance Industries Limited Navi Mumbai,
India

Yogendra Shastri
Department of Chemical Engineering,
Indian Institute of Technology Bombay,
Powai, Mumbai, India

G. Venkata Subhash
Reliance Technology Group
Reliance Industries Limited Ghansoli,
India

Hrishikesh A. Tavanandi
Department of Food Engineering CSIR-
Central Food Technological Research
Institute (CSIR-CFTRI) Mysuru, India

Ashiwin Vadiveloo
Algae R & D Centre
School of Veterinary and
Life Sciences
Murdoch University
Murdoch, Australia

Section 1

Biomolecules from Microalgae for Commercial Production

1 Biomolecules from Microalgae for Commercial Applications

Meghna Rajvanshi, Uma Shankar Sagaram,
G. Venkata Subhash, G. Raja Krishna
Kumar, Chitranshu Kumar, Sridharan
Govindachary, Santanu Dasgupta

CONTENTS

1.1 INTRODUCTION

Algae are a diverse group of photosynthetic microorganisms, encompassing both prokaryotic blue-green algae and uni- and multicellular eukaryotic micro- and macroalgae. Macroalgae are well utilized by humans as sources of food, fertilizers, etc. (Garcia-Vaquero and Hayes 2016). Microalgae, though not as popular as macroalgae, can sequester atmospheric CO_2 using the sun's energy and grow rapidly both in fresh and sea water and produce more biomass per unit area than vascular plants. *Chlorophyta, Chrysophyta, Cyanophyta* and *Bacillariophyta* are generally known as green, golden, blue-green algae and diatoms, respectively, and are the most prominent microalgal divisions, encompassing ≈0.2 million species (Norton, Melkonian, and Andersen 1996). Microalgae are highly versatile due to their unique ability to grow under varied environmental conditions. Their capability to sequester CO_2 from the atmosphere and grow in marine water in non-arable lands makes them even more lucrative for commercial applications. It not only reduces the dependence on fresh water for cultivation but also helps in CO_2 mitigation. Many microalgal species are rich sources of biomolecules viz. proteins, carbohydrates, lipids, vitamins, minerals, pigments and other small molecules. Potential health benefits offered by these bioactive compounds make microalgae highly attractive candidates for use in various industrial applications, including food/feed, nutraceuticals, pharmaceuticals, cosmetics, thickening agents and ecofriendly biofuels, to name the few (Figure. 1.1).

Achieving adequate nutrition for all living organisms is a growing global concern with the ever-increasing population and diminishing resources. In such a scenario, algae as a rich source of protein offer an attractive alternative to meet the population's protein requirement and thus help solve the protein deficiency faced by human society. Similarly, algal proteins are a sustainable alternative protein source for the aqua and animal feed industry. Further, this chapter speaks about the use of microalgae as sugar feedstock for bioethanol production and various functional carbohydrates, like exopolysaccharides, β-glucan and sulfated polysaccharides, and their potential utility in the nutraceutical industry. The next section focusses on various pigments produced from microalgae, like chlorophyll, astaxanthin, carotenoids, phycocyanin, etc., and their application in the food, cosmetics, nutraceutical and pharmaceutical industries. The subsequent section describes the use of algal lipids for commercial biodiesel production and its potential use as a viable source of omega-3 (ω-3) fatty acids. The last section discusses microalgae as a source of bioactive compounds and peptides that have numerous pharmaceutical and therapeutic applications.

1.2 PROTEINS FROM MICROALGAE

Proteins are complex macromolecules, which have an essential structural, functional and regulatory role in the body. Being an essential nutrient, and due to growing awareness about the role of proteins in health and nutrition, demand for protein is steadily increasing. It is projected that the demand for animal-derived protein will be doubled by 2050, when the world population will reach 9.7 billion (Henchion et al. 2017; United Nations 2015). However, animal-derived protein raises concerns related to greenhouse gas emissions, extensive land use, animal welfare and health

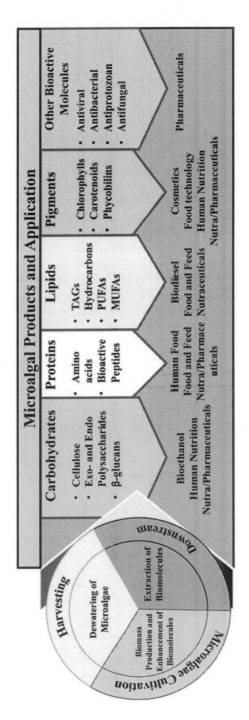

FIGURE 1.1 Microalgal biomolecules and key applications.

effects associated with high levels of saturated fatty acids and cholesterol. Therefore, alternative protein sources such as land and aquatic plants, insect protein, cell culture–based meat and single-cell proteins are constantly being explored to meet the demands of a growing population (Barka and Blecker 2016; Henry et al. 2015). Microalgae are one of the most-sought-after alternative protein sources, due to their high growth rates, lack of competition for arable land and water, no negative impact on the environment, high protein content and presence of health-boosting bioactive compounds (Borowitzka 2013b). Changing trends in dietary habits from omnivores to vegan also strengthen the prospects of microalgal-based protein. It is estimated that microalgal protein may cover 18% of the protein market by the middle of this century (Caporgno and Mathys 2018). A few examples of microalgal species that are a rich source of protein are *Chlamydomonas rheinhardii* (48%), *Chlorella pyrenoidosa* (57%), *Chlorella vulgaris* (51%–58%), *Scenedesmus obliquuus* (50%–56%), *Scenedesmus quadricauda* (47%), *Spirulina maxima* (60%–71%) and *Spirulina platensis* (46%–63%) (Guedes, Sousa Pinto, and Xavier Malcata 2015). In addition, microalgal protein has high nutritional value due to the presence of essential amino acids (AAs), which meets the World Health Organization (WHO) requirement. Also, the composition of microalgal proteins is comparable to other high-protein sources, like soybeans, eggs, meat and milk (Bleakley and Hayes 2017). Further, several microalgae species like *Arthrospira, Dunaliella, Chlorella, Haematococcus, Porphyridium cruentum, Crypthecodinium cohnii* and *Schizochytrium* have been granted Generally Recognized as Safe (GRAS) status by the U.S. Food and Drug Administration (FDA) (Caporgno and Mathys 2018; García, de Vicente, and Galán 2017), rendering their safe use in food. Due to these nutritional qualities, microalgae are highly attractive candidates for food and feed applications.

1.2.1 MICROALGAL PROTEIN PRODUCTS

Microalgal protein is categorized into whole-cell protein (whole microalgal cell), protein concentrates, isolates, hydrolysates and bioactive peptides, depending on the processing involved (Soto-Sierra, Stoykova, and Nikolov 2018). For a detailed discussion on extraction and separation of proteins from microalgae, refer to the review by Soto-Sierra, Stoykova, and Nikolov (2018). However, this aspect is briefly discussed here for understanding purposes only. Whole-cell protein is protected in intact cells but may not be bioavailable because intracellular protein is not readily accessible to digestive enzymes. The presence of cellulose and other complex carbohydrates in the microalgae cell wall makes it hard to digest after consumption. Disruption/digestion of the cell wall by mechanical or chemical means can significantly improve the availability of the protein. Alternatively, heat treatment appears to improve the digestibility of protein by partial denaturation and breakdown of proteins into smaller peptides (Bleakley and Hayes 2017). In the case of *Spirulina*, digestibility is not a major concern, as the cell wall is devoid of cellulose and hence does not require additional processing steps (Dillon, Phuc, and Dubacq 1995); still, its digestibility could be further improved by enzymatic hydrolysis. Therefore, it is recommended that algal supplements comprising whole-cell protein in the form of powder, tablets, granules, etc., should be devoid of the cell wall and pulverized for improved digestibility. For the preparation of protein concentrates and isolates, soluble protein from the cells is released by either a single method or a combination of mechanical (bead milling, high-pressure homogenization, ultrasonication, microwave and pulse electric field) or non-mechanical (enzymatic and chemical) cell disruption methods (Günerken et al. 2015; Demuez et al. 2015). The selection and effectiveness of disruption methods highly depend on the cell wall composition. For example, microalgae like *Scenedesmus*, *Nannochloropsis* and a few *Chlorella* species contain algaenan in their cell wall (Allard and Templier 2000; Gelin et al. 1997, 1999), which is highly resistant to enzymatic digestion, and hence an appropriate mechanical disruption method needs to be devised. Extracted soluble protein obtained post-removal of cell debris by centrifugation is further selectively purified and concentrated either by using different protein precipitation methods like acid, heat or ethanol, or by membrane-based tangential flow filtration (TFF). The concentrated proteins (60%–95% protein content) prepared following the process noted earlier have relatively higher digestibility compared to whole-cell proteins (Soto-Sierra, Stoykova, and Nikolov 2018). Protein hydrolysates (70%–95% protein content), on the other hand, are obtained by enzymatic hydrolysis of whole algal cells or extracted protein. The hydrolysis process breaks down the protein into smaller peptides and substantially improves the digestibility and bioavailability. *C. vulgaris* and *S. platensis* untreated biomass showed 35% and 50% apparent digestibility, respectively. However, the protein hydrolysate from the two strains obtained after treatment with pancreatin improved the digestibility substantially to 70% and 97%, respectively. This observation illustrates that digestibility can be enhanced significantly and matched to the levels of commercially available proteins like casein (Kose et al. 2017). Protein hydrolysates containing peptides >10 to 20 kDa are preferred for the desirable functional properties of the protein (Soto-Sierra, Stoykova, and Nikolov 2018). Peptides ranging between 2 and 5 kDa are called bioactive peptides and will be discussed in a later section. Various proteolytic enzymes have been successfully used for the hydrolysis of proteins to improve digestibility. For example, Morris et al. (2008) have shown through comparative enzymatic hydrolysis of proteins from a *Chlorella* biomass that papain and pancreatin were equally efficient in hydrolysis, followed by trypsin, bromelain from *Bacillus subtilis* preparation and pepsin, respectively. Also it was observed that proteins hydrolyze readily when an ethanol-extracted biomass is used, as ethanol removes lipids, pigments and other hydrophobic molecules, resulting in improved enzyme substrate interaction (Morris et al. 2008). A similar finding was earlier reported in *Scenedesmus* and *Chlorella* extracted with solvents including ethanol, ethyl acetate, acetone, chloroform and pentane (Tchorbanov and Bozhkova 1988).

1.2.2 Microalgal Protein in Human Nutrition

Microalgae have been used as a protein source in human nutrition since ancient times. Asian countries like Korea, Japan and China commonly use them as part of their diet (Rath 2012). Out of several algal species, *Spirulina, Nostoc, Chlorella* and *Aphanizomenon*, which are protein rich, are commonly consumed (Caporgno and Mathys 2018). *Spirulina* gained importance, as it has been highly consumed over the past several decades. It is a non-heterocystous, filamentous cyanobacterium usually grown in high-alkaline water that restricts the growth of other contaminating organisms. *Spirulina* can accumulate up to 70% protein of its dry weight, which is almost double the best plant protein source: soy flour (35% crude protein) (Dillon, Phuc, and Dubacq 1995). It is also a rich source of B-vitamins, free radical scavenging phycobiliproteins, γ-linoleic acid, calcium and iron (Khan, Bhadouria, and Bisen 2005). Thus, it provides holistic health benefits, which include anti-hypertension, anti-hyperlipidaemia, anti-hyperglycaemia, anti-oxidant, anti-inflammatory and renal protective functions. Due to these immense health benefits, *Spirulina* is considered a "superfood" by the WHO and used as a nutrient-dense compact food by the National Aeronautics and Space Administration (NASA) (Khan, Bhadouria, and Bisen 2005; Bleakley and Hayes 2017). Based on the AA profile of *Spirulina*, it is considered a complete protein, as all the essential AAs are present (Dillon, Phuc, and Dubacq 1995). However, levels of lysine and methionine are low (20%–30%) compared to WHO guidelines, which can be compensated for by complementary foods.

Chlorella is another important microalga strain, which is commonly used as food supplements. *Chlorella pyrenoidosa*, a fresh water alga, not only is a rich source of protein but also is believed to offer several health benefits, like a reduction in high blood pressure, cholesterol and glucose levels. It is also known to enhance the immunity of the body. The essential AA index of *C. pyrenoidosa* is greater than soy protein, indicating better nutritional properties compared to soy protein (Waghmare et al. 2016). *Chlorella vulgaris*, with 51% to 58% protein content, is another strain with high potential as an alternative protein source (Barka and Blecker 2016). Because *Chlorella* has a rigid cell wall, it is tough to digest in its natural form, and hence biomass processing is required to ensure better availability of nutrients.

Apart from a superior AA profile, another important aspect of proteins from algae utilized as a food source is their protein quality. According to the Food and Agriculture Organization (FAO) and WHO, the protein quality is determined by the essential AA profile (which should match the requirements of the body), its digestibility and the bioavailability of the AAs (Gurevich 2014). Multiple methods are available to estimate protein quality: protein efficiency ratio (PER), biological value (BV), net protein utilization (NPU), protein digestibility corrected amino acid score (PDCAAS) and digestible indispensable amino acid scores (DIAAS). Since 1989, PDCAAS has been the most recommended method by the FAO-WHO for protein quality evaluation. The PDCAAS of egg, whey and soy proteins are in the range of 0.9 to 1. In 2013, the FAO-WHO proposed replacing PDCAAS with DIAAS to address individual dietary AA digestibility (Lee et al. 2016). DIAAS can overcome the limitations of PDCAAS; however, the applicability of DIAAS will largely depend on sufficient research data in favor of DIAAS (Leser 2013).

1.2.3 MICROALGAL PROTEIN IN AQUA FEED

Similar to human nutrition, there is a growing requirement for alternative protein sources in aqua and animal feed. Fish meal is the primary protein source in commercial aqua feed. However, the continuous increase in fish meal price, increasing demand and decline in fish meal production have forced the search for alternative protein sources (Hemaiswarya et al. 2011). Microalgae can be a suitable replacement for fish meal due to its high protein content and presence of all essential AAs (Brown et al. 1997). Apart from high-quality protein, microalgae also can provide poly-unsaturated fatty acids (PUFAs) and carotenoids, which add important value to the feed. *Chlorella, Nannochloropsis, Chaetoceros, Isochrysis, Pavlova* and *Tetraselmis* are some of the commonly used species in aquaculture. Numerous studies have revealed the comparable growth performance of various aquaculture animals when fish meal was replaced to a varied extent with algal proteins. This strengthens the potential use of algae as a novel protein source in the aquaculture industry (Conceição et al. 2010; Kiron et al. 2012). However, current microalgae use is restricted to nursery feed (fresh microalgae) due to challenges related to digestibility in the formulated feed. Skrede et al. (2011) have shown variable digestibility of *Nannochloropsis oceanica* (35.5%), *Phaeodactylum tricornutum* (79.9%) and *Isochrysis galbana* (18.8%) in carnivorous mink, a model organism used for salmon (Skrede et al. 2011). Nonetheless, the nutritional value of the feed could be improved by using a combination of algal species and improving digestibility of the cell wall to ensure better availability of nutrients (Spolaore et al. 2006; Raja et al. 2004).

1.2.4 ACCEPTANCE OF MICROALGAL PRODUCTS

Odor, color and fishy smell are the key obstacles in the acceptance of microalgae as food. Acceptance requires innovative food product development, where the smell can be masked, and the color is not a hindrance. For example, *Chlorella* has been added to yogurt (Jin-Kyoung and Shin-In 2005) and cheeses (Jeon 2006); likewise, *A. platensis* was incorporated into cookies to increase the protein and fiber content (Abd El Baky, El Baroty, and Ibrahem 2015). Pasta and smoothies are other options for the incorporation of microalgal protein, where different colors are easily acceptable (Caporgno and Mathys 2018). Functional properties, like foaming, gelling, solubility and texture of a protein could be enhanced through chemical and enzymatic modifications, thus making algal proteins more suitable for incorporation into a wider variety of food products. For instance, succinic anhydride, through the acylation reaction, improves the solubility of the protein. Similarly, disulfide bond cleavage using sodium sulfite improves the viscosity and adhesive strength of soya protein (Hettiarachchy and Ziegler 1994). Similar strategies need to be applied to microalgae proteins for improving the acceptance levels. The scalability of protein extraction is another bottleneck, as the technology is in its infancy. Out of the many technologies available, pulse electric field, ultrasound and membrane technology have been considered options for large-scale operations (Bleakley and Hayes 2017).

1.3 MICROALGAE AS A SOURCE OF CARBOHYDRATES

Carbohydrates constitute a significant fraction of the algae biomass. Microalgal carbohydrates consist of mono- and disaccharides, polysaccharides for energy storage (starch and glycogen) and structural polysaccharides (cellulose, hemicellulose and pectin) (Tang et al. 2016). Some microalgae and cyanobacteria accumulate carbohydrate storage as high as >40% and ~70%, respectively, of their body weight

(Choi, Nguyen, and Sim 2010; Branyikova et al. 2011; Aikawa et al. 2012). Using thermochemical and biological conversion technologies, microalgae carbohydrates were shown to be suitable feedstocks for the production of renewable fuels (Markou, Angelidaki, and Georgakakis 2012), food, feed and bulk chemicals (Wijffels, Barbosa, and Eppink 2010).

1.3.1 Polysaccharides as Feedstock for Fermentation

In the past decade, microalgae have gained significant consideration as feedstock for bioethanol production through fermentation, probably due to the evolution of production technologies and the ability of microalgae to accumulate substantial amounts of carbohydrates (Harun, Danquah, and Forde 2010; Menetrez 2012; John et al. 2011; Markou, Angelidaki, and Georgakakis 2012). Typically, algae accumulate carbohydrates as cell wall polysaccharides such as cellulose, hemicellulose, mannose or xylans, or storage polysaccharides such as starch, laminaran and glycogen (cyanobacteria). Fermentable sugars can be produced from both cellular and cell wall polysaccharides (Wang, Liu, and Wang 2011). Some examples of microalgae strains that are commonly proposed for bioethanol production include *Chlorella* sp., *Chlamydomonas reinhardtii*, *Chlorococcum* sp., *Scenedesmus* sp. and *Spirulina fusiformis* (Bhalamurugan, Valerie, and Mark 2018).

Considerable research efforts were dedicated to developing carbohydrates as feedstock for fermentative bioethanol production. Although a major effort has been made to optimize pre-treatment of feedstock for fermentation (Choi, Nguyen, and Sim 2010; Dong et al. 2016; Harun et al. 2011; Miranda, Passarinho, and Gouveia 2012), a few successful efforts were reported with improved feed quality by enhancing the carbohydrate content in the biomass (Branyikova et al. 2011; Aikawa et al. 2012). Biomass pre-treatment with the enzymes α-amylase and amyloglucosidase (Choi, Nguyen, and Sim 2010), NaOH (Miranda, Passarinho, and Gouveia 2012) and H_2SO_4 (Harun et al. 2011), along with optimized temperature, resulted in improved recovery of starch from the biomass, thereby enhancing the availability of feedstock for fermentative production of ethanol. On the other hand, Branyikova et al. (2011) reported that the starch accumulation of 8.5% dry weight (DW) was obtained at a light intensity of 215 μmol/m²s¹ in a fresh water *Chlorella* species, which was enhanced to 40% of DW at a mean light intensity of 330 μmol/m²s¹. Also, it was noticed that the phase of the cell cycle plays a key role in starch accumulation; the starch content before cell division was ~45% DW, and after the division, the starch content dropped to 13% DW (Branyikova et al. 2011). In addition, Aikawa et al. (2012), demonstrated that the glycogen content of *Arthrospira platensis* could be enhanced by optimizing light intensity and nitrogen supply, thus enhancing the concentration of carbohydrates and the feedstock quality (Aikawa et al. 2012). This suggests that careful cultivation and harvesting practices need to be adopted if algae biomass is the intended feedstock for conversion to ethanol and related fermentation processes. Even with the significant progress in research related to feedstock quality and conversion technology, the biomass treatment and conversion costs have remained key limiting factors in cellulosic ethanol production (Carriquiry, Du, and Timilsina 2011).

1.3.2 FUNCTIONAL POLYSACCHARIDES

Apart from feedstock in biofuel application, carbohydrates of microalgae origin are known to possess structural and functional roles. Several green, brown and red microalgae are known to produce polysaccharides with economic importance; however, very few of these compounds were established as commercial products because of the availability of alternative compounds from plants and macroalgae (Borowitzka 2013b). For example, xanthan gum and guar gum are produced from bacteria and guar, respectively, and carrageen, alginate, gels, agar, etc., are produced from macroalgae (Bixler and Porse 2011). Apart from macroalgae, some microalgae—for example, *Botryococcus braunii*, *Porphyridium cruentum* and cyanobacteria (*Geitlerinema* sp., *Oscillatoria* sp., *Spirulina maxima*, *Spirulina platensis*, *Synechocystis* sp., *Synechococcus* sp., *Cyanothece* sp. etc)—are also known to synthesize exopolysaccharides (EPS), which either can be cell surface associated or are released into the medium (Donot et al. 2012). EPS from microalgae are gaining attention due to potential applications as industrial gums, bioflocculants, soil conditioners and biosorbants (Li, Harding, and Liu 2001). A group of carbohydrates from microalgae origin that deserves a special mention is Sulphur-containing polysaccharides, which have long been characterized for their properties and potential application as anti-oxidant and therapeutic agents (Guedes, Amaro, and Malcata 2011; Amaro, Guedes, and Malcata 2011). These sulfated polysaccharides of therapeutic application are discussed in the later section.

1.3.3 β-GLUCAN

A notable microalgal polysaccharide molecule that gained significant market share is β-glucan. It is a cell wall component formed through the linking of D-glucose units through β-glycosidic bonds (Zhu, Du, and Xu 2016). It gained a lot of attention because of its well-proven bioactive properties such as anti-tumor, anti-obesity, anti-inflammatory, anti-osteoporotic, anti-allergic and immunomodulating activities (Bashir and Choi 2017). Crops like barley and oats are rich sources of β-glucan, containing ~20 and 8%, respectively. However, microalgal species like *Euglena gracialis, Chlorella pyrenoidosa* and *Nannochloropsis* also are potent sources of β-glucan. *Nannochloropsis* has been reported to contain 3% to 6% β-glucan. Reported structures of β-glucan in microalgae are β-1,3 glucan, β-1,3/1,6 glucan and β-1,6/1,3 glucan. β-glucan in *Isochrysis galbana* and *Chlorella pyrenoidosa* is branched, whereas it is linear (β-1,3 glucan) in *Euglena gracilis* and called paramylon (Rojo et al. 2017). *E. gracilis* has been used by Kemin Industries to produce β-glucan commercially for human and animal nutrition, and products are sold under the brand name BitaVia Pure and BitaVia Complete for human nutrition and Aleta for animal nutrition.

1.4 PIGMENTS

Microalgae and cyanobacteria are diversified based on their pigments, which have taxonomical importance in algal systematics. Based on their physical and chemical properties, pigments are broadly categorized into fat-soluble chlorophylls and carotenoids present in thylakoids and water-soluble phycobilins in the cytoplasm. Chlorophyll *a* and phycocyanin are major pigments in algae and cyanobacteria, respectively. The carotenoids—chlorophyll *b*, *c*, *d* and *e*—are accessory pigments. Pigments of different colors can absorb light, transfer the energy to reaction centers and convert them into chemical energy through photosynthesis. Pigment accumulation in microalgae is mainly influenced by light, temperature, pH, salinity and nutrition (Begum et al. 2016). The anti-oxidant property of pigments is commonly used in pharmaceutical and nutraceutical industries, along with their use as a natural dye in food and cosmeceuticals (Poonam and Nivedita 2017; Rath 2012). Hexa Research has estimated that the natural food color market size was USD 1.5 billion in 2017 and will reach USD 2.5 billion by 2025 (Michelle 2018).

1.4.1 CHLOROPHYLLS

Chlorophyll *a* is a green pigment universally present in microalgae, cyanobacteria and green plants. Chlorophyll *b* is present in *Chlorophyta* and *Euglenophyta,* whereas chlorophyll *c*, *d* and *e* are found in diatoms and act as accessory pigments in photosynthesis. Notably, the anti-oxidant properties of chlorophyll are used in health care and as a colorant in the food industry. For example, its therapeutic use reduces the risk of colorectal cancer (Balder et al. 2006). Chlorophyll-related compounds (CRC) are obtained by natural catabolism or are derived chemically or thermally. The major CRC used in many therapeutics are pheophytin (lack of Mg atom), chlorophyllide or chlide (lacking phytol tail) and pheophorbide or pho (lacking both phytol tail and Mg atom) (Hsu et al. 2014). Historically, chlorophyll-rich foods are consumed by humans, yet there is limited literature on the bioavailability of chlorophyll, and it is assumed to be non-absorbable by humans (Ferruzzi and Blakeslee 2007). However, Chao et al. (2018) have reported that upon consumption of spinach, chlorophyll rapidly converted into CRC and was detected in the blood after three hours as pheophytin and pheophorbide and thus can be absorbed by the human body (Chao et al. 2018). Chlorophyllin, another derivative of chlorophyll, acts as a chemopreventive agent in human colon cancer cells and as a deodorant in personal hygiene. Copper-chlorophyllin (Cu-Chl) and sodium copper chlorophyllin (SCC) are chemically derived by replacing the magnesium ion with copper through acid treatment. The FDA has approved SCC as a food colorant derived from a plant source. SCC is also used in wound healing and for the control of calcium oxalate crystals (Ferruzzi and Blakeslee 2007). In addition, porphyrin-based photosensitizers are used in cancer therapy (Sternberg, Dolphin, and Brückner 1998). Microalgae can potentially be a reliable source of chlorophyll globally because of its year-round availability.

1.4.2 CAROTENOIDS (CAROTENES AND XANTHOPHYLLS)

Carotenoids are present in microalgae and cyanobacteria. There are more than 750 structurally defined carotenoids reported. In photosynthesis, approximately 30 types of carotenoids are involved in absorbing light energy of different wavelengths that chlorophyll cannot absorb and eventually transferring to chlorophyll (Alvarez et al. 2014; Takaichi 2011). Carotenoids also play a role as a non-photochemical quenching agent that prevents damage to the cells caused by excess light energy. Carotenes are classified into α, β, γ, δ, ε and lycopenes, which lack oxygen in their structure, and xanthophylls are oxygenated carotenoids—namely lutein, astaxanthin, zeaxanthin, canthaxanthin, neoxanthin, violaxanthin, flavoxanthin and α- and β-cryptoxanthin. The global carotenoid market was estimated to be USD 1.24 billion in 2016 and is predicted to reach USD 1.53 billion by 2021, at a compound annual growth rate (CAGR) of 3.78% (Rohan 2016).

ß-carotene

ß-carotene is a common precursor of vitamin A and can be supplied to human nutrition through plant sources. *Dunaliella salina* is a commonly cultivated microalga for ß-carotene (40% 9-*cis* and 50% all-*trans* stereoisomers) and can accumulate up to 14% of ß-carotene inside the cells. ß-carotene, also known as pro–vitamin A or retinol, serves as a nutritional supplement, preventing cataracts and night blindness. Additionally, the anti-oxidant property of ß-carotene has manifold applications in anti-aging, anti-cancer immunity, skin diseases, arthritis, coronary heart diseases and various forms of cancer (Begum et al. 2016; Milledge 2011). In 2015, the global ß-carotene market size was USD 432.2 million, and revenue generated from algae-derived ß-carotene was more than 35% of the total revenue (GrandViewResearch 2016). In 1979, the FDA granted GRAS status to ß-carotene, and many countries extensively use ß-carotene as a food colorant and pharmaceutical and cosmetic ingredient (USDA 2011).

Astaxanthin

Astaxanthin (3,3'-dihydroxy-b, b-carotene-4,4'-dione) is produced by some micro- and macro- green algae, plants, fungi and bacteria. In microalgae, *Haematococcus pluvialis, Neochloris wimmeri, Chlorella zofingiensis, Chlorococcum* sp., *Protosiphon botryoides, Scotiellopsis oocystiformis* and *Scenedesmus vacuolatus* are reported to produce astaxanthin (Orosa et al. 2001). Among all of these, *H. pluvialis* is considered a commercially viable strain because it can accumulate up to about 5% of dry weight as astaxanthin. The anti-oxidant property of astaxanthin, which is commonly represented by oxygen radical absorbance capacity (ORAC) (2,822,200 µ mol TE/100 G), is a great deal higher than the ORAC of vitamin E and vitamin C (Superfoodly 2018). Further, the anti-cancer, anti-inflammatory, anti-diabetic, photoprotective, skin damage protection and DNA repair properties of astaxanthin have increased the demand in global pharmaceutical and nutraceutical industries. Commercially, astaxanthin is currently produced from *H. pluvialis*, yeast fermentation and chemical synthesis. Natural astaxanthin derived from *H. pluvialis* is esterified (>95%) and ideal for human applications like dietary supplements, cosmetics and food. Yeast accumulates ~0.5% astaxanthin by dry weight. Chemically, astaxanthin is mainly synthesized by the reaction of C15 phosphonium salts with C10 dialdehyde (Wittig reaction), or by hydroxylation of canthaxanthin, or by oxidation of zeaxanthin. Chemically synthesized astaxanthin is an unesterified or free form with low ORAC, and hence suspected to be less effective. However, chemically synthesized astaxanthin dominates the global market due to the lower cost of synthesis (Nguyen 2013). Astaxanthin is widely used as a food colorant and in aqua culture to feed salmon and crustaceans like shrimp, crab, lobster and krill (Miyashita and Hosokawa 2019). Global astaxanthin production was USD 615 million in 2016. Astaxanthin—Global Market Outlook (2017–2023) estimates that world astaxanthin production will be USD 1226 million in 2023, with a CAGR of 10.3% because of increasing demand in cosmetics, nutraceuticals and health care products (Wood 2017). Demand for natural astaxanthin derived from *H. pluvialis* was ~55% in 2017 by the global nutraceutical market, and it is predicted to be 190 Metric Ton by 2024 (Wood 2018). The FDA has granted GRAS status to astaxanthin extracted from *H. pluvialis* by the supercritical CO_2 method and advised a permissible human consumption dosage up to 12 mg/day to 24 mg/day for a maximum of 30 days (Davinelli, Nielsen, and Scapagnini 2018). Market analysts opined that natural astaxanthin market (USD 7000/kg) growth can go higher because of expanding end-use applications. However, the labor-intensive cultivation and production processes of natural astaxanthin and adulteration with cheap, chemically synthesized (USD 2000/kg) or yeast-fermented astaxanthin (USD 2500/kg) are major obstacles in the sustained growth of the natural astaxanthin market (Nguyen 2013).

Lutein

Lutein is an oxygenated carotenoid that is present in microalgae like *Chlorella sorokiniana, C. protothecoides, C. pyrenoidosa, C. vulgaris, Dunaliella salina, Chlamydomonas reinhardtii. Haematococcus, Murielopsis, Scenedesmus ameriensis* and many fruits, vegetables and flowers, especially marigold petals. Microalgal-based lutein production has many advantages over land plants because of the high lutein content, ranging from 0.5% to 4.6% DW basis, five to ten times higher growth rate, use of non-arable land, sea water cultivation, year-round harvest and availability with other co-value–added products from algae (Del Campo et al. 2001; Lin, Lee, and Chang 2015). Lutein is used as a food color in the poultry industry to give the yellowish color to the egg yolk and is given in feed to improve the color of broiler chicken skin. The anti-oxidant property of lutein plays an ameliorative role in degenerative human diseases, such as age-related macular degeneration (AMD), cataracts and in skin health. In addition, lutein reduced the risk of developing colon cancer and is protective against breast cancer and head-and-neck cancer. Newer applications are emerging in cosmetics, skin care and as an antioxidant (Sun et al. 2016; Buscemi et al. 2018). The diverse biological functions of lutein and its wide therapeutic uses make it a high-value product with huge market potential. The approximate market price of lutein is USD 500/kg (unpublished data). The lutein market was about USD 263.8 million in 2017 and is projected to reach USD 357.7 million in the next five years, growing at a CAGR of 6.3% (Rohan 2017). The European Food Safety Authority (EFSA) recommends 1 mg/kg of body weight as the daily intake of lutein for adults, and for infants 250 μg/L, whereas the Joint Expert Committee on Food Additives advises 2 mg/kg for adults (Buscemi et al. 2018). However, clinical reports suggest that ~6 mg of lutein is required per day to reduce AMD in adults (Ranard et al. 2017). Currently, marigold petals are exploited for commercial production of lutein. Despite the many advantages of microalgae as a lutein source, they have not been used for commercial production. A few attempts were made in tubular reactors and open pond cultivation for sustainable production of lutein from algae, but more developments are needed to increase the overall lutein content, lower cultivation costs and develop cost-effective processes for cell disruption and extraction (Ho et al. 2014; Borowitzka 2013b).

Other Carotenoids

Canthaxanthin is a secondary carotenoid, which is produced at the end of the growth phase in several green algae and cyanobacteria. Canthaxanthin is used as a food colorant, and it improves the color of chicken skins, egg yolks, salmon and trout when added in animal feed. In addition, canthaxanthin is used in cosmetics and medications. In the United States, the quantity of canthaxanthin consumption permitted is up to 30 mg per pound of solid or semi-solid foods—the EFSA recommends ~0.3 mg/kg average daily intake, but not in Australia and New Zealand (Administration 2018; Koller, Muhr, and Braunegg 2014). Violaxanthin is another carotenoid that is produced by microalgae like *Dunaliella tertiolecta* and *Botryococcus braunii*. It is used as a food colorant in Australia and New Zealand but not in the EU and United States. Violaxanthin also exhibits a strong anti-proliferative activity on human mammary cancer cell lines, suggesting potential therapeutic use in treating human mammary cancers (Koller, Muhr, and Braunegg 2014). Fucoxanthin is one of the most abundant carotenoids present in diatoms like *Phaeodactylum tricornutum*, *Cylindrotheca closterium* and macroalgae. Fucoxanthin has anti-oxidant activity, anti-inflammatory effect, anti-cancer activity, anti-obese effect and anti-diabetic activity (Kim et al. 2012). The global fucoxanthin market is projected to grow at a CAGR of 3.2% through 2022 and is estimated nearly 700 tons annually. Traditionally, fucoxanthin is extracted from seaweed, which contains only ~0.01% fucoxanthin. Recent studies have shown that the most promising algae for fucoxanthin production are the diatom *Phaeodactylum tricornutum* (fucoxanthin—15.33 mg/g DW) (Kim et al. 2012). Algatech of Israel has recently launched Fucovital, containing 3% natural fucoxanthin oleoresin extracted from *P. tricornutum*. Fucovital has been granted New Dietary Ingredient Notification (NDIN) status from the FDA.

1.4.3 Phycobilisomes

Phycocyanin, allophycocyanin and phycoerythrin are collectively called phycobilisomes. Among these, C-phycocyanin is present in red algae, *Cryptomonads* and *Cyanelles*. *Spirulina platensis* and *Porphyridium* are extensively used for the commercial production of phycocyanin and phycoerythrin, respectively. Phycocyanin is a unique pigment and has an exhaustive list of commercial applications in food, nutraceuticals, pharmaceuticals, textile and printing dye industries, and newer applications are emerging frequently. In addition, it is used as a fluorescent probe in cytometry and as an immunological analysis and DNA staining agent in research and diagnostic applications (Singh, Kuddus, and Thomas 2009). Phycocyanin also selectively inhibits the activity of cyclooxygenase-2 in cancer cells and MCF-7 (Michigan Cancer Foundation) in human breast cancer cells (Hosseini, Khosravi-Darani, and Mozafari 2013; Eriksen 2008). The FDA and EFSA recommend up to 3 to 10 g daily of *S. platensis* for human health and have approved phycocyanin as a colorant in the food and confectionary industry (Seyidoglu, Inan, and Aydin 2017). In 2018, the global phycocyanin market was valued at USD 112 million, and by the end of 2025, it is estimated to reach more than USD 232 million, growing at a CAGR of 7.6% (*Phycocyanin to Show Consistent Penetration Owing to Use in Multiple Food and Beverages Applications* 2018). Phycoerythrin is a red protein-pigment complex from the light-harvesting phycobiliprotein family. It is produced from *Cyanobacteria, Cryptomonads* and the red alga *Porphyridium cruentum*. Phycoerythrin is used as a colorant in ice creams, confectionaries, milk products and candies (Dufossé et al. 2005). Also, R-phycoerythrin is used as a fluorescence-based indicator to detect the presence of cyanobacteria and for labeling antibodies in flow cytometry (Sekar and Chandramohan 2008; De Rosa, Brenchley, and Roederer 2003).

1.5 MICROALGAL LIPIDS

In microalgae, energy is stored in the form of starch or triacylglycerols (TAGs), which can be used as feedstock to produce bioethanol and biodiesel, respectively. Certain long-chain PUFAs, especially ω-3 FAs, are considered alternative sources of metabolic ingredients in functional food formulations (Kumari et al. 2013; Mendis and Kim 2011). In general, microalgal species are able to accumulate about 25% to 35% lipids, with some species accumulating lipids as high as 70% of DW. These widespread species belong to the genera *Porphyridium, Dunaliella, Isochrysis, Nannochloropsis, Tetraselmis, Phaeodactylum, Chlorella* and *Schizochytrium*. Microalgal lipids with 14 to 20 carbons (FAs) are used for biodiesel production, and 20+ C PUFAs are used as health food supplements.

Lipid production or synthesis in microalgae is influenced by environmental factors and nutrient medium composition. Light illumination, temperature and other chemical stimuli such as pH, salinity, mineral salts and nitrogen starvation are the main factors shown to influence lipid accumulation in microalgae. Among these, nitrogen limitation is the most well studied and appears to be a reliable strategy for increasing the lipid content. Under nitrogen-stressed conditions, the major biochemical steps involved in the lipid biosynthesis pathway in microalgae are i.) conversion of acetyl-Co A to malonyl-CoA that occurs in chloroplast, ii.) carbon chain elongation of the FAs, and iii.) TAG formation (Venkata Subhash et al. 2017). Biosynthetic pathways of FA and TAG synthesis in algae have been less studied compared to higher plants. However, recent genomic studies and bioinformatics analyses have helped to predict genes encoding proteins involved in membrane and storage lipid biogenesis in microalgae (Harwood and Guschina 2009).

Microalgal lipids consist of phospholipids, glycolipids (glycosyl glycerides) and non-polar glycerolipids (neutral lipids). Chain length and degree of unsaturation in microalgal lipids are significantly higher than lipids from plants. Phospholipids (PL) represent 10% to 20% of total lipids in algae (Dembitsky and Rozentsvet 1990; Dembitsky 1996) and are located in extra-chloroplast membranes in significant amounts and maintain the structural integrity of the photosynthetic apparatus. Glycolipids contain 1,2-diacyl-*sn*—glycerol moiety with mono- or oligosaccharide groups at the *sn*-3 position of the glycerol backbone. Monogalactosyldiacylglycerol (MGDG), digalactosyldiacylglycerol (DGDG), sulfolipids and sulfoquinovosyldiacylglycerol (SQDG) are the typical glycolipids. Glycolipids are strictly restricted to the thylakoid membranes of the chloroplast. Glycerolipids consists of a glycerol backbone to which acyl groups (hydrophobic) are esterified to either one, two or all three positions. Acyl groups may be saturated or unsaturated, forming mono-, di- and TAG. TAG is the neutral lipids that accumulate in algae as a storage and energy reservoir (Fan et al. 2007; Kulikova and Khotimchenko 2000). Apart from TAGs, many oleaginous algae exhibit the potential to accumulate long-chain PUFAs, namely arachidonic acid, eicosapentaenoic acid (EPA) and docosahexaenoic acid (DHA). *Parietochloris incisa* accumulates arachidonic acid; *Phaeodactylum tricornutum*, *Porphyridium cruentum*, *Nitzschia laevis* and *Nannochloropsis* sp. accumulate EPA; *Pavlova lutheri* accumulates both arachidonic acid and EPA; and *Schizochytrium mangrovei* and *Isochrysis galbana* accumulate DHA (Chen, Jiang, and Chen 2007; Khozin-Goldberg and Boussiba 2011; Patil et al. 2007). Long-chain PUFAs offer several health benefits; for instance, they contribute in reducing chances of cardiovascular diseases, diabetes, hypertension and autoimmune diseases. DHA plays a positive role in visual and neural health. Arachidonic acid and EPA are precursors of bioregulators, such as prostaglandins, thromboxane and other eicosanoids, which influence inflammatory processes and immune reactions (Calder and Grimble 2002). Lang et al. (2011) screened about 2,071 strains, and their study revealed that taxonomic groups belonging to glaucophytes, rhohophytes, eustigmatophytes and phaeophytes are rich sources of arachidonic acid and EPA. Similarly, other taxonomic groups belonging to haptophytes and dinophytes are rich in EPA and DHA, euglenoids in arachidonic acid and DHA, xanthophytes in arachidonic acid and cryptophytes in EPA (Lang et al. 2011).

1.5.1 MICROALGAL LIPIDS FOR BIODIESEL PRODUCTION

Currently, vegetable oils harvested from soybeans, rapeseed (Georgogianni et al. 2009), canola (Thanh et al. 2010), sunflower seeds (Harrington and D'Arcy-Evans 1985), corn (Bi, Ding, and Wang 2010) and palm (El-Araby et al. 2018) are the main plant feedstocks for biodiesel production. Alternative feedstocks from the non-edible vegetable oils jatropha (Suresh, Jawahar, and Richard 2018), karanja (Das et al. 2009), mahua (Godiganur, Suryanarayana Murthy, and Reddy 2009) and polanga (Sahoo et al. 2009) are also used for biodiesel production, as their FA composition is suitable for biodiesel production. In contrast, the high oil content and suitable FA composition of microalgae make them highly promising organisms, which have the potential to replace petroleum-based fuel completely without adversely affecting the food supply. Thus, microalgal lipids are even superior to existing oil crops (Venkata Subhash et al. 2017). FA composition significantly affects the fuel properties of algal biodiesel. For instance, lower saturated FAs result in better cold temperature properties. Similarly, the cloud point and pour properties of biodiesel increase due to the presence of long-chain saturated fatty esters. Therefore, lipid content and composition are critical factors that must be considered in the selection of microalgae species/strains to produce biofuel.

Though microalgae appear to be a promising feedstock for biodiesel due to their high lipid content and fast growth rate, several critical issues influence the production. For example, the size of microalga is too small, so the lipid cannot be easily extracted by the conventional mechanical methods (such as expelling) used for obtaining oil from oil crops. In addition, intensive dewatering of the microalgal biomass is still required to achieve better oil yield. Besides, the perception of bio-diesel from microalgal lipids is currently poor due to several biological and techni-cal restrictions and high production costs when compared to that of fossil diesel. Successful commercialization of biodiesel from microalgae depends on successfully addressing these challenges.

1.5.2 OMEGA-3 FATTY ACIDS

Omega-3 (ω-3) FAs are known as essential FAs because the human body can-not synthesize them. In addition, they stimulate many health benefits, like anti-inflammatory, reduce the risk of blood clotting, lower the TAG level, reduce blood pressure, lower the risk of diabetes and some cancers, etc, (Sahin, Tas, and Altindag 2018; Lian et al. 2010; Xie, Jackson, and Zhu 2015). EPA (C20:5, ω-3) and DHA (C22:6, ω-3) are considered typical ω-3 FAs, and their traditional source is fish oil. The fish oil may not be a sustainable option to meet the future growing demand for EPA and DHA, as it has already reached maximum global production. Moreover, the quality of fish-derived oil is not consistent and depends on the season and location (Lenihan-Geels, Bishop, and Ferguson 2013). To overcome such limitations, marine algae have attracted attention as a potentially novel source for ω-3 FAs (Gupta, Barrow, and Puri 2012). Algae belonging to the genera *Schizochytrium*, *Ulkenia* and *Crypthecodinium* are used for the production of DHA (Barclay, Meager, and Abril 1994; Borowitzka 2013a; Klok et al. 2014). Traditionally, *Phaeodactylum tri-cornutum*, *Nannochloropsis* and *Nitzchia* are used for EPA production (Wen and Chen 2003). Microalgae as novel sources of ω-3 FAs could eliminate many of the taste and odor problems associated with fish and discard the shortcomings of the fish oil–based process. Commercial production of microalgal PUFAs appears to be more realistic than the production of biodiesel. DSM Nutritional Products use the microalgae *Crypthecodinium cohnii* to make their DHASCO oil by fermentation for the infant formula market. DSM also uses another strain, *Schizochytrium* spp., to produce oil containing DHA:EPA in 2:1 ratio, which matches high-quality fish oils. This product is marketed as Life's Omega, with minimum DHA and EPA levels of 24% and 12%, respectively (Winwood 2013). However, the production of EPA is still restricted to the laboratory scale. The relatively low accumulated biomass and EPA, the slow growth rate of algae, lack of efficient cell disruption methods and expen-sive downstream processes have hindered the industrial production of EPA from algae (Ji, Ren, and Huang 2015). Thus, for the successful commercialization of EPA production from algae, there is a need for bioprospecting for algae species with fast growth rates, high biomass content and high oil accumulating capability. This needs to be complemented by the development of efficient and economical downstream processes for the extraction and purification of EPA with fewer steps and a lower level of solvents. Recently, in 2018, DSM and Evonik established a joint venture to produce EPA and DHA from marine algae for animal nutrition. This EPA and DHA–rich algae oil is being marketed as an alternative to fish oil and will be ready for sale in 2019.

1.6 BIOACTIVE COMPOUNDS FROM MICROALGAE IN PHARMACEUTICAL AND THERAPEUTIC APPLICATIONS

Microalgal research has gained considerable attention lately in relation to pharmaceutical and therapeutic applications due to the ability of microalgae to produce several bioactive metabolites such as sterols, polyphenols, polysaccharides, carotenoids and PUFAs. These compounds exhibit several pharmacological activities such as anti-bacterial, anti-fungal, anti-parasitic, anti-coagulant and anti-tumor activities (Olasehinde, Olaniran, and Okoh 2017; Mimouni et al. 2012). Screening of algae for potential pharmaceutical applications and antibiotics began in the 1950s but received greater attention in the last decade (Borowitzka 2014). The advancements in screening and analytical technology have led to the discovery of many bioactive compounds from microalgae. Due to space limitations, only microalgae sterols, microalgae extract (and subsequently purified products) and bioactive peptides from microalgae of pharmaceutical or therapeutic importance will be discussed in detail in this section. For recent detailed reviews on microalgae compounds of pharmacological significance, please refer to (Falaise et al. 2016; Sathasivam et al. 2017; de Vera et al. 2018).

1.6.1 MICROALGAE STEROLS

Sterols from plant sources, commonly called phytosterols, are well known to reduce cholesterol and prevent cardiovascular disorders in humans. Whereas PUFAs derived from microalgae and their roles in promoting human health have been extensively studied, research on sterols from microalgae is poorly represented. However, there is accumulating evidence that phytosterols from microalgae possess anti-inflammatory, anti-cancer, anti-atherogenic and anti-oxidative activities and offer protection against nervous system disorders (Luo, Su, and Zhang 2015). Microalgae produce several types of phytosterols, like brassicasterol, sitosterol and stigmasterol, based on the taxonomic classification (Volkman 2003). The amounts of sterols can vary depending on the physiological condition (Fábregas et al. 1997). Microalgae of the members of Chlorophyceae (green algae), Phaeophyceae (brown algae) and Rhodophyceae (red algae) are known to produce sterols (Lopes et al. 2013). Recently, *Pavlova lutheri*, *Tetraselmis* sp. M8 and *Nannochloropsis* sp. BR2 were identified as the highest microalgal phytosterol producers (0.4%—2.6 % DW). Further, a two-fold increase (5.1% w/w) in phytosterol accumulation was achieved in *P. lutheri* by modifying salinity, nutrient and cultivation duration (Ahmed, Zhou, and Schenk 2015). The plant sterols that are reduced to the stanols (sterols saturated at C-5) are commonly used in vegetable oils and are considered effective in lowering plasma cholesterol (Piironen et al. 2000). Studies suggest that the production of phytosterols by microalgae is comparable to or higher than plant sources, and hence it may be more advantageous to use microalgae as an alternative phytosterol source.

Phytosterols from *Dunaliella tertiolecta* and *Chlorella vulgaris* have been extensively studied for their anti-cancer and anti-inflammatory activities. In mice studies, it was shown that sterols from *C. vulgaris*, 7-dehydroporiferasterol peroxide and 7-oxocholesterol showed effective anti-inflammatory activities on TPA

(12-*O*-tetradecanoylphorbol-13-acetate), a potent tumor promoter. Also, ergosterol peroxide from *C. vulgaris* remarkably inhibited tumor progression by TPA and DMBA (7,12-dimethylbenz [*a*] anthracene), an immunosuppressor and tumor initiator (Yasukawa et al. 1996). In a similar study, phytosterols extracted from *D. tertiolecta* were orally administered in rats, and an *in vivo* neuromodulatory action was observed in selective brain areas (Francavilla et al. 2012). In addition, a mixture of 7-dehydroporiferasterol and ergosterol purified from *D. tertiolecta* provoked a suppressive effect on cell proliferation and a reduction of pro-inflammatory cytokines production, suggesting that purified extract from *D. tertiolecta* reduced inflammation-related immune reactions in sheep (Caroprese et al. 2012). Further, purified phytosterols from *D. tertiolecta* showed an anti-proliferative effect on sheep peripheral blood mononuclear cells, regardless of their chemical composition and concentration (Ciliberti et al. 2017).

In addition to cholesterol lowering, anti-inflammatory and anticancer properties, phytosterols from microalgae possess anti-microbial ability. Saringosterol, from Chilean brown alga, has been shown to possess growth inhibitory activity against *Mycobacterium tuberculosis*, the causative agent of tuberculosis. Saringosterol was also identified in some species of green microalgae from the Prasinophyceae class and can be an alternative source (Volkman 2003). These studies clearly indicate the therapeutic applications of phytosterols; however, the safety and commercial feasibility of microalgae as a source of bioactive molecules in medical applications remain to be fully explored.

1.6.2 EXTRACTS AND PURIFIED COMPOUNDS OF MICROALGAL ORIGIN

A wide range of bioactivities, such as anti-oxidant, anti-biotic, anti-viral, anti-cancer, anti-inflammatory and anti-hypotensive, were reported from extracts and purified molecules of microalgae (Borowitzka 2013b). *In vitro* screening studies conducted by the National Cancer Institute have revealed that 10% of 300 fresh water algae and 5% of 532 marine and fresh water strains showed anti-bacterial and anti-fungal activity, respectively (Borowitzka 2014). There is substantial evidence of anti-bacterial activity of microalgae against bacteria belonging to human pathogens and in the food industry (Shannon and Abu Ghannam 2016). Ganesh Sanmukh et al. (2014) reported that methanol and ethyl acetate extracts of *Chlorella* and *Chlamydomonas* species exhibited anti-bacterial activity against *Staphylococcus* spp. and *Escherichia coli* (Ganesh Sanmukh et al. 2014). Likewise, by using the disc diffusion method, ethanol and methanol extracts of several fresh water microalgae were shown to have a growth inhibitory effect against *E. coli, Staphylococcus aureus,* and *Salmonella typhi* (Prakash, Antonisamy, and Jeeva 2011). In addition, methanol extracts from marine microalga, *Dunaliella salina*, and fresh water microalga, *Pseudokirchneriella subcapitata*, were tested on bacteria responsible for human external otitis. Both algal extracts inhibited the growth of *S. aureus, Pseudomonas aeruginosa, E. coli,* and *Klebsiella* spp. with a minimum inhibitory concentration between 1.4×10^9 and 2.2×10^{10} cells/mL (Pane et al. 2015). In a similar manner, different solvent and aqueous extracts of *D. salina* suppressed the growth of *E. coli* and *S. aureus* (Mendiola et al. 2008). Interestingly, methanol extracts from one of the *Nannochloris* sp. strains, SBL1 isolated from the Red Sea, significantly reduced the viability of a parasitic species, *Leishmania infantum*, the causal agent of infantile leishmaniasis. Extracts from other microalgae strains from the Red Sea, *Picochlorum* sp. SBL2 and *Nannochloris* sp. SBL4, showed anti-tumor activity by significantly reducing the tumoral cells (HepG2 and HeLa) but with reduced toxicity against the non-tumoral murine stromal (S17) cells (Pereira et al. 2015).

Microalgae also display anti-fungal activity, but there are fewer published reports on the fungicidal activity of microalgae when compared to bactericidal activity. In a large-scale screening study to identify new anti-fungal agents from culture filtrates and solvent extracts of 132 marine and 400 fresh water microalgal cultures, Kellam et al. (1988) found that marine microalgae species were better in comparison to fresh water species. Characterization of anti-fungal compounds from several microalgae revealed interesting trends in the specificity of their action (Kellam et al. 1988). High-molecular-weight polysaccharides from the diatom *Chaetoceros lauderi* (Walter and Mahesh 2000) and gambieric acids from the dinoflagellate *Gambierdiscus toxicus* (Nagai et al. 1993) showed inhibitory activity against dermatophytes and molds, but no activity was detected against the yeasts *Candida albicans* or *Saccharomyces cerevisiae*. In contrast, the diatom *Thalassiothrix frauenfeldii* showed activity against molds and yeasts but not against dermathophytes (Walter and Mahesh 2000). The solvent and aqueous extracts from *D. salina* exhibited activity against both yeast and filamentous forms of the fungi *C. albicans* and *Aspergillus niger*, respectively (Mendiola et al. 2008). Other compounds of microalgae origin that have shown anti-fungal activity are pigments like chlorophyll-a, chlorophyll-b and β-carotene from *Chlorococcum humicola* (Bhagavathy, Sumathi, and Jancy Sherene Bell 2011); phycobiliproteins from *Porphyridium aerugineum* (Najdenski et al. 2013); and polyene-polyhydroxy metabolites (amphidinols) from the dinoflagellate *Amphidinium klebsii* (Satake et al. 1991).

In addition to anti-bacterial and anti-fungal activities, screening of microalgae and cyanobacteria for anti-viral compounds is rapidly increasing. The most common anti-viral compounds extracted from microalgae are polysaccharides. One of the notable molecules isolated from the cyanobacteria *Spirulina (Arthrospira platensis)* is spirulan, a sulphated polysaccharide. Spirulan and a spirulan-like molecule, both from *S. platensis*, have shown potent anti-viral activity against the human immunodeficiency virus type 1 (HIV-1) and the herpes simplex virus type 1 (HSV-1) without showing cytotoxic effects (Hayashi et al. 1996; Rechter et al. 2007). Other important polysaccharides that were demonstrated to possess anti-viral activity are naviculan from the diatom *Navicula directa*, which showed activity against several enveloped viruses such as influenza virus type A (IFV-A) or HIV-1 or HSV-1 (Lee et al. 2006), or A1 and A2 from *Cochlodinium polykrikoides* (Hasui et al. 1995). Similarly, another sulphated polysaccharide, p-KG03 extracted from *Gyrodinium impudicum*, exhibited activity against several strains of influenza viruses but was not active against HSV-1 and HSV-2 (Yim et al. 2004; Kim et al. 2011). In addition to polysaccharides, short-chain FAs and palmitic and α-linoleic acids from *D. salina* were shown to reduce the infectivity of HSV-1 (Santoyo et al. 2012). Although the mechanisms of action of these anti-viral compounds are not clear yet, the polysaccharides have been proposed to act by blocking the internalization of the virus into the host cells or inhibiting the binding or suppressing of DNA replication and protein synthesis (Ahmadi et al. 2015); however, there is no clear mention on modes of action of FA and pigments.

Like humans, aquatic animals are frequently affected by bacteria, fungi and viruses, which leads to significant production losses in terms of aquaculture. Aquaculture is another area in which considerable efforts have been put forth to identify microalgal compounds for controlling aquaculture pathogens or diseases. Several reports of microalgae used as whole cells or aqueous/solvent extracts or purified compounds, such as PUFAs or pigments, have shown promising results in controlling the target pathogens such as *Vibrio* and *Pseudomonas* species (Walter and Mahesh 2000; Desbois, Mearns-Spragg, and Smith 2009; González-Davis et al. 2012; Das et al. 2005; Gastineau, Hardivillier et al. 2012). Interestingly, because microalgae occur in the food chain of fish and shrimp, it was shown that the anti-bacterial activity could be successfully attained by feeding processed or live microalgae directly to fish larval stages or through brine shrimp (Marques et al. 2006; Austin and Day 1990; Austin, Baudet, and Stobie 1992). Moreover, micro-algae have also shown some promise in controlling fungal and viral pathogens of aquaculture. Fungi belonging to the genera *Fusarium* and *Aspergillus* are the com-monly occurring contaminants in aquaculture. *Fusarium moniliforme*, which causes black gill disease in shrimps, was shown to be controlled by organic extracts of the green microalgae *Scenedesmus quadricauda* and *Chlorella pyrenoidosa* (Abedin and Taha 2008), whereas *Aspergillus fumigatus*, which produces toxins in marine bivalves, was shown to be controlled by compounds produced by *Chaetocerus laud-eri, Gambierdiscus toxicus* and *Chlorella vulgaris* (Walter and Mahesh 2000; Nagai et al. 1993; Ghasemi et al. 2007). Though several menacing viruses—such as infec-tious pancreatic necrosis virus (IPNV), affecting a wide range of finfish; white spot syndrome virus (WSSV) and yellow head virus (YHV), causing significant losses in shrimp culture; and the ostreid herpesvirus-1 (OsHV-1), causing high mortality in marine bivalves (Ahne 1994; Crane and Hyatt 2011)—are common in aquacul-ture, very few studies have been reported on the anti-viral activity of microalgae in controlling these pathogens. In one study, several exo- and endo-cellular extracts of various microalgae—*Dunaliella tertiolecta, Ellipsoidon* sp., *Isochrysis galbana, Chlorella autotrophica* and *Porphyridium cruentum*—were tested and shown to inhibit the infection of viral hemorrhagic septicemia virus (VHSV) *in vitro* in epi-thelioma papulosum cyprinid (EPC) cells (Fabregas et al. 1999).

1.6.3 Bioactive Peptides of Microalgae Origin

As described in earlier sections, microalgae have been recognized as an excellent source of protein. Typically, protein hydrolysates from microalgae are nutritionally beneficial, as they have better digestibility (Frokjaer 1994; Kose and Oncel 2015; Anahite Ovando et al. 2016; Morris et al. 2008) and higher nitrogen (N) retention (Boza et al. 1995), when compared to unhydrolyzed biomass. Protein hydrolysates are the result of hydrolysis and contain peptides of different sizes and molecular weight distributions, depending upon the degree of hydrolysis (Guadix 2000). In comparison to the native form, protein hydrolysates exhibit superior physical properties, such as emulsification, foaming and solubility (Medina et al. 2015). More importantly, the resultant peptides, which are inactive within the native sequence of the parent protein, gain biological activity (commonly called "bioactive" peptides) and show a positive impact on several physiological functions. It was demonstrated that bioactive peptides exhibit a broad spectrum of functions such as anti-oxidant (Karawita 2007; Sheih, Wu, and Fang 2009), anti-cancer (Sheih et al. 2009), anti-hypertensive (Murray and FitzGerald 2007; E. Merchant, A. Andre, and Sica 2002), immune-modulatory (Morris et al. 2007), hepatic-protective (Kang et al. 2012) and anti-coagulant activities (Athukorala et al. 2007). Vo et al. (2013), isolated two bio-active peptides (each consisting of six AA residues) from the enzymatic hydrolysis of *Spirulina maxima* that displayed anti-inflammatory and anti-allergic activity (Vo, Ryu, and Kim 2013). Similarly, Nguyen et al. (2013) isolated and characterized a bio-active oligopeptide consisting of Met-Pro-Asp-Trp from deoiled *N. oculata* hydro-lysate, which promoted osteoblast differentiation in MG-63 and D1 cells and hence was proposed to provide a possibility for treating bone diseases (Nguyen et al. 2013). Sheih et al. (2009) isolated and conducted detailed characterization of a purified hendeca-peptide, Val-Glu-Cys-Tyr-Gly-Pro-Asn-Arg-Pro-Gln-Phe, from the pepsin hydrolysate of protein waste produced as a by-product of algae essence from *C. vul-garis* (Sheih, Fang, and Wu 2009). This peptide exhibited potent inhibitory activity against angiotensin-converting enzyme (ACE) and was resistant to *in vitro* digestion by gastrointestinal enzymes. Moreover, the purified peptide was stable at tempera-tures from 40 to 100 °C and a wide pH range of 2 to 12, indicating its possible role in pharmaceutical applications.

Apart from bioactive peptides derived from protein hydrolysis, most organisms are known to produce small proteins or peptides known as anti-microbial peptides (AMPs), which are considered part of the innate immune system and act as the first line of defense (Zhang and Gallo 2016). Several micro and macro flora and fauna such as fish, bacteria, fungi, sponges, crabs, jellyfish, etc., produce AMPs (Kang, Seo, and Park 2015). Surprisingly, though microalgae occupy a significant portion of marine life, there is no notable report of AMPs from microalgae. However, cyano-bacteria are known to produce several linear and cyclic peptides that show bioactivity in native forms (Welker and Von Döhren 2006; Berry et al. 2008). For example, fila-mentous cyanobacteria such as *Lyngbya*, *Symploca* and *Oscillatoria* were reported to produce short peptides of non-ribosomal origin (Tan 2013). Some of these pep-tides exhibit allelopathic effects against pathogenic bacteria as a mechanism to gain an ecological advantage (Shannon and Abu Ghannam 2016; Kearns and Hunter 2001), whereas a few peptides possess potent therapeutic activity (Sathasivam et al. 2017). For example, several linear and cyclic peptides and depsipeptides isolated from cyanobacteria possess protease inhibitory activity and are used for the treat-ment of pulmonary emphysema, strokes and coronary artery occlusions (Skulberg 2007). Similarly, depsipeptides like cyanopeptolin, oscillapeptin, micropeptin, microcystilide and nostocyclin possess inhibitory activity against enzymes such as plasmin, thrombin, trypsin and chymotrypsin and hence offer possible application in pharmaceuticals (Patterson 1996).

Taken together, these studies conducted with extracts from microalgae strongly suggest that microalgae can act as a promising source for a variety of compounds that have pharmaceutical and therapeutic importance. However, it should be noted that in very few studies the extracts were characterized well enough to identify the specific compounds responsible for bioactivity. One such compound that has been well described is a blue pigment, marennine, produced by diatoms belonging to the genus *Haslea* (Gastineau et al. 2014). The range of activity of purified extra-cellular marennine was extensively verified in a dose-dependent manner against pathogenic *Vibrio tasmaniensis* LGP32 species, and it was observed that 10 µg/ mL concentration of marennine slowed down the growth of tested *Vibrio* species and 1 mg/ mL totally inhibited growth (Gastineau, Pouvreau et al. 2012). In a different study, Herrero et al. (2006) reported anti-microbial activity of pressurized liquid extracts from *D. salina* containing as many as 15 different volatile compounds and several FAs (mainly palmitic, alpha-linolenic and oleic acids) (Mendiola et al. 2008; Herrero et al. 2006). In the volatile compounds fraction, β-cyclocitral, alpha- and β-ionone, neophytadiene and phytol were identified, all of which have previously been described as anti-microbial agents.

The production of different bioactive molecules by microalgae might be primarily species specific; however, the activity of a given anti-microbial compound also depends on the target species. For example, bacterial membranes might be the target for some microalgal compounds, though the exact mechanism of the anti-bacterial action is not fully understood. The phycobiliproteins from the red microalgae *Porphyridium* species exhibited clear antagonistic activity against the gram-positive bacteria *S. aureus* and *Streptoccocus pyogenes* but it had no influence against the gram-negative bacteria *E. coli* and *P. aeruginosa* (Najdenski et al. 2013). In contrast, EPA, an FA isolated from *Phaeodactylum tricornutum*, is active against a range of both gram-negative and gram-positive bacteria, including multi-drug-resistant *S. aureus* at micromolar concentrations (Desbois, Mearns-Spragg, and Smith 2009).

In general, the anti-microbial activity of microalgae is attributed to several compounds belonging to a variety of chemical classes such as acetogenins, indoles, phenols, FAs, terpenes and volatile halogenated hydrocarbons, and the attempts to characterize/identify these new compounds have not advanced (Amaro, Guedes, and Malcata 2011). In addition, the bioactive pool of molecules produced from a given species might vary significantly depending on the growth or environmental conditions and the extraction methodology. For example, the biological activity of microalgal compounds was mainly found on non-aqueous extracts, indicating that bioactive compounds in microalgae are mostly hydrophobic and can be preferably extracted with organic solvents (Prakash, Antonisamy, and Jeeva 2011; Cannell, Owsianka, and Walker 1988; Pane et al. 2015; Pereira et al. 2015). In studies by Pereira et al. (2015), among the microalgae strains tested from the Red Sea collection, *Picochlorum* sp. SBL2 was observed to produce the highest total phenolic content, and the major compounds identified were coumaric, gallic acids and salicylic acids. All other strains tested were shown to produce β-carotene, neoxanthin, lutein violaxanthin and zeaxanthin, whereas canthaxanthin was only observed in *Picochlorum* sp. SBL2 (Pereira et al. 2015), indicating the specificity in producing a particular set of bioactive molecules by certain specific microalgae.

As described earlier, in many of the cited reports, the activity of almost all microalgae extracts might be due to the presence of more than one compound. It is common to perceive that at least in some cases, the bioactivity might be a synergistic effect of multiple molecules. Hence, there is a need for significant scientific efforts in the areas of extraction, fractionation and characterization of individual bioactive molecules in a dose-dependent manner to develop a pathogen- or disorder-specific pharmaceutical or therapeutic activity for potential commercialization. Between 1965 and 2006, around 18,500 new compounds were isolated from marine sources, and yet around 97% of these compounds have not been characterized (Blunt et al. 2008). Considering their extreme physiological state, marine microalgae possess the additional advantage of substantial metabolic plasticity, and their secondary metabolism can easily be stimulated by external or applied stresses such as nitrogen depletion (Guedes, Amaro, and Malcata 2011). Hence, microalgae, being an important component of the marine environment, present a unique opportunity to discover novel metabolites of economic importance.

1.7 CONCLUSION AND FUTURE PERSPECTIVES

Microalgae offer unequaled potential to provide a variety of sustainable products to humankind. In addition, they contribute to economic growth, mitigate elevated CO_2 levels and reduce greenhouse gas emissions. They have demonstrated the potential to produce biofuels, proteins for food/feed and aquaculture, bioethanol, anti-oxidants, pigments, ω-3 FAs and many medicinal biomolecules, as discussed earlier. However, an array of key challenges remain to be addressed before the full potential of algae can be unleashed at a commercial scale. For instance, most of the recent progress in algal cultivation, crop protection, dewatering and harvesting is a result of numerous efforts carried out to develop commercially viable biofuels from microalgae to replace fossil fuels. However, despite the large efforts and significant advancements in technology, several logjams remain to be overcome before biofuels from algae become tangible.

Although the challenges and bottlenecks will vary according to the product desired, the major problems that obstruct the transition of microalgae from pilot to commercial scale are:

1. Lack of robust algal strains for the desired products to sustainably produce copious amounts of biomass on a large commercial scale.
2. Development of cost-effective cultivation systems and economic and low-energy-consuming harvesting and dewatering/drying processes for treating the microalgal biomass. For instance, large photobioreactors (PBRs) are required to cultivate algae in a clean condition to produce proteins or bioactive biomolecules.
3. Lack of scalable extraction and purification methods, particularly for the purification of target nutraceutical compounds from the production strains.
4. Risk of a large environmental footprint due to lack of recycling of water, energy and nutrients.
5. Inability to compete with synthetic or plant-derived products due to low levels of target compounds, high downstream recovery costs and lack of multiple products from one strain.

6. Lack of genetic engineering tools for manipulation of microalgal strains to achieve high-level production of target compounds and limited knowledge of metabolic pathways, proteins, metabolites and genomes due to very few "-omics" studies to date.
7. Regulatory, safety and consumer acceptance concerns, as many microalgae strains produce toxins, and there is lack of awareness among consumers.
8. Lack of innovative food products where microalgae or their components can be used.

Some of the issues regarding strain suitability and robustness are already being addressed by more intensive and intelligent bioprospecting, keeping in mind the product and cultivation conditions. For instance, in-depth bioprospecting and screening have established the potential for algae in producing several bioactive compounds and peptides. Similarly, screening of high-yielding, robust or seasonal algal strains and optimization of growth parameters to enhance algal pigment production should help improve the commercial feasibility of pigment production from algae. More in-depth "-omics," systems and synthetic biology studies are required to shed light on pathways and processes regulating protein synthesis, cell division, signaling and synthesis of nutraceuticals and other bioactive metabolites. Algal molecular biology and genetic engineering are still in its infancy, and much work is required to develop genetic transformation and genome engineering tool kits for algae to enable gene editing/overexpression to produce high titers of desired proteins or biomolecules, just like in yeast or *E. coli*. Further, in a way to improve the economics of algal biomass, industries and research groups must come up with the "biorefinery concept" where multiple products from the algal feedstock are harvested to make the process economically attractive.

In today's world, there is an ever-increasing demand for environmentally friendly, healthy and sustainable bioinspired green products by replacing synthetic and petroleum-derived products. In such a scenario, microalgae, using the power of photosynthesis, unequivocally holds an upper hand and offers a plethora of biological products with unprecedented benefits for human health and society. Microalgal biofuels may be still a distance away from economic reality, but microalgae as a platform for commercial production of proteins, novel pigments, bioactive compounds and recombinant vaccines/proteins is well poised to disrupt the world food, feed, aquaculture and nutraceutical markets.

REFERENCES

Abd El Baky, H. H., G. S. El Baroty, and E. A. Ibrahem. 2015. "Functional characters evaluation of biscuits sublimated with pure phycocyanin isolated from *spirulina* and *spirulina* biomass." *Nutricion Hospitalaria* 32 (1):231–241. doi: 10.3305/nh.2015.32.1.8804.

Abedin, R. M. A., and H. M. Taha. 2008. "Antibacterial and antifungal activity of cyanobacteria and Green microalgae. Evaluation of medium components by Plackett-Burman design for antimicrobial activity of Spirulina platensis." *Global Journal of Biochemistry and Biotechnology* 3:22–31.

Administration, U.S. Food and Drug. 2018. Guidance for Industry: Estimating Dietary Intake of Substances in Food.

Ahmadi, A., S. Zorofchian, S. A. Bakar, and K. Zandi. 2015. "Antiviral potential of algae polysaccharides isolated from marine sources: A review." *BioMed Research International*:1–10.

Ahmed, F., W. Zhou, and P. M. Schenk. 2015. "Pavlova lutheri is a high-level producer of phytosterols." *Algal Research* 10:210–217. doi: 10.1016/j.algal.2015.05.013.

Ahne, W. 1994. "Viral infections of aquatic animals with special reference to Asian aquaculture." *Annual Review of Fish Diseases* 4:375–388. https://doi.org/10.1016/0959-8030(94)90036-1.

Aikawa, S., Y. Izumi, F. Matsuda, T. Hasunuma, J. S. Chang, and A. Kondo. 2012. "Synergistic enhancement of glycogen production in Arthrospira platensis by optimization of light intensity and nitrate supply." *Bioresource Technology* 108:211–215.

Allard, B., and J. Templier. 2000. "Comparison of neutral lipid profile of various trilaminar outer cell wall (TLS)-containing microalgae with emphasis on algaenan occurrence." *Phytochemistry* 54 (4):369–380. doi: 10.1016/S0031-9422(00)00135-7.

Alvarez, R., B. Vaz, H. Gronemeyer, and A. R. de Lera. 2014. "Functions, therapeutic applications, and synthesis of retinoids and carotenoids." *Chemical Reviews* 114 (1):1–125.

Amaro, H. M., A. C. Guedes, and F. X. Malcata. 2011. "Antimicrobial activities of microalgae: An invited review." *Science Against Microbial Pathogens: Communicating Current Research and Technological Advances*: 1272–1280.

Anahite O. C., J. De Carvalho, G. Pereira, P. Jacques, V. Thomaz-Soccol, and C. Soccol. 2016. "Functional properties and health benefits of bioactive peptides derived from Spirulina: A review." *Food Reviews International* 34 (1):31–51.

Athukorala, Yasantha, Ki-Wan Lee, S. J. Kim, and You-Jin Jeon. 2007. "Anticoagulant activity of marine green and brown algae collected from Jeju Island in Korea." *Bioresource Technology* 98 (9):1711–1716.

Austin, B., and J. G. Day. 1990. "Inhibition of prawn pathogenic Vibrio spp. By a commercial spray-dried preparation of Tetraselmis suecica." *Aquaculture* 90 (3–4):389–392.

Austin, B., E. Baudet, and M. Stobie. 1992. "Inhibition of bacterial fish pathogens by Tetraselmis suecica." *Journal of Fish Diseases* 15 (1):55–61. doi: 10.1111/j.1365-2761.1992.tb00636.x.

Balder, H. F., J. Vogel, M. C. Jansen, M. P. Weijenberg, P. A. van den Brandt, S. Westenbrink, R. van der Meer, and R. A. Goldbohm. 2006. "Heme and chlorophyll intake and risk of colorectal cancer in the Netherlands cohort study." *Cancer Epidemiology, Biomarkers and Prevention* 15 (4):717–725.

Barclay, W. R., K. M. Meager, and J. R. Abril. 1994. "Heterotrophic production of long chain omega-3 fatty acids utilizing algae and algae-like microorganisms." *Journal of Applied Phycology* 6 (2):123–129.

Barka, A., and C. Blecker. 2016. "Microalgae as a potential source of single-cell proteins." *Biotechnology, Agronomy, Society and Environment* 20:427–436.

Bashir, K., M. Imran, and J.-S. Choi. 2017. "Clinical and physiological perspectives of β-glucans: The past, present, and future." *International Journal of Molecular Sciences* 18 (9):1906. doi: 10.3390/ijms18091906.

Begum, H., and Fatimah M. D. Yusoff, Sanjoy Banerjee, Helena Khatoon, and Mohamed Shariff. 2016. "Availability and utilization of pigments from microalgae." *Critical Reviews in Food Science and Nutrition* 56 (13):2209–2222. doi: 10.1080/10408398.2013.764841.

Berry, J. P., M. Gantar, Mario H. Perez, Gerald Berry, and Fernando G. Noriega. 2008. "Cyanobacterial toxins as allelochemicals with potential applications as algaecides, herbicides and insecticides." *Marine Drugs* 6 (2):117–146. doi: 10.3390/md20080007.

Bhagavathy, S., P. Sumathi, and I. Jancy Sherene Bell. 2011. "Green algae Chlorococcum humicola-a new source of bioactive compounds with antimicrobial activity." *Asian Pacific Journal of Tropical Biomedicine* 1 (1, Supplement):S1-S7. doi: 10.1016/S2221-1691(11)60111-1.

Bhalamurugan, G. L., O. Valerie, and L. Mark. 2018. "Valuable bioproducts obtained from microalgal biomass and their commercial applications: A review." *Environmental Engineering Research* 23 (3):229–241. doi: 10.4491/eer.2017.220.

Bi, Y., D. Ding, and D. Wang. 2010. "Low-melting-point biodiesel derived from corn oil via urea complexation." *Bioresource Technology* 101 (4):1220–1226. doi: S0960-8524(09)01242-5 [pii] 10.1016/j.biortech.2009.09.036.

Bixler, H. J., and H. Porse. 2011. A decade of change in the seaweed hydrocolloids industry. 23.

Bleakley, S., and M. Hayes. 2017. "Algal Proteins: Extraction, Application, and Challenges Concerning Production." *Foods (Basel, Switzerland)* 6 (5):33. doi: 10.3390/foods6050033.

Blunt, J. W., B. R. Copp, W. P. Hu, M. H. Munro, P. T. Northcote, and M. R. Prinsep. 2008. "Marine natural products." *Natural Product Reports* 25 (1):35–94.

Borowitzka, M. A. 2013a. "High-value products from microalgae—their development and commercialisation." *Journal of Applied Phycology* 25 (3):743–756.

Borowitzka, M. A. 2013b. High-value products from microalgae—Their development and commercialisation. *Journal of Applied Phycology* 25 (3):743–756.

Borowitzka, M. A. 2014. "Pharmaceuticals from Algae." In *Fundamentals in Biotechnology, BIOTECHNOLOGY: Special Processes for Products, Fuel and Energy*, Volume VII, edited by Horst W. Doelle, J. Stefan Rokem, and Marin Berovic. 94–117. EOLSS Publishers, Oxford, UK.

Boza, Julio J., Olga Martínez-Augustin, Luis Baró, M. Dolores Suarez, and Angel Gil. 1995. "Protein v. enzymic protein hydrolysates. Nitrogen utilization in starved rats." *British Journal of Nutrition* 73 (1):65–71. doi: 10.1079/bjn19950009.

Branyikova, I., B. Marsalkova, J. Doucha, T. Branyik, K. Bisova, V. Zachleder, and M. Vitova. 2011. "Microalgae–novel highly efficient starch producers." *Biotechnology and Bioengineering* 108 (4):766–776. doi: 10.1002/bit.23016.

Brown, M. R., S. W. Jeffrey, J. K. Volkman, and G. A. Dunstan. 1997. "Nutritional properties of microalgae for mariculture." *Aquaculture* 151 (1):315–331. doi: 10.1016/S0044-8486(96)01501-3.

Buscemi, S., D. Corleo, F. Di Pace, M. L. Petroni, A. Satriano, and G. Marchesini. 2018. "The Effect of Lutein on Eye and Extra-Eye Health." *Nutrients* 10 (9).

Calder, P. C., and R. F. Grimble. 2002. "Polyunsaturated fatty acids, inflammation and immunity." *European Journal of Clinical Nutrition* 56 (Suppl 3):S14–S19. doi: 10.1038/sj.ejcn.1601478.

Cannell, R. J. P., A. M. Owsianka, and J. M. Walker. 1988. "Results of a large-scale screening programme to detect antibacterial activity from freshwater algae." *British Phycological Journal* 23 (1):41–44. doi: 10.1080/00071618800650051.

Caporgno, M. P., and A. Mathys. 2018. "Trends in Microalgae Incorporation Into Innovative Food Products With Potential Health Benefits." *Frontiers in Nutrition* 5:58–58. doi: 10.3389/fnut.2018.00058.

Caroprese, M., M. Albenzio, M. G. Ciliberti, M. Francavilla, and A. Sevi. 2012. "A mixture of phytosterols from Dunaliella tertiolecta affects proliferation of peripheral blood mononuclear cells and cytokine production in sheep." *Veterinary Immunology and Immunopathology* 150 (1–2):27–35. doi: S0165-2427(12)00293-0 [pii] 10.1016/j.vetimm.2012.08.002.

Carriquiry, M. A., X. Du, and G. R. Timilsina. 2011. "Second generation biofuels: Economics and policies." *Energy Policy* 39 (7):4222–4234. doi: 10.1016/j.enpol.2011.04.036.

Chao, Pi-Yu, M.-Y. Huang, W.-D. Huang, K.-H. R. Lin, S.-Y. Chen, and C.-M. Yang. 2018. "Study of chlorophyll-related compounds from dietary spinach in human blood." *Notulae Botanicae Horti Agrobotanici Cluj-Napoca* 46 (2). doi: 10.15835/nbha46210918.

Chen, G.-Q., Y. Jiang, and F. Chen. 2007. "Fatty acid and lipid class composition of the eicosapentaenoic acid-producing microalga, Nitzschia laevis." *Food Chemistry* 104 (4):1580–1585. doi: 10.1016/j.foodchem.2007.03.008.

Choi, S. P., M. T. Nguyen, and S. Jun Sim. 2010. "Enzymatic pretreatment of Chlamydomonas reinhardtii biomass for ethanol production." *Bioresource Technology* 101 (14):5330–5336. doi: 10.1016/j.biortech.2010.02.026.

Ciliberti, M. G., M. Francavilla, S. Intini, M. Albenzio, R. Marino, A. Santillo, and M. Caroprese. 2017. "Phytosterols from Dunaliella tertiolecta Reduce Cell Proliferation in Sheep Fed Flaxseed during Post Partum." *Marine Drugs* 15 (7):216. doi: 10.3390/md15070216.

Conceição, L. E. C., M. Yúfera, P. Makridis, S. Morais, and M. T. Dinis. 2010. "Live feeds for early stages of fish rearing." *Aquaculture Research* 41 (5):613–640. doi: 10.1111/j.1365-2109.2009.02242.x.

Crane, M., and A. Hyatt. 2011. "Viruses of fish: An overview of significant pathogens." *Viruses* 3 (11):2025.

Das, B., J. Pradhan, P. Pattnaik, B. R. Samantaray, and S. Samal. 2005. "Production of antibacterials from the freshwater alga *Euglena viridis* (Ehren)." *World Journal of Microbiology and Biotechnology* 21 (1):45–50.

Das, L. M., D. K. Bora, S. Pradhan, M. K. Naik, and S. N. Naik. 2009. "Long-term storage stability of biodiesel produced from Karanja oil." *Fuel* 88 (11):2315–2318. doi: 10.1016/j.fuel.2009.05.005.

Davinelli, S., M. E. Nielsen, and G. Scapagnini. 2018. "Astaxanthin in Skin Health, Repair, and Disease: A Comprehensive Review." *Nutrients* 10 (4).

De Rosa, S. C., J. M. Brenchley, and M. Roederer. 2003. "Beyond six colors: A new era in flow cytometry." *Nature Medicine* 9:112. doi: 10.1038/nm0103-112. www.nature.com/articles/nm0103-112#supplementary-information.

de Vera, C. R., G. D. Crespin, A. H. Daranas, S. M. Looga, K. E. Lillsunde, P. Tammela, M. Perala, V. Hongisto, J. Virtanen, H. Rischer, C. D. Muller, M. Norte, J. J. Fernandez, and M. L. Souto. 2018. "Marine microalgae: Promising source for new bioactive compounds." *Marine Drugs* 16 (9). doi: md16090317 [pii] 10.3390/md16090317.

Del Campo, J. A., H. Rodríguez, J. Moreno, M. Á. Vargas, J. Rivas, and M. G. Guerrero. 2001. "Lutein production by Muriellopsis sp. in an outdoor tubular photobioreactor." *Journal of Biotechnology* 85 (3):289–295.

Dembitsky, V. M. 1996. "Betaine ether-linked glycerolipids: Chemistry and biology." *Progress in Lipid Research* 35 (1):1–51.

Dembitsky, V. M., and O. A. Rozentsvet. 1990. "Phospholipid composition of some marine red algae." *Phytochemistry* 29 (10):3149–3152. doi: 10.1016/0031-9422(90)80175-G.

Demuez, M., A. Mahdy, E. Tomás-Pejó, C. González-Fernández, and M. Ballesteros. 2015. "Enzymatic cell disruption of microalgae biomass in biorefinery processes." *Biotechnology and Bioengineering* 112 (10):1955–1966. doi: 10.1002/bit.25644.

Desbois, A. P., A. Mearns-Spragg, and V. J. Smith. 2009. "A fatty acid from the diatom phaeodactylum tricornutum is antibacterial against diverse bacteria including multi-resistant staphylococcus aureus (MRSA)." *Marine Biotechnology* 11 (1):45–52. doi: 10.1007/s10126-008-9118-5.

Dillon, J. C., A. P. Phuc, and J. P. Dubacq. 1995. "Nutritional value of the alga spirulina." In *Plants in Human Nutrition*, edited by A. P. Simopoulos, 32–46. Switzerland: Karger.

Dong, T., E. P. Knoshaug, R. Davis, L. M. L. Laurens, S. V. Wychen, P. T. Pienkos, and N. Nagle. 2016. "Combined algal processing: A novel integrated biorefinery process to produce algal biofuels and bioproducts." *Algal Research* 19:316–323. doi: 10.1016/j.algal.2015.12.021.

Donot, F., A. Fontana, J. C. Baccou, and S. Schorr-Galindo. 2012. "Microbial exopolysaccharides: Main examples of synthesis, excretion, genetics and extraction." *Carbohydrate Polymers* 87 (2):951–962. doi: 10.1016/j.carbpol.2011.08.083.

Dufossé, L., P. Galaup, A. Yaron, S. M. Arad, P. Blanc, K. N. C. Murthy, and G. A. Ravishankar. 2005. "Microorganisms and microalgae as sources of pigments for food use: A scientific oddity or an industrial reality?" *Trends in Food Science & Technology* 16 (9):389–406. doi: 10.1016/j.tifs.2005.02.006.

El-Araby, R., A. Amin, A. K. El Morsi, N. N. El-Ibiari, and G. I. El-Diwani. 2018. "Study on the characteristics of palm oil—biodiesel—diesel fuel blend." *Egyptian Journal of Petroleum* 27 (2):187–194. doi: 10.1016/j.ejpe.2017.03.002.

Eriksen, N. T. 2008. "Production of phycocyanin–a pigment with applications in biology, biotechnology, foods and medicine." *Applied Microbiology Biotechnology* 80 (1):1–14.

Fabregas, J., D. García, M. Fernandez-Alonso, A. I. Rocha, P. Gomez-Puertas, J. M. Escribano, A. Otero, and J. Coll. 1999. "In vitro inhibition of the replication of haemorrhagic septicaemia virus (VHSV) and African swine fever virus (ASFV) by extracts from marine microalgae." *Antiviral Research* 44 (1):67–73.

Fábregas, J., J. Arán, E. D. Morales, T. Lamela, and A. Otero. 1997. "Modification of sterol concentration in marine microalgae." *Phytochemistry* 46 (7):1189–1191. doi: 10.1016/S0031-9422(97)80009-X.

Falaise, C., C. Francois, M. A. Travers, B. Morga, J. Haure, R. Tremblay, F. Turcotte, P. Pasetto, R. Gastineau, Y. Hardivillier, V. Leignel, and J. L. Mouget. 2016. "Antimicrobial compounds from eukaryotic microalgae against human pathogens and diseases in aquaculture." *Marine Drugs* 14 (9).

Fan, K. W., Y. Jiang, Y. W. Faan, and F. Chen. 2007. "Lipid characterization of mangrove thraustochytrid–Schizochytrium mangrovei." *Journal of Agricultural and Food Chemistry* 55 (8):2906–2910. doi: 10.1021/jf070058y.

Ferruzzi, Mario G., and Joshua Blakeslee. 2007. "Digestion, absorption, and cancer preventative activity of dietary chlorophyll derivatives." *Nutrition Research* 27 (1):1–12. doi: 10.1016/j.nutres.2006.12.003.

Francavilla, M., M. Colaianna, M. Zotti, M. G. Morgese, P. Trotta, P. Tucci, S. Schiavone, V. Cuomo, and L. Trabace. 2012. "Extraction, characterization and in vivo neuromodulatory activity of phytosterols from microalga Dunaliella tertiolecta." *Current Medicinal Chemistry* 19 (18):3058–3067. doi: CMC-EPUB-20120420-009 [pii].

Frokjaer, S. 1994. "Use of hydrolysates for protein supplementation." *Food Technology* 48 (10):86–88.

Ganesh S., S., B. Bruno, U. Ramakrishnan, K. Khairnar, S. Sandhya, and W. Paunikar. 2014. "Bioactive compounds derived from microalgae showing antimicrobial activities." *Journal of Aquaculture Research and Development* 5:22.

García, J. L., M. de Vicente, and B. Galán. 2017. "Microalgae, old sustainable food and fashion nutraceuticals." *Microbial Biotechnology* 10 (5):1017–1024.

Garcia-Vaquero, M., and M. Hayes. 2016. "Red and green macroalgae for fish and animal feed and human functional food development." *Food Reviews International* 32 (1):15–45. doi: 10.1080/87559129.2015.1041184.

Gastineau, R., F. Turcotte, J.-B. Pouvreau, M. Morançais, J. Fleurence, E. Windarto, F. S. Prasetiya, S. Arsad, P. Jaouen, M. Babin, L. Coiffard, C. Couteau, J.-F. Bardeau, B. Jacquette, V. Leignel, Y. Hardivillier, I. Marcotte, N. Bourgougnon, R. Tremblay, J.-S. Deschênes, H. Badawy, P. Pasetto, N. Davidovich, G. Hansen, J. Dittmer, and J.-L. Mouget. 2014. "Marennine, promising blue pigments from a widespread Haslea diatom species complex." *Marine Drugs* 12 (6):3161–3189. doi: 10.3390/md12063161.

Gastineau, R., J.-B. Pouvreau, C. Hellio, M. Morançais, J. Fleurence, P. Gaudin, N. Bourgougnon, and J.-L. Mouget. 2012. "Biological activities of purified marennine, the blue pigmentresponsible for the greening of oysters." *Journal of Agricultural and Food Chemistry* 60 (12):3599–3605.

Gastineau, R., Y. Hardivillier, V. Leignel, N. Tekaya, M. Morançais, J. Fleurence, N. Davidovich, B. Jacquette, P. Gaudin, C. Hellio, N. Bourgougnon, and J.-L. Mouget. 2012. "Greening effect on oysters and biological activities of the blue pigments produced by the diatom Haslea karadagensis (Naviculaceae)." *Aquaculture* 368–369:61–67.

Gelin, F., J. K. Volkman, C. Largeau, S. Derenne, J. S. Sinninghe Damsté, and J. W. De Leeuw. 1999. "Distribution of aliphatic, nonhydrolyzable biopolymers in marine microalgae." *Organic Geochemistry* 30 (2):147–159. doi: 10.1016/S0146-6380(98)00206-X.

Gelin, F., I. Boogers, A. A. M. Noordeloos, J. S. S. Damste, R. Riegman, and J. W. De Leeuw. 1997. "Resistant biomacromolecules in marine microalgae of the classes eustigmatophyceae and chlorophyceae: Geochemical implications." *Organic Geochemistry* 26 (11):659–675. doi: 10.1016/S0146-6380(97)00035-1.

Georgogianni, K. G., A. P. Katsoulidis, P. J. Pomonis, and M. G. Kontominas. 2009. "Transesterification of soybean frying oil to biodiesel using heterogeneous catalysts." *Fuel Processing Technology* 90 (5):671–676. doi: 10.1016/j.fuproc.2008.12.004.

Ghasemi, Y., A. Moradian, A. Mohagheghzadeh, S. Shokravi, and H. M. Mohammad. 2007. "Antifungal and antibacterial activity of the microalgae collected from paddy fields of Iran: Characterization of antimicrobial activity of chroococcus dispersus." *Journal of Biological Sciences* 7 (6):904–910.

Godiganur, S., C. H. S. Murthy, and R. P. Reddy. 2009. "6BTA 5.9 G2-1 Cummins engine performance and emission tests using methyl ester mahua (Madhuca indica) oil/diesel blends." *Renewable Energy* 34 (10):2172–2177. doi: 10.1016/j.renene.2008.12.035.

González-Davis, O., E. Ponce-Rivas, M. D. P. Sánchez-Saavedra, M.-E. Muñoz-Márquez, and W. H. Gerwick. 2012. "Bioprospection of microalgae and cyanobacteria as biocontrol agents against vibrio campbellii and their use in white shrimp litopenaeus vannamei culture." *Journal of the World Aquaculture Society* 43 (3):387–399. doi: 10.1111/j.1749-7345.2012.00567.x.

GrandViewResearch. 2016. Beta-Carotene Market Analysis By Source (Algae, Fruits & Vegetables, & Synthetic), By Application (Food & Beverages, Dietary Supplements, Cosmetics, & Animal Feed) And Segment Forecasts To 2024.

Guadix, A. 2000. "Technological processes and methods of control in the hydrolysis of proteins." *Ars Pharmaceutica* 41:79–89.

Guedes, A. C., H. M. Amaro, and F. X. Malcata. 2011. "Microalgae as sources of high addedvalue compounds–a brief review of recent work." *Biotechnology Progress* 27 (3):597–613. doi: 10.1002/btpr.575.

Guedes, A., I. S. Pinto, and F. X. Malcata. 2015. *Application of Microalgae Protein to Aquafeed* (Chapter 8). In *Hand Book of Marine Microalgae. Biotechnology Advances*, 93–125. doi: 10.1016/B978-0-12-800776-1.00008-X.

Günerken, E., E. D'Hondt, M. H. M. Eppink, L. Garcia-Gonzalez, K. Elst, and R. H. Wijffels. 2015. "Cell disruption for microalgae biorefineries." *Biotechnology Advances* 33 (2):243–260. doi: 10.1016/j.biotechadv.2015.01.008.

Gupta, A., C. J. Barrow, and M. Puri. 2012. "Omega-3 biotechnology: Thraustochytrids as a novel source of omega-3 oils." *Biotechnology Advances* 30 (6):1733–1745.

Gurevich, P. 2014. Protein quality- The 4 most important metrices. *Labdoor Magazine*.

Harrington, K. J., and C. D'Arcy-Evans. 1985. "A comparison of conventional and in situ methods of transesterification of seed oil from a series of sunflower cultivars." *Journal of the American Oil Chemists' Society* 62 (6):1009–1013. doi: 10.1007/BF02935703.

Harun, R., M. K. Danquah, and G. M. Forde. 2010. "Microalgal biomass as a fermentation feedstock for bioethanol production." *Journal of Chemical Technology & Biotechnology* 85 (2):199–203. doi: 10.1002/jctb.2287.

Harun, R., W. S. Y. Jason, T. Cherrington, and M. K. Danquah. 2011. "Exploring alkaline pre-treatment of microalgal biomass for bioethanol production." *Applied Energy* 88 (10):3464–3467. doi: 10.1016/j.apenergy.2010.10.048.

Harwood, J. L., and I. A. Guschina. 2009. "The versatility of algae and their lipid metabolism." *Biochimie* 91 (6):679–684. doi: S0300-9084(08)00313-1 [pii] 10.1016/j.biochi.2008.11.004.

Hasui, M., M. Matsuda, K. Okutani, and S. Shigeta. 1995. "In vitro antiviral activities of sulfated polysaccharides from a marine microalga (Cochlodinium polykrikoides) against human immunodeficiency virus and other enveloped viruses." *International Journal of Biological Macromolecules* 17 (5):293–297. doi: 10.1016/0141-8130(95)98157-T.

Hayashi, T., K. Hayashi, M. Maeda, and I. Kojima. 1996. "Calcium Spirulan, an inhibitor of enveloped virus replication, from a blue-green alga spirulina platensis." *Journal of Natural Products* 59 (1):83–87. doi: 10.1021/np960017o.

Hemaiswarya, S., R. Raja, R. Ravi Kumar, V. Ganesan, and C. Anbazhagan. 2011. "Microalgae: A sustainable feed source for aquaculture." *World Journal of Microbiology and Biotechnology* 27 (8):1737–1746.

Henchion, M., M. Hayes, A. M. Mullen, M. Fenelon, and B. Tiwari. 2017. "Future protein supply and demand: Strategies and factors influencing a sustainable equilibrium." *Foods* 6 (53):1–21. doi: 10.3390/foods6070053.

Henry, M., L. Gasco, G. Piccolo, and E. Fountoulaki. 2015. "Review on the use of insects in the diet of farmed fish: Past and future." *Animal Feed Science and Technology* 203:1–22.

Herrero, M., E. Ibanez, A. Cifuentes, G. Reglero, and S. Santoyo. 2006. "Dunaliella salina microalga pressurized liquid extracts as potential antimicrobials." *Journal of Food Protection* 69 (10):2471–2477.

Hettiarachchy, N. S., and G. R. Ziegler. 1994. *Protein Functionality in Food Systems.* New York: CRC press.

Ho, S. H., M. C. Chan, C. C. Liu, C. Y. Chen, W. L. Lee, D. J. Lee, and J. S. Chang. 2014. "Enhancing lutein productivity of an indigenous microalga *Scenedesmus obliquus* FSP-3 using light-related strategies." *Bioresource Technology* 152:275–282.

Hosseini, S. M., K. Khosravi-Darani, and M. R. Mozafari. 2013. "Nutritional and medical applications of spirulina microalgae." *Mini-Reviews in Medicinal Chemistry* 13 (8):1231–1237.

Hsu, C. Y., T. H. Yeh, M. Y. Huang, S. P. Hu, P. Y. Chao, and C. M. Yang. 2014. "Organ-specific distribution of chlorophyll-related compounds from dietary spinach in rabbits." *Indian Journal of Biochemistry Biophysics* 51 (5):388–395.

Jeon, J.-K. 2006. "Effect of chlorella addition on the quality of processed cheese." *Journal of the Korean Society of Food Science and Nutrition* 35:373–377.

Ji, X.-J., L.-J. Ren, and H. Huang. 2015. "Omega-3 biotechnology: A green and sustainable process for omega-3 fatty acids production." *Frontiers in Bioengineering and Biotechnology* 3:158. doi: 10.3389/fbioe.2015.00158.

Jin-Kyoung, G., and P. Shin-In. 2005. "Sensory property and keeping quality of curd yoghurt added with loquat (Eriobotrya Japonica Lindley) extract." *The Korean Journal of Food and Nutrition* 18.

John, R. P., G. S. Anisha, K. M. Nampoothiri, and A. Pandey. 2011. "Micro and macroalgal biomass: A renewable source for bioethanol." *Bioresource Technology* 102 (1):186–193. doi: 10.1016/j.biortech.2010.06.139.

Kang, H. K., C. H. Seo, and Y. Park. 2015. "Marine peptides and their anti-infective activities." *Marine Drugs* 13 (1):618–654. doi: 10.3390/md13010618.

Kang, K.-H., Z.-J. Qian, B. M. Ryu, F. Karadeniz, D. Kim, and S.-K. Kim. 2012. "Antioxidant peptides from protein hydrolysate of microalgae navicula incerta and their protective effects in Hepg2/CYP2E1 cells induced by ethanol." *Phytotherapy Research* 26 (10):1555–1563. doi: 10.1002/ptr.4603.

Karawita, R. 2007. "Protective effect of enzymatic extracts from microalgae against DNA damage induced by H_2O_2."*Marine Biotechnology (New York)* 9:479–490.

Kearns, K. D., and M. D. Hunter. 2001. "Toxin-producing anabaena flos-aquae induces settling of chlamydomonas reinhardtii, a competing motile alga." *Microbial Ecology* 42 (1):80–86.

Kellam, J., S., R. J. P. Cannell, Ania Owsianka, and J. M. Walker. 1988. "Results of a large-scale screening programme to detect antifungal activity from marine and freshwater microalgae in laboratory culture." *British Phycological Journal* 23 (1):45–47.

Khan, Z., P. Bhadouria, and P. S. Bisen. 2005. "Nutritional and therapeutic potential of spirulina." *Current Pharmaceutical Biotechnology* 6 (5):373–379.

Khozin-Goldberg, I., and S. Boussiba. 2011. "Concerns over the reporting of inconsistent data on fatty acid composition for microalgae of the genus Nannochloropsis (Eustigmatophyceae)." *Journal of Applied Phycology* 23 (5):933–934.

Kim, M., J. H. Yim, S.-Y. Kim, H. S. Kim, W. G. Lee, S. Kim, P.-S. Kang, and C.-K. Lee. 2011. "In vitro inhibition of influenza a virus infection by marine microalga-derived sulfated polysaccharide P-Kg03." *Antiviral Research* 93 (2):253–259.

Kim, S. M., Y. J. Jung, O. N. Kwon, K. H. Cha, B. H. Um, D. Chung, and C. H. Pan. 2012. "A potential commercial source of fucoxanthin extracted from the microalga Phaeodactylum tricornutum." *Applied Biochemistry and Biotechnology* 166 (7):1843–1855.

Kiron, V., W. Phromkunthong, M. Huntley, I. Archibald, and G. Scheemaker. 2012. "Marine microalgae from biorefinery as a potential feed protein source for Atlantic salmon, common carp and whiteleg shrimp." *Aquaculture Nutrition* 18 (5):521–531. doi: 10.1111/j.1365-2095.2011.00923.x.

Klok, A. J., P. P. Lamers, D. E. Martens, R. B. Draaisma, and R. H. Wijffels. 2014. "Edible oils from microalgae: Insights in TAG accumulation." *Trends in Biotechnology* 32 (10):521–528. doi: 10.1016/j.tibtech.2014.07.004.

Koller, M., A. Muhr, and G. Braunegg. 2014. "Microalgae as versatile cellular factories for valued products." *Algal Research* 6:52–63. doi: 10.1016/j.algal.2014.09.002.

Kose, A., and S. S. Oncel. 2015. "Properties of microalgal enzymatic protein hydrolysates: Biochemical composition, protein distribution and FTIR characteristics." *Biotechnology Reports* 6:137–143. doi: 10.1016/j.btre.2015.02.005.

Kose, A., M. O. Ozen, M. Elibol, and S. S. Oncel. 2017. "Investigation of in vitro digestibility of dietary microalga Chlorella vulgaris and cyanobacterium Spirulina platensis as a nutritional supplement." *3 Biotech* 7 (3):170. doi: 10.1007/s13205-017-0832-4.

Kulikova, I. V., and S. V. Khotimchenko. 2000. "Lipids of different thallus regions of the brown algaSargassum miyabei from the sea of Japan." *Russian Journal of Marine Biology* 26 (1):54–57.

Kumari, P., M. Kumar, C. R. K. Reddy, and B. Jha. 2013. "3 — Algal lipids, fatty acids and sterols." In *Functional Ingredients from Algae for Foods and Nutraceuticals*, edited by H. Domínguez, 87–134. Cambridge: Woodhead Publishing Limited.

Lang, I., L. Hodac, T. Friedl, and I. Feussner. 2011. "Fatty acid profiles and their distribution patterns in microalgae: A comprehensive analysis of more than 2000 strains from the SAG culture collection." *BMC Plant Biology* 11 (1):124. doi: 10.1186/1471-2229-11-124.

Lee, J.-B., K. Hayashi, M. Hirata, E. Kuroda, E. Suzuki, Y. Kubo, and T. Hayashi. 2006. "Antiviral sulfated polysaccharide from navicula directa, a diatom collected from deep-sea water in Toyama Bay." *Biological and Pharmaceutical Bulletin* 29 (10):2135–2139.

Lee, W. T. K., R. Weisell, J. Albert, D. Tomé, A. V. Kurpad, and R. Uauy. 2016. "Research approaches and methods for evaluating the protein quality of human foods proposed by an FAO Expert Working Group in 2014." *The Journal of Nutrition* 146 (5):929. doi: 10.3945/jn.115.222109.

Lenihan-Geels, G., K. S. Bishop, and L. R. Ferguson. 2013. "Alternative sources of omega-3 fats: Can we find a sustainable substitute for fish?" *Nutrients* 5 (4):1301–1315. doi: nu5041301 [pii] 10.3390/nu5041301.

Leser, S. 2013. "The 2013 FAO report on dietary protein quality evaluation in human nutrition: Recommendations and implications." *Nutrition Bulletin* 38 (4):421–428. doi: 10.1111/nbu.12063.

Li, P., S. Harding, and Z. Liu. 2001. "Cyanobacterial exopolysaccharides: Their nature and potential biotechnological applications." *Biotechnology and Genetic Engineering Reviews* 18 (1):375–404.

Lian, M., H. Huang, L. Ren, X. Ji, J. Zhu, and L. Jin. 2010. "Increase of docosahexaenoic acid production by Schizochytrium sp. through mutagenesis and enzyme assay." *Applied Biochemistry Biotechnology* 162 (4):935–941. doi: 10.1007/s12010-009-8865-8.

Lin, J. H., D. J. Lee, and J. S. Chang. 2015. "Lutein production from biomass: Marigold flowers versus microalgae." *Bioresource Technology* 184:421–428.

Lopes, G., C. Sousa, P. Valentão, and P. B. Andrade. 2013. "Sterols in algae and health." *Bioactive Compounds from Marine Foods: Plant and Animal Sources*, 173–191.

Luo, X., P. Su, and W. Zhang. 2015. "Advances in microalgae-derived phytosterols for functional food and pharmaceutical applications." *Marine Drugs* 13 (7):4231–4254. doi: 10.3390/md13074231.

Markou, G., I. Angelidaki, and D. Georgakakis. 2012. "Microalgal carbohydrates: An overview of the factors influencing carbohydrates production, and of main bioconversion technologies for production of biofuels." *Applied Microbiology Biotechnology* 96 (3):631–645.

Marques, A., T. Huynh, P. Sorgeloos, and P. Bossier. 2006. "Use of microalgae and bacteria to enhance protection of gnotobiotic Artemia against different pathogens." *Aquaculture* 258 (1–4):116–126

Medina, C., M. Rubilar, C. Shene, S. Torres, and M. Verdugo. 2015. "Protein fractions with techno-functional and antioxidant properties from *nannochloropsis gaditana* microalgal biomass." *Journal of Biobased Materials and Bioenergy* 9 (4):417–425.

Mendiola, J. A., S. Santoyo, A. Cifuentes, G. Reglero, E. Ibanez, and F. J. Senorans. 2008. "Antimicrobial activity of sub- and supercritical CO_2 extracts of the green alga Dunaliella salina." *Journal of Food Protection* 71 (10):2138–2143.

Mendis, E., and S. K. Kim. 2011. "Present and future prospects of seaweeds in developing functional foods." *Advances in Food and Nutrition Research* 64:1–15. doi: B978-0-12-387669-0.00001-6 [pii] 10.1016/B978-0-12-387669-0.00001-6.

Menetrez, M. Y. 2012. "An overview of algae biofuel production and potential environmental impact." *Environmental Science & Technology* 46 (13):7073–7085.

Merchant, E. R., C. A. Andre, and D. Sica. 2002. "Chlorella supplementation for controlling hypertension: A clinical evaluation." *Alternative and Complementary Therapies* 8 (6).

Michelle, T. 2018. "Natural Food Colors Market Size to Reach USD 2.5 Billion by 2025: Hexa Research." *Cision PR Newswire.*

Milledge, J. 2011. "Commercial application of microalgae other than as biofuels: A brief review."*Reviews in Environmental Science and Bio/Technology* 10 (1):31–41.

Mimouni, V., L. Ulmann, V. Pasquet, M. Mathieu, L. Picot, G. Bougaran, J. P. Cadoret, A. Morant-Manceau, and B. Schoefs. 2012. "The potential of microalgae for the production of bioactive molecules of pharmaceutical interest." *Current Pharmaceutical Biotechnology* 13 (15):2733–2750.

Miranda, J. R., P. C. Passarinho, and L. Gouveia. 2012. "Pre-treatment optimization of *Scenedesmus obliquus* microalga for bioethanol production." *Bioresource Technology* 104:342–348. doi: 10.1016/j.biortech.2011.10.059.

Miyashita, K., and M. Hosokawa. 2019. "Carotenoids as a nutraceutical therapy for visceral obesity." In *Nutrition in the Prevention and Treatment of Abdominal Obesity* (Second Edition), edited by R. R. Watson. 459–477. Cambridge: Academic Press.

Morris, H. J., A. Almarales, O. Carrillo, and R. C. Bermúdez. 2008. "Utilisation of *Chlorella vulgaris* cell biomass for the production of enzymatic protein hydrolysates." *Bioresource Technology* 99 (16):7723–7729. doi: 10.1016/j.biortech.2008.01.080.

Morris, H. J., O. Carrillo, A. Almarales, R. C. Bermúdez, Y. Lebeque, R. Fontaine, G. Llauradó, and Y. Beltrán. 2007. "Immunostimulant activity of an enzymatic protein hydrolysate from green microalga Chlorella vulgaris on undernourished mice." *Enzyme and Microbial Technology* 40 (3):456–460. doi: 10.1016/j.enzmictec.2006.07.021.

Murray, B., and R. FitzGerald. 2007. "Angiotensin converting enzyme inhibitory peptides derived from food proteins: Biochemistry, bioactivity and production." *Current Pharmaceutical Design* 13 (8):773–791.

Nagai, H., Y. Mikami, K. Yazawa, T. Gonoi, and T. Yasumoto. 1993. "Biological activities of novel polyether antifungals, gambieric acids A and B from a marine dinoflagellate Gambierdiscus toxicus." *The Journal of Antibiotics* 46 (3):520–522.

Najdenski, H. M., L. G. Gigova, I. I. Iliev, P. S. Pilarski, J. Lukavský, I. V. Tsvetkova, M. S. Ninova, and V. K. Kussovski. 2013. "Antibacterial and antifungal activities of selected microalgae and cyanobacteria." *International Journal of Food Science & Technology* 48 (7):1533–1540. doi: 10.1111/ijfs.12122.

Nguyen, K. 2013. "Astaxanthin: A comparative case of synthetic vs. natural production." *Chemical and Biomolecular Engineering Publications and Other Works* 1:1–11.

Nguyen, T., Q. Zhong-Ji, V.-T. Nguyen, I.-W. Choi, S.-J. Heo, C. Oh, D.-H. Kang, G. H. Kim, and W.-K. Jung. 2013. "Tetrameric peptide purified from hydrolysates of biodiesel byproducts of Nannochloropsis oculata induces osteoblastic differentiation through MAPK and Smad pathway on MG-63 and D1 cells." *Process Biochemistry* 48 (9):1387–1394.

Norton, T. A., M. Melkonian, and R. A. Andersen. 1996. "Algal biodiversity." *Phycologia* 35 (4):308–326. doi: 10.2216/i0031-8884-35-4-308.1.

Olasehinde, T. A., A. O. Olaniran, and A. I. Okoh. 2017. "Therapeutic potentials of microalgae in the treatment of Alzheimer's disease." *Molecules* 22 (3).

Orosa, M., J. Fernandez-Valero, C. Herrero, and J. Abalde. 2001. "Comparison of the accumulation of astaxanthin in Haematococcus pluvialis and other green microalgae under N-starvation and high light conditions." *Biotechnology Letters* 23 (13):1079–1085.

Pane, G., G. Cacciola, E. Giacco, G. L. Mariottini, and E. Coppo. 2015. "Assessment of the antimicrobial activity of algae extracts on bacteria responsible of external otitis." *Marine Drugs* 13 (10):6440–6452. doi: 10.3390/md13106440.

Patil, V., T. Källqvist, E. Olsen, G. Vogt, and H. R. Gislerød. 2007. "Fatty acid composition of 12 microalgae for possible use in aquaculture feed." *Aquaculture International* 15 (1):1–9.

Patterson, G. M. L. 1996. "Biotechnological applications of cyanobacteria." *Journal of Scientific and Industrial Research* 55:669–684.

Pereira, H., L. Custodio, M. J. Rodrigues, C. B. de Sousa, M. Oliveira, L. Barreira, R. Neng Nda, J. M. Nogueira, S. A. Alrokayan, F. Mouffouk, K. M. Abu-Salah, R. Ben-Hamadou, and J. Varela. 2015. "Biological activities and chemical composition of methanolic extracts of selected autochthonous microalgae strains from the Red Sea." *Marine Drugs* 13 (6):3531–3549. doi: md13063531 [pii] 10.3390/md13063531.

Phycocyanin to Show Consistent Penetration Owing to Use in Multiple Food and Beverages Applications. Future Market Insights 2018.

Piironen, V., D. G. Lindsay, T. A. Miettinen, J. Toivo, and A.-M. Lampi. 2000. "Plant sterols: Biosynthesis, biological function and their importance to human nutrition." *Journal of the Science of Food and Agriculture* 80 (7):939–966. doi: 10.1002/(SICI)1097-0010(20000515)80:7<939::AID-JSFA644>3.0.CO;2-C.

Poonam, S., and S. Nivedita. 2017. "Industrial and biotechnological applications of algae: A review." *Journal of Advances in Plant Biology* 1 (1):1–25. doi: 10.14302/issn.2638-4469.japb-17-1534.

Prakash, J. W., J. M. Antonisamy, and S. Jeeva. 2011. "Antimicrobial activity of certain fresh water microalgae from Thamirabarani River, Tamil Nadu, South India." *Asian Pacific Journal of Tropical Biomedicine* 1 (2, Supplement):S170–173. doi: 10.1016/S2221-1691(11)60149-4.

Raja, R., C. Anbazhagan, S. Senthil, D. Lakshmi, and R. Rengasamy. 2004. "Studies on Dunaliella salina (Volvocales, Chlorophyta) under laboratory conditions."*Seaweeds Research utilisation Journal* 26 (1–2):127–146.

Ranard, K. M., S. Jeon, E. S. Mohn, J. C. Griffiths, E. J. Johnson, and J. W. Erdman, Jr. 2017. "Dietary guidance for lutein: Consideration for intake recommendations is scientifically supported." *European Journal of Nutrition* 56 (Suppl 3):37–42.

Rath, B. 2012. "Commercial and industrial applications of micro algae—A review." *Journal of Algal Biomass Utilization* 3 (4):89–100.

Rechter, S., T. König, S. Auerochs, S. Thulke, H. Walter, H. Dörnenburg, C. Walter, and M. Marschall. 2007. "Antiviral activity of Arthrospira-derived spirulan-like substances." *Antiviral Research* 72 (3):197–206.

Rohan. *Carotenoids Market worth 1.53 Billion USD by 2021* 2016. Available from www.marketsandmarkets.com/PressReleases/carotenoid.asp.

Rohan. *Lutein Market worth 357.7 Million USD by 2022* 2017. Available from www.marketsandmarkets.com/PressReleases/lutein.asp.

Rojo, A., L. Ibarra, M.-B. Jm, G. Velasco, M. Martínez-Téllez, M.-J. Ma, N.-S. M, and Q.-Z. D. 2017. "Potential of Nannochloropsis in beta glucan production." In *Nannochloropsis: Biology, Biotechnological, Potential and Challenges (Edition 1)*, 181–225, edited by M. Jan and P. Kazik. New York: Nova Sciences Publishers.

Sahin, D., E. Tas, and U. H. Altindag. 2018. "Enhancement of docosahexaenoic acid (DHA) production from Schizochytrium sp. S31 using different growth medium conditions." *AMB Express* 8 (7):1–8.

Sahoo, P. K., L. M. Das, M. K. G. Babu, P. Arora, V. P. Singh, N. R. Kumar, and T. S. Varyani. 2009. "Comparative evaluation of performance and emission characteristics of jatropha, karanja and polanga based biodiesel as fuel in a tractor engine." *Fuel* 88 (9):1698–1707. doi: 10.1016/j.fuel.2009.02.015.

Santoyo, S., L. Jaime, M. Plaza, M. Herrero, I. Rodriguez-Meizoso, E. Ibáñez, and G. Reglero. 2012. "Antiviral compounds obtained from microalgae commonly used as carotenoid sources." *Journal of Applied Phycology* 24 (4):731–741.

Satake, M., M. Murata, T. Yasumoto, T. Fujita, and H. Naoki. 1991. "Amphidinol, a polyhydroxy-polyene antifungal agent with an unprecedented structure, from a marine dinoflagellate, Amphidinium klebsii." *Journal of the American Chemical Society* 113 (26):9859–9861. doi: 10.1021/ja00026a027.

Sathasivam, R., R. Radhakrishnan, A. Hashem, and E. F. Abd_Allah. 2017. "Microalgae metabolites: A rich source for food and medicine." *Saudi Journal of Biological Sciences*. doi: 10.1016/j.sjbs.2017.11.003.

Sekar, S., and M. Chandramohan. 2008. "Phycobiliproteins as a commodity: Trends in applied research, patents and commercialization." *Journal of Applied Phycology* 20 (2):113–136. doi: 10.1007/s10811-007-9188-1.

Seyidoglu, N., S. Inan, and C. Aydin. 2017. "A prominent superfood: *Spirulina platensis*." In *Superfood and Functional Food*, edited by N. Shiomi and V. Waisundara, IntechOpen, DOI: 10.5772/66118. Available from: https://www.intechopen.com/books/superfood-and-functional-food-the-development-of-superfoods-and-their-roles-as-medicine/a-prominent-superfood-spirulina-platensis.

Shannon, E., and N. A. Ghannam. 2016. "Antibacterial derivatives of marine algae: An overview of pharmacological mechanisms and applications." *Marine Drugs* 14 (4):81.

Sheih, I. C., T. Fang, T.-K. Wu, and P.-H. Lin. 2009. "Anticancer and antioxidant activities of the peptide fraction from algae protein waste."*Journal of Agricultural and Food Chemistry* 58 (2):1202–1207.

Sheih, I. C., T. J. Fang, and T.-K. Wu. 2009. "Isolation and characterisation of a novel angiotensin I-converting enzyme (ACE) inhibitory peptide from the algae protein waste." *Food Chemistry* 115 (1):279–284. doi: 10.1016/j.foodchem.2008.12.019.

Sheih, I. C., T.-K. Wu, and T. J. Fang. 2009. "Antioxidant properties of a new antioxidative peptide from algae protein waste hydrolysate in different oxidation systems." *Bioresource Technology* 100 (13):3419–3425. doi: 10.1016/j.biortech.2009.02.014.

Singh, P., M. Kuddus, and G. Thomas. 2009. "An efficient method for extraction of C-phycocyanin from Spirulina sp. and its binding affinity to blood cells, nuclei and genomic DNA." *International Research Journal of Biotechnology* 1 (5):080–085.

Skrede, M., K. R. Ahlstrøm, H. Gislerød, and M. Øverland. 2011. "Evaluation of microalgae as sources of digestible nutrients for monogastric animals." *Journal of Animal and Feed Sciences* 20 (1):131–142.

Skulberg, O. M. 2007. "Bioactive chemicals in microalgae" (Chapter 30). In *Handbook of Microalgal Culture. Biotechnology and Applied Phycology*, 485–512. Hoboken, NJ: Blackwell Publishing. doi: 10.1002/9780470995280.ch30.

Soto-Sierra, L., P. Stoykova, and Z. L. Nikolov. 2018. "Extraction and fractionation of microalgae-based protein products." *Algal Research* 36:175–192. doi: 10.1016/j. algal.2018.10.023.

Spolaore, P., C. Joannis-Cassan, E. Duran, and A. Isambert. 2006. "Commercial applications of microalgae." *Journal of Bioscience and Bioengineering* 101 (2):87–96. doi: 10.1263/ jbb.101.87.

Sternberg, E. D., D. Dolphin, and C. Brückner. 1998. "Porphyrin-based photosensitizers for use in photodynamic therapy." *Tetrahedron* 54 (17):4151–4202. doi: 10.1016/ S0040-4020(98)00015-5.

Sun, Z., T. Li, Z. G. Zhou, and Y. Jiang. 2016. Microalgae as a source of lutein: Chemistry, biosynthesis, and carotenogenesis. *Advances in Biochemical Engineering/Biotechnology* 153:37–58.

Superfoodly. *Astaxanthin Supplements* 2018. Available from www.superfoodly.com/ orac-value/astaxanthin/.

Suresh, M., C. P. Jawahar, and A. Richard. 2018. "A review on biodiesel production, combustion, performance, and emission characteristics of non-edible oils in variable compression ratio diesel engine using biodiesel and its blends." *Renewable and Sustainable Energy Reviews* 92:38–49. doi: 10.1016/j.rser.2018.04.048.

Takaichi, S. 2011. "Carotenoids in algae: Distributions, biosyntheses and functions." *Marine Drugs* 9 (6):1101–1118.

Tan, L. T. 2013. "Pharmaceutical agents from filamentous marine cyanobacteria." *Drug Discovery Today* 18 (17–18):863–871.

Tang, Y., J. N. Rosenberg, P. Bohutskyi, G. Yu, M. J. Betenbaugh, and F. Wang. 2016. "Microalgae as a feedstock for biofuel precursors and value-added products: Green fuels and golden opportunities." *BioResources* 11 (1).

Tchorbanov, B., and M. Bozhkova. 1988. "Enzymatic hydrolysis of cell proteins in green algae Chlorella and Scenedesmus after extraction with organic solvents." *Enzyme and Microbial Technology* 10 (4):233–238. doi: 10.1016/0141-0229(88)90072-5.

Thanh, L. T., K. Okitsu, Y. Sadanaga, N. Takenaka, Y. Maeda, and H. Bandow. 2010. "Ultrasound-assisted production of biodiesel fuel from vegetable oils in a small scale circulation process." *Bioresource Technology* 101 (2):639–645. doi: 10.1016/j. biortech.2009.08.050.

United Nations, Department of Economic and Social Affairs, Population Division. 2015. World Population Prospects: The 2015 Revision, Key Findings and Advance Tables. edited by United Nations: Department of Economic and Social Affairs. New York.

USDA. 2011. β-Carotene Handling/Processing. Technical Services Branch for the USDA National Organic Program.

Venkata S. G., M. Rajvanshi, B. Navish Kumar, S. Govindachary, V. Prasad, and S. Dasgupta. 2017. "Carbon streaming in microalgae: Extraction and analysis methods for high value compounds." *Bioresource Technology* 244:1304–1316. doi: 10.1016/j.biortech.2017.07.024.

Vo, T.-S., B. Ryu, and S.-K. Kim. 2013. "Purification of novel anti-inflammatory peptides from enzymatic hydrolysate of the edible microalgal Spirulina maxima." *Journal of Functional Foods* 5 (3):1336–1346. doi: 10.1016/j.jff.2013.05.001.

Volkman, J. K. 2003. "Sterols in microorganisms." *Applied Microbiology Biotechnology* 60 (5):495–506. doi: 10.1007/s00253-002-1172-8.

Waghmare, A. G., M. K. Salve, J. G. LeBlanc, and S. S. Arya. 2016. "Concentration and characterization of microalgae proteins from Chlorella pyrenoidosa." *Bioresources and Bioprocessing* 3 (1):16. doi: 10.1186/s40643-016-0094-8.

Walter, C. S., and R. Mahesh. 2000. "Antibacterial and antifungal activities of some marine diatoms in culture." *International Journal of Medical Science* 29 (3):238–242.

Wang, X., X. Liu, and G. Wang. 2011. "Two-stage hydrolysis of invasive algal feedstock for ethanol fermentation." *Journal of Integrative Plant Biology* 53 (3):246–252.

Welker, M., and H. Von Döhren. 2006. "Cyanobacterial peptides—Nature's own combinatorial biosynthesis." *FEMS Microbiology Reviews* 30 (4):530–563. doi: 10.1111/j.1574-6976.2006.00022.x.

Wen, Z.-Y., and F. Chen. 2003. "Heterotrophic production of eicosapentaenoic acid by microalgae." *Biotechnology Advances* 21 (4):273–294.

Wijffels, R. H., M. J. Barbosa, and M. H. M. Eppink. 2010. "Microalgae for the production of bulk chemicals and biofuels." *Biofuels, Bioproducts and Biorefining* 4 (3):287–295. doi: 10.1002/bbb.215.

Winwood, R. J. 2013. "Recent developments in the commercial production of DHA and EPA rich oils from micro-algae." *Oilseeds & Fats Crops and Lipids* 20 (6):D604.

Wood, L. 2017. *Global $1.22 Billion Astaxanthin Market Outlook 2016–2023 — Rising Demand for Cosmetic Products, Customers Varying Preferences Towards Healthcare & Nutraceutical Products—Research and Markets* 2017. Available from www.research andmarkets.com/research/439dvs/astaxanthin.

Wood, L. 2018. Global *Astaxanthin Market 2014–2017 & 2017–2023: Analysis & Forecasts by Sources, Technologies and Applications—ResearchAndMarkets.com* 2018. Available from www.researchandmarkets.com/research/c7pvhk/forecast_to?w=4.

Xie, D., E. N. Jackson, and Q. Zhu. 2015. "Sustainable source of omega-3 eicosapentaenoic acid from metabolically engineered Yarrowia lipolytica: From fundamental research to commercial production." *Applied Microbiology Biotechnology* 99 (4):1599–1610.

Yim, J. H., S. Kim, S. H. Ahn, C. K. Lee, K. T. Rhie, and H. K. Lee. 2004. "Antiviral effects of sulfated exopolysaccharide from the marine microalga gyrodinium impudicum strain KG03." *Marine Biotechnology (New York)* 6 (1):17–25.

Zhang, L.-J., and R. L. Gallo. 2016. "Antimicrobial peptides." *Current Biology* 26 (1):R14–R19. doi: 10.1016/j.cub.2015.11.017.

Zhu, F., B. Du, and B. Xu. 2016. "A critical review on production and industrial applications of beta-glucans." *Food Hydrocolloids* 52:275–288. doi: 10.1016/j.foodhyd.2015.07.003.

2 Promising Future Products from Microalgae for Commercial Applications

F. C. Thomas Allnutt

CONTENTS

2.1   BACKGROUND AND DRIVING FORCES

The commercial-scale algal biorefinery being run for the sole production of algal biofuels probably will never be the future of the industry. More likely is that as the conditions improve for biofuels, algal biorefineries will be engineered to produce biofuels in synergy with a portfolio of bioproducts to take advantage of the availability of the complex mixture of biomolecules present in the algal biomass in order to make the overall plant more profitable.

> At least until oil prices return to near their pre-August 2014 levels, or carbon emissions reductions are rewarded through higher carbon pricing in a global climate disruption mitigation policy, primary strategies for bioenergy production from algae will need to rely on a multi-product biorefinery approach.
>
> **(Laurens, Chen-Glasser, and McMillan 2017)**

This quote reflects on the current opportunities and challenges facing a successful microalgal biofuels biorefinery. These authors also emphasize the importance that co-products are likely to play in any successful commercialization of large-scale microalgal biofuels. Current energy markets and the regulatory climate surrounding biofuels erect significant hurdles to successful commercialization of microalgae-derived biofuels based solely on a dedicated microalgal biorefinery (i.e., no additional bioproducts are produced).

Although technologies are in place to produce renewable and sustainable microalgal fuels from microalgal lipids or biomass (Chen et al. 2015, Leite, Abdelaziz, and Hallenbeck 2013, Beal et al. 2012, NAABB 2014), additional technical advances need to be combined with stimulation from governmental policies in order to accelerate algal biofuel's commercial success. Some of the required technical advances include improved microalgal production technologies (e.g., higher lipid or biomass yields), innovative new approaches to incorporate co-products in the production train (e.g., improved harvesting and flexible processing), and improved production strains that deliver the primary product and target high-value co-products for optimal value generation.

There are those who have already written the obituary for algal biofuels, giving a eulogy citing the large amounts of funding already expended by governments and corporations to support microalgal biofuel research, hype that was overly optimistic in the yields of algal biofuels, and "the huge learning curve between what takes place in a lab and commercial production" (Rapier 2018, Cohen 2019). Nonetheless, there continues to be optimism that microalgae will eventually contribute to the overall health of the planet (Kite-Powell 2018).

The combination of a huge variety of potential products from microalgae and processing that makes complete use of all the biomass produced will most likely have microalgal biorefineries increasingly producing both bioproducts and biofuel, although the biofuel could be a secondary rather than primary product of the biorefinery, as market conditions dictate.

Putting in place a cost for carbon emissions to create pull from the marketplace is an example of a policy approach that could stimulate the commercial deployment of microalgal biofuels. The Europeans are leading the way by providing carbon dioxide emission allowances (European Union Allowances, or EUAs) to be bought and sold in an attempt to reduce carbon emissions through monetization of emission credits. It is promising that Canada recently introduced a carbon cap and trade format (Ambasta and Buonocore 2018), and groups like the European Bioeconomy Alliance (EUBA 2019), the Minnesota Bioeconomy Coalition (GPI 2019), and the Algae Biomass Organization (ABO 2019) continue to nudge the world toward a more sustainable bioeconomy. Microalgae have the potential to play a significant role in this emerging bioeconomy (Karan et al. 2019).

Many of the early startup algal biofuels companies from the 2000s have closed or shifted their focus to non-fuel algal products as the price of oil declined from a high of approximately US\$160 to US\$35 a barrel in 2018. Today's microalgae-based companies are algal bioproducts companies that are commercially viable through a shift in focus to making (or developing) a wide variety of microalgae-derived products, which encompass feeds, foods, nutraceuticals, cosmetics, and printer inks.

The first chapter in this book provided a comprehensive overview of potential products that can be produced from microalgae (Rajvanshi et al. 2019), as have a number of recent review articles (Chew et al. 2017, Foley, Beach, and Zimmerman 2011, Spolaore et al. 2006); the full range of potential products will not be discussed in detail in this chapter. This chapter's focus will be on how algal companies are progressing toward commercialization of specific biofuels and bioproducts, as well as potential pathways to commercially relevant microalgal biofuels. The challenge for future commercial scale biorefineries will be 1) to produce a flexible and commercially viable portfolio of products that are responsive to market needs and 2) to transform all the biomass produced into useful products to achieve maximal value. Some future algal biorefineries may not even produce biofuels and could therefore be called biofactories; however, no distinction between the two will be made in this chapter.

2.1.1 Chapter Objectives

In this chapter, recently introduced or in-development microalgal products, as well as emerging methods and tools that will enable the development of commercially relevant biofuels and co-products derived from microalgae, will be discussed. The daunting challenge of integrating one or several commercially viable co-products in a microalgal biorefinery in a manner that is flexible, sustainable, and commercially relevant will be discussed as an overview of the approach needed rather than in detail. Downstream processing will vary according to the production strain, products to be produced, and the regulations governing the control of both native and engineered production strains. Downstream processing is discussed in detail in other chapters in this book.

This chapter discusses: 1) products that address large markets that would absorb the production output generated by a commercial-scale algal biorefinery with minimal market disruption; 2) smaller market products are likely to be derived from a side stream taken from the biorefinery production or processing train; and 3) products that take advantage of the existing biorefinery culturing, harvesting, and processing infrastructure to make a synergistic set of products using different organisms and processes. The last two categories address smaller markets that would be saturated when using the entire scaled production of a commercial scale microalgal production plant as feedstock.

The chapter also provides a brief overview of algal biotechnology to better understand what can be done to develop novel products from algal biorefineries, as well as to improve the production of biofuels and co-products from engineered strains. Additionally, the chapter will look at research and commercial activities to identify new microalgal products, as well as identify products of possible future value, whether directly coupled to a biofuel microalgal biorefinery or biofactories.

2.1.2 CURRENT MICROALGAL PRODUCTS MARKET

In 2018, the algae-based products market was estimated to be US$3.98 billion (MarketResearch 2018). A large part of this market was driven by algae-based nutraceutical products, largely long-chain polyunsaturated omega-3–based lipids (LC-PUFAs), such as docosahexaenoic acid (DHA) and eicosapentaenoic acid (EPA) (Gyekye 2018). The algae-based products market is projected to reach US$5.17 billion by 2023 (MarketResearch 2018) and is dominated by large companies such as Cargill (United States), DIC Corporation (Japan), DSM (Netherlands), BASF (Germany), Kerry Ingredients & Flavors (Ireland), EID-Parrys (India), Corbion (Netherlands), Roquette Freres (France), and Fenchem Biotech (China). However, there are smaller, well-established companies, such as ALGA Technologies (Malaysia); Taiwan Chlorella Manufacturing Company (China); BlueBioTech (Germany); AlgaTechnologies Ltd. (Israel); Cyanotech, Cellana, Ingredion (United States); and Nature Beta Technologies (Israel), that are continuing to produce products and drive innovation in the algal products market sector.

A reflection of the continuing interest in algal products is the proliferation of small microalgae-based companies producing products from microalgae, such as Euglena Ltd. (Japan); Algal Scientific, Algenol, Valensa International, and Heliae Development (United States); and AstaReal Co. LTD (Japan), and the list continues to expand.

Although the advancement of algal biofuels has been constrained by a decrease in the price of fossil fuels and the lack of a regulatory or policy push (to generate market pull), the use of microalgae-based products for food, animal feeds, and nutraceuticals is predicted to drive the scaled-up production of a variety of algal strains in the foreseeable future (Gyekye 2018).

2.1.3 ALGAL BIOREFINERIES' DEPENDENCE ON CO-PRODUCTS

Because the market is not currently favorable for a dedicated microalgal biofuel biorefinery (Laurens, Chen-Glasser, and McMillan 2017), methods to improve harvesting and processing that retain the desired co-products need to be developed and customized to accommodate a portfolio of microalgae-derived co-products to create a profitable, multiproduct algal biorefinery (Lam et al. 2018, Allnutt and Kessler 2015, Haznedaroglu et al. 2016). These algal biorefineries could be based on phototrophic, heterotrophic, or mixed trophic state approaches.

The microalgal industry needs to develop bioproducts that allow the maximization of value for the algal biomass produced at the commercial scale biorefinery. This is reflected in a statement from Wijffels and Barbosa: "The main objective is to reduce production costs and energy requirements while maximizing lipid productivity and to increase the biomass value by making use of all algal biomass components" (Wijffels and Barbosa 2010). When considering all microalgae-derived products, their statement should be modified to broaden its focus from lipid productivity to product productivity—including the co-products, or even energy production, strictly from the biomass (or residual biomass after the extraction of valuable co-products).

Large-Scale Production and Cultivation—Impact on Viability of Bioproducts from Microalgae

Large-scale production of algal biomass (nearly unialgal) will provide the industry with a unique opportunity to develop co-products that have heretofore been impossible due to the high cost and low availability of the algal biomass needed as a feedstock for their production. Such products could be secondary metabolites, structural components, pigments, and sterols that are too expensive when the biomass is produced for a single purpose, but they could be commercially relevant as a co-product produced from residual biomass/materials funneled into a portfolio of products. Again, it should be noted that the downstream processing will be critical to maintaining these valuable products of interest.

Although the availability of low-cost algal feedstock is a significant advantage, one issue with making co-products as an integral part of the microalgal biorefinery will be to match the scale needed for the production of biofuels with the potential markets for all co-product(s). As an algal biofuel biorefinery will be making thousands (hopefully millions) of barrels of biofuel, markets for co-products made from all the residual biomass, or even from a side stream in the production process, would generate quantities of the product that might exceed the market demand and thereby lower the value of that product. If the sales of these co-products are not balanced against their demand (or potential demand) in the marketplace, they will quickly flood the market and decrease the value of the co-product to the biorefinery, thereby cutting into the value proposition of the integrated biorefinery.

Carotenoids are an example of this problem. Carotenoids currently are a very high-value co-product for the nutraceutical and food industries and continue to be a focus as an alternative product for the biorefinery (Panis and Carreon 2016, Yuan et al. 2011, Long et al. 2017). The current market for natural astaxanthin (at ~US$7,000/kg) and for synthetic astaxanthin (~US$2,000/kg) is high, having a total market of a few hundred tons per year for all astaxanthin. Potentially, a single commercial-scale algal biorefinery making natural astaxanthin as a co-product could flood the existing market and drastically reduce the value of the pigment produced to the overall cost structure of the biorefinery, thus making it less economically viable.

This is in contrast to a co-product, such as cattle feed, that is based on the residual biomass after oil extraction (lipid-extracted microalgae, or LEA), where there is a huge potential market (in terms of mass) for such a product that is unlikely to be saturated due to an increasing demand for protein for people that is putting increased pressure on agriculture to deliver it (Henchion et al. 2017).

The potential for the development of useful co-products from microalgal biorefineries is high due to the huge biological diversity in algae, which have a possibility of up to a million species (Guiry 2012). There are several ways to group potential products based on their position in the production process. A conceptually useful grouping is 1) products that are extracted from algal biomass or spent medium at a production scale that addresses markets that can absorb huge quantities of the product (e.g., animal feed or protein products), 2) products that are extracted from the algal biomass that address small markets but are high-value products (e.g., astaxanthin or lutein), and 3) products that are engineered into the algal cell to address either of the first two categories.

One way to synchronize co-product production with the markets addressed is to design a plant to be able to produce a portfolio of products, whose production can be adjusted to meet changing market demands. Such an approach would maximize value while reducing risk to the overall biorefinery cost structure (Chew et al. 2017).

2.1.4   INNOVATION—TOOLS TO IDENTIFY CO-PRODUCTS
AND OPTIMIZE PRODUCTION

A commercial-scale algal biorefinery would provide a rich source of complex biomolecules that could be used as a feedstock for the production of commercially relevant bioproducts based on these biomolecules or the bioconversion of these biomolecules into a different product. Products could be generated during the production process (e.g., by extraction from the biomass or spent medium), during processing (e.g., through a side stream removed during the processing), as an unprocessed side stream where the co-product is extracted and then the residual materials are returned to the processing stream, or generated from the residual biomass after the extraction of lipid (either directly or indirectly). The downstream processing to produce co-products must be synchronized with the main product being produced (i.e., such that the processing does not destroy the co-product or the biofuel).

Having previously worked for microalgae-centric companies, a major hurdle encountered was the cost of producing biomass feedstock to produce a specific product at a commercially relevant price. Having a commercial-scale biorefinery, where large amounts of biomass are produced routinely, would provide the opportunity to harvest a side stream of biomass or to use partially processed materials as feedstock to extract high-value products. Such routine availability of inexpensive algal feedstocks would enable the production of products that had been previously limited due to the high cost of a dedicated algal biomass feedstock. Potential product examples are functional and structural polysaccharides, phytosterols, and bioactive peptides (Rajvanshi et al. 2019). As an added benefit, any residual materials from co-product production could be returned to the biorefinery process train for use downstream, either for the production of additional co-products or bioenergy. The overall goal must be to generate the most value from the biomass produced.

2.2 LARGE MARKET CO-PRODUCTS

Large market co-products are those products produced by direct extraction or produced indirectly from the algal biomass or residuals generated at the commercial-scale algal biorefinery. These products must address markets large enough to absorb the quantities of co-products being produced. Such co-products need to be in synchrony with the markets, or they risk reducing their value, which will negatively affect the economics of the biorefinery.

2.2.1 ANIMAL FEEDS

The use of the algal biomass or the LEA in animal feeds would provide a source of protein, carbon, and other nutrients to the feed industry that, as a co-product of biofuel production, would be available in large quantities and at competitive pricing. However, the use of inexpensive fertilizers and/or wastewaters high in organic nutrients to nourish the microalgae during production could lead the biorefinery biomass to be more suitable as an animal feed or feed supplement than for direct human consumption.

Both whole algal biomass and LEA have been looked at for use as an animal feed supplement (Gatrell et al. 2014, Tibbetts and Fredeen 2017). This effort is encouraged by the agriculture industry's push to use phytogenic additives in animal feeds in a shift from synthetic materials, both by choice and through regulatory push (Karaskova, Suchy, and Strakova 2015).

Ruminants—The lipid, protein, and carbohydrate content of the residual biomass from the algal biorefinery makes it ideal for use as a component in ruminant feeds (Lodge-Ivey, Tracey, and Salazar 2013, 2014, Stokes et al. 2015). Significant research has been done with algal biomass, and the work is continuing. An example is the recent study with *Scenedesmus*, a microalga being considered for biofuel production, which was recently evaluated for its suitability as a ruminant feed additive (Tibbetts and Fredeen 2017). Both whole-cell biomass and LEA were evaluated *in vitro* using dairy cattle rumen fluid. The LEA provided an effective replacement of protein sources when using the LEA inclusion rates up to 50% and found that the whole cell (high-lipid)–based diets had no significant difference than control feeds. They concluded that this alga had value in the ruminant feed market. Another recent study looked at LEA from *Chlorella* spp. and *Nannochloropsis salina* using an *in vitro* model (Lodge-Ivey, Tracey, and Salazar 2013, 2014). Although the data were noisy, the research demonstrated the potential LEA has as a component of ruminant feeds. Studies like these continue to demonstrate that algal biomass, as well as delipidated residuals from various potential algal production strains, can be viable components for ruminant feed. Another recent *in vitro* study looked at the digestibility of a variety of different microalgae and concluded that all the microalgae tested were digestible in the ruminant model test and that broken cells were more digestible than whole cells (Wild, Steingass, and Rodehutscord 2018). These authors concluded that microalgae have promise as a protein source for use in ruminant animal feeds.

The partially delipidated biomass of *Prototheca moriformis*, a heterotrophic alga, was tested for its digestibility in sheep (Stokes et al. 2015). They found that the biomass was highly digestible and attractive to sheep in terms of palatability and nutrient content. Their experiments did demonstrate that the small particle size and the rates of inclusion were important considerations for optimal use in ruminant diets.

Once large-scale algal biorefineries come online, the use of their residual biomass could make microalgae-derived materials important ingredients in ruminant feed formulations and could contribute positively to the overall cost balance of the plant while generating sustainable and renewable animal feed products.

Monogastrics—Although most monogastric organisms will have difficulty digesting the complex carbohydrates (e.g. sporopollenin and cellulose) that make up the cell wells of many algal strains (He, Dai, and Wu 2016, Rodrigues and da Silva Bon 2011), algal protein, lipids, and other nutrients derived from the residual biomass are likely to play a role in formulations for monogastric nutrition when large-scale biorefineries are commercially active and provide competitive pricing for these ingredients.

A recent *in vitro* study with *Scenedesmus*, using both delipidated biomass (high-protein) and whole-cell biomass, found simulated porcine digestion was very effective for the delipidated biomass but low for the whole-cell material (Tibbetts and Fredeen 2017). The authors considered that further studies are needed to understand why whole cells were not digested (e.g., cell rupture might be required to release the nutrients or expose the cellular components for enzymatic action). Another example is a study evaluating the use of *Chlorella* sp. (heterotrophically grown) in layers, which provided improved egg quality (Karaskova, Suchy, and Strakova 2015). One of the main issues preventing the widespread use of *Chlorella* biomass in chicken feeds is, again, the current high price of the biomass.

The use of *Staurosira* sp. whole algal biomass or LEA in swine studies showed no adverse impact of either formulation when added at up to 6.6% of the whole algal biomass or 7.2% of the LEA added to the feed (Gatrell et al. 2014).

Production of biomass as a side product of a biorefinery could reduce that hurdle and encourage more extensive use of microalgal biomass in feeds.

Aquaculture—Whole algae have long been used in aquaculture feeds as a rich source of protein, lipids, and other nutritive factors (Ronquillo, Fraser, and McConkey 2012, Su, Su, and Liao 2001), as well as in live algal cultures, or "green water," for larval culture (Su, Su, and Liao 2001). Continuing innovation in the use of microalgae to replace live feeds in larval culture (Arney et al. 2015), as well as the LC-PUFAs from fishmeal and fish oil (Chauton et al. 2015), will provide more data to support algal supplement use in aquaculture.

A recent review of the use of microalgae grown on wastewater generated by agricultural processing (e.g., slaughter houses) as a source of food for aquaculture is interesting in several aspects (Maizatul et al. 2017). The review points out that wastewaters from some industrial applications and wastewater treatment plants contain pathogens and heavy metals, making them unsuitable for use in aquaculture. However, other wastewaters, such as agricultural processing wastes, are low in pathogens and metals but high in nutrients useful for algal growth. This is positive for the possibilities of biorefineries using some of these recycled, high-nutrient waters, resulting in the residual biomass being applicable to aquaculture feeds if attention is paid to eliminate harmful components in the wastewaters used.

Residual biomass and whole microalgae are also being investigated to enable the increased use of plant-based feeds, such as soy-based products, to reduce the overall cost of feed, as well as to lower the need for marine-based feedstocks (i.e., fishmeal and fish oil). Some recent work on microalgal strains demonstrated that a *Chlorella* or *Tetraselmis* biomass would improve the health of zebrafish fed diets containing 50% soy meal and less than 50% of the fishmeal than in the normal diet (Bravo-Tello et al. 2017). These authors traced this improvement in fish health to improved control of gut inflammation due to soy antigens and an improved mucus layer in the intestinal tract.

2.2.2 Bioplastics and Biopolymers

Microalgae have been studied for their potential use in the production of bioplastics (Hempel, Bozarth et al. 2011, Karan et al. 2019, Zeller et al. 2013, Rahman and Miller 2017). A recent review also points out the need for renewable feedstocks for the production of bioplastic, with microalgae cited as a potential source (Karan et al. 2019). Direct use of microalgae for the production of bioplastic or biopolymers can take several forms: 1) the biomass or residual materials from the biorefinery (e.g., LEA) can be part of a blend of materials, where the algae contribute structural support or elasticity; and 2) where the microalgae are genetically engineered to produce components that can be converted directly to a biopolymer.

Blend Ingredient

Algal species being considered for use in biorefineries (*Chlorella, Spirulina, Nannochloropsis*, and *Botryococcus*) have already been shown to be useful when blended with plastic polymers, such as polypropylene, polyvinyl chloride, and polyethylene (Rahman and Miller 2017). Similar studies are continuing, but what is lacking is an inexpensive algal feedstock. Therefore, there is a validated and pending potential market for excess biomass and residuals from the biorefinery for use in biopolymer production.

A recent study evaluated blends of *Spirulina platensis* and *Chlorella vulgaris* biomass with glycerol (Zeller et al. 2013). They optimized the blends (biomass to glycerol) and produced test materials using thermochemical molding. They found differences in the physical characteristics of the bioplastics produced from blends when different algae were blended with polyethylene, and both algae produced bioplastics suitable for molding into plastic.

Direct Production

A direct approach produces the bioplastic or biopolymer in the microalga through genetic modification.

In one study, the diatom *Phaeodactylum tricornutum* was molecularly engineered to contain the pathway to produce poly-3-hydroxy butyrate (PHB), which is a substrate for the production of renewable bioplastics (Hempel, Bozarth et al. 2011). The genes for ketothiolase, acetoacetyl-CoA reductase, and PHB synthase from the bacterium *Ralstonia eutropha* were introduced into the alga, allowing production of PHB in this microalga. This transgenic diatom produced more than 10% of its biomass as PHB, demonstrating that this alga could be a production system for the feedstocks for bioplastic production. However, additional product optimization will be required to make this economically feasible, and the conversion of the remaining 80% of the biomass into useful co-products is a necessity.

There continues to be an active interest in the production of biobased polymers from microalgae. The announcement that the California Center for Algae Biotechnology was awarded US$2 million to develop monomers in microalgae for the production of polyurethane is a good example of this continuing activity (Gyekye 2019).

Indirect Production

Production of bioplastics and biopolymers indirectly involves using the algal biomass or residuals after lipid extraction as a feedstock for bioconversion or chemical conversion to useful polymers or intermediates. The commercial indirect production of bioplastics or biopolymers using algal biomass as feedstock could be enabled if the biorefinery could deliver the feedstock inexpensively.

Indirect production of polyhydroxyalkanoates (PHA) by bacterial action using algal biomass as a renewable feedstock is a potential approach to making renewable bioplastics. A recent publication using macroalgal biomass and the halophilic archaea *Haloferax mediterranei* demonstrated the ability of using a macroalga, *Ulva* sp., as feedstock for PHA production (Ghosh et al. 2019); a similar process could be developed for using biomass produced in the microalgal biorefinery and/or residual biomass, because lipid extraction leaves the structural carbohydrates.

Lipid as Feedstock

Algal lipids, if available at economical price points, would have value as a feedstock for the production of many different products, such as lubricants, surfactants, and other chemicals, now produced from fossil fuels. A recent example of this approach is the conversion of algae-derived methylated fatty acid esters to useful hydrophobic intermediates (Seo et al. 2018).

Complex Carbohydrates

The residual biomass from lipid extraction contains a large amount of cell wall material that has potential as a feedstock for the production of carbohydrate-based products and materials. The cell walls of the microalgae vary widely and can contain alginate, algaenan, cellulose, hemicellulose, sporopollenin, and other complex carbohydrates (He, Dai, and Wu 2016, Macfie and Welbourn 2000, Rodrigues and da

Silva Bon 2011, Zych et al. 2009). Although some approaches advocate for enzymatic conversion of the residual cell wall material to simple carbohydrates and the conversion of the monomers and small multimers to alcohol through fermentation (Sander and Murthy 2009), use of these complex materials for the production of fibers, which is already being attempted for macroalgae (Pauli 2010), would be an interesting approach that could deliver high value from the microalgal residual biomass.

2.2.3 BIOMASS AS FEEDSTOCK

Utilization of algal biomass as a substrate for microbial production systems makes sense in the context of an algal biorefinery or from wild harvests of algae—where the cost of biomass production is either low, such as the use of residual material (such as LEA from a biorefinery), or for the price of collection (such as an algal bloom that occurs periodically—the bloom that occurred in China just prior to the 2008 Olympics is a great example). The advantage of using a biorefinery's residual biomass rather than wild algal blooms is that a biorefinery would provide a constant feedstock (both in composition and availability) generated as the biorefinery makes other products in its production portfolio. An example of where this feedstock could be important was discussed in the previous section on bioplastics from microalgae.

The efforts to use microalgal residual carbohydrates to produce bioethanol were discussed in a recent review (Tang et al. 2016). Both high-carbohydrate-containing biomass and LEA have been investigated for applicability in the production of bioethanol. Although both applications need pretreatment to release smaller fermentable sugars, these materials have been demonstrated as ready sources for bioethanol production (Rodrigues and da Silva Bon 2011).

A recent announcement by BASF on its efforts to make renewable methanol needs a renewable feedstock for its organisms to convert to methanol (BASF 2018). This would be an example of the indirect utilization of excess biomass or residual biomass being converted into a useful chemical intermediate.

Another example of an indirect approach comes from the efforts of the company LanzaTech (United States). They are using proprietary acetogenic microbes to produce acetic acid from CO and CO_2 (and other industrial gases) as the energy and carbon sources (Simpson 2017) and then feeding the acetate to microalgae to make LC-PUFAs (mainly DHA). This work is being done in cooperation with the IOC-DBT Center for Advanced Bio-Energy Research (a joint group with Indian Oil Corporation Ltd. and Indian Dept. of Biotechnology) (LanzaTech 2014). Their production organism is a *Clostridium autoethanogenum* strain (LanzaTech 2018) for which they have developed an extensive set of synthetic tools to custom-produce a variety of different products using "gas fermentation"—a process described by them as using gases rather than sugars as their energy and carbon sources.

2.2.4 HETEROTROPHIC PRODUCTION

Algae are capable of photoautotrophic, heterotrophic, and mixotrophic growth (Girard et al., 2014, O'Grady and Morgan 2011, Zhou et al. 2012), offering a diverse set of potential growth conditions and feedstocks. The algal products industry is currently making many products from heterotrophic aerobic fermentation systems, such as vitamins from *Chlorella* (Running, Huss, and Olson 1994), LC-PUFAs from *Cryptosporidium cohnii* and *Schizochytrium sp.* (DSM and others), and algal food ingredients produced from *Auxenochlorella protothecoides* and *Prototheca* (Corbion).

The use of heterotrophic microalgae to produce biofuels from feedstocks such as sugar and agriculture and food processing wastewaters provides an indirect way to capture the sun's energy then conversion of this chemical energy into biofuel. This is a potential approach to biofuel production in the microalgal biorefinery, but it is limited by the price of sugar and the cost of fermentation equipment. The use of residual biomass to produce sugars that can then be converted to biofuel or bioproducts would further diversify the microalgal biorefinery.

2.2.5 BIOFERTILIZERS AND AGRICULTURAL PRODUCTS

The residual products—left over after all the most valuable products are extracted—can be used as a biofertilizer (Koller, Muhr, and Braunegg 2014), which, although not a huge profit driver, will make the overall process more environmentally friendly. The company Heliae (United States) is selling and developing products from algae, such as a soil amendment (PhycoTerra) and plant stimulators, based on a more direct application of algal extracts than the use of the residual biomass in composted form.

2.3 SMALL-MARKET HIGH-VALUE CO-PRODUCTS

Small-market high-value co-products are produced by removing or recovering them from the algal biomass that is generated for algal biofuel production or, alternatively, are produced on-site at the biorefinery, utilizing excess infrastructure capacity (e.g., unused inoculation train capacity, photobioreactors, or bags) to make additional co-products and generate more value for the biorefinery.

2.3.1 ISOLATED PURIFIED NATURAL PRODUCTS

Microalgae have long been studied for their potential to be the source of useful bioproducts. Among the potential products identified are antibacterial agents, antifungal agents, antiviral agents, and nutraceuticals—the list is expansive (Rasala and Mayfield 2015). It has been suggested that the biomass generated could be a source of useful bioactive compounds (Bohutskyi et al. 2015). The efforts at bioprospecting from different algal strains is already in progress—for instance, looking for drug candidates (de Vera et al. 2018) and bioactive peptides (Giordano et al. 2018). Subsequent to the development of these bioprospecting tools, microalgal products are already in the marketplace or on the verge of introduction.

β-glucan. β-1,3-glucan has been isolated from the unicellular euglenoid *Euglena gracilis* and is currently being marketed as Algamune β-1,3-glucan (Algal Scientific, United States). Other companies are also developing similar products, such as Euglena Co. Ltd. (Japan). The product is being marketed as a nutritional supplement that improves the immune system and optimizes growth and performance in animals. Studies have been done showing its bioactivity in a variety of potential uses in medical applications such as dermatitis (Sugiyama et al. 2010) and hepatoprotection (Sugiyama et al. 2009).

Vitamins. Ascorbic acid is known to be excreted by different algal species such as *Auxenochlorella protothecoides, Prototheca* sp., and *Prototheca moriformis* during aerobic heterotrophic growth (Running, Huss, and Olson 1994). This could also be a co-product in phototrophic systems if the processing systems could capture the more dilute product. Vitamin E and pro-vitamin A (e.g., β-carotene) are other products already in the marketplace.

Carotenoids. Skin protection using carotenoids such as β-carotene and zeaxanthin from microalgae has been demonstrated, due mostly to their protection against reactive oxygen species (Wang et al. 2015) and could be a potential co-product from the lipid extraction process, as they tend to co-purify with lipids. Fucoxanthin is used in creams for improved skin protection (Rodriguez-Luna et al. 2018).

Polysaccharides. The bioactive polysaccharide sacran was discovered in the cyanobacterium *Aphanothece sacrum* and has anti-inflammatory properties (Ngatu et al. 2012). It was found to inhibit allergic dermatitis in mouse models.

Novel Proteins and Peptides. Proteomic evaluation of *Nannochloropsis gaditana* biomass produced under model biorefinery conditions was examined for novel proteins that could be mined as co-products (Fernández-Acero et al. 2019). This organism is being evaluated as a biofuel and biorefinery strain (Fernández-Acero et al. 2019, Jinkerson, Radakovits, and Posewitz 2013, Radakovits et al. 2012, Sanchez-Silva et al. 2013). Fernández-Acero and colleagues used proteomics techniques to look for products with commercial potential that are produced by *N. gaditana.* They found a number of potentially useful enzymes (e.g., ATP synthase, catalase, and phosphoenolpyruvate carboxykinase), bioactive compounds (e.g., "prohibitin", "resistance to phytophthora", and ferredoxin) while characterizing >600 potential protein targets of interest (Fernández-Acero et al. 2019). When coupled to a biorefinery producing *N. gaditana* biomass and tailoring the downstream processing to capture some of these proteins, the cost of the biomass would no longer be a limiting factor for producing these high-value products.

The company Checkerspot (United States) is engineering materials from microalgae using fermentation as the production process, although there is not much public information on their efforts (Checkerspot 2019). However, a review of their approach has been published that indicates they are fermenting microalgae from the microalgae class Trebouxiophyceae to produce high-performance materials (Gyekye 2018). Because many well-known and industrially relevant algal species are from genera (e.g., *Auxenochlorella, Chlorella, Parachlorella,* and *Botryococcus*) in this class, it promises to be an interesting venture to address advanced materials needs in specific markets.

Antineoplastic Agents. Large-scale screening for antineoplastic agents has been done for cyanobacteria that has led to the identification of many potentially useful compounds (Patterson et al. 1991). For example, from the cyanobacterium *Oscillatoria acutissima* two antineoplastic agents were described: acutiphycin and 20,21-didehydroactutiphycin (Barchi, Moore, and Paatterson 1984).

2.3.2 Heterotrophic Production of High-Value Bioproducts

Although heterotrophic algae have been contemplated for biofuel production (see Section 2.2.4), they have been restrained in terms of commercialization due to the cost of the sugar needed as feedstock. However, the use of heterotrophic algae, where appropriate for market conditions, to make a small-market, high-value bioproduct is already a reality.

Innovative products such as flour and fat replacement products made from heterotrophic microalgae are already on the market through Corbion (Netherlands) and Roquette Freres, SA (France). But new formulations with microalgae are appearing at companies such as the startup Algama (France and United States), which is producing a vegan mayonnaise containing microalgae (GoodSpoon 2019, Crawford 2018).

Corbion (Netherlands) has built a renewables business through development and acquisition. It purchased TerraVia's (formerly Solazyme) assets out of bankruptcy and has been continuing the marketing and development of products derived from heterotrophic microalgae such as *Auxenochlorella (Chlorella) prototheocoides*. They are marketing food and cosmetic products based on microalgal biomass. The food ingredients have a large potential market and are focused on food additives and algal oils. Their Algavia whole microalgae ingredients are marketed as making low-fat foods taste richer and healthier. They also market their AlgaWise algal oils that have high omega-9 content and are marketed as ingredients for improved healthy foods. Other oils derived from microalgae are their Thrive cooking oil, which provides high-temperature stability for frying, and AlgaPrime DHA for a sustainable alternative to fish oil in human nutrition.

2.3.3 Food and Nutraceuticals from the Photosynthetic Biorefinery

Although less likely to be produced from residual biomass, foods and nutraceuticals based on microalgae can be a valuable bioproduct produced based on a side stream from the biorefinery, using the facilities for algal growth and processing already in place in the biorefinery, or advanced processing technologies to assure quality.

Microalgae biomass products grown in open ponds and outside photobioreactors are already being marketed by companies around the world for human nutrition. Examples of this are *Arthrospira (Spirulina) platensis, Chlorella* sp., *Aphanizomenon* sp., and *Haematococcus pluvialis*, which have been used as food for many years (Spolaore et al. 2006) and are now sold as dried pellets and powders and extracts for nutraceutical applications by companies such as Cyanotech, Inc. (United States) and AlgaTechnologies Ltd. (Israel).

Bulk protein isolates from either a residual biomass or a side stream of the production of a biorefinery would also provide a valuable addition to human nutrition. At least a few well-studied species, such as *Chlorella* and *Arthrospira*, are known to produce high-quality protein and have amino acid profiles that are well balanced and suitable for human nutrition (Becker 2007).

Lipids. Qualitas Health has microalgae farms in Texas where it produces algal products. Their focus is on omega-3 LC-PUFAs. They sell nutraceuticals under the

Almega PL trademark that focus on omega-3s for eye health, joint health, and cardiac health using the trade name alGeepa (Qualtitas 2019). AlmegaPL, their EPA product containing 25% EPA and 15% polar lipids, is microalgae derived, but the company does not disclose the strains they are using for this process (Qualitas 2019).

DSM Company (The Netherlands) produces oil that is used as an infant formula additive, as well as a nutraceutical sourced from the heterotrophic production of *Cryphecodinium cohnii* and *Schizochytrium* sp. These products came out of the acquisition of Martek Biosciences and Omega Tech, both early pioneers in algal product development (Gladue and Behrens 1999, Barclay et al. 1998).

2.3.4 INNOVATIVE PRODUCTS

While commercialization of biofuels foundered, many new companies sprouted up with innovative products that utilized microalgae in novel ways. Some of these companies are described here as examples of where the algal products market is now or is heading in the near future.

Printer Ink. Microalgae as a substitute for petroleum-based inks is an example of a product that is already in development by a company called Living Ink Technologies, LLC (United States). They have been funded by the US National Science Foundation's Small Business Innovation Research (SBIR) program (NSF Award #1758587) and have prototype products with promise as potential substitutes for petroleum-based printer inks. As scaled-up algal biomass production accelerates in the microalgal biorefinery, the potential to use pigments that might be extracted prior to destructive processing could reduce the cost of microalgae-derived printer inks and make this company's approach a disruptive high-value co-product.

Microalgae-Based Foams. Materials incorporating microalgae-derived materials have been developed by Algix (United States) and are being marketed as Bloom Algal Foam. This licensed technology takes advantage of the proteins and other molecules in the algal biomass to provide flexibility to an ethylene vinyl acetate–based foam.

Renewable Chemicals and Precursors

The high value product area closest to being experimentally demonstrated and deployed at scales compatible with those of fuels is the production of oleochemical products (e.g. surfactants, lubricants, additives and bio-polymers) from algal oils.

(Laurens, Chen-Glasser, and McMillan 2017)

Electrospun Fibers. The cyanobacterium *Spirulina* sp. LEB 18 produces a number of different bioproducts, including PHB and phenolic compounds (Coelho et al. 2015). Electrospun fibers using *Spirulina*-derived materials were made and evaluated for their use in food packaging (Kuntzler et al. 2018). Prior work was done to create microfibers based on algal protein concentrates isolated from *Botryococcus braunii* residual biomass from a biorefinery pilot through electrospinning (Verdugo, Lim, and Rubilar 2014). These researchers created electrospun microalgal protein concentrate fibers through a formulation that incorporated poly(ethylene oxide) at a low pH that contained up to 93% of the microalgal protein content by weight. They demonstrated the ability for *B. braunii* residual protein wastes to be processed into useful materials could have many applications.

2.4   BIOPRODUCTS LEVERAGING THE BIOREFINERY INFRASTRUCTURE

Microalgae represent a huge and diverse group of organisms (Guiry 2012) that have been and continue to be mined to discover new biomolecules or biomaterials and generate new methods to manipulate microalgae to produce valuable products. With the expertise and infrastructure put in place in an algal biorefinery, it is probable that small-market high-value bioproducts could be generated with minimal additional infrastructure costs (in terms of capital expenditure). This will allow the biorefinery to produce a flexible portfolio of microalgae-based products that could offset the overall costs of the algal biofuel, allowing more rapid commercialization. Some of these products will be discovered by prospecting for naturally occurring biomolecules (e.g., novel peptides), engineering metabolic pathways (e.g., overproduction of specific products), collection of novel materials from spent medium (e.g., extracellular polysaccharides, enzymes and proteins), and resistant polysaccharides from the residual biomass (e.g., sporopollenin).

2.4.1   BIOREFINERY INFRASTRUCTURE SUPPORTS BIOPRODUCT PRODUCTION

The unique capabilities presented by a large-scale algal production plant open up a commercially viable pathway where the existing biorefinery infrastructure can be leveraged into new products. Excess capacity for large-scale phototrophic biomass production of different strains could be engineered into the plant design, allowing managers to expand their portfolio quickly to respond to market opportunities. Coupling phototrophic and heterotrophic production capabilities, as contemplated in mixed trophic state production systems (Bohutskyi et al. 2014, Rismani-Yazdi et al. 2014), would expand these capabilities to provide an even wider array of products (via fermentation).

2.4.2   MINING SPENT MEDIUM

One unknown of phototrophic algal production systems is what can be recovered from the spent medium. Potentially millions of gallons of spent medium will be produced at a commercial scale biorefinery, which must then be treated and either reused for further algal growth or released into the environment once the biological oxygen demand meets the required discharge standard. Some research has been carried out to analyze the potential of spent medium to derive additional products. A review by Liu and colleagues summarizes some of these potential products (Liu, Pohnert, and Wei 2016). Among the potential co-products are extracellular polysaccharides, some of which have already demonstrated a variety of bioactivities, such as antiviral activity, antitumor activity, immunomodulatory activity, metal chelating activity, and many other potentially useful applications (Fan, Han, and Xu 2001, Misurcova et al. 2012, Engene et al. 2011, Koller, Muhr, and Braunegg 2014, Plaza et al. 2009).

An unknown variety of extracellular proteins are produced by microalgal species that could have commercial value. These include proteins such as extracellular carbonic anhydrase, proteases, phenol oxidases, protease inhibitors, pigments, organic acids (e.g., ascorbic acid, lactic acid, 5-amionlevulinic acid, and glycolic acid), extracellular lipids, hydrocarbons, and phytohormones. Extracellular polysaccharides (EPS) have been found for many algal species, such as *Arthrospira platensis* and *Porphyridium cruentum*, and could offer a huge potential product if the downstream processing can be done cost-effectively (Zille et al. 2009, Patel et al. 2013).

The list provided in the Liu review (Liu, Pohnert, and Wei 2016) is extensive and yet cannot encompass all of the potential co-products available from mining the extracellular metabolites produced by industrial microalgae systems; this is mostly due to the variety of different production strains and varied culture conditions. The extracellular compounds found in the spent medium from both phototrophic and heterotrophic growth systems offer a difficult downstream processing challenge, but potentially, due to the huge volumes of spent medium generated by the microalgal biorefinery, are an underexploited resource for novel and high-value co-product generation.

2.4.3 TRANSGENIC PRODUCTION PLATFORMS

Engineered high-value co-products are designed and produced using molecular methods and could benefit from the use of the existing biorefinery infrastructure. For instance, cultivation systems for phototrophic and/or heterotrophic microalgae growth could be designed with excess capacity that would be filled with microalgae producing specialty co-products, and the harvesting and processing systems would have synergy with the biofuel production process, lowering overall manufacturing costs for the portfolio of products.

Microalgae have the ability to be biofactories for recombinant produced proteins (Gimpel et al. 2015, Rasala and Mayfield 2015, Specht, Miyake-Stoner, and Mayfield 2010). Genetic systems have been developed for using both heterotrophic (Talebi et al. 2013) and autotrophic (Anil et al. 2016, Guo et al. 2013, Lauersen et al. 2013, Rasala et al. 2014, Rosenberg et al. 2008) algal growth to produce recombinant proteins. How this can be incorporated into biorefineries remains an unsolved question (see the section on challenges). However, emerging recombinant products are being developed that demonstrate the potential of microalgae in the marketplace; some examples of recent activity follow.

The majority of molecular engineering in microalgae has been carried out with the unicellular, flagellated, haploid green alga *Chlamydomonas reinhardtii* (Allen et al. 2007, Rosales-Mendoza, Paz-Maldonado, and Soria-Guerra 2012, Lauersen et al. 2013, Oey et al. 2013, Berger et al. 2014, Jiang et al. 2014, Bertalan et al. 2015, Shamriz and Ofoghi 2018, Specht, Miyake-Stoner, and Mayfield 2010). Due to the algal molecular biology community's concentration on *C. reinhardtii*, this alga has been promoted as a potential production organism (Rasala et al. 2014, Rasala and Mayfield 2015). However, alternative algal genetic systems based on oleaginous or rapidly growing algae have been developed that may provide viable alternatives for transgenic production platforms (Daboussi et al. 2014, Siaut et al. 2007, Anil et al. 2016).

A substantial body of this work with *C. reinhardtii* focused on chloroplastic expression utilizing this alga's ability to grow on acetate as a way to select for microalgae mutated in the photosynthetic process (Shamriz and Ofoghi 2018, Bertalan et al. 2015, el-Sheekh 2000, Mayfield, Franklin, and Lerner 2003, Tran et al. 2009). The chloroplastic expression can be problematic due to the lack of post-translational modification in proteins expressed in the chloroplast (Vanier et al. 2015). These *C. reinhardtii* efforts have increased the knowledge of how to work with this alga, spurring more rapid development of genetic systems for algal strains more fitting for use in a biorefinery (e.g., high oil producing or rapid photosynthetic growth rates). These are a necessity for using molecular tools to improve the economics of algal biofuels and increase the co-product diversity of the biorefinery product portfolio.

Synthetic Meat. Certain *C. reinhardtii* molecular tools are well developed and have a strong impact on the development of molecularly engineered products from microalgae. Triton Algae Innovations (United States) is developing products based on *C. reinhardtii* (a single-celled green microalga). They issued a press release (Triton 2018) that the biomass they are producing is classified as Generally Regarded as Safe (GRAS) through self-affirmation, thereby concluding that it is safe for use in feeds and foods. However, the US Food and Drug Administration's (FDA) Select Committee on GRAS Substances (SCOGS) database does not list *C. reinhardtii* as an approved product. Triton published a study on the toxicology of the *C. reinhardtii* biomass (Murbach et al. 2018) and has a substantial body of publications that support their conclusions (Triton 2019a). This is a company with strong fundamental science that is moving toward commercial reality.

Triton is also developing a portfolio of interesting products. This includes a meat alternative using a recombinant yellow *C. reinhardtii* to produce meat proteins (leghemoglobin and myoglobin) that are combined with algal biomass to make a meat alternative.

Milk Proteins. Triton has also expressed the milk protein osteopontin with potential as both a nutraceutical product and as an anti-inflammatory treatment. This product uses recombinant *C. reinhardtii* to make colostrum proteins that are important to the early nutrition of mammals (Triton 2019b).

Genetic Systems. Recently, there has been an uptick in the work being done to develop genetic systems for a wider variety of algal strains that may be important to the production of new co-products for the algal biorefinery (Urtubia, Betanzo, and Vasquez 2016, Kumar et al. 2016). A review of the current status of the genetic engineering of microalgae useful for biofuels and biobased chemicals has recently been published (Ng et al. 2017).

Through the work of many laboratories and the US Department of Energy's Joint Genome Project[1] many algal strains have had partial or whole genomes sequenced (Rosenberg et al. 2008, Kumar et al. 2016). These include the oleaginous diatom *Phaeodactylum tricornutum* (Siaut et al. 2007, Thamatrakoln et al. 2012), the Eustigmatophyte alga *Nannochloropsis oculata (Chen et al. 2008)*, the unicellular green alga *C. reinhardtii* so important to the development of genetic systems for algae (Merchant et al. 2007), the marine-centric diatom important to

---

[1] https://genome.jgi.doe.gov/algae/algae.info.html

aquaculture *Thalassiosira pseudonana* (Armbrust et al. 2004), and the heterotrophic oleaginous green alga *Auxenochlorella protothecoides* (Anil et al. 2016) as a few examples of algae important to biorefinery platforms in development.

Microalgal Production vs. Other Microbial Systems. One of the advantages of the microalgal system is the huge and varied library of genetic diversity available. There are estimated to be about 75,000 known algal species, with an unknown number yet to be discovered (Guiry 2012). Microalgae have different trophic capabilities (phototrophic to heterotrophic), temperature tolerances (hot springs to ice), secondary metabolite production, toxin production, nutrient mobilization (e.g., siderophores), cell wall composition, and beyond. However, once a useful product is discovered, a decision must be made whether production of the target molecule in microalgae is more valuable than the production of the product in an alternative production system that may provide more simple scale-up and isolation (e.g., yeast or bacteria).

An example of this decision is the production of a novel restriction enzyme found initially in cyanobacteria (e.g., AnaI) that is now produced commercially in recombinant bacteria and not in the source organism *Anabaena*. The reasoning behind the choice of a bacterial system over the original species is that the methods for expression of a single protein in bacteria are easy and have relatively high expression rates with existing industrial technologies. Production of more complicated molecules, such as some secondary metabolites from microalgae or pigments with specifically positioned chromophores with high value, could be a challenge to make in recombinant expression systems. However, if these compounds can be easily produced by synthetic chemistry, often the algal strain used to discover the compound is not used in the production of the final product. This reduces the number of potential algal products that will be commercialized in synergy with the algal biorefinery.

Many previous studies have been done looking for bioactive compounds produced by microalgae (Liu, Pohnert, and Wei 2016). It is likely that this bioprospecting, when successful, will lead to biomolecules that will be synthesized rather than isolated from the algal species (Ravishankar et al. 2006). The scaled production of algal biomass for a biorefinery opens an alternative approach to making some of these bioactives through a readily available biomass feedstock if efficient isolation protocols can be developed. Residual materials could then be fed back to the biorefinery for other uses.

Using *Phaeodactylum tricornutum* (a diatom) as the expression system, the triterpenes betulin and its precursor, lupeol, were produced as a means of making these high-value compounds (originally from higher plants) (D'Adamo et al. 2018).

Recent work on making renewable chemicals and precursors using engineered cyanobacteria is an example of a potential parallel bioproduct for the biorefinery using excess capacity in its enclosed photobioreactors. An example is the recent engineering of *Synechococcus elongatus* PCC 7942 to produce heparosan, an unsulfonated polysaccharide with applications ranging from cosmetic to pharmaceutical applications (Sarnaik et al. 2019). One use of this product is as a precursor for the synthesis of heparin.

Genome Editing. A recent review of the progress in genome editing and its potential impact on the production of biofuels and co-products details the present status based on CRISP-Cas9, TALEN, and ZFN methods (Ng et al. 2017). Recent work in a combined effort of Synthetic Genomics and ExxonMobil (both in the United

States) used genome editing to double the yield of lipid in *Nannochloropsis gaditana* (Ajjawi et al. 2017). They applied the CRISPR-Cas9 genome editing method to shut down a transcription factor that provided increased production of lipid under nutrient-replete conditions. This is a critical finding, because most oleaginous microalgae either grow fast with little lipid accumulation or need to be stressed (e.g., nitrogen limitation), which reduces the growth rate but stimulates lipid accumulation (Rismani-Yazdi et al. 2012).

Medical Applications

Vaccine Production. Using heterotrophic alga, recombinant expression of a viral antigen(s) is proposed as a vaccine production strategy (Gregory et al. 2013, Rasala and Mayfield 2015). Recently, the heterotrophic alga *Schizochytrium* sp. was used to express recombinant Zika virus glycoprotein antigens, and an *E. coli* enterotoxin subunit was used as an adjuvant to stimulate mucosal immunity (Márquez-Escobar, Bañuelos-Hernández, and Rosales-Mendoza 2018).

Recombinant Therapeutic Proteins. The systems are well developed for several algal strains to express recombinant proteins in microalgae in the chloroplast (Taunt, Stoffels, and Purton 2018). Some of these proteins, such as human growth hormone, endolysins, and immunotoxins, have strong potential for use in human medial applications.

Solarvest BioEnergy, Inc. (Vancouver, Canada) is a microalgae technology company whose algal-based production platform provides it with an extremely flexible system capable of being adapted to produce numerous products—from omega-3 fatty acids to human therapeutic proteins. The company has successfully demonstrated (news release dated March 16, 2015) the expression of bone morphogenetic protein (BMP), a high-value therapeutic protein, to accelerate bone growth. In addition, the company's platform has successfully produced recombinant viral antigens (immune-stimulating proteins) and cecropins (antimicrobial peptides). The company has completed a feasibility study for the expression of cannabidiol oil (CBD) and tetrahydrocannabinol (THC) as a way to produce CBD and THC in sterile bioreactors.

Antibody Production. Human antibodies have been produced in *Phaeodactylum tricornutum* (Hempel, Lau et al. 2011, Hempel and Maier 2012). A study was done on the glycosylation patterns of these humanized monoclonal antibodies to hepatitis B virus surface antigen (HbsAg) to understand the potential value of this approach in microalgae to the production of these very high-value biotherapeutics (Vanier et al. 2015). These authors showed that *P. tricornutum* could produce glycoproteins that carried high mannose type glycans that could either be secreted or retained in the endoplasmic reticulum. Although the antibodies produced hold significant promise and were shown to have *in vivo* activity, additional work on the glycosylation system of *P. tricornutum* will be required to make fully humanized antibodies.

The technical and physical infrastructure that will be available at the commercial-scale biorefinery has the capacity to broaden the product portfolio possible through contained growth of a number of these high-value products that could be game changing in terms of the economics surrounding these facilities.

2.4.3.1   Challenges for Transgenic Bioproducts

Genetically engineered microalgal strains face significant challenges as they seek to enter the marketplace in order to overcome safety concerns, as well as emotional concerns tied to the use of transgenic microalgae, especially in open cultivation systems, which are often applied in phototrophic biorefineries. There have been several different attempts to judge scientifically how safe the use of genetically engineered strains would be (Henley et al. 2013, Snow and Smith 2012, Austin 2009). There is not a consensus, as the conclusions are not uniform in their assessment of the safety of the microalgal biorefinery based on genetically engineered microalgae. Nonetheless, conditional approvals have been granted in the United States for several operations where these strains were used in controlled cultivation systems (Rosenberg et al. 2008). The societal and emotional resistance to the use of genetically engineered strains will be harder to overcome and must be dealt with by careful science and open communication of these results in a transparent fashion with the wider society.

The cutting-edge genome editing technologies under development (e.g., CRISPR-Cas9, TALEN, and ZFN) pose a dilemma that is still being evaluated by the regulatory agencies and advocacy groups (Sprink et al. 2016), as these technologies do not fall neatly into previous transgenic organism categories and are capable of very precise control of expression of existing genes without the addition of foreign nucleic acid (Strack 2019).

2.5   CONCLUSIONS

A blue wave is coming to the biobased economy where natural and engineered organisms (marine and fresh water) will provide the means to manufacture useful and large-scale products based on microalgae. These microalgae-based bioproducts will be produced in biorefineries that are not solely biofuel based, due to the competitive nature of the energy market. However, the huge biodiversity represented by the many algal strains under investigation, the improved methods for the manufacture of biomass, and improved processing methods promise to enhance the availability of unique biobased microalgae-derived products in the near future through direct (phototrophic) production of algal feedstocks and indirect conversion of biobased feedstocks by algal fermentation to produce useful products. Ironically, as the scaled-up production of materials and feeds based on algal biomass or extracted materials occurs, algal biofuels may end up being co-products of the algal biofactories as they provide renewable and sustainable products to the marketplace.

2.6   REFERENCES CITED

ABO. 2019. "About the Algae Bioass Organization." https://algaebiomass.org/about/who-we-are/philosophy-purpose/.

Ajjawi, I., J. Verruto, M. Aqui, L. B. Soriaga, J. Coppersmith, K. Kwok, L. Peach, E. Orchard, R. Kalb, W. Xu, T. J. Carlson, K. Francis, K. Konigsfeld, J. Bartalis, A. Schultz, W. Lambert, A. S. Schwartz, R. Brown, and E. R. Moellering. 2017. "Lipid production in

Nannochloropsis gaditana is doubled by decreasing expression of a single transcriptional regulator." *Nat Biotech* 35:647. doi: 10.1038/nbt.3865 www.nature.com/articles/nbt.3865#supplementary-information.

Allen, M. D., J. A. del Campo, J. Kropat, and S. S. Merchant. 2007. "FEA1, FEA2, and FRE1, encoding two homologous secreted proteins and a candidate ferrireductase, are expressed coordinately with FOX1 and FTR1 in iron-deficient *Chlamydomonas reinhardtii.*" *Eukaryot Cell* 6 (10):1841–1852. doi: 10.1128/EC.00205-07.

Allnutt, F. C. T., and B. A. Kessler. 2015. "Harvesting and downstream processing—and their economics." In *Biomass and Biofuels from Microalgae*, edited by N. R. Hoheimani, M. P McHenry, K. de Boer, and P. Bahri, 251–273. Switzerland: Springer.

Ambasta, A., and J. J. Buonocore. 2018. "Carbon pricing: A win-win environmental and public health policy." *Can J Public Health* 109 (5–6):779–781. doi: 10.17269/s41997-018-0099-5.

Anil, K., P. Zoee, S. Clayton, L. P. Bradley, A. C. Daniel, T. S. Richard, and F. C. T. Allnutt. 2016. "Molecular tools for bioengineering eukaryotic microalgae." *Curr Biotech* 5 (2):93–108. doi: 10.2174/2211550105666160127002147.

Armbrust, E. V., J. A. Berges, C. Bowler, B. R. Green, D. Martinez, N. H. Putnam, S. Zhou, A. E. Allen, K. E. Apt, M. Bechner, M. A. Brzezinski, B. K. Chaal, A. Chiovitti, A. K. Davis, M. S. Demarest, J. C. Detter, T. Glavina, D. Goodstein, M. Z. Hadi, U. Hellsten, M. Hildebrand, B. D. Jenkins, J. Jurka, V. V. Kapitonov, N. Kroger, W. W. Lau, T. W. Lane, F. W. Larimer, J. C. Lippmeier, S. Lucas, M. Medina, A. Montsant, M. Obornik, M. S. Parker, B. Palenik, G. J. Pazour, P. M. Richardson, T. A. Rynearson, M. A. Saito, D. C. Schwartz, K. Thamatrakoln, K. Valentin, A. Vardi, F. P. Wilkerson, and D. S. Rokhsar. 2004. "The genome of the diatom Thalassiosira pseudonana: Ecology, evolution, and metabolism." *Science* 306 (5693):79–86. doi: 10.1126/science.1101156.

Arney, B., W. Liu, I. P. Forster, R. S. McKinley, and C. M. Pearce. 2015. "Feasibility of dietary substitution of live microalgae with spray-dried *Schizochytrium* sp. or *Spirulina* in the hatchery culture of juveniles of the Pacific geoduck clam (*Panopea generosa*)." *Aquaculture* 444 (0):117–133. doi: 10.1016/j.aquaculture.2015.02.014.

Austin, A. 2009. "Transgenic algae pose environmental risks." *Biomass Magazine.*

Barchi, J. J., R. E. Moore, and G. M. L. Patterson. 1984. "Acutiphycin and 20,21-didehydroacutiphycin, new antineoplastic agents from the cyanophyte *Oscillatoria acutissima.*" *J Am Chem Soc* 106 (26):8193–8197. doi: 10.1021/ja00338a031.

Barclay, W., R. Abril, P. Abril, C. Weaver, and A. Ashford. 1998. "Production of docosahexaenoic acid from microalgae and its benefits for use in animal feeds." *World Rev Nutr Dietetics* 83:61–76.

BASF. 2018. "BASF announce renewable methanol production." Biofuels International. https://biofuels-news.com/display_news/14140/basf_announce_renewable_methanol_production/. (Accessed February 18, 2019).

Beal, C. M., R. E. Hebner, M. E. Webber, R. S. Ruoff, A. F. Seibert, and C. W. King. 2012. "Comprehensive evaluation of algal biofuel production: Experimental and target results." *Energies* 5 (6):1943–1981.

Becker, E. W. 2007. "Micro-algae as a source of protein." *Biotech Adv* 25 (2):207–210. doi: 10.1016/j.biotechadv.2006.11.002.

Berger, H., O. Blifernez-Klassen, M. Ballottari, R. Bassi, L. Wobbe, and O. Kruse. 2014. "Integration of carbon assimilation modes with photosynthetic light capture in the green alga *chlamydomonas reinhardtii.*" *Mol Plant.* doi: 10.1093/mp/ssu083.

Bertalan, I., M. C. Munder, C. Weiss, J. Kopf, D. Fischer, and U. Johanningmeier. 2015. "A rapid, modular and marker-free chloroplast expression system for the green alga *Chlamydomonas reinhardtii.*" *J Biotechnol* 195:60–66. doi: 10.1016/j.jbiotec.2014.12.017.

Bohutskyi, P., K. Liu, L. K. Nasr, N. Byers, J. N. Rosenberg, G. A. Oyler, M. J. Betenbaugh, and E. J. Bouwer. 2015. "Bioprospecting of microalgae for integrated biomass production and phytoremediation of unsterilized wastewater and anaerobic digestion centrate." *Appl Micro Biotech* 99 (14):6139–6154. doi: 10.1007/s00253-015-6603-4.

Bohutskyi, P., T. Kula, B. A. Kessler, Y. Hong, E. J. Bouwer, M. J. Betenbaugh, and F. C. T. Allnutt. 2014. "Mixed trophic state production process for microalgal biomass with high lipid content for generating biodiesel and biogas." *BioEnergy Res* 1–12. doi: 10.1007/s12155-014-9453-5.

Bravo-Tello, K., N. Ehrenfeld, C. J. Solís, P. E. Ulloa, M. Hedrera, M. Pizarro-Guajardo, D. Paredes-Sabja, and C. G. Feijóo. 2017. "Effect of microalgae on intestinal inflammation triggered by soybean meal and bacterial infection in zebrafish." *PLOS One* 12 (11):e0187696. doi: 10.1371/journal.pone.0187696.

Chauton, M. S., K. I. Reitan, N. H. Norsker, R. Tveterås, and H. T. Kleivdal. 2015. "A techno-economic analysis of industrial production of marine microalgae as a source of EPA and DHA-rich raw material for aquafeed: Research challenges and possibilities." *Aquaculture* 436 (0):95–103. doi: 10.1016/j.aquaculture.2014.10.038.

Checkerspot. 2019. "A materials innovation company." www.checkerspot.com/.

Chen, H., T. Qiu, J. Rong, C. He, and Q. Wang. 2015. "Microalgal biofuel revisted: An informatics-based analysis of developments to date and future prospects." *Applied Energy* 155: 585–598.

Chen, H. L., S. S. Li, R. Huang, and H.-J. Tsai. 2008. "Conditional production of a functional fish growth hormone in the transgenic line of *Nannochloropis oculata* (Eustigmatophyceae)." *J Phycol* 44 (3):768–776.

Chew, K. W., J. Y. Yap, P. L. Show, N. H. Suan, J. C. Juan, T. C. Ling, D. J. Lee, and J. S. Chang. 2017. "Microalgae biorefinery: High value products perspectives." *Bioresour Technol* 229:53–62. doi: 10.1016/j.biortech.2017.01.006.

Coelho, V. C., C. K. de Silva, A. L. Terra, J. A. V. Costa, and M. G. de Morais. 2015. "Polyhydroxybutyrate production by *Spirulina* sp. LEB 18 grown under different nutrient concentrations." *African J Micro Res* 9 (24):1586–1594. doi: 10.5897/AJMR2015.7530.

Cohen, B. 2019. National Algae Association's 2018 Algae Year in Review.

Crawford, E. 2018. Algama is using microalgae to disrupt how food is made—starting with mayo. *FOOD Navigator USA*.

D'Adamo, S., G. Schiano di Visconte, G. Lowe, J. Szaub-Newton, T. Beacham, A. Landels, M. J. Allen, A. Spicer, and M. Matthijs. 2018. "Engineering the unicellular alga *Phaeodactylum tricornutum* for high-value plant triterpenoid production." *Plant Biotech J*. doi: 10.1111/pbi.12948.

Daboussi, F., S. Leduc, A. Marechal, G. Dubois, V. Guyot, C. Perez-Michaut, A. Amato, A. Falciatore, A. Juillerat, M. Beurdeley, D. F. Voytas, L. Cavarec, and P. Duchateau. 2014. "Genome engineering empowers the diatom *Phaeodactylum tricornutum* for biotechnology." *Nat Commun* 5:3831. doi: 10.1038/ncomms4831.

de Vera, C. R., G. Diaz Crespin, A. H. Daranas, S. M. Looga, K. E. Lillsunde, P. Tammela, M. Perala, V. Hongisto, J. Virtanen, H. Rischer, C. D. Muller, M. Norte, J. J. Fernandez, and M. L. Souto. 2018. "Marine microalgae: Promising source for new bioactive compounds." *Mar Drugs* 16 (9). doi: 10.3390/md16090317.

el-Sheekh, M. M. 2000. "Stable chloroplast transformation in *Chlamydomonas reinhardtii* using microprojectile bombardment." *Folia Microl* 45 (6):496–504.

Engene, N., E. C. Rottacker, J. Kastovsky, T. Byrum, H. Choi, M. H. Ellisman, J. Komarek, and W. H. Gerwick. 2011. "*Moorea producta* gen. nov., sp. nov. and *Moorea bouillonii* comb. nov., tropical marine cyanobacteria rich in bioactive secondary metabolites." *Int J Syst Evol Micro*. doi: ijs.0.033761–0 [pii]10.1099/ijs.0.033761-0.

EUBA. 2019. "EUBA: About Us." European Bioeconomy Alliance. https://bioeconomyalliance.eu/node/82.

Fan, X., L. Han, and N. Xu. 2001. "Screening of antitumor compounds and the activity substances from marine algae." *Studia marina sinica/Haiyang Kexue Jikan. Qingdao [Stud. Mar. Sin./Haiyang Kexue Jikan]* (43):120–128.

Fernández-Acero, F. J., F. Amil-Ruiz, M. J. Durán-Peña, R. Carrasco, C. Fajardo, P. Guarnizo, C. Fuentes-Almagro, and R. A. Vallejo. 2019. "Valorisation of the microalgae *Nannochloropsis gaditana* biomass by proteomic approach in the context of circular economy." *Journal of Proteomics* 193:239–242. doi: 10.1016/j.jprot.2018.10.015.

Foley, P. M., E. S. Beach, and J. B. Zimmerman. 2011. "Algae as a source of renewable chemicals: Opportunities and challenges." *Green Chem* 13 (6):1399–1405. doi: 10.1039/C1GC00015B.

Gatrell, S., K. Lum, J. Kim, and X. G. Lei. 2014. "Non-ruminant nutrition symposium: Potential of defatted microalgae from the biofuel industry as an ingredient to replace corn and soybean meal in swine and poultry diets." *J. Anim. Sci.* 92:1306–1314.

Ghosh, S., R. Gnaim, S. Greiserman, L. Fadeev, M. Gozin, and A. Golberg. 2019. "Macroalgal biomass subcritical hydrolysates for the production of polyhydroxyalkanoate (PHA) by *Haloferax mediterranei.*" *Bioresour Technol* 271:166–173. doi: 10.1016/j.biortech.2018.09.108.

Gimpel, J. A., J. S. Hyun, N. G. Schoepp, and S. P. Mayfield. 2015. "Production of recombinant proteins in microalgae at pilot greenhouse scale." *Biotechl Bioeng* 112 (2):339–345. doi: 10.1002/bit.25357.

Giordano, D., M. Costantini, D. Coppola, C. Lauritano, L. Nunez Pons, N. Ruocco, G. di Prisco, A. Ianora, and C. Verde. 2018. "Biotechnological applications of bioactive peptides from marine sources." *Adv Micro Physiol* 73:171–220. doi: 10.1016/bs.ampbs.2018.05.002.

Girard, J.-M., M.-L. Roy, M. B. Hafsa, J. Gagnon, N. Faucheux, M. Heitz, R. Tremblay, and J.-S. Deschênes. "Mixotrophic cultivation of green microalgae *Scenedesmus obliquus* on cheese whey permeate for biodiesel production." *Algal Res* (0). doi: 10.1016/j.algal.2014.03.002.

Gladue, R. M., and P. W. Behrens. 1999. DHA-containing nutritional compositions and methods for their production. In *WO 09906585*: Martek Biosciences Corp.

GoodSpoon. 2019. "Gourmet eggless condiments." The Good Spoon. http://thegoodspoon foods.com/

GPI. 2019. "Minnesota Policy Leaders Continue to Support Bioeconomy Production Incentive." www.betterenergy.org/blog/minnesota-policy-leaders-reaffirm-support-bioeconomy-production-incentive/.

Gregory, J. A., A. B. Topol, D. Z. Doerner, and S. Mayfield. 2013. "Alga-produced cholera toxin-Pfs25 fusion proteins as oral vaccines." *Appl Environ Micro* 79 (13):3917–3925. doi: 10.1128/aem.00714-13.

Guiry, M. D. 2012. "How many species of algae are there?" *J Phycol* 48 (5):1057–1063. doi: 10.1111/j.1529-8817.2012.01222.x.

Guo, S.-L., X.-Q. Zhao, Y. Tang, C. Wan, Md A. Alam, S.-H. Ho, F.-W. Bai, and J.-S. Chang. 2013. "Establishment of an efficient genetic transformation system in *Scenedesmus obliquus.*" *J Biotech* 163 (1):61–68. doi: 10.1016/j.jbiotec.2012.10.020.

Gyekye, L. 2018. "The green revolution will be blue: Harvesting algae for the bio-economy." *Bio-Based World Quarterly* 12:10–11.

Gyekye, L. 2019. US researchers awarded $2m to advance algae-based bio-polymers. *Bio-Based World News*.

Haznedaroglu, B. Z., H. Rismani-Yazdi, F. C. T. Allnutt, D. Reeves, and J. Peccia. 2016. "Chapter 18: Algal biorefiney for high-value platform chemicals." In *Platform Chemistry Biorefinery, Future Green Chemistry*, edited by K. Brar, J. Sarma, and K. Pakshirajan, 333–360. Amsterdam, The Netherlands: Elsevier. Original edition, June 18, 2016.

He, X., J. Dai, and Q. Wu. 2016. "Identification of Sporopollenin as the Outer Layer of Cell Wall in Microalga *Chlorella protothecoides.*" *Front Micro* 7:1047. doi: 10.3389/fmicb.2016.01047.

Hempel, F., J. Lau, A. Klingl, and U. G. Maier. 2011. "Algae as protein factories: Expression of a human antibody and the respective antigen in the diatom *Phaeodactylum tricornutum.*" *PLOS ONE* 6 (12):e28424. doi: 10.1371/journal.pone.0028424.

Hempel, F, and U. G. Maier. 2012. "An engineered diatom acting like a plasma cell secreting human IgG antibodies with high efficiency." *Micro Cell Factories* 11 (1):126. doi: 10.1186/1475-2859-11-126.

Hempel, F., A. S. Bozarth, N. Lindenkamp, A. Klingl, S. Zauner, U. Linne, A. Steinbuchel, and U. G. Maie. 2011. "Microalgae as bioreactors for bioplastic production." *Micro Cell Factories* 10:81. doi: 10.1186/1475-2859-10-81.

Henchion, M., M. Hayes, A. M. Mullen, M. Fenelon, and B. Tiwari. 2017. "Future protein supply and demand: Strategies and factors influencing a sustainable equilibrium." *Foods* 6 (7):53.

Henley, W. J., R. W. Litaker, L. Novoveska, C. S. Duke, H. D. Quemada, and R. T. Sayre. 2013. "Initial risk assessment of genetically modified (GM) microalgae for commodity-scale biofuel cultivation." *Algal Res* 2 (1).

Jiang, W., A. J. Brueggeman, K. M. Horken, T. M. Plucinak, and D. P. Weeks. 2014. "Successful transient expression of Cas9 and single guide RNA genes in *Chlamydomonas reinhardtii.*" *Eukaryot Cell* 13 (11):1465–1469. doi: 10.1128/EC.00213-14.

Jinkerson, R. E., R. Radakovits, and M. C. Posewitz. 2013. "Genomic insights from the oleaginous model alga *Nannochloropsis gaditana.*" *Bioengineered* 4 (1):37–43. doi: 10.4161/bioe.21880.

Karan, H., C. Funk, M. Grabert, M. Oey, and B. Hankamer. 2019. "Green bioplastics as part of a circular bioeconomy." *Trends Plant Sci* doi: 10.1016/j.tplants.2018.11.010.

Karaskova, K., P. Suchy, and E. Strakova. 2015. "Current use of phytogenic feed additives in animal nutrition: A review." *Czech J Amin Sci* 60 (12):521–530.

Kite-Powell, J. 2018. See how algae could change our world. *Forbes.*

Koller, M., A. Muhr, and G. Braunegg. 2014. "Microalgae as versatile cellular factories for valued products." *Algal Res* 6:52–63. doi: 10.1016/j.algal.2014.09.002.

Kumar, A., Z. Perrine, C. Srroff, B. Postier, D. A. Coury, R. T. Sayre, and F. C. T. Allnutt. 2016. "Molecular tools for bioengineering eukaryotic microalgae." *Curr Biotech* 5 (5):93–108.

Kuntzler, S. G., A. C. A. Almeida, J. A. V. Costa, and M. G. Morais. 2018. "Polyhydroxybutyrate and phenolic compounds microalgae electrospun nanofibers: A novel nanomaterial with antibacterial activity." *Int J Biol Macromol* 113:1008–1014. doi: 10.1016/j.ijbiomac.2018.03.002.

Lam, G. P., M. H. Vermue, M. H. M. Eppink, R. H. Wijffels, and C. van den Berg. 2018. "Multi-product microalgae biorefineries: From concept towards reality." *Trends Biotech* 36 (2):216–227. doi: 10.1016/j.tibtech.2017.10.011.

LanzaTech. 2014. Capturing Carbon to Feed the World. In *LanzaTech offers a sustainable alternative*: LanzaTech.

LanzaTech. 2018. "Ancient life to sustain future progress." www.lanzatech.com/innovation/technical-overview/.

Lauersen, K. J., H. Berger, J. H. Mussgnug, and O. Kruse. 2013. "Efficient recombinant protein production and secretion from nuclear transgenes in *Chlamydomonas reinhardtii.*" *J Biotech* 167 (2):101–110. doi: 10.1016/j.jbiotec.2012.10.010.

Laurens, L. M. L., M. Chen-Glasser, and J. D. McMillan. 2017. "A perspective on renewable bioenergy from photosynthetic algae as feedstock for biofuels and bioproducts." *Algal Res* 24:261–264.

Leite, G. B., A. E. M. Abdelaziz, and P. C. Hallenbeck. 2013. "Algal biofuels: Challenges and opportunities." *Bioresource Technol* 145 (0):134–141. doi: 10.1016/j.biortech.2013.02.007.

Liu, L., G. Pohnert, and D. Wei. 2016. "Extracellular metabolites from industrial microalgae and their biotechnological potential." *Mar Drugs* 14 (10). doi: 10.3390/md14100191.

Lodge-Ivey, S. L., L. N. Tracey, and A. Salazar. 2013. "The utility of lipid extracted algae as a protein source in forage or starch-based ruminant diets." *J Animal Sci* doi: 10.2527/jas.2013-7027.

Lodge-Ivey, S. L., L. N. Tracey, and A. Salazar. 2014. "Ruminant nutrition symposium: The utility of lipid extracted algae as a protein source in forage or starch-based ruminant diets." *J Animal Sci* 92 (4):1331–1342. doi: 10.2527/jas.2013-7027.

Long, X., X. Wu, L. Zhao, J. Liu, and Y. Cheng. 2017. "Effects of dietary supplementation with *Haematococcus pluvialis* cell powder on coloration, ovarian development and antioxidation capacity of adult female Chinese mitten crab, *Eriocheir sinensis*." *Aquaculture* 473:545–553. doi: 10.1016/j.aquaculture.2017.03.010.

Macfie, S. M., and P. M. Welbourn. 2000. "The cell wall as a barrier to uptake of metal ions in the unicellular green alga *Chlamydomonas reinhardtii* (*chlorophyceae*)." *Arch Envir Cont Tox* 39 (4):413–419.

Maizatul, A. Y., R. M. S. R. Mohamed, A. A. Al-Gheethi, and M. K. A. Hashim. 2017. "An overview of the utilisation of microalgae biomass derived from nutrient recycling of wet market wastewater and slaughterhouse wastewater." *Int Aquat Res* 9:177–193. doi: 10.1007/s40071-017-0168-z.

MarketResearch. 2018. Algae products market by application & source—Global forecast 2023. Markets and Markets Research Private Ltd.

Márquez-Escobar, V. A., B. Bañuelos-Hernández, and S. Rosales-Mendoza. 2018. "Expression of a Zika virus antigen in microalgae: Towards mucosal vaccine development." *J Biotech* 282:86–91. doi: 10.1016/j.jbiotec.2018.07.025.

Mayfield, S. P., S. E. Franklin, and R. A. Lerner. 2003. "Expression and assembly of a fully active antibody in algae." *Proc Natl Acad Sci USA* 100 (2):438–442. doi: 10.1073/pnas.0237108100.

Merchant, S. S., S. E. Prochnik, O. Vallon, E. H. Harris, S. J. Karpowicz, G. B. Witman, A. Terry, A. Salamov, L. K. Fritz-Laylin, L. Marechal-Drouard, W. F. Marshall, L. H. Qu, D. R. Nelson, A. A. Sanderfoot, M. H. Spalding, V. V. Kapitonov, Q. Ren, P. Ferris, E. Lindquist, H. Shapiro, S. M. Lucas, J. Grimwood, J. Schmutz, P. Cardol, H. Cerutti, G. Chanfreau, C. L. Chen, V. Cognat, M. T. Croft, R. Dent, S. Dutcher, E. Fernandez, H. Fukuzawa, D. Gonzalez-Ballester, D. Gonzalez-Halphen, A. Hallmann, M. Hanikenne, M. Hippler, W. Inwood, K. Jabbari, M. Kalanon, R. Kuras, P. A. Lefebvre, S. D. Lemaire, A. V. Lobanov, M. Lohr, A. Manuell, I. Meier, L. Mets, M. Mittag, T. Mittelmeier, J. V. Moroney, J. Moseley, C. Napoli, A. M. Nedelcu, K. Niyogi, S. V. Novoselov, I. T. Paulsen, G. Pazour, S. Purton, J. P. Ral, D. M. Riano-Pachon, W. Riekhof, L. Rymarquis, M. Schroda, D. Stern, J. Umen, R. Willows, N. Wilson, S. L. Zimmer, J. Allmer, J. Balk, K. Bisova, C. J. Chen, M. Elias, K. Gendler, C. Hauser, M. R. Lamb, H. Ledford, J. C. Long, J. Minagawa, M. D. Page, J. Pan, W. Pootakham, S. Roje, A. Rose, E. Stahlberg, A. M. Terauchi, P. Yang, S. Ball, C. Bowler, C. L. Dieckmann, V. N. Gladyshev, P. Green, R. Jorgensen, S. Mayfield, B. Mueller-Roeber, S. Rajamani, R. T. Sayre, P. Brokstein, I. Dubchak, D. Goodstein, L. Hornick, Y. W. Huang, J. Jhaveri, Y. Luo, D. Martinez, W. C. Ngau, B. Otillar, A. Poliakov, A. Porter, L. Szajkowski, G. Werner, K. Zhou, I. V. Grigoriev, D. S. Rokhsar, and A. R. Grossman. 2007. "The Chlamydomonas genome reveals the evolution of key animal and plant functions." *Science* 318 (5848):245–250. doi: 10.1126/science.1143609.

Misurcova, L., S. Skrovankova, D. Samek, J. Ambrozova, and L. Machu. 2012. "Health benefits of algal polysaccharides in human nutrition." *Adv Food Nutr Res* 66:75–145. doi: 10.1016/b978-0-12-394597-6.00003-3.

Murbach, T. S., R. Glavits, J. R. Endres, G. Hirka, A. Vertesi, E. Beres, and I. P. Szakonyine. 2018. "A toxicological evaluation of *Chlamydomonas reinhardtii*, a green algae." *Int J Toxicol* 37 (1):53–62. doi: 10.1177/1091581817746109.

NAABB. 2014. National Alliance for Advanced Biofuels and Bioproducts—Final Report Synopsis. DOE.

Ng, I.-S., S.-I. Tan, P.-H. Kao, Y.-K. Chang, and J.-S. Chang. 2017. "Recent developments on genetic engineering of microalgae for biofuels and bio-based chemicals." 12 (10):1600644. doi: 10.1002/biot.201600644.

Ngatu, N. R., M. K. Okajima, M. Yokogawa, R. Hirota, M. Eitoku, B. A. Muzembo, N. Dumavibhat, M. Takaishi, S. Sano, T. Kaneko, T. Tanaka, H. Nakamura, and N. Suganuma. 2012. "Anti-inflammatory effects of sacran, a novel polysaccharide from *Aphanothece sacrum*, on 2,4,6-trinitrochlorobenzene-induced allergic dermatitis *in vivo*." *Ann Allergy Asthma Immunol* 108 (2):117–122. doi: 10.1016/j.anai.2011.10.013.

O'Grady, J., and J. A. Morgan. 2011. "Heterotrophic growth and lipid production of *Chlorella prototothecoides* on glycerol." *Bioprocess Biosyst Eng* 34 (1):121–125. doi: 10.1007/s00449-010-0474-y.

Oey, Oey, M., I. L. Ross, E. Stephens, J. Steinbeck, J. Wolf, K. A. Radzun, J. Kugler, A. K. Ringsmuth, O. Kruse, and B. Hankamer. 2013. "RNAi knock-down of LHCBM1, 2 and 3 increases photosynthetic H2 production efficiency of the green alga *Chlamydomonas reinhardtii*." *PLoS One* 8 (4):e61375. doi: 10.1371/journal.pone.0061375.

Panis, G., and J. R. Carreon. 2016. "Commercial astaxanthin production derived by green alga *Haematococcus pluvialis*: A microalgae process model and a techno-eco nomic assessment all through production line." *Algal Res* 18:175–190. doi: 10.1016/j.algal.2016.06.007.

Patel, A. K., C. Laroche, A. Marcati, A. V. Ursu, S. Jubeau, L. Marchal, E. Petit, G. Djelveh, and P. Michaud. 2013. "Separation and fractionation of exopolysaccharides from *Porphyridium cruentum*." *Bioresource Technol* 145:345–350. doi: 10.1016/j.biortech.2012.12.038.

Patterson, G. M. L., C. L. Baldwin, C. M. Bolis, F. R. Caplan, H. Karuso, L. K. Larsen, I. A. Levine, R. E. Moore, C. S. Nelson, K. D. Tschappat, G. D. Tuang, E. Furusawa, S. Furusawa, T. R. Norton, and R. B. Raybourne. 1991. "Antineoplastic activity of culture blue-green algae (Cyanophyta)." *J Phycol* 27 (4):530–536. doi: 10.1111/j.0022-3646.1991.00530.x.

Pauli, G. 2010. Fibers from Algae. *The Blue Economy*.

Plaza, M., M. Herrero, A. Cifuentes, and E. Ibanez. 2009. "Innovative natural functional ingredients from microalgae." *J AgriFood Chem* 57 (16):7159–7170. doi: 10.1021/jf901070g.

Qualitas. 2019. "almega PL." www.qualitas-health.com/almega-pl.

Qualtitas. 2019. "Solar power strength: The power of algae." www.algeepa.com/algeepa.

Radakovits, R., R. E. Jinkerson, S. I. Fuerstenberg, H. Tae, R. E. Settlage, J. L. Boore, and M. C. Posewitz. 2012. "Draft genome sequence and genetic transformation of the oleaginous alga *Nannochloropis gaditana*." *Nat Commun* 3:686. doi: 10.1038/ncomms1688.

Rahman, A., and C. D. Miller. 2017. "Microalgae as a source of bioplastics." In *Algal Green Chemistry: Recent Progress in Biotechnology*, edited by R. P. Rastogi, D Madamwar and A Pandey, 121–136. Amsterdam, The Netherland: Elseviers.

Rajvanshi, M., U. S. Sagaram, G. V. Subhask, R. Kumar, C Kumar, S. Govindachary, and S. Dasgupta. 2019. "Biomolecules from microalgae for commercial applications." In *Sustainable Downstream Processing of Microalgae for Industrial Application*. Boca Raton, FL: CRC Press.

Rapier, R. 2018. An algal biofuel obituary. *Forbes*.

Rasala, B. A., S. S. Chao, M. Pier, D. J. Barrera, and S. P. Mayfield. 2014. "Enhanced genetic tools for engineering multigene traits into green algae." *PLoS One* 9 (4):e94028. doi: 10.1371/journal.pone.0094028.

Rasala, B. A., and S. P. Mayfield. 2015. "Photosynthetic biomanufacturing in green algae; production of recombinant proteins for industrial, nutritional, and medical uses." *Photosynth Res* 123 (3):227–239. doi: 10.1007/s11120-014-9994-7.

Ravishankar, G. A., R. Manoj Sarada, B. S. Kamath, and K. K. Namitha. 2006. "Food applications of algae." In *Food Biotechnology*, edited by K Shetty, G Paliyath, A Pometto and R. E. Levin, 508–541. Boca Raton, FL: Taylor & Francis Group.

Rismani-Yazdi, H., K. H. Hampel, C. D. Lane, B. A. Kessler, N. M. White, K. M. Moats, and F. C. T. Allnutt. 2014. "High-productivity lipid production using mixed trophic state cultivation of *Auxenochlorella* (*Chlorella*) *protothecoides*." *BioprocessBiosystems Eng*:1–12. doi: 10.1007/s00449-014-1303-5.

Rismani-Yazdi, H., B. Z. Haznedaroglu, C. Hsin, and J. Peccia. 2012. "Transcriptomic analysis of the oleaginous microalga *Neochloris oleoabundans* reveals metabolic insights into triacylglyceride accumulation." *Biotechnol Biofuels* 5 (1):74. doi: 10.1186/1754-6834-5-74.

Rodrigues, M. A., and E. P. da Silva Bon. 2011. "Evaluation of *Chlorella* (Chlorophyta) as source of fermentable sugars via cell wall enzymatic hydrolysis." *Enzyme Res* 2011:405603. doi: 10.4061/2011/405603.

Rodriguez-Luna, A., J. Avila-Roman, M. L. Gonzalez-Rodriguez, M. J. Cozar, A. M. Rabasco, V. Motilva, and E. Talero. 2018. "Fucoxanthin-containing cream prevents epidermal hyperplasia and UVB-induced skin erythema in mice." *Mar Drugs* 16 (10). doi: 10.3390/md16100378.

Ronquillo, J. D., J. Fraser, and A.-J. McConkey. 2012. "Effect of mixed microalgal diets on growth and polyunsaturated fatty acid profile of European oyster (*Ostrea edulis*) juveniles." *Aquaculture* 360–361 (0):64–68. doi: 10.1016/j.aquaculture.2012.07.018.

Rosales-Mendoza, S., L. M. Paz-Maldonado, and R. E. Soria-Guerra. 2012. "*Chlamydomonas reinhardtii* as a viable platform for the production of recombinant proteins: Current status and perspectives." *Plant Cell Rep* 31 (3):479–494. doi: 10.1007/s00299-011-1186-8.

Rosenberg, J. N., G. A. Oyler, L. Wilkinson, and M. J. Betenbaugh. 2008. "A green light for engineered algae: Redirecting metabolism to fuel a biotechnology revolution." *Curr OpBiotech* 19 (5):430–436. doi: 10.1016/j.copbio.2008.07.008.

Running, J. A., R. J. Huss, and P. T. Olson. 1994. "Heterotrophic production of ascorbic acid by microalgae." *J Appl Phycol* 6:99–104.

Sanchez-Silva, L., D. López-González, A. M. Garcia-Minguillan, and J. L. Valverde. 2013. "Pyrolysis, combustion and gasification characteristics of *Nannochloropsis gaditana* microalgae." *Bioresource Tech* 130:321–331. doi: 10.1016/j.biortech.2012.12.002.

Sander, K., and G. S. Murthy. 2009. *Enzymatic Degradation of Microalgal Cell Walls*. St. Joseph, MI: ASABE.

Sarnaik, A., M. H. Abernathy, X. Han, Y. Ouyang, K. Xia, Y. Chen, B. Cress, F. Zhang, A. Lali, R. Pandit, R. J. Linhardt, Y. J. Tang, and M. A. G. Koffas. 2019. "Metabolic engineering of cyanobacteria for photoautotrophic production of heparosan, a pharmaceutical precursor of heparin." *Algal Res* 37:57–63. doi: 10.1016/j.algal.2018.11.010.

Seo, E. J., Y. J. Yeon, J. H. Seo, J. H. Lee, J. P. Bongol, Y. Oh, J. M. Park, S. M. Lim, C. G. Lee, and J. B. Park. 2018. "Enzyme/whole-cell biotransformation of plant oils, yeast derived oils, and microalgae fatty acid methyl esters into n-nonanoic acid, 9-hydroxynonanoic acid, and 1,9-nonanedioic acid." *Bioresour Technol* 251:288–294. doi: 10.1016/j.biortech.2017.12.036.

Shamriz, S., and H. Ofoghi. 2018. "Engineering the chloroplast of *Chlamydomonas reinhardtii* to express the recombinant PfCelTOS-Il2 antigen-adjuvant fusion protein." *J Biotech* 266:111–117. doi: 10.1016/j.jbiotec.2017.12.015.

Siaut, M., M. Heijde, M. Mangogna, A. Montsant, S. Coesel, A. Allen, A. Manfredonia, A. Falciatore, and C. Bowler. 2007. "Molecular toolbox for studying diatom biology in *Phaeodactylum tricornutum*." *Gene* 406 (1–2):23–35. doi: 10.1016/j.gene.2007.05.022.

Simpson, S. 2017. "The LanzaTech process is driving innovation." www.energy.gov/sites/prod/files/2017/07/f35/BETO_2017WTE-Workshop_SeanSimpson-LanzaTech.pdf (Accessed February 18, 2019).

Snow, A. A., and V. H. Smith. 2012. "Genetically engineered algae for biofuels: A key role for ecologists." *Bioscience* 62 (8):765–768.

Specht, E., S. Miyake-Stoner, and S. Mayfield. 2010. "Micro-algae come of age as a platform for recombinant protein production." *Biotechnol Lett* 32 (10):1373–1383. doi: 10.1007/s10529-010-0326-5.

Spolaore, P., C. Joannis-Cassan, E. Duran, and A. Isambert. 2006. "Commercial applications of microalgae." *J Biosci Bioeng* 101 (2):87–96. doi: 10.1263/jbb.101.87.

Sprink, T., D. Eriksson, J. Schiemann, and F. Hartung. 2016. "Regulatory hurdles for genome editing: Process- vs. product-based approaches in different regulatory contexts." *Plant Cell Reports* 35 (7):1493–1506. doi: 10.1007/s00299-016-1990-2.

Stokes, R. S., M. L. Van Emon, D. D. Loy, and S. L. Hansen. 2015. "Assessment of algae meal as a ruminant feedstuff: Nutrient digestibility in sheep as a model species." *J Animal Sci* 93 (11):5386–5394. doi: 10.2527/jas.2015-9583.

Strack, R. 2019. "Precision genome editing." *Nat Methods* 16 (1):21. doi: 10.1038/s41592-018-0286-6.

Su, H. M., M. S. Su, and I. C. Liao. 2001. "The Culture and Use of Microalgae for Larval Rearing in Taiwan." Conference Joint Taiwan-Australia Aquaculture and Fisheries Resources and Management Forum, 199 Hou-Ih Road Keelung 202 Taiwan, November 2–8, 1998.

Sugiyama, A., S. Hata, K. Suzuki, E. Yoshida, R. Nakano, S. Mitra, R. Arashida, Y. Asayama, Y. Yabuta, and T. Takeuchi. 2010. "Oral administration of paramylon, a beta-1,3-D-glucan isolated from *Euglena gracilis* Z inhibits development of atopic dermatitis-like skin lesions in NC/Nga mice." *J Vet Med Sci* 72 (6):755–763.

Sugiyama, A., K. Suzuki, S. Mitra, R. Arashida, E. Yoshida, R. Nakano, Y. Yabuta, and T. Takeuchi. 2009. "Hepatoprotective effects of paramylon, a beta-1, 3-D-glucan isolated from *Euglena gracilis* Z, on acute liver injury induced by carbon tetrachloride in rats." *J Vet Med Sci* 71 (7):885–890.

Talebi, A. F., M. Tohidfar, M. Tabatabaei, A. Bagheri, M. Mohsenpor, and S. K. Mohtashami. 2013. "Genetic manipulation, a feasible tool to enhance unique characteristic of *Chlorella vulgaris* as a feedstock for biodiesel production." *Mol Biol Rep* 40 (7):4421–4428. doi: 10.1007/s11033-013-2532-4.

Tang, Ying, J. N. Rosenberg, P. Bohutskyi, G Yu, M. J. Betenbaugh, and F. Wang. 2016. "Microalgae as a feedstock for biofuel precursors and value-added products: Green fuels and golden opportunities." *BioResources* 11 (1):2850–2885.

Taunt, H. N., L. Stoffels, and S. Purton. 2018. "Green biologics: The algal chloroplast as a platform for making biopharmaceuticals." *Bioengineered* 9 (1):48–54. doi: 10.1080/21655979.2017.1377867.

Thamatrakoln, K., O. Korenovska, A. K. Niheu, and K. D. Bidle. 2012. "Whole-genome expression analysis reveals a role for death-related genes in stress acclimation of the diatom *Thalassiosira pseudonana*." *Environ Micro* 14 (1):67–81. doi: 10.1111/j.1462-2920.2011.02468.x.

Tibbetts, S. M., and A. H. Fredeen. 2017. "Nutritional evaluation of whole and lipid-extracted biomass of the microalga *Scenedesmus* sp. AMDD for animal feeds: Simulated ruminal fermentation and in vitro monogastric digestibility." *Curr Biotech* 6 (3):264–272. doi: 10.2174/2211550105666160906123939.

Tran, M., B. Zhou, P. L. Pettersson, M. J. Gonzalez, and S. P. Mayfield. 2009. "Synthesis and assembly of a full-length human monoclonal antibody in algal chloroplasts." *Biotechnol Bioeng* 104 (4):663–673. doi: 10.1002/bit.22446.

Triton. 2018. San Diego algae products company hits key regulatory milestone.

Triton. 2019a. "*Chlamydomonas* as a safe to consume green algae." www.tritonai.com/publications/.

Triton. 2019b. "Triton colostrum proteins." www.tritonai.com/tritons-milk-proteins/ (Accessed February 18, 2019).

Urtubia, H. O., L. B. Betanzo, and M. Vasquez. 2016. "Microalgae and cyanobacteria as green molecular factories: Tools and perspectives." In *Algae—Organisms for Imminent Biotechnology*, 1–27.

Vanier, G., F. Hempel, P. Chan, M. Rodamer, D. Vaudry, U. G. Maier, P. Lerouge, and M. Bardor. 2015. "Biochemical characterization of human anti-hepatitis B monoclonal antibody produced in the microalgae *Phaeodactylum tricornutum*." *PLoS One* 10 (10):e0139282. doi: 10.1371/journal.pone.0139282.

Verdugo, M., L.-T. Lim, and M. Rubilar. 2014. "Electrospun protein concentrate fibers from microalgae residual biomass." *J Polym Environ* 22:373–383.

Wang, H. M., C. C. Chen, P. Huynh, and J. S. Chang. 2015. "Exploring the potential of using algae in cosmetics." *Bioresour Technol* 184:355–362. doi: 10.1016/j.biortech.2014.12.001.

Wijffels, R. H., and M. J. Barbosa. 2010. "An outlook on microalgal biofuels." *Science* 329 (5993):796–799. doi: 10.1126/science.1189003.

Wild, K. J., H. Steingass, and M. Rodehutscord. 2018. "Variability of *in vitro* ruminal fermentation and nutritional value of cell-disrupted and non-disrupted microalgae for ruminants." *GCB Bioenergy* 11:345–359. doi: 10.1111/gcbb.12539.

Yuan, J. P., J. Peng, K. Yin, and J. H. Wang. 2011. "Potential health-promoting effects of astaxanthin: A high-value carotenoid mostly from microalgae." *Mol Nutr & Food Res* 55 (1):150–165. doi: 10.1002/mnfr.201000414.

Zeller, M. A., R. Hunt, A. Jones, and S. Sharma. 2013. "Bioplastics and their thermoplastic blends from *Spirulina* and *Chlorella* microalgae." 130 (5):3263–3275. doi: 10.1002/app.39559.

Zhou, W., B. Hu, Y. Li, M. Min, M. Mohr, Z. Du, P. Chen, and R. Ruan. 2012. "Mass cultivation of microalgae on animal wastewater: A sequential two-stage cultivation process for energy crop and omega-3-rich animal feed production." *Appl Biochem Biotech* 168:348–363. doi: 10.1007/s12010-012-9779-4.

Zille, A., S. Pereira, P. Tamagnini, P. Moradas-Ferreira, E. Micheletti, and R. De Philippis. 2009. "Complexity of cyanobacterial exopolysaccharides: Composition, structures, inducing factors and putative genes involved in their biosynthesis and assembly." *FEMS Micro Rev* 33 (5):917–941. doi: 10.1111/j.1574-6976.2009.00183.x %J

Zych, M., J. Burczyk, M. Kotowska, A. Kapuścik, A. Banaś, A. Stolarczyk, K. Termińska-Pabis, S. Dudek, and S. Klasik. 2009. "Differences in staining of the unicellular algae *Chlorococcales* as a function of algaenan content." *Acta Agronomica Hungarica* 57 (3):377–381. doi: 10.1556/AAgr.57.2009.3.12.

3 Cultivation of Microalgae for the Production of Biomolecules and Bioproducts at an Industrial Level

Preeti Mehta, Kritika Singh, Ravi Prakash Gupta, Anshu Shankar Mathur

CONTENTS

3.1 INTRODUCTION

Every year, total oil consumption in the United States is approximately 7 billion barrels, which elucidates 22% demand worldwide (Silva et al. 2014). In India, total oil consumption exceeded 16.9 million tons, according to the oil ministry's Petroleum Planning and Analysis Cell. Biofuel, a carbon-neutral fuel derived from renewable sources, has the potential to replace petroleum. Intensive research activities were started for developing renewable energy resources such as first-generation and second-generation biofuels. Amidst the world vitality emergency and real issues related with first- and second-generation biofuels, researchers have focused on third-generation biofuels derived from microorganisms. Among them, microalgae are considered a promising alternative towards a societal economy (Borowitzka 2013; Brownbridge et al. 2014; Nagaranjan et al. 2013).

Microalgae are the fastest-growing phototrophic organism. It has also the ability to sequester CO_2 and grow in a wide variety of aquatic habitats and store oil in much higher density (around 50 wt% of biomass). Algae do not require arable land and have a short life cycle. Therefore, they can be harvested throughout the year. They can be cultivated autotrophically, mixotrophically, and/or heterotrophically in a broad spectrum of temperature, pH, salt concentration, and light intensities (Amer et al. 2011). Based on the category of species and its cultivation conditions, algae produce the different concentrations of lipids that are converted into biofuels, including biodiesel or bioethanol. Algal biomass for only biofuels production is not cost-effective at commercial scale. Commercialization of biofuel requires selective biorefining of biomass, which is still a challenge (Vanthoor-Koopmans et al. 2013, Davis et al. 2011, Delrue et al. 2012, Louw et al., 2016). The evaluated cost of a barrel of algae biofuel production in open and closed systems is US$300 to 2600, which is 10 to 30 times the production cost of petroleum (Hannon et al. 2010).

Microalgae have attracted considerable interest worldwide for being a potential source of a variety of value-added bio-products in addition to biofuels. It yields a variety of pigments, proteins, vitamins, polysaccharides, carbohydrates, antioxidants, and lipid molecules (Figure 3.1). Due to its considerable application in the

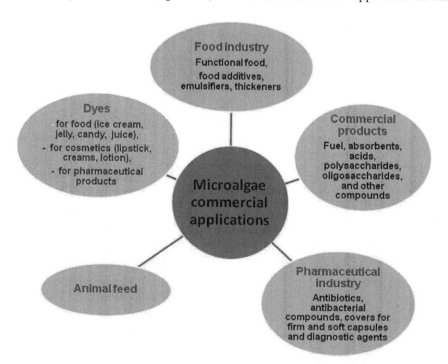

FIGURE 3.1 Commercial applications of microalgae.

biopharmaceutical, neutraceutical, and renewable energy industry, the industrial-scale cultivation of microalgae to produce bio-products and bio-molecules has expanded adequately over the last decades. Some microalgae are also capable of producing stable isotopes such as ^{13}C, ^{15}N, and ^{2}H, which are used in the manufacturing of various products. To improve the process economics, the concept of using microalgae as a part of the biorefinery model has been adopted at large-scale. This chapter briefly reviews the currently available microalgae-derived high-value bio-products and bio-molecules and their commercial significance and development.

3.2 CURRENT STATUS OF MICROALGAE PRODUCTION: CULTIVATION SYSTEMS AND GROWTH PARAMETERS

3.2.1 CULTIVATION SYSTEMS

The success of large-scale commercial production of microalgae depends on an effective large-scale cultivation system. The microalgal biomass is costly to produce, even though numerous efforts are in progress to accomplish cost-efficient modes for mass cultivation of microalgae at commercial scale. Different cultivation systems (horizontal tubular photo-bioreactor, vertically stacked tubular photo-bioreactor, flat panels photo-bioreactor, and open raceway pond) for biomass production have been designed, and this continues to be a gradual process conducted on a largescale (Ruiz et al. 2016; Silva et al. 2014).

Paddle-wheel, open, shallow raceway ponds are the simplest and best choice for the cultivation of microalgae because this requires half the initial investment than closed systems. Every unit of a raceway pond covers a region of several hundred to a couple of thousand square meters. In raceway ponds, researchers use pH statistics for the sufficient supply of carbon dioxide and also to determine the optimum pH for the culture. The raceway pond reactor also has some shortcomings, such as less productivity. The culture is also more prone to contamination in the raceway pond reactor. A recently developed advanced closed photo-bioreactor provides a better alternative for the cultivation of microalgal strains without contamination under controlled operational conditions. These photo-bioreactors are either flat or tubular and are easy to design, construct, and operate. Although closed systems involve a larger investment, they have high productivity and better quality of algal biomass. Commercial cultivation in a closed system also includes some drawbacks, such as overheating, buildup of oxygen, etc. Although a huge amount of research for improvement in the designs of photo-bioreactors is going on, discussion on the best cultivation is a work in progress, because none of the systems seem to shine out from others (Louw et al. 2016). Table 3.1 summarizes some of the main differences between open and closed culture systems.

3.2.2 CULTIVATION PARAMETERS

Various factors, such as light (based on intensity), nutrient source (including trace elements), temperature, pH, agitation, and CO_2, affect the growth of microalgae. These parameters should be controlled to reproduce the desired set of results (Figure 3.2).

TABLE 3.1

A Comparison of Open and Closed Large-scale Culture Systems for Microalgae

	Closed systems	Open systems
Contamination control	Easy	Difficult
Operation mode	Continuous	Batch or fed-batch
Area/volume ratio	High	Low
Culture density	High	Low
Process control	Easy	Difficult
Investment	High	Low
Operation cost	High	Low
Scale-up	Difficult	Easy
Photo efficiency	Excellent	Poor

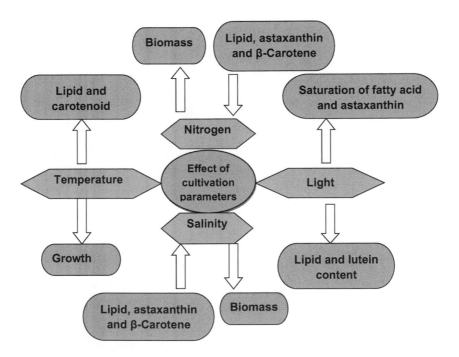

FIGURE 3.2   Effect of different cultivation factors on accumulation of various parameters.

Temperature is one of the important factors. The optimum growth temperature generally varies from 20 to 30°C. Numerous algal species can withstand temperatures above and below this range. However, variations and changes in temperature are known to give low biomass productivity. pH plays a crucial role in algal growth. Most algal species grow well in the pH range of 6.8 to 7.86. During the cultivation, the pH rises in the culture medium. This has been attributed to the accumulation of -OH ions during photosynthesis. In acidic conditions (pH <5), the dissolved inorganic carbon is CO_2, and under alkaline conditions, microalgae will easily absorb the CO_2 from the atmosphere and produce more biomass.

The nutrient requirement may vary for different microalgae species, but the basic requirements remain the same. The key nutrients are C, O_2, H_2, P, and nitrogen. The quantity of available N_2 in the culture media directly affects cell growth. Nitrogen depletion can reduce growth and biomass productivity, but it has a positive effect on lipids and carbohydrates. The micronutrients Mo, Co, Fe, Mg, K, Mn, B, and Zn are needed in small quantities, but they have a strong influence on the enzymatic activity of the algal cells. Light duration and intensity directly affect the photosynthetic efficiency of microalgae. They also affect the biochemical constitution of microalgae and biomass yield. Light and dark periods both are needed for algal photosynthesis. The light period is needed for the production of energy compounds such as adenosine triphosphate (ATP) and nicotinamide adenine dinucleotide phosphate (NADPH). In the dark cycle, there is the production of C skeletons using the energy reservoirs produced during the light phase. Most studies have shown that 16 hours of light and 8 hours of dark is most suitable for algal growth. Mixing and aeration provide equal distribution of all the available nutrients, air, and CO_2 in the culture. An appropriate mixing system not only provides efficient gas exchange and light penetration; it also enables nutrient dissolution. A well-facilitated system also prevents the biomass from settling and aggregating.

3.3   BIOREFINERY APPROACH

Biorefinery is an industrial process in which the biomass is converted into a variety of bio-chemicals and energy products. This concept corresponds to the oil industry, in which a number of products are manufactured from the same oil. As a totally integrated and versatile process, this approach provides a varied range of materials to give a number of different products, including food and feed (Milledge, 2012). The biorefinery process consists of the pre-treatment of components and the further conversion through utilization and recycling of the components. Efforts are made to widely use raw materials and reduce resource losses. As is already known, microalgae can be used to generate a variety of bio-products, based on the microalgal compositions (Figure 3.3).

Microalgae are highly rich in lipids, proteins, carbohydrates, and various valuable compounds (Chauhan and Pathak 2010; Pulz et al. 2004). The lipids from these cells can be easily used for biodiesel production, and the carbohydrates can be utilized for bioethanol production (Brennan and Owende 2009). Methane and bio-hydrogen can be produced via anaerobic digestion of carbohydrates, proteins, and fats. Various other products such as bio-butanol, bio-oil, syngas, jet fuel, etc., can be produced via thermo-chemical, chemical, and biochemical conversion processes. Microalgae are the ideal source for the production of a variety of high-value products, which are utilized on a daily basis by a wide variety of people. The co-production of various products along with the primary product of interest also decreases the final cost of production. For example, when our product of interest is a specific fatty acid, the

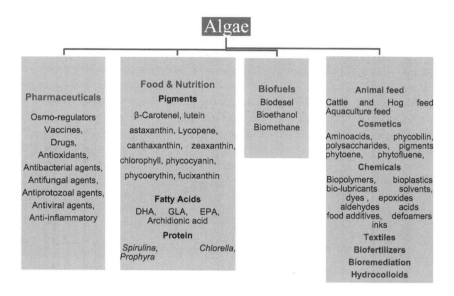

FIGURE 3.3 Various value-added bioproducts and biomolecules from algae.

fatty acids are purified from the biomass, but the supernatant can also be utilized at the same moment for the extraction of proteins, thereby producing two commercial products from the same amount of culture and thus decreasing the cost of production.

3.4 VARIOUS BIO-MOLECULES AND BIO-PRODUCTS

3.4.1 CAROTENOIDS

Carotenoids are soluble natural pigments, which are an important component of the photosynthetic process in algae (Pulz and Gross 2004). They play a crucial role in the food, feed, cosmetic, and bio-pharmacy industry (Dufosse et al. 2005; Enzing et al. 2014). They also have efficient antioxidant activity, which makes them therapeutic for strokes, aging, diabetes, etc (Lv et al. 2015; Munir et al. 2013). β-Carotene is the first commercialized pigment molecule. It is the precursor molecule for vitamin A synthesis in humans. It has an important role for the well-being of eyes and the immune system (Costa 2003). It also saves the membrane lipids from oxidation, which is related to very chronic diseases such as cancer, atherosclerosis, Parkinson's disease, and cardiovascular diseases. *Dunaliella salina* is a very rich source of β-carotene, and its first pilot plant was set in USSR, Taiwan-based industry in 1960. *Dunaliella*, when grown under high light intensity and salinity, produces a higher amount of β-carotene. Nitrogen deficiency is also known to enhance its accumulation. It is able to tolerate temperatures around 40°C. Because of the extreme survival conditions, the high cultivation cost is not associated with the scale-up of this alga. The cultivation and downstream processing used in the industry is explained in Figure 3.4.

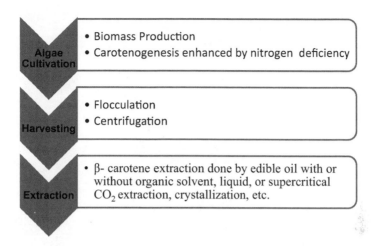

FIGURE 3.4 Production and extraction of Beta-carotene.

Astaxanthin is becoming popular as an abundant source of vitamin E because of its natural antioxidant powers. The main demand for astaxanthin is as a source for pigmentation in the aquaculture industry (Patterson et al. 1994). But it is also being used in the nutraceutical, cosmetics, food, and feed industries (Kim and Pangestuti 2011). It has also been reported as a promising source in human health. Natural astaxanthin was found effective against functional dyspepsia and various DNA alterations caused by ultraviolet light A (UVA) in human skin fibroblasts. *Haematococcus pluvialis* has emerged as a very crucial organism for astaxanthin production along with its utilization for the biorefinery (Lorenz and Cysewski 2000). This alga, when grown under stressed conditions, accumulates astaxanthin. These stress conditions also induce deposition of triglycerides, due to the coexistence of metabolites obtained from the same feedstock, which lowers the cost of production. Its ability to utilize various carbon sources opens the possibility of utilization of various waste streams as a feedstock. The other nutrients in the form of energy supplied can be replenished astaxanthin and triglycerides in space and time; simultaneously, both products can be produced from the anaerobic digestion process. This process is elucidated in Figure 3.5.

Lutein, another important carotenoid, is also accumulated by *Chlorella* sp., *Muriellopsis* sp., *Scenedesmus* sp., and *Chlamydomonas* sp. Its main function is to maintain the structure and proper functioning of the eye.

3.4.2 PROTEINS

Various species of microalgae produce significant quantities of amino acids and proteins. These are used in the food industry. It also improves the immune system of the human body. Some species can produce proteins equivalent to the proteins present in eggs or meat. They are also known to have comparatively high nutritional value and reduce cholesterol levels by activating cholecystokinin. Another protein-rich microalga is *Arthrospira* (Sánchez et al. 2003). Important enzymatic effects have also been

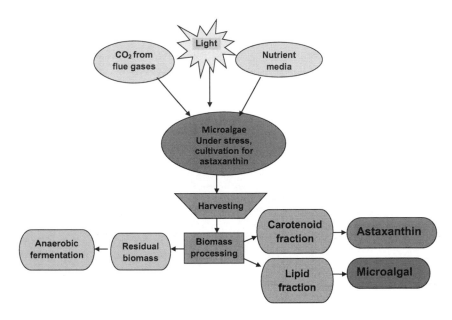

FIGURE 3.5   Production of astaxanthin.

observed from microalgal proteins. Microcolin-A, an immunosuppressive agent, is accumulated by *Lyngbya majuscula*. Cyanovirin is known for its antiviral activities against HIV. *M. aeruginosa* is also known to produce a number of important amino acids. Superoxide dismutase (SOD), known to protect against oxidative damage, is produced by *Anabeana* and *Porphyridium*, and *Isochyris galbana* produces the indispensable enzyme carbonic anhydrase, which has a pivotal role in the conversion of carbon dioxide into carbonic acid and bicarbonate.

3.4.3   FATTY ACIDS

Polyunsaturated fatty acids (PUFAs) are significant for tissue integrity and have beneficial health effects. Omega-3 and omega-6 fatty acids are crucial for humans, but our body is not able to make these fatty acids. Thus, external intake is essential. Natural PUFAs can be obtained from fish. But there is a risk of heavy metal contamination in these, along with several drawbacks, such as unpleasant odor, taste, and low oxidative stability. Thus, algae appear as an alternative natural source for PUFA production (Kyle 2005; Wynn et al. 2010). It may be noted that PUFA in fish is actually originated from algae (Rasmussen et al. 2008; Ratledge 2004). Docosahexaenoic acid (DHA) is the only commercially available algal PUFA. It is a major structural fatty acid in the brain. It is also important for eye and brain development in infants and provides cardiovascular support to adults. DHA is known to improve the body's immune system and protects from various allergic diseases and eosinophilic disorders (Barclay et al. 2010). The DHA oil production process is discussed in Figure 3.6.

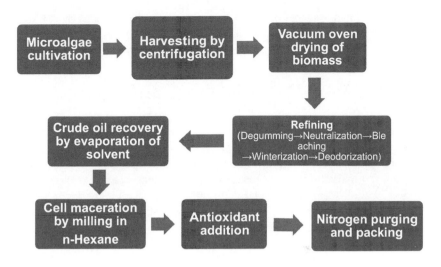

FIGURE 3.6 Strategy for DHA oil production in DSM.

3.4.4 Carbohydrates

Agar is a multiplex of polysaccharides found in the cell walls of marine red algae. Agar molecules have apliable shape, which protects the algae from stress caused by moving sea water (Arad and Levy-Ontman 2010). Agar is made up of agarose and agaropectin. It is widely used in the bakery industry as a thickening agent (Muller and Alegre 2007). For commercial alar extraction, *Gracilaria* and *Gelidium* are used. Carrageenan is a natural sulphated galactan polysaccharides, which is used in the food and pharmaceuticalindustries. A number of red algal strains (e.g., *Chondrus* sp., *Gigartina* sp., *Hypnea* sp., and *Furcellaran*sp.) are used for the extraction of carrageenan. Alginates are structural phyco-colloids used in the food, cosmetics, and pharmaceutical industries (Cardozo and Guaratini 2007). It is known to provide a uniform appearance due to its moisture retention property. In the medicine industry, it is used as a pill disintegrator because of its biodegradable nature. Several brown algal species such as *Macrocystis pyrifera, Ascophyllum nodosum, Laminaria hyperborean,* etc., are used for alginate production. Fucoidan is made up of L-fructose sugar, and molecular weights are in the 13 to 950 kDa range. It has anticoagulant, antitumor, antivirus, and antioxidant capacities, making it an attractive choice for applications in the pharmaceutical field (Deng and Chow 2010). It has been shown that fucoidans are known to inhibit virus proliferation.

3.4.5 Biopolymers

Algae are known as a source of biopolymers after biofuel generation to reduce the cost of the production of the biofuels. A variety of plastics, such as hybrid plastics, cellulose-based plastics, polylactic acid, and biopolyethylene, can be obtained from algal biomass. Polyhydroxyalkanoates (PHA) are widely used in the packaging industry. Being biodegradable in nature, a number of materials such as PHA, poly-lactides, aliphatic polyesters, polysaccharides, and co-polymers are under development (Philippis et al. 2011).

3.5   MARKET VALUE AND DEMAND

With the increasing interest in microalgae-derived products, the market value and demand for these products are showing an increasing trend. Several companies are producing various products. The United States occupies the highest place in the production of microalgal products, followed by China and Japan. The products that are dominating the market are β-carotene, astaxanthin, eicosapentaenoic acid (EPA)/DHA, and ethanol. As far as the volume is concerned, the algal market is estimated to reach 26,849.11 tons by 2022. The market value of these products is expected to reach USD 1,128,000. A number of private companies are producing these algae-derived products (Table 3.2).

The green alga *Chlorella* is renowned as a human health supplement and has a market value of USD 44/kg. The major companies dominating the *Chlorella* market are Yaeyama Shokusan Co. Ltd. (Japan) and Maypro Industries, Inc., and Taiwan Chlorella Manufacturing Co. Ltd. (Taiwan). Carotenoids are assessed to achieve

TABLE 3.2
Companies Producing Algae-derived Products

Products	Names of companies
Dried algae, β-carotene, biofuels	Cognis Nutrition and Health, Muradel, Origin Oil Bio-Fuels Pvt. Ltd.
Dried algae, astaxanthin, algal paste	Algae Can's, Pond Technology, Algabloom International
Dried algae, β-carotene, EPA/DHA	Yunnan Ginko Asta Biotech Co. Ltd., Fuji Chemical Industry Co. Ltd., Cyanotech Corporation, Far East Algae Ind Co. Ltd.
EPA/DHA and other dietary supplements	Aleor, Roquette (France)
Dried algae, β-carotene, EPA/DHA, Astaxanthin	Breen BiotechGmbH, Greeenovation Blue Biotech GmbH
Dried algae, β-carotene, astaxanthin	NB LaboratoriesPvt. Ltd., Global Green Company Ltd., Parry Nutraceuticals, Energy Algae, Jovialis, Sateera Nutria Biotech
EPA/DHA	Aquaflow Bionomic Corporation, SeaDragon
Astaxanthin	Asta Real, BioReal
Dried algae, astaxanthin, EPA/DHA ethanol	Algal Biosciences Algenol, Aurora Algae, Inc., Cynotech, Terra Via, Sapphire Energy, Solazyme, Inc., Solix BioSystems, Algae Systems, Algae to Omega Holdings, Algoil Energy, Aquatic Energy, Global Green Solutions, Greener Bio-Energy

USD 1.53 billion by 2021, at a compound annual growth rate (CAGR) of 3.78% from 2016 to 2021. In carotenoids, production of β-carotene is the most flourishing business attributed to its high adequacy and therapeutic properties. It sells at a cost of USD 300 to 1500/kg. Following it is astaxanthin, another highly esteemed carotenoid, whose average market price is USD 2500/kg. Various fatty acids are being derived from algal sources, especially alpha-lipoic acid (ALA), DHA, and EPA. Among them, DHA dominates the market, whereas ALA is the quickest blooming omega-3, with a CAGR of 15.8% from 2016 to 2022. By 2020, it is forecasted that PUFA demand will increase to 241,000 metric tons, and to fulfil this demand microalgal products need to surface with a lesser production cost (Bermudez et al. 2010).

3.6 CONCLUSIONS

Microalgae have received tremendous attention for their various high-value products. It is a very abundant source of several metabolites and molecules, which are not readily available and are even rather expensive. But the constraints lie in the cultivation, harvesting, and downstream processing to make it economically feasible at an industrial level. Cultivation at a largescale requires proper maintenance, aeration, agitation, and light supply, which is otherwise difficult to maintain in open systems. These factors can be maintained in closed systems, but with additional costs. There are several methods of harvesting, but a proper method with low capital input is still being devised. The bio-molecules of interest need to be recovered either from the biomass or the culture media that is obtained after biomass removal. These products have a huge demand in the market due to several associated benefits and use. So, attention needs to be diverted away from only biodiesel production with microalgae to producing co-products along with biodiesel. This integrated approach would meet the market demand for other value-added products. This integration would also decrease the cost of production, thereby attracting more market demand for the commodities. The products, which are expensive and not being utilized by a wide customer range, will also become accessible with this co-integration.

REFERENCES

Amer, L., B. Adhikari and J. Pellegrino. 2011. "Technoeconomic analysis of five algae-to-biofuels processes of varying complexity." *Bioresource Technology* 102:9350–9359.

Arad, S., and O. Levy-Ontman. 2010. "Red microalgal cell-wall polysaccharides: Biotechnological aspects." *Curr Opin Biotechnol* 21:358–364.

Barclay, W., C. Weaver, and J. Metz. 2010. "Development of docosahexaenoic acid production technology using *Schizochytrium*: Historical perspective and update." In *Single Cell Oils. Microbial and Algal Oils*, edited by Z. Cohen, C. Ratledge, 75–96. Urbana: AOCSs Press.

Bermudez, S. P., and I. Aguilar-Hernandez et al. 2010. "Extraction and purification of high-value metabolites from algae: Essential lipids, astaxanthin and phycobiliproteins." *Microbial Biotechnology* 8(2):190–209.

Borowitzka, M. A. 2013. "Energy from algae: A short history." In *Algae for Biofuels and Energy*, edited by M. A. Borowitzka, and N. R. Moheimani, 1–15. Dordrecht: Springer.

Borowitzka, M. A. 2013. "High-value products from algae—their development and commercialisation." *JApplPhycol* 25:743–756.

Brennan, L., and P. Owende. 2009. "Biofuels from algae—A review of technologies for production, processing, and extractions of biofuels and co-products." *Renewable and Sustainable Energy Reviews* 14 (2):557–577.

Brownbridge, G., P. Azadi and A. Smallbone et al. 2014. "The future viability of algae-derived biodiesel under economic and technical uncertainties." *Bioresource Technology* 151:166–173. doi: 10.1016/j.biortech.2013.10.062.

Cardozo, K. H. M., and T. Guaratini et al. 2007. "Metabolites from algae with economical impact." *Comparative Biochem Physiol Toxicology Pharmacol* 146 (1–2):60–78.

Chauhan, U. K., and N. Pathak. 2010. "Effect of different conditions on the production of chlorophyll by *Spirulina platensis*." *J Algal Biomass Utln* 1 (4):89–99.

Costa, P. J. 2003. Method of producing beta-carotene by means of mixed culture fermentation using (+) and (–) strains of *Blakesleatrispora*. European Patent Application 1,367,131.

Davis, R., A. Aden and P. T. Pienkos. 2011. "Techno-economic analysis of autotrophic algae for fuel production." *Applied Energy* 88:3524–3531. doi: 10.1016/j.apenergy.2011.04.018.

Delrue, F., P. Setier, and C. Sahut et al. 2012. "An economic, sustainability, and energetic model of biodiesel production from algae." *Bioresource Technology* 111:191–200.

Deng, R., and T. J. Chow. 2010. "Hyperlipidemic, antioxidant and antiinflammatory activities of algae Spirulina." *Cardiovasc Ther.* 28:33–45.

Dufosse, L., P. Galaupa and Y. Anina. 2005. "Microorganisms and algae as sources of pigments for food use: A scientific oddity or an industrial reality?" *Trends in Food Science & Technology* 16:389–406.

Enzing, C., M. Ploeg and M. Barbosa. 2014. Algae based products for the food and feed sector: An outlook for Europe. In *JRC Scientific and Policy Reports*, edited by M. Vigani, C. Parisi, and E. Rodr_iguezCerezo.https://ec.europa.eu/jrc/sites/default/files/final version_online_ipts_jrc_85709.pdf.

Hannon, M., J. Gimpel, and M. T. Miller et al. 2010. Biofuels from algae: Challenges and potential." Biofuels 1(5):763–784.

Kim, S. K., and R. Pangestuti. 2011. "Biological properties of cosmeceuticals derived from marine algae." In *Marine Cosmeceuticals*, edited by S. K. Kim, 191–200. Boca Raton: CRC Press.

Kyle, D. J. 2005. "The future development of single cell oils." In *Single Cell Oils*, edited by Z. Cohen, and C. Ratledge, 239–248. Urbana: AOCS.

Lorenz, R. T., G. R. Cysewski. 2000. Commercial potential for *Haematococcus* algae as a natural source of astaxanthin." Trends Biotechnol 18:160–167.

Louw, T. M., Griffiths M. J., Jones S. M., and Harrison S. T. 2016. "Techno-economics of Algal Biodiesel." In *Algae Biotechnology. Green Energy and Technology*, edited by F. Bux, and Y. Chisti, 111–141. Cham: Springer.

Lv, J., X. Yang, and H. Ma et al. 2015. The oxidative stability of algae oil (*Schizochytrium aggregatum*) and its antioxidant ability after gastrointestinal digestion: Relationship with constituents." *Eur J Lipid Sci Technol* 117 (12),1928–1939.

Milledge, J. J. 2012. "Algae: Commercial potential for fuel, food and feed." *Food Science and Technology* 26 (1):28–31.

Muller, J. M., and R. M. Alegre. 2007. "Alginate production by *Pseudomonas mendocina* in a stirred draft fermenter." *World J Microbiol Biotechnol* 23 (5):691–695.

Munir, N., N. Sharif, and S. Naz. 2013. "Algae: A potent antioxidant source." *Sky J Microbiol Res* 1(3):22–31.

Nagarajan, S., S. K. Chou, and S. Cao et al. 2013. "An updated comprehensive techno-economic analysis of algae biodiesel." *Biores Technol* 145:150–156.

NREL U.S. Department of energy. National Algal Biofuels Technology Review. 2016 https://energy.gov/eere/bioenergy/downloads/2016-national-algal-biofuels-technology-review.

Patterson, G. M. L., L. K. Larsen, and R. E. Moore. 1994. "Bioactive natural products from blue-green algae." *J Appl Phycol* 6:151–157.

Philippis, R., G. Colica, and E. Micheletti. 2011. "Exopolysaccharide producing cyanobacteria in heavy metal removal from water: Molecular basis and practical applicability of the biosorption process." *Appl Microbiol Biotechnol* 92:697–708.

Pulz, O., and W. Gross. 2004. "Valuable products from biotechnology of algae." *Appl Microbiol Biotechnol* 65 (6):635–648.

Rasmussen, H. E., K. R. Blobaum, and Y. K. Park et al. 2008. "Lipid extract of *Nostoc commune var. sphaeroides* Kützing." *J Nutr*138:476–481.

Ratledge, C. 2004. "Fatty acid biosynthesis in microorganisms being used for single cell oil production." *Biochimie* 86:807–815.

Sánchez, M., J. C. Bernal, and C. Rozo. 2003. "*Spirulina* (arthrospira): An edible microorganism: A review." Univ Sci 8:7–24.

Silva, C., E. Soliman, G. Cameron, Fabiano, L. A. D. Seider, E. H. Dunlop and A. Kimi Coaldrake. 2014. Commercial-scale biodiesel production from algae. *Ind. Eng. Chem. Res* 53:5311–5324.

Wynn, J., P. Behrens, and A. Sundararajan. 2010. "Production of single cell oils from Dinoflagellates." In *Single Cell Oils. Microbial and Algal Oils*, edited by Z. Cohen, and C. Ratledge, 115–129. Urbana: AOCSs Press.

Yasukawa, K., T. Akihisa, and H. Kanno et al. 1996. "Inhibitory effects of sterols isolated from *Chlorella vulgaris* on 12-*O*-tetradecanoylphorbol-13-acetate-Induced inflammation and tumor promotion in mouse skin." *Biol Pharm Bull* 19:573–576.

4 Industrial Scope with High-Value Biomolecules from Microalgae

Chetan Paliwal, Asha A. Nesamma,
Pannaga P. Jutur

CONTENTS

4.1 INTRODUCTION

The complexity in the earth's ecosystem has metamorphosed in the last century, leading to major environmental multifaceted problems, including non-availability of sustainable food and energy sources, rampant climatic changes/global warming, degrading of land resources, and loss in biodiversity (Arora et al. 2018). Nevertheless, a sustainable solution could be achieved through an efficient biochemical process, such as photosynthesis, capable of converting atmospheric CO_2 to industrially relevant high-value biomolecules (HVBs) (Nath et al. 2015). In this context, microalgae offers several advantages due to their higher photosynthetic efficiencies, accounting for ~40% of atmospheric CO_2 assimilation, thus reducing global warming, as well as an ability to thrive in non-arable/off-shore areas and mimic any stress conditions by altering their cellular pathways (Brennan and Owende 2010, Markou and Nerantzis 2013, Li et al. 2015, Jutur, Nesamma, and Shaikh 2016). The potential use of microalgae in any industrial biorefinery will be a step forward for the production of a low or/and zero carbon footprint technology (Wichuk, Brynjolfsson, and Fu 2014, Hariskos and Posten 2014).

Microalgae are regarded as promising feedstocks for biofuels and other HVBs, namely carotenoids, long-chain polyunsaturated fatty acids (LC-PUFAs), polysaccharides, phycobiliproteins, and therapeutic proteins for their use in the food and feed industries, bioactive pharmaceuticals, nutraceuticals, functional foods, and bio-fuels (de la Noue and de Pauw 1988, Pulz and Gross 2004, Wijffels and Barbosa 2010, Pangestuti and Kim 2011, Vanthoor-Koopmans et al. 2013). Furthermore, these microalgae-based high-value renewables have been proven to be effective in the reduction of gastric ulcers, constipation, anemia, hypertension, diabetes, cardio-circulatory problems, and coronary diseases (Spolaore et al. 2006, Gouveia 2014). These biomolecules are produced in different growth phases, under specific time course conditions, leading to phenotypic and metabolism changes. Co-extraction of HVBs such as eicosapentaenoic acid (EPA), docosahexaenoic acid (DHA), vitamin E (a-tocopherol), and arachidonic acid (AA) may further enhance their industrial scope and commercial value of microalgae (Dewapriya and Kim 2014, Nesamma, Shaikh, and Jutur 2015).

Microalgae, depending upon the species, can accumulate carbohydrates and/or lipids up to 50% of their dry weight, and its genetic diversity corresponds to the variety of habitats in which algae exists, but our knowledge about their organismal and biochemical diversity is limited. Apart from biofuel precursors, some species can also accumulate specific high-value secondary metabolites like pigments, vita-mins, and carotenoids which has implications for their potential applications in the nutraceutical, pharmaceutical, cosmetic, food, and feed industries (Skjanes, Rebours, and Lindblad 2013, Paliwal et al. 2015). Unfortunately, the economic viability of microalgae-based biofuels and associated HVBs are still not feasible and sustainable (Clarens et al. 2011). The economics of producing HVBs in microalgae are promis-ing due to inexpensive cultivation costs, where media would be priced around \$0.002 L^{-1} and the production cost for algal facilities would be a fraction compared to other models, that is, \$0.05 g^{-1} in plants and would be approximately \$150 g^{-1} in mamma-lian cells (Barrera and Mayfield 2013). The cost-effectiveness and high productivities of these HVBs make microalgae a sustainable candidate for industrial production.

In this chapter, our aim is to provide insights on various HVBs that can be pro-duced from microalgae, which may be a significant precursor to relevant industries. It is also essential to know the market scenario and key hurdles involved in develop-ing a sustainable biorefinery model for these HVBs with a factor of cost-economics adding to its advantage.

4.2 HIGH-VALUE BIOMOLECULES FROM MICROALGAE

Algal research was extensively predominant for the production of biofuel precursors; however, the process at this juncture seems to be expensive, thus making them eco-nomically non-feasible (Misra et al. 2014). Nonetheless, recently the trend has been diverted to the production of cost-effective HVBs simultaneously with new emerg-ing technologies (Borowitzka 2013, Skjanes, Rebours, and Lindblad 2013, Gomaa, Al-Haj, and Abed 2016, Chew et al. 2017, Zhao et al. 2019). Certain HVBs with potential industrial scope are discussed next.

4.2.1 POLYUNSATURATED FATTY ACIDS

PUFAs, or ω-3/6 fatty acids, such as DHA (22:6) and EPA (20:5) have gained more attention due to their associated health benefits such as combatting anti-inflammatory disorders, enhancing brain development and neural signaling, and combatting aging (Robertson et al. 2013, Rosales-Mendoza 2016, Zárate et al. 2017). These PUFAs have a characteristic chain length of 18 to 22 carbons containing two or more double bonds and classified as ω-3 and ω-6 derived biosynthetically via linoleic acids (18:2n-2) and α-linoleic acids (18:3n-3), respectively (Li et al. 2014). Fish oil is the primary source of EPA and/or DHA, but due to the scarcity of wild fishes and marine pollution, there is a need for an alternative sustainable source for PUFA production which can also be extracted from microalgae, namely *Schizochytrium, Chlorella pyrenoidosa, Isochrysis galbana*, and *Crypthecodinium* (El Abed et al. 2008, Winwood 2013, Matos et al. 2017). Henceforth, microalgae as the primary producers are able to synthesize PUFAs, but their regulatory pathways are still not entirely known (Guiheneuf et al. 2015, Guiheneuf, Schmid, and Stengel 2015). Mostly microalgae tend to increase the ratio of unsaturation to saturation when the temperature declines; however, these responses are species specific (Neidleman 1987, Olofsson et al. 2012).

Microalgal oils extracted from *Tetraselmis* and *Nannochloropsis* sp. have shown higher antioxidant properties due to the co-derivatization of high-value carotenoids and polyphenols (Gangl et al. 2015). Also, a few species of microalgae, like *Schizochytrium, Aurantiochytrium, Phaeodactylum*, and *Nannochloropsis*, can produce DHA and/or EPA in the desired ratio along with high growth rates, which may be amicable for industrial-scale production (Table 4.1) (Paliwal et al. 2017,

TABLE 4.1
EPA and DHA Contents of Potential Microalgae Cultivated in Their Respective Conditions

Strains	Medium	Conditions	EPA (mg/100 g)	DHA (mg/100 g)	References
Chlorella vulgaris (green)	Sorokin and Krauss medium	25 °C; 150 $\mu Em^{-2} s^{-1}$; stationary phase harvesting; bubbling	19 ± 1	16 ± 1	(Batista et al. 2013)
Chlorella vulgaris (orange)	Sorokin and Krauss medium	25 °C; 150 $\mu Em^{-2} s^{-1}$; stationary phase harvesting followed by nitrogen starvation and 30% NaCl addition at 1000 $\mu Em^{-2} s^{-1}$; bubbling	39 ± 1	80 ± 1	
Haematococcus pluvialis	Bold basal modified medium	25 °C; 150 $\mu Em^{-2} s^{-1}$; stationary phase harvesting followed by nitrogen starvation and 2% NaCl addition at 1000 $\mu Em^{-2} s^{-1}$; bubbling	579 ± 6	–	

(Continued)

TABLE 4.1 (*Continued*)

Strains	Medium	Conditions	EPA (mg/100 g)	DHA (mg/100 g)	References
Diacronema vlkianum	Wallerstein and Míquel medium (3:1, in filtered seawater with 35% salt)	18 °C; 150 µEm^{-2} s^{-1}; stationary phase harvesting; bubbling	3212±57	836±41	
Isochrysis galbana	Wallerstein and Míquel medium (3:1, in filtered seawater with 35% salt)	18 °C; 150 µEm^{-2} s^{-1}; stationary phase harvesting; bubbling	4875±108	1156±40	
Chaetoceros sp.	F/2-enriched artificial seawater (f/2AW, pH 8.5) medium	25 °C; 50 µmol photons m^{-2} s^{-1}; 7 days; bubbling	3200	–	(Gong et al. 2013)
Phaeodactylum tricornutum	F/2-enriched artificial seawater (f/2AW, pH 8.5) medium	25 °C; 50 µmol photons m^{-2} s^{-1}; 7 days; bubbling	2500	–	
Pavlova viridis	F/2-enriched artificial seawater (f/2AW, pH 8.5) medium	25 °C; 50 µmol photons m^{-2} s^{-1}; 11 days; bubbling	2500	–	
Nannochloropsis oculata	F/2-enriched artificial seawater (f/2AW, pH 8.5) medium	25 °C; 50 µmol photons m^{-2} s^{-1}; 7 days; bubbling	1000	–	
Schizochytrium sp. S31	100 g/L glycerol +14.28 g/L yeast extract + salts	28 °C; 2 d 1 h; baffled flask; 200 rpm	–	16880	(Chang et al. 2013)
Attheya septentrionalis	Walne's enriched with 80% seawater	22% salinity; 50/200 and day 5; 1% CO$_2$ bubbling	7100	–	(Steinrucken et al. 2018)

Hamilton et al. 2015). Production of EPA and DHA has been commercially available from microalgae such as *Phaeodactylum tricornutum*, *Nannochloropsis* sp., and *Crypthecodinium cohnii* (Hamilton et al. 2014). However, due to certain technological factors, more versatile strategies need to be employed to improve PUFA productivities (O'Neill and Kelly 2017).

These microalgae use either conventional desaturation and elongation biosynthetic pathway or the polyketide synthase (PKS) pathway (anaerobic conditions) for LC-PUFA biosynthesis (Kapase, Nesamma, and Jutur 2018). The *Schizochytrium* sp. uses both pathways, depending upon the substrate concentration in the media (Li, Chen et al. 2018, Geng et al. 2019, Yin et al. 2019). Therefore, there is a specific mechanism to check the gene expression of the pathway regulating the levels of crucial enzymes involved in desaturation and/or elongation reaction of fatty acid (FA) biosynthesis, along with acyltransferase activity, so that higher PUFA levels are achieved, but these are also species specific (Liang et al. 2018, Yin et al. 2018, Li, Meng et al. 2018)

Genetic engineering is one of the best strategies through which we can improve and achieve the production of PUFAs at commercial scale. To demonstrate the enhancement of PUFAs, a few studies have expressed elongase, malic, and desaturase enzymes from *Osterococcus tauri* and *Phaeodactylum tricornutum* (Hamilton et al. 2014, Peng et al. 2014, Degraeve-Guilbault et al. 2017, Velmurugan and Deka 2018, Stukenberg et al. 2018, D'Adamo et al. 2019). Recently, a new strategy has been employed as an alternative to genetic engineering, that is, the use of chemical modulators, where certain small chemicals enhance the overall growth, as well as lipid productivities including PUFA content in *Schizochytrium* sp. (Wang et al. 2018, Sahin, Tas, and Altindag 2018, Sun et al. 2018).

4.2.2 CAROTENOIDS

Carotenoids are natural hydrophobic yellow/red pigments with 40-carbon structures classified as the carotenes (non-oxygenated molecules) and the xanthophylls (oxygenated molecules), ranging between 600 and 700 different molecules, namely α- and β-carotene, astaxanthin, lycopene, lutein, and zeaxanthin (Table 4.2) (Siepelmeyer et al. 2016, Schieber and Weber 2016). The global market estimates for the carotenoids are projected to reach ~2.0 billion USD by 2022, and is growing at

---

TABLE 4.2

Some Important Carotenoids From Different Potential Microalgae With Industrial Scope

Strains	Medium	Conditions	Carotenoids (mg/g dw)	References
Dunaliella salina UTEX 2538	F/2 media	10 MJm^{-2} d^{-1}; two stages first grow 5 days in full strength medium, then 6 days in N free F/2	97.0 (β-carotene; after second stage)	(Prieto, Pedro Canavate, and Garcia-Gonzalez 2011)
Chlorella protothecoides CS-41	modified Basal medium with 40 g/L glucose	28 °C; pH 6.6; 480 rpm	4.58 (Lutein)	(Shi, Zhang, and Chen 2000)

(Continued)

TABLE 4.2 (*Continued*)

Strains	Medium	Conditions	Carotenoids (mg/g dw)	References
Haematococcus pluvialis	Rudic's medium	25 °C; 75 µmol photons m^{-2} s^{-1} continuous; N free/ distilled water with CO$_2$ (1.5% v/v; continuous)	30.07 (Astaxanthin)	(Imamoglu, Dalay, and Sukan 2009)
Phaeodactylum tricornutum	Conway medium	20 °C; 2,500 lux; continuous air 5 L/ min; 5 days	15.33 (Fucoxanthin)	(Kim et al. 2012)

a compound annual growth rate (CAGR) of 5.7% from 2017 to 2022 (McWilliams 2018). In microalgae, these molecules play significant roles in the process of photosynthesis, such as light harvesting, photoprotection, free radical scavenging, excess energy dissipation, and structure stabilization (Koller, Muhr, and Braunegg 2014). Microalgae such as *Dunaliella salina*, *Tetraselmis suecica*, *Isochrysis galbana*, and *Pavlova salina* have been exploited for commercial production of carotenoids due to their association with various health benefits, for example, in the prevention of age-related macular degeneration, cataract formation, cancers, rheumatoid arthritis, muscular dystrophy, and cardiovascular and chronic diseases (Kleinegris et al. 2011, de Los Reyes et al. 2016, Diprat et al. 2017, Bhalamurugan, Valerie, and Mark 2018, Yi et al. 2018, Ortiz and Ferruzzi 2019, Cezare-Gomes et al. 2019, Sathasivam and Ki 2019). Carotenoids are also essential biomolecules used in chemotaxonomy and therapeutics, besides being beneficial in the food and cosmetic industry as a natural colorant (Ghosh et al. 2015, Paliwal et al. 2015, Paliwal et al. 2016).

Currently, the bulk production of industrially relevant carotenoids are achieved by chemical synthesis, though only small amounts are obtained from either plants or algae naturally (Sathasivam and Ki 2018, Huang et al. 2018, Ambati et al. 2018). Studies with β-carotene isolated from *Dunaliella salina* demonstrated the reduction of plasma cholesterol levels and atherogenesis in mice (Gangl et al. 2015), having a commerical value of ~0.6 USD g^{-1} (Borowitzka 2013). Another well-known example of a industrial carotenoid extracted from the freshwater green alga *Haematococcus pluvialis* is astaxanthin, which has been used as a feed supplement due to its high antioxident properties (Ambati et al. 2014), with a market value of ~1.8 USD g^{-1} (Panis and Carreon 2016, Shah et al. 2016). Therefore, to enhance the production of carotenoids at industrial scale the use of molecular tools such as the genetic and pathway engineering of regulatory networks would be better strategies in developing the efficient microalgal cell factories (Ng et al. 2017).

4.2.3 MICROALGAL PROTEINS

Microalgae are consumed as a source of food due to their high protein content. However, non-pigmented protein from microalgae is still poorly valorized (Ursu et al. 2014). The protein content in *Chlorella vulgaris* is exceptionally higher at about 50% to 58% of dry weight and contains all the necessary amino acids (Bleakley and Hayes 2017). Specialty proteins like phycobiliproteins are a class of fluorescent pigmented proteins that exist only in algae, especially in red and blue-green algae, with a wide range of biological applications (Borowitzka 2013). Phycobiliproteins are classified as phycoerythrin (PE, pink-purple, λ_{max} = 540– to 70 nm), phycocyanin (PC, blue, λ_{max} = 610– to 20 nm), phycoerythrocyanin (PEC, orange, λ_{max} = 560 to 600 nm), and allophycocyanin (APC, bluish-green, λ_{max} = 650 to 655 nm) (Dufossé 2018). Besides their relevance to the food and cosmetic industries, these biomolecules are used as antibody and receptor labels, biosensors, and biomedicines (Paliwal et al. 2017). The commercial value of pure R-phycoerythrin obtained from *Porphyridium cruentum* would be ~30 to 150 USD mg^{-1} depending upon the purity of the phyco-biliproteins (Chaloub et al. 2015, Cuellar-Bermudez et al. 2015). Microalgal-based bioactive peptides are extensively used in the treatment of certain diseases for its immunomodulatory- and cholesterol-lowering effects (Ibanez and Cifuentes 2013, Singh et al. 2017).

4.2.4 TOCOPHEROLS

Tocopherols are another class of lipid-soluble antioxidant biomolecules synthesized only in photosynthetic organisms (Maeda et al. 2005). They are amphipathic molecules made up of a chromanol polar head group attached with a phytyl chain (hydrophobic) and play a key role in inhibiting lipid peroxidation due to reactive oxygen species, along with participating in the signalling cascades and expression of genes involved in cellular proliferation (Wawrzyniak et al. 2013). α-tocopherol is the most abundant tocopherol and exhibits the highest *in vivo* antioxidant activity and functions in rendering membrane stability and electron transport chains. Tocopherol is widely used as a dietary supplement to prevent oxidative damages caused by stress or pollution, and it also has hypocholestemic, antimutagenic, and cardioprotective health benefits (Kottuparambil, Thankamony, and Agusti 2019).

4.2.5 POLYSACCHARIDES

Polysaccharides are generally employed in the industry as gelling or thickening agents. Microalgae species are also recognized for their richness and diversity of polysaccharides, such as chitosan, laminarin, alginate, and fucoidan (Jutur, Nesamma, and Shaikh 2016), which are essential in cosmeceutical products. Edible algae contain more than 50% fiber content, either laminarin (brown algae) or floridean starch (red algae), making them difficult to be digested by humans (Ibanez and Cifuentes 2013). Certain species of algae—mainly diatoms, which are a rich source of sulfated polysaccharides—have shown functional properties in antitumor, immune-modulation, antiviral, and other therapeutics (Xiao and Zheng 2016). Today, paramylon, a β-1-3-glucan-type polysaccharide obtained from microalgae *Euglena* sp., is the most prominent polysaccharide used due to its dietary properties, which has significant cytokine-related immunopotent and stimulating effects (Kottuparambil, Thankamony, and Agusti 2019). Microalgae produce exopolysaccharides (EPS) to avoid erosion from their natural habitat and clump together to form biofilms (Angelaalincy et al. 2017).

4.3 COMMERCIAL RELEVANCE OF HIGH-VALUE BIOMOLECULES

Numerous microalgal-based products are available in the commercial market, mostly in the cosmetic industries as anti-aging or skin conditioning agents. For example, Depollutine (Givaudan Active Beauty), derived from *Phaeodactylum tricornutum* peptidic extract, is used for fighting wrinkles. Another biomolecule, Solasta Astaxanthin (Solix Algredients), uses a natural astaxanthin extract from *Haematococcus pluvialis* to reduce wrinkles and improve skin elasticity and moisture. Pepha-Tight (DSM) is marketed as an energizing agent, and it contains *Nannochloropsis oculata* extract blended with polysaccharide fractions to give exceptional skin tightening and firming properties, and the company has used *Dunaliella salina* as an energizing agent promoting dermal cell growth and energy metabolism. There are several more microalgae extract–based products described in Table 4.3.

The demand for nutraceutical products and dietary supplements from microalgae has increased in the past decades because of their unique nutrition content. Some of the microalgae are a rich source of PUFAs like EPA and DHA, whereas others are rich in phytonutrients. There are several marketed microalgae-based nutraceutical products (Table 4.3) designed for dietary purposes, like Zanthin, from the Indian company Parry Nutraceuticals, which contains natural

---

TABLE 4.3

List of Companies Involved in the Commercial Production of HVBs and Their Products From Algae

Supplier / Company	Product	Microalgae	Properties	References
Givaudan Active Beauty	Hydrintense	*Porphyridium cruentum*	Moisturizing agent	www.givaudan.com/
	Grevilline	*Skeletonema costatum*	Soothing agent	
	Mariliance	*Rhodosorus marinus*	Neuro-soothing agent	
	Depollutine	*Phaeodactylum tricornutum*	Anti-aging agent	
NaturZell (BASM blue International)	Naturnutritive Cosmetic	*Nannochloropsis gaditana*	Anti-aging, anti-inflammatory, UV protecting and nourishing agent	
	Acqualift	*Porphyridium* sp.	Healing and hydrating agent	

Supplier / Company	Product	Microalgae	Properties	References
DSM	Pepha-Age	*Scenedesmus rubescens*	Anti-aging and sunscreen agent	www.dsm.com/
	Pepha-Ctive	*Dunaliella salina*	Energizing agent	
	Pepha-Tight	*Nannochloropsis oculata*	Skin tightening and firming agent	
	DSM-NP Life's DHA plus EPA	*Schizochytrium sp.*	EPA and DHA for nutrition	
	DSM-NP Life's DHA and GCI	*Crypthecodinium cohnii*	Nutrients DHA Algae 35% Oil	
Phytomer	Phytomer	*Chlorella vulgaris*	Neutralize inflammation and improve skin	www.phytomer.fr/
Euglena Co. Ltd.	Euglena Bar	*Euglena* sp.	Dietary cookies with low glycemic index, high tocopherols	www.euglena.jp/
Parry Nutraceuticals	Zanthin	*Haematococcus pluvialis*	3S, 3'S Astaxanthin, powerful antioxidant, anti-inflammatory	www.eidparry.com/

astaxanthin from *Haematococcus pluvialis,* and Euglena Co. Ltd. produces several *Euglena*-based products rich in paramilon (Kottuparambil, Thankamony, and Agusti 2019).

4.4   CONCLUSIONS

Although industrial production of HVBs from microalgae is still not feasible and to increase the economic sustainability of large-scale processes, all the microalgal biomass constituents should be completely valorized (i.e., aiming towards a multi-product biorefinery). However, this scenario is too expensive to pursue at this stage. All downstream procedures should be cost-effective in terms of the overall production of industrial HVBs ranging from start-to-end technologies. However, significant technological advancements are in progress, but overall understanding of the biology of these photosynthetic organisms will be beneficial for engineering these strains for the production of biomass, biofuels, and biomolecules (B^3) (Figure 4.1). In conclusion, the development of a sustainable process for the complete valorization of microalgal biomass will lead to a change in the transition from the present low-scale profile of specialty biomolecules to a futuristic high-scale production of bulk commodities from microalgae.

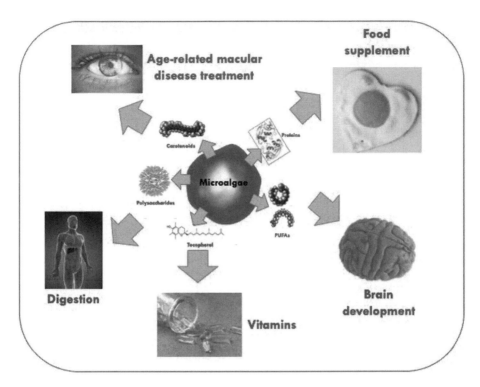

FIGURE 4.1 Schematic overview for the production of different high-value biomolecules with industrial scope from microalgae.

REFERENCES

Ambati, R. R., D. Gogisetty, R. G. Aswathanarayana, S. Ravi, P. N. Bikkina, L. Bo, and S. Yuepeng. 2018. "Industrial potential of carotenoid pigments from microalgae: Current trends and future prospects." *Crit Rev Food Sci Nutr*:1–22. doi: 10.1080/10408398.2018.1432561.

Ambati, R. R., S. M. Phang, S. Ravi, and R. G. Aswathanarayana. 2014. "Astaxanthin: Sources, extraction, stability, biological activities and its commercial applications–a review." *Mar Drugs* 12 (1):128–152. doi: 10.3390/md12010128.

Angelaalincy, M., N. Senthilkumar, R. Karpagam, G. G. Kumar, B. Ashokkumar, and P. Varalakshmi. 2017. "Enhanced extracellular polysaccharide production and self-sustainable electricity generation for PAMFCs by *Scenedesmus* sp. SB1." *ACS Omega* 2 (7):3754–3765. doi: 10.1021/acsomega.7b00326.

Arora, N. K., T. Fatima, I. Mishra, M. Verma, J. Mishra, and V. Mishra. 2018. "Environmental sustainability: Challenges and viable solutions." *Environ Sustain* 1 (4):309–340. doi: 10.1007/s42398-018-00038-w.

Barrera, D. J., and S. P. Mayfield. 2013. "High-value recombinant protein production in microalgae." In *Handbook of Microalgal Culture: Applied phycology and biotechnology.* 2nd ed., edited by A. Richmond and Q. Hu, 532–544. Chichester (UK): John Wiley & Sons, Ltd.

Batista, A. P., L. Gouveia, N. M. Bandarra, J. M. Franco, and A. Raymundo. 2013. "Comparison of microalgal biomass profiles as novel functional ingredient for food products." *Algal Res* 2 (2):164–173. doi: 10.1016/j.algal.2013.01.004.

Bhalamurugan, G. L., O. Valerie, and L. Mark. 2018. "Valuable bioproducts obtained from microalgal biomass and their commercial applications: A review." *Environ Eng Res* 23 (3):229–241. doi: 10.4491/eer.2017.220.

Bleakley, S., and M. Hayes. 2017. "Algal proteins: Extraction, application, and challenges concerning production." *Foods* 6 (5):33. doi: 10.3390/foods6050033.

Borowitzka, M. A. 2013. "High-value products from microalgae-their development and commercialisation." *J Appl Phycol* 25 (3):743–756. doi: 10.1007/s10811-013-9983-9.

Brennan, L., and P. Owende. 2010. "Biofuels from microalgae—A review of technologies for production, processing, and extractions of biofuels and co-products." *Renew Sust Energ Rev* 14 (2):557–577. doi: 10.1016/j.rser.2009.10.009.

Cezare-Gomes, E. A., L. D. C. Mejia-da-Silva, L. S. Perez-Mora, M. C. Matsudo, L. S. Ferreira-Camargo, A. K. Singh, and J. C. M. de Carvalho. 2019. "Potential of microalgae carotenoids for industrial application." *Appl Biochem Biotechnol*:1–33. doi: 10.1007/s12010-018-02945-4.

Chaloub, R. M., N. M. S. Motta, S. P. de Araujo, P. F. de Aguiar, and A. F. da Silva. 2015. "Combined effects of irradiance, temperature and nitrate concentration on phycoerythrin content in the microalga *Rhodomonas* sp. (Cryptophyceae)." *Algal Res* 8:89–94. doi: 10.1016/j.algal.2015.01.008.

Chang, G., N. Gao, G. Tian, Q. Wu, M. Chang, and X. Wang. 2013. "Improvement of docosahexaenoic acid production on glycerol by *Schizochytrium* sp. S31 with constantly high oxygen transfer coefficient." *Bioresour Technol* 142:400–406. doi: 10.1016/j.biortech.2013.04.107.

Chew, K. W., J. Y. Yap, P. L. Show, N. H. Suan, J. C. Juan, T. C. Ling, D. J. Lee, and J. S. Chang. 2017. "Microalgae biorefinery: High value products perspectives." *Bioresour Technol* 229:53–62. doi: 10.1016/j.biortech.2017.01.006.

Clarens, A. F., H. Nassau, E. P. Resurreccion, M. A. White, and L. M. Colosi. 2011. "Environmental impacts of algae-derived biodiesel and bioelectricity for transportation." *Environ Sci Technol* 45 (17):7554–7560. doi: 10.1021/es200760n.

Cuellar-Bermudez, S. P., I. Aguilar-Hernandez, D. L. Cardenas-Chavez, N. Ornelas-Soto, M. A. Romero-Ogawa, and R. Parra-Saldivar. 2015. "Extraction and purification of high-value metabolites from microalgae: Essential lipids, astaxanthin and phycobiliproteins." *Microbial Biotechnology* 8 (2):190–209. doi: 10.1111/1751-7915.12167.

D'Adamo, S., G. Schiano di Visconte, G. Lowe, J. Szaub-Newton, T. Beacham, A. Landels, M. J. Allen, A. Spicer, and M. Matthijs. 2019. "Engineering the unicellular alga *Phaeodactylum tricornutum* for high-value plant triterpenoid production." *Plant Biotechnol J* 17 (1):75–87. doi: 10.1111/pbi.12948.

de la Noue, J., and N. de Pauw. 1988. "The potential of microalgal biotechnology: A review of production and uses of microalgae." *Biotechnol Adv* 6 (4):725–770.

de Los Reyes, C., M. J. Ortega, A. Rodriguez-Luna, E. Talero, V. Motilva, and E. Zubia. 2016. "Molecular characterization and anti-inflammatory activity of galactosylglycerides and galactosylceramides from the microalga *Isochrysis galbana*." *J Agric Food Chem* 64 (46):8783–8794. doi: 10.1021/acs.jafc.6b03931.

Degraeve-Guilbault, C., C. Brehelin, R. Haslam, O. Sayanova, G. Marie-Luce, J. Jouhet, and F. Corellou. 2017. "Glycerolipid characterization and nutrient deprivation-associated changes in the green picoalga *Ostreococcus tauri*." *Plant Physiol* 173 (4):2060–2080. doi: 10.1104/pp.16.01467.

Dewapriya, P., and S.-K. Kim. 2014. "Marine microorganisms: An emerging avenue in modern nutraceuticals and functional foods." *Food Research International* 56:115–125. doi: 10.1016/j.foodres.2013.12.022.

Diprat, A. B., T. Menegol, J. F. Boelter, A. Zmozinski, M. G. Rodrigues Vale, E. Rodrigues, and R. Rech. 2017. "Chemical composition of microalgae *Heterochlorella luteoviridis* and *Dunaliella tertiolecta* with emphasis on carotenoids." *J Sci Food Agric* 97 (10):3463–3468. doi: 10.1002/jsfa.8159.

Dufossé, L. 2018. "Microbial pigments from bacteria, yeasts, fungi, and microalgae for the food and feed Industries." In *Natural and Artificial Flavoring Agents and Food Dyes*, edited by Alexandru Mihai Grumezescu and Alina Maria Holban, 113–132. London: Academic Press.

El Abed, M. M., B. Marzouk, M. N. Medhioub, A. N. Helal, and A. Medhioub. 2008. "Microalgae: A potential source of polyunsaturated fatty acids." *Nutr Health* 19 (3):221–226. doi: 10.1177/026010600801900309.

Gangl, D., J. A. Zedler, P. D. Rajakumar, E. M. Martinez, A. Riseley, A. Wlodarczyk, S. Purton, Y. Sakuragi, C. J. Howe, P. E. Jensen, and C. Robinson. 2015. "Biotechnological exploitation of microalgae." *J Exp Bot* 66 (22):6975–6990. doi: 10.1093/jxb/erv426.

Geng, L., S. Chen, X. Sun, X. Hu, X. Ji, H. Huang, and L. Ren. 2019. "Fermentation performance and metabolomic analysis of an engineered high-yield PUFA-producing strain of *Schizochytrium* sp." *Bioprocess Biosyst Eng* 42 (1):71–81. doi: 10.1007/s00449-018-2015-z.

Ghosh, T., C. Paliwal, R. Maurya, and S. Mishra. 2015. "Microalgal Rainbow Colours for Nutraceutical and Pharmaceutical Applications." In *Plant Diversity, Organization, Function and Improvement*, edited by B. Bahadur, M. V. Rajam, L. Sahijram and K. V. Krishnamurthy, 777–791. New Delhi: Springer India. http://dx.doi.org/10.1007/978-81-322-2286-6_32.

Gomaa, M. A., L. Al-Haj, and R. M. Abed. 2016. "Metabolic engineering of cyanobacteria and microalgae for enhanced production of biofuels and high-value products." *J Appl Microbiol* 121 (4):919–931. doi: 10.1111/jam.13232.

Gong, Y., X. Guo, X. Wan, Z. Liang, and M. Jiang. 2013. "Triacylglycerol accumulation and change in fatty acid content of four marine oleaginous microalgae under nutrient limitation and at different culture ages." *J Basic Microbiol* 53 (1):29–36. doi: 10.1002/jobm.201100487.

Gouveia, L. 2014. "From tiny microalgae to huge biorefineries." *Oceanography: Open Access* 2 (1):1–8. doi: 10.4172/2332-2632.1000120.

Guiheneuf, F., M. Schmid, and D. B. Stengel. 2015. "Lipids and fatty acids in algae: Extraction, fractionation into lipid classes, and analysis by gas chromatography coupled with flame ionization detector (GC-FID)." *Methods Mol Biol* 1308:173–190. doi: 10.1007/978-1-4939-2684-8_11.

Guiheneuf, F., V. Mimouni, G. Tremblin, and L. Ulmann. 2015. "Light intensity regulates LC-PUFA incorporation into lipids of *Pavlova lutheri* and the final desaturase and elongase activities involved in their biosynthesis." *J Agric Food Chem* 63 (4):1261–1267. doi: 10.1021/jf504863u.

Hamilton, M. L., J. Warwick, A. Terry, M. J. Allen, J. A. Napier, and O. Sayanova. 2015. "Towards the industrial production of omega-3 long chain polyunsaturated fatty acids from a genetically modified diatom *Phaeodactylum tricornutum*." *PLoS One* 10 (12):e0144054. doi: 10.1371/journal.pone.0144054.

Hamilton, M. L., R. P. Haslam, J. A. Napier, and O. Sayanova. 2014. "Metabolic engineering of *Phaeodactylum tricornutum* for the enhanced accumulation of omega-3 long chain polyunsaturated fatty acids." *Metab Eng* 22:3–9. doi: 10.1016/j.ymben.2013.12.003.

Hariskos, I., and C. Posten. 2014. "Biorefinery of microalgae—opportunities and constraints for different production scenarios." *Biotechnol J* 9 (6):739–752. doi: 10.1002/biot.201300142.

Huang, W., Y. Lin, M. He, Y. Gong, and J. Huang. 2018. "Induced high-yield production of zeaxanthin, lutein, and beta-carotene by a mutant of *Chlorella zofingiensis*." *J Agric Food Chem* 66 (4):891–897. doi: 10.1021/acs.jafc.7b05400.

Ibanez, E., and A. Cifuentes. 2013. "Benefits of using algae as natural sources of functional ingredients." *J Sci Food Agric* 93 (4):703–709. doi: 10.1002/jsfa.6023.

Imamoglu, E., M. C. Dalay, and F. V. Sukan. 2009. "Influences of different stress media and high light intensities on accumulation of astaxanthin in the green alga *Haematococcus pluvialis.*" *N Biotechnol* 26 (3–4):199–204. doi: 10.1016/j.nbt.2009.08.007.

Jutur, P. P., A. A. Nesamma, and K. M. Shaikh. 2016. "Algae-derived marine oligosaccharides and their biological applications." *Front Mar Sci* 3 (83). doi: 10.3389/fmars.2016.00083.

Kapase, V. U., A. A. Nesamma, and P. P. Jutur. 2018. "Identification and characterization of candidates involved in production of OMEGAs in microalgae: A gene mining and phylogenomic approach." *Prep Biochem Biotechnol*:1–10. doi: 10.1080/10826068.2018.1476886.

Kim, S. M., Y. J. Jung, O. N. Kwon, K. H. Cha, B. H. Um, D. Chung, and C. H. Pan. 2012. "A potential commercial source of fucoxanthin extracted from the microalga *Phaeodactylum tricornutum.*" *Appl Biochem Biotechnol* 166 (7):1843–1855. doi: 10.1007/s12010-012-9602-2.

Kleinegris, D. M., M. Janssen, W. A. Brandenburg, and R. H. Wijffels. 2011. "Continuous production of carotenoids from *Dunaliella salina.*" *Enzyme Microb Technol* 48 (3):253–259. doi: 10.1016/j.enzmictec.2010.11.005.

Koller, M., A. Muhr, and G. Braunegg. 2014. "Microalgae as versatile cellular factories for valued products." *Algal Research* 6:52–63. doi: 10.1016/j.algal.2014.09.002.

Kottuparambil, S., R. L. Thankamony, and S. Agusti. 2019. "*Euglena* as a potential natural source of value-added metabolites. A review." *Algal Res* 37:154–159. doi: 10.1016/j.algal.2018.11.024.

Li, H. Y., Y. Lu, J. W. Zheng, W. D. Yang, and J. S. Liu. 2014. "Biochemical and genetic engineering of diatoms for polyunsaturated fatty acid biosynthesis." *Mar Drugs* 12 (1):153–166. doi: 10.3390/md12010153.

Li, T., M. Gargouri, J. Feng, J. J. Park, D. Gao, C. Miao, T. Dong, D. R. Gang, and S. Chen. 2015. "Regulation of starch and lipid accumulation in a microalga *Chlorella sorokiniana.*" *Bioresour Technol* 180:250–257. doi: 10.1016/j.biortech.2015.01.005.

Li, Z., T. Meng, X. Ling, J. Li, C. Zheng, Y. Shi, Z. Chen, Z. Li, Q. Li, Y. Lu, and N. He. 2018. "Overexpression of malonyl-CoA: ACP transacylase in *Schizochytrium* sp. to improve polyunsaturated fatty acid production." *J Agric Food Chem* 66 (21):5382–5391. doi: 10.1021/acs.jafc.8b01026.

Li, Z., X. Chen, J. Li, T. Meng, L. Wang, Z. Chen, Y. Shi, X. Ling, W. Luo, D. Liang, Y. Lu, Q. Li, and N. He. 2018. "Functions of PKS genes in lipid synthesis of *Schizochytrium* sp. by gene disruption and metabolomics analysis." *Mar Biotechnol (NY)* 20 (6):792–802. doi: 10.1007/s10126-018-9849-x.

Liang, Y., Y. Liu, J. Tang, J. Ma, J. J. Cheng, and M. Daroch. 2018. "Transcriptomic profiling and gene disruption revealed that two genes related to PUFAs/DHA biosynthesis may be essential for cell growth of *Aurantiochytrium* sp." *Marine Drugs* 16 (9):310.

Maeda, H., Y. Sakuragi, D. A. Bryant, and D. Dellapenna. 2005. "Tocopherols protect Synechocystis sp. strain PCC 6803 from lipid peroxidation." *Plant Physiol* 138 (3):1422–1435. doi: 10.1104/pp.105.061135.

Markou, G., and E. Nerantzis. 2013. "Microalgae for high-value compounds and biofuels production: A review with focus on cultivation under stress conditions." *Biotechnol Adv* 31 (8):1532–1542. doi: 10.1016/j.biotechadv.2013.07.011.

Matos, J., C. Cardoso, N. M. Bandarra, and C. Afonso. 2017. "Microalgae as healthy ingredients for functional food: A review." *Food Funct* 8 (8):2672–2685. doi: 10.1039/c7fo00409e.

McWilliams, A. 2018. The global market for carotenoids. In *BCC Research, www.bccre search.com/market-research/food-and-beverage/the-global-market-for-carotenoids-fod025f.html* (Accessed January 20, 2019).

Misra, R., A. Guldhe, P. Singh, I. Rawat, and F. Bux. 2014. "Electrochemical harvesting process for microalgae by using nonsacrificial carbon electrode: A sustainable approach for biodiesel production." *Chem Eng J* 255:327–333. doi: 10.1016/j.cej.2014.06.010.

Nath, K., M. M. Najafpour, R. A. Voloshin, S. E. Balaghi, E. Tyystjärvi, R. Timilsina, J. J. Eaton-Rye, T. Tomo, H. G. Nam, H. Nishihara, S. Ramakrishna, J-R. Shen, and S. I. Allakhverdiev. 2015. "Photobiological hydrogen production and artificial photosynthesis for clean energy: From bio to nanotechnologies." *Photosynth Res* 126 (2):237–247. doi: 10.1007/s11120-015-0139-4.

Neidleman, S. L. 1987. "Effects of temperature on lipid unsaturation." *Biotechnol Genet Eng Rev* 5:245–268.

Nesamma, A. A., K. M. Shaikh, and P. P. Jutur. 2015. "Genetic engineering of microalgae for production of value-added ingredients." In *Handbook of Marine Microalgae*, edited by Se-Kwon Kim, 405–414. Boston, London: Academic Press.

Ng, I. S., S. I. Tan, P. H. Kao, Y. K. Chang, and J. S. Chang. 2017. "Recent developments on genetic engineering of microalgae for biofuels and bio-based chemicals." *Biotechnol J* 12 (10). doi: 10.1002/biot.201600644.

O'Neill, E. C., and S. Kelly. 2017. "Engineering biosynthesis of high-value compounds in photosynthetic organisms." *Crit Rev Biotechnol* 37 (6):779–802. doi: 10.1080/0738 8551.2016.1237467.

Olofsson, M., T. Lamela, E. Nilsson, J. P. Bergé, V. Del Pino, P. Uronen, and C. Legrand. 2012. "Seasonal variation of lipids and fatty acids of the microalgae *Nannochloropsis oculata* grown in outdoor large-scale photobioreactors." *Energies* 5 (5):1577.

Ortiz, D., and M. G. Ferruzzi. 2019. "Identification and quantification of carotenoids and tocochromanols in sorghum grain by high-performance liquid chromatography." *Methods Mol Biol* 1931:141–151. doi: 10.1007/978-1-4939-9039-9_10.

Paliwal, C., I. Pancha, T. Ghosh, R. Maurya, K. Chokshi, S. V. Vamsi Bharadwaj, S. Ram, and S. Mishra. 2015. "Selective carotenoid accumulation by varying nutrient media and salinity in *Synechocystis* sp. CCNM 2501." *Bioresour Technol* 197:363–368. doi: 10.1016/j.biortech.2015.08.122.

Paliwal, C., M. Mitra, K. Bhayani, S. V. V. Bharadwaj, T. Ghosh, S. Dubey, and S. Mishra. 2017. "Abiotic stresses as tools for metabolites in microalgae." *Bioresour Technol* 244 (Pt 2):1216–1226. doi: 10.1016/j.biortech.2017.05.058.

Paliwal, C., T. Ghosh, B. George, I. Pancha, R. Maurya, K. Chokshi, A. Ghosh, and S. Mishra. 2016. "Microalgal carotenoids: Potential nutraceutical compounds with chemotaxonomic importance." *Algal Research* 15:24–31. doi: 10.1016/j.algal.2016.01.017.

Paliwal, C., T. Ghosh, K. Bhayani, R. Maurya, and S. Mishra. 2015. "Antioxidant, anti-nephrolithe Activities and *in vitro* digestibility studies of three different cyanobacterial pigment extracts." *Mar Drugs* 13 (8):5384–5401. doi: 10.3390/md13085384.

Pangestuti, R., and S. K. Kim. 2011. "Neuroprotective effects of marine algae." *Mar Drugs* 9 (5):803–818. doi: 10.3390/md9050803.

Panis, G., and J. R. Carreon. 2016. "Commercial astaxanthin production derived by green alga *Haematococcus pluvialis*: A microalgae process model and a techno-economic assessment all through production line." *Algal Research* 18:175–190. doi: 10.1016/j.algal.2016.06.007.

Peng, K.T., C.N. Zheng, J. Xue, X.-Y. Chen, W. D. Yang, J. S. Liu, W. Bai, and H.-Y. Li. 2014. "Delta 5 fatty acid desaturase upregulates the synthesis of polyunsaturated fatty acids in the marine diatom *Phaeodactylum tricornutum*." *J Agric Food Chem* 62 (35):8773–8776. doi: 10.1021/jf5031086.

Prieto, A., J. Pedro Canavate, and M. Garcia-Gonzalez. 2011. "Assessment of carotenoid production by *Dunaliella salina* in different culture systems and operation regimes." *J Biotechnol* 151 (2):180–185. doi: 10.1016/j.jbiotec.2010.11.011.

Pulz, O., and W. Gross. 2004. "Valuable products from biotechnology of microalgae." *Appl Microbiol Biotechnol* 65 (6):635–648. doi: 10.1007/s00253-004-1647-x.

Robertson, R., F. Guihéneuf, B. Schmid, D. Stengel, G. Fitzgerald, P. Ross, and C. Stanton. 2013. "Algae-derived polyunsaturated fatty acids: Implications for human health." In *Polyunsaturated Fatty Acids: Sources, Antioxidant Properties and Health Benefits*, edited by Angel Catalá, 1–54. Hauppauge, NY: Nova Sciences Publishers, Inc.

Rosales-Mendoza, S. 2016. *Algae-based Biopharmaceuticals*, 1–166. Switzerland: Springer International Publishing.

Sahin, D., E. Tas, and U. H. Altindag. 2018. "Enhancement of docosahexaenoic acid (DHA) production from *Schizochytrium* sp. S31 using different growth medium conditions." *AMB Express* 8 (1):7. doi: 10.1186/s13568-018-0540-4.

Sathasivam, R., and J. S. Ki. 2019. "Differential transcriptional responses of carotenoid biosynthesis genes in the marine green alga *Tetraselmis suecica* exposed to redox and non-redox active metals." *Mol Biol Rep.* doi: 10.1007/s11033-018-04583-9.

Sathasivam, R., and J.-S. Ki. 2018. "A review of the biological activities of microalgal carotenoids and their potential use in healthcare and cosmetic industries." *Mar Drugs* 16 (1):26.

Schieber, A., and F. Weber. 2016. "Carotenoids." In *Handbook on Natural Pigments in Food and Beverages*, edited by R. Carle and R. M. Schweiggert, 101–123. Cambridge: Woodhead Publishing Limited.

Shah, M. M., Y. Liang, J. J. Cheng, and M. Daroch. 2016. "Astaxanthin-producing green microalga *Haematococcus pluvialis*: From single cell to high value commercial products." *Front Plant Sci* 7:531. doi: 10.3389/fpls.2016.00531.

Shi, X., X. Zhang, and F. Chen. 2000. "Heterotrophic production of biomass and lutein by *Chlorella protothecoides* on various nitrogen sources." *Enzyme Microb Technol* 27 (3–5):312–318.

Siepelmeyer, A., A. Micka, A. Simm, and J. Bernhardt. 2016. "Nutritional biomarkers of aging." In *Molecular Basis of Nutrition and Aging*, edited by Marco Malavolta and Eugenio Mocchegiani, 109–120. San Diego: Academic Press.

Singh, R., P. Parihar, M. Singh, A. Bajguz, J. Kumar, S. Singh, V. P. Singh, and S. M. Prasad. 2017. "Uncovering potential applications of cyanobacteria and algal metabolites in biology, agriculture and medicine: Current status and future prospects." *Front Microbiol* 8:515. doi: 10.3389/fmicb.2017.00515.

Skjanes, K., C. Rebours, and P. Lindblad. 2013. "Potential for green microalgae to produce hydrogen, pharmaceuticals and other high value products in a combined process." *Crit Rev Biotechnol* 33 (2):172–215. doi: 10.3109/07388551.2012.681625.

Spolaore, P., C. Joannis-Cassan, E. Duran, and A. Isambert. 2006. "Commercial applications of microalgae." *J Biosci Bioeng* 101 (2):87–96. doi: 10.1263/jbb.101.87.

Steinrucken, P., S. K. Prestegard, J. H. de Vree, J. E. Storesund, B. Pree, S. A. Mjos, and S. R. Erga. 2018. "Comparing EPA production and fatty acid profiles of three *Phaeodactylum tricornutum* strains under western Norwegian climate conditions." *Algal Res* 30:11–22. doi: 10.1016/j.algal.2017.12.001.

Stukenberg, D., S. Zauner, G. Dell' Aquila, and U. G. Maier. 2018. "Optimizing CRISPR/Cas9 for the diatom *Phaeodactylum tricornutum*." *Front Plant Sci* 9:740. doi: 10.3389/fpls.2018.00740.

Sun, X. M., L. J. Ren, X. J. Ji, and H. Huang. 2018. "Enhancing biomass and lipid accumulation in the microalgae *Schizochytrium* sp. by addition of fulvic acid and EDTA." *AMB Express* 8 (1):150. doi: 10.1186/s13568-018-0681-5.

Ursu, A. V., A. Marcati, T. Sayd, V. Sante-Lhoutellier, G. Djelveh, and P. Michaud. 2014. "Extraction, fractionation and functional properties of proteins from the microalgae *Chlorella vulgaris*." *Bioresour Technol* 157:134–139. doi: 10.1016/j.biortech.2014.01.071.

Vanthoor-Koopmans, M., R. H. Wijffels, M. J. Barbosa, and M. H. Eppink. 2013. "Biorefinery of microalgae for food and fuel." *Bioresour Technol* 135:142–149. doi: 10.1016/j.biortech.2012.10.135.

Velmurugan, N., and D. Deka. 2018. "Transformation techniques for metabolic engineering of diatoms and haptophytes: Current state and prospects." *Appl Biochem Biotechnol* 102 (10):4255–4267. doi: 10.1007/s00253-018-8925-5.

Wang, K., T. Sun, J. Cui, L. Liu, Y. Bi, G. Pei, L. Chen, and W. Zhang. 2018. "Screening of chemical modulators for lipid accumulation in *Schizochytrium* sp. S31." *Bioresour Technol* 260:124–129. doi: 10.1016/j.biortech.2018.03.104.

Wawrzyniak, A., M. Gornicka, J. Hamulka, M. Gajewska, M. Drywien, J. Pierzynowska, and A. Gronowska-Senger. 2013. "alpha-Tocopherol, ascorbic acid, and beta-carotene protect against oxidative stress but reveal no direct influence on p53 expression in rats subjected to stress." *Nutr Res* 33 (10):868–875. doi: 10.1016/j.nutres.2013.07.001.

Wichuk, K., S. Brynjolfsson, and W. Fu. 2014. "Biotechnological production of value-added carotenoids from microalgae: Emerging technology and prospects." *Bioengineered* 5 (3):204–208. doi: 10.4161/bioe.28720.

Wijffels, R. H., and M. J. Barbosa. 2010. "An outlook on microalgal biofuels." *Science* 329 (5993):796–769. doi: 10.1126/science.1189003.

Winwood, R. J. 2013. "Recent developments in the commercial production of DHA and EPA rich oils from micro-algae." *OCL* 20 (6):D604.

Xiao, R., and Y. Zheng. 2016. "Overview of microalgal extracellular polymeric substances (EPS) and their applications." *Biotechnol Adv* 34 (7):1225–1244. doi: 10.1016/j.biotechadv.2016.08.004.

Yi, Z., Y. Su, M. Xu, A. Bergmann, S. Ingthorsson, O. Rolfsson, K. Salehi-Ashtiani, S. Brynjolfsson, and W. Fu. 2018. "Chemical mutagenesis and fluorescence-based high-throughput screening for enhanced accumulation of carotenoids in a model marine diatom *Phaeodactylum tricornutum*." *Mar Drugs* 16 (8). doi: 10.3390/md16080272.

Yin, F. W., D. S. Guo, L. J. Ren, X. J. Ji, and H. Huang. 2018. "Development of a method for the valorization of fermentation wastewater and algal-residue extract in docosa-hexaenoic acid production by *Schizochytrium* sp." *Bioresour Technol* 266:482–487. doi: 10.1016/j.biortech.2018.06.109.

Yin, F. W., S. Y. Zhu, D. S. Guo, L. J. Ren, X. J. Ji, H. Huang, and Z. Gao. 2019. "Development of a strategy for the production of docosahexaenoic acid by *Schizochytrium* sp. from cane molasses and algae-residue." *Bioresour Technol* 271:118–124. doi: 10.1016/j.biortech.2018.09.114.

Zárate, R., N. el Jaber-Vazdekis, N. Tejera, J. A. Pérez, and C. Rodríguez. 2017. "Significance of long chain polyunsaturated fatty acids in human health." *Clin Transl Med* 6 (1):25. doi: 10.1186/s40169-017-0153-6.

Zhao, Y., H. P. Wang, B. Han, and X. Yu. 2019. "Coupling of abiotic stresses and phyto-hormones for the production of lipids and high-value by-products by microalgae: A review." *Bioresour Technol* 274:549–556. doi: 10.1016/j.biortech.2018.12.030.

Section 2

Sustainable Approach for Industrial-Scale Operations

5 Sustainable Pre-treatment Methods for Downstream Processing of Harvested Microalgae

Hrishikesh A. Tavanandi, A. Chandralekha Devi, K. S. M. S. Raghavarao

CONTENTS

5.1 INTRODUCTION

The demand for cosmetic, biopharmaceutical and nutraceutical (of natural origin instead of synthetic) products is increasing because of increased consumer awareness (Ottman 2011; Gernaey et al. 2012). Microalgae are a major source of phycobiliproteins, chlorophylls, carotenoids, polyphenols and lipids. Attention to microalgae has increased in the 21st century after problems related to the basic needs of modern human beings, such as health problems, food security, oil depletion and global warming, came to the fore. Due to progress in biotechnology, it has become possible for us to exploit the capability of microalgae to address all these problems to some extent.

A significant amount of research is in progress on the possibilities of increasing the bioactive compounds in microalgae during their cultivation. But if we are not able to efficiently extract them, merely increasing the bioactive compounds in the biomass is a futile exercise. For instance, it can be observed that estimation of the yield in reports related to the purification of C-phycocyanin (C-PC) (Patil et al. 2006; Patil and Raghavarao 2007; Chethana et al. 2015) is on the basis of the content of C-PC in the primary extract/crude. However, the overall yield of the process is quite a bit less compared to the total C-PC content in the biomass, as the extraction efficiency of most of the conventional cell disruption methods is ~50% to 60%. Additionally, the cell wall structure of the microalgae is complex and needs more understanding to achieve higher extraction efficiency (Gerken et al. 2013). The efficiency of its disruption is influenced by different process parameters, such as the thickness, rigidity and composition of the cell wall, which in turn varies greatly based on many factors, such as growth phase, culture conditions and time of harvesting (Safi et al. 2014b; Günerken et al. 2015). Longer processing times and harsher operating conditions denature the shear-sensitive biomolecules (Gharibzahedi et al. 2013; Maran et al. 2015). In addition, disruption of cells for a longer time results in the formation of fine cell debris and the release of contaminant proteins, which pose problems during further downstream processing. The conditions used to disrupt cells also depend on the location of the biomolecules. Hence, the cell disruption conditions should be standardized precisely so as to achieve the extraction of exactly the desired biomolecule (selectivity) (Fonseca and Cabral 2002). Such efficient disruption (ideal condition) will release maximum target biomolecules (without compromising purity) and have a minimum amount of fine cell debris while preserving its integrity (Follows et al. 1971).

Accordingly, this chapter aims to provide useful information on various mechanical and non-mechanical cell disruption methods, their mechanism of operation, advantages and disadvantages, their application for microalgal cell disruption, recent advancements and future perspectives.

5.2 CELL DISRUPTION METHODS

Current algal cell disruption methods can be classified into two categories: mechanical and non-mechanical. A few of the most common mechanical and non-mechanical methods are presented in Figure 5.1.

5.2.1 MECHANICAL METHODS

A few conventionally used mechanical methods are milling, homogenization and ultrasonication. Mechanical methods are simple and cost-effective and can be universally applied to all microalgae species (Kim et al. 2013). The cells are broken by physical force.

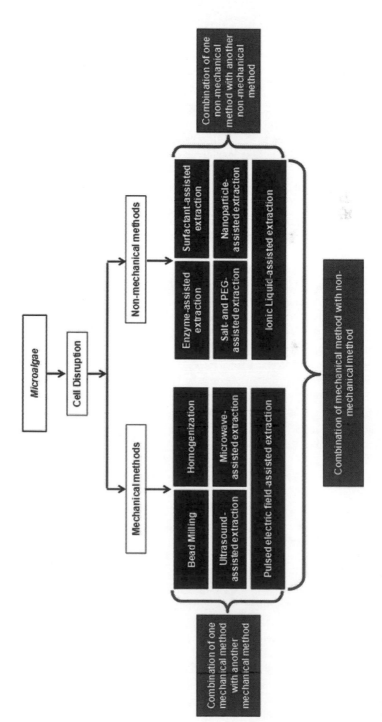

FIGURE 5.1 Different methods for cell disruption of microalgae.

5.2.1.1   Bead Milling

Bead mills are considered one of the most effective methods for microalgal biomass disruption. Disruption occurs because of the impact of grinding beads against the cells.

Agitated bead and shaking vessel are the two basic types of bead mills. The former is the most commonly used. It consists of a fixed vessel containing a rotating agitator of different designs (rings or eccentric or concentric disks). The designs and detailed configurations are reported elsewhere (Middelberg 1995). The vessel will be filled with biomass and beads. The cell disruption is achieved due to shear forces or compaction (MacNeill et al. 1985; Melendres et al. 1992; Bunge et al. 1992), during which the energy transfer occurs from the rotating shaft to cells passing through beads (MacNeill et al. 1985; Currie et al. 1972; Garrido et al. 1994; Halim et al. 2012; Lee et al. 2012; Postma et al. 2015). During this transfer of energy, a good amount of energy is transformed into heat because of friction between the shaft, beads and cells. This undesired increase in medium temperature is expected to degrade thermolabile biomolecules. In order to overcome this problem, an effective cooling mechanism is required. The combination of the collision, agitation and grinding of the beads produces a more effective disruption process.

The disruption efficiency, energy consumption and overall cost depend on the load filling (Ricci-Silva et al. 2000; Mei et al. 2005), diameter, shape and composition of beads, rotational speed (Van Gaver and Huyghebaert 1991; Melendres et al. 1992), design of the agitator and vessel, feed flow rate, dry cell weight (DCW) concentration and rheological properties of the feed (Lee et al. 2012). Low-density beads (e.g., glass) are preferred for low-viscous media (Hedenskog et al. 1969; Schuette and Kula 1990; Doucha and Lívanský 2008; Lee et al. 2012; Günerken et al. 2015). The higher the hardness and density of the bead, the higher the disruption efficiency; therefore, these are used for media with high viscosity. Beads made of titanium carbide, zirconium oxide or zirconia-silica (Seetharam and Sharma 1991) are the most commonly used for high-viscosity media. Bead loading with the volume fraction of ~0.5 and bead diameter of 0.5 mm are considered optimal. It is reported that extraction efficiency is directly proportional to the bead diameter in the range of 0.1 to 0.5 mm; it has a negative effect when increased above 0.5 mm (Doucha and Lívanský 2008), and hence the same is considered optimal for microalgal disruption (Lee et al. 2012). A few of the case studies on the extraction of biomolecules from microalgae using bead milling are presented in Table 5.1.

Bead mills offer many advantages such as high efficiency, low labor intensity, easy scale-up and commercial-scale equipment is readily available (Bierau et al. 1999; Jahanshahi et al. 2002; Show et al. 2015; Günerken et al. 2015). Separation of beads from the medium is based on gravity and does not take extra energy (Lee et al. 2012). Most of the works are reported on the extraction of lipids, which are stable at higher temperatures (Shen et al. 2009). For instance, *Scenedesmus dimorphus* biomass subjected to bead milling before hexane extraction resulted in 4 times higher yield compared to intact biomass. Similarly, 3.5 times higher yield was observed in the case of *Chlorella protothecoides*. However, this equipment is not energy efficient because of inefficient energy transfer to the individual cells from the rotating

TABLE 5.1
Mechanical Methods for the Extraction of Biomolecules From Microalgae

Species	Target Biomolecule	Operating Condition	Efficiency	Reference
Bead milling				
Tetraselmis sp.	Protein	Beads: ceramic Bead diameter: 0.3–0.6 mm Bead filling: 65% DCW concentration: 12% Feed flow rate: 1.5 l/min Retention time: 30 min	Extraction efficiency: 21%	(Schwenzfeier et al. 2011)
Scenedesmus quadricauda (fresh and spray dried biomass)	Disrupted biomass	Beads: Ballotini Bead filling: 33% DCW concentration: 5% Retention time: 5 min Agitator speed: 2800 rpm	Cell disruption efficiency: 55% for fresh biomass and 87% for spray dried biomass	(Hedenskog et al. 1969)
Chlorella sp.	Disrupted biomass	Beads: glass Bead diameter: 0.42–0.58 mm Bead filling: 82% DCW concentration: 10.7% Feed flow rate: 3 kg/h	Cell disruption efficiency: 99.9% and 90.2% for *Chlorella* and bacteria, respectively.	(Doucha and Lívanský 2008)
High-pressure homogenization				
Nannochloropsis sp.	Protein	Pressure: 1500 bar Passes: 6 DCW concentration: 1%	Extraction efficiency: 91%	(Grimi et al. 2014)
Chlorococcum sp.	Disrupted biomass	Pressure: 850 bar Passes: 4 DCW concentration: 0.85%	Cell disruption efficiency: 90% 83% of colony diameter reduction after first pass	(Halim et al. 2012)
Nannochloropsis oculata	Disrupted biomass	Pressure: 2100 bar Passes: 3 DCW concentration: 0.015–0.023%	Cell disruption efficiency: 100%	(Samarasinghe et al. 2012)

(Continued)

TABLE 5.1 (*Continued*)

Species	Target Biomolecule	Operating Condition	Efficiency	Reference
Nannochloropsis oculata	Lipid	Pressure: 125 MPa, pH: 6.0 Extraction solvent: Petroleum ether	Petroleum ether was observed to be the best solvent	(Shene et al. 2016)
High-speed homogenization				
Arthrospira platensis	C-Phycocyanin	Speed: 10,000 rpm Pre-soaking time: 120 mins S/L ratio: 1:6 Extraction time: 10 mins	Extraction efficiency: 44%	(Tavanandi et al. 2018a)
Arthrospira platensis	Allophycocyanin	Speed: 10,000 rpm Pre-soaking time: 120 mins S/L ratio: 1:6 Extraction time: 10 mins	Extraction efficiency: 48%	(Tavanandi et al. 2018b)
Phaeodactylum tricornutum	Antioxidant	Speed: 14,000 rpm DCW concentration: 0.12% Extraction time: 0.5 mins Extraction solvent: EtOH (or MetOH)/water solvent (1:1, v/v)	30 mg and 22.5 mg equivalent ascorbic acid/l antioxidant activity for ethyl alcohol and methanol, respectively	(Guedes et al. 2013)
Nannochloropsis sp.	Lipid	Speed: 10,000 rpm DCW concentration: 6% Extraction time: 1 min	Extraction efficiency: 78%	(Wang and Wang 2012)
Nannochloropsis sp.	Lipid	Speed: 12,000 rpm S/L ratio: 1:50 (g/ml) DCW concentration: 2%	Extraction efficiency: 38%	(Rajesh Kumar Balasubramanian et al. 2013)
Ultrasonication				
Arthrospira platensis	C-Phycocyanin	Pre-soaking time: 120 mins Amplitude: 50%, S/L ratio: 1:6 Ultrasonication time: 2.5 mins followed by freezing and thawing	Extraction efficiency: 92%	(Tavanandi et al. 2018a)

Species	Target Biomolecule	Operating Condition	Efficiency	Reference
Arthrospira platensis	Allophycocyanin	Pre-soaking time: 120 mins Amplitude: 50%, S/L ratio: 1:6 Ultrasonication time: 3.5 mins followed by 4 cycles of freezing and thawing	Extraction efficiency: 94% Sequential extraction of C-PC and A-PC achieved	(Tavanandi et al. 2018b)
Stichococcus sp.	Chlorophyll a	Power: 70 W, Cycles: 3 cycles of 90 s with 5 min breaks	Local heat caused degradation of chlorophyll a	(Schumann et al. 2005)
Arthrospira platensis	Total chlorophylls	Pre-soaking time: 120 mins Amplitude: 50%, S/L ratio: 1:6 Ultrasonication time: 3.5 mins followed by 4 cycles of freezing and thawing Low humidity dried spent biomass followed by ethanol extraction at 50°C.	Extraction efficiency: 80% Low humidity dried spent biomass was able to perform at par with the freeze dried sample	(Tavanandi and Raghavarao 2019)
Scenedesmus obliquus	Fermentable sugars	Power: 200 W Cycles: 5 cycles of 30 s with 10 min breaks DCW concentration: 7–10%	Yield: 0.025 equal gram of glucose/ gram biomass Complex sugars were converted to fermentable sugars	(Miranda et al. 2012)
Nannochloropsis salina	Anaerobic digestion and Biogas from treated Biomass	Power: 200 W Ultrasonication time: 45 s Ultrasonication frequency: 30 kHz	21% decrease in biogas production in comparison with untreated biomass	(Schwede et al. 2011)
Scenedesmus dimorphus	Lipid	Power: 100 W Ultrasonication time: 2 mins Cycles: 2	Extraction efficiency: 21%	(Shen et al. 2009)
Chlorella vulgaris	Lipid	Power: 600 W Cycles: 34 cycles with 5 second breaks	5.11-fold more yield than intact cells	(Zheng et al. 2011)

(Continued)

TABLE 5.1 (*Continued*)

Species	Target Biomolecule	Operating Condition	Efficiency	Reference
Nannochloropsis sp.	Lipid	Power: 400 W Ultrasonication time: 5 min Ultrasonication frequency: 24 kHz Extraction solvent: Ethanol and dimethyl sulfoxide	Higher yields achieved	(Parniakov et al. 2015)
Microwave				
Arthrospira platensis	Phycocyanin	Microwave power: 90W Extraction time: 25 mins pH: 7	Extraction Yield: 4.5 mg/mL Highest purity of 1.27	(Vali Aftari et al. 2015)
Dunaliella tertiolecta	Pigments	Microwave power: 50 W Extraction temperature: 56 °C Extraction time: 5 min DCW concentration: 0.16% Extraction solvent: acetone	Chlorophyll-a yield: 4.5 µg/l Chlorophyll-b yield: 1.4 µg/l, β-carotene yield: 1.3 µg/l	(Pasquet et al. 2011)
Chlorella vulgaris	Lipid	Microwave frequency: 2450 MHz Extraction temperature: 100 °C Extraction time: 5 min	3.875-fold more extraction than untreated cells	(Zheng et al. 2011)
Scenedesmus obliquus	Lipid	Microwave power: 1.2 kW Microwave frequency: 2.45 GHz Extraction temperature: 95 °C Extraction time: 30 min DCW concentration: 7.6% Extraction solvent: Hexane	Recovery:77%	(Sundar Balasubramanian et al. 2011)
Nannochloropsis sp.	Lipid	Microwave power: 0.5 kW Extraction temperature: 65 °C	Microwaves induced permeabilization contribute up to	(Teo and Idris 2014)

Species	Target Biomolecule	Operating Condition	Efficiency	Reference
		Extraction solvent: Chloroform, dichloromethane and sodium sulfate	45% of the total yield compared to microwave induced temperature increase	

Pulsed electric field

Species	Target Biomolecule	Operating Condition	Efficiency	Reference
Arthrospira platensis	Phycocyanin	Electric field strength: 25 kW/cm, Extraction temperature: 40 °C Extraction time: 150 µs	152 mg/g dry biomass translating into extraction efficiency of 70%	(Martínez et al. 2016)
Nannochloropsis salina	Protein	Electric field strength: 15.44–30.89 kWh/kg Extraction temperature: 37 °C (overflow) DCW concentration: 0.0545–0.109% Extraction solvent: Water and Methanol	Four-fold more extraction with water than methanol extraction of untreated cells	(M Coustets et al. 2013)
Synechocystis PCC 6803	Lipid	Electric field strength: 59.67–239 kWh/kg Extraction temperature: 36–54 °C (overflow) DCW concentration: 0.03%	Cell disruption efficiency: 99%	(Sheng et al. 2012)
Auxenochlorella protothecoides	Lipid	Electric field strength: 0.42–0.63 kWh/kg DCW concentration: 10% Extraction solvent: Ethanol	Recovery: 70%	(Eing et al. 2013)
Ankistrodesmus falcatus	Lipid	Electric field strength 5.8 kWh/kg DCW concentration: 0.19% Extraction solvent: Methanol-ethyl acetate	Two times more Yield than untreated biomass	(Zbinden et al. 2013)

shaft. Dissipation of energy into heat results in an increase in the temperature of the media and demands energy-intensive cooling (only in the case of thermolabile biomolecules) and accessories such as cooling jackets, which again adds to the cost. Not much work is reported for application of bead milling for the extraction of thermolabile biomolecules such as C-PC and allophycocyanin (A-PC). Recently, bead milling was employed for the extraction of thermolabile proteins from *Chlorella vulgaris*, where lower extraction efficiency was observed when experiments were performed at 35 °C (Postma et al. 2015). Because this method is shear based, it is non-selective and results in very fine cell debris (Middelberg 1995), which poses a difficulty in downstream processing.

5.2.1.2   Homogenization

i) High-Pressure Homogenization

High-pressure homogenizers (HPHs) are the most preferred equipment for the industrial-scale cell disruption of microalgae. The initial attempts of HPH for the cell disruption of microorganisms was performed on yeast cells (Brookman 1974; Engler 1985; Middelberg 1993). The equipment consists of positive displacement pumps (one or two), homogenizer valve, valve seat and impact ring (Middelberg 1995; Kelly and Muske 2004). The pump forces the feed under high pressure to hit a ring inside the valve seat. The force of the impact is determined by the pressure exerted by the pump. The cell disruption occurs due to a combination of multiple effects, such as (a) mechanical effects because of the high pressure impact of cells on the impact ring (Middelberg 1995; Kelly and Muske 2004), including turbulence (Engler and Robinson 1981); (b) shear stress (Miller et al. 2002); (c) bursting of gas bubbles (because of the sudden decrease in pressure) inside the cells (Clarke et al. 2010); and (d) collapse of cavitation bubbles due to decreased flow velocities (Save et al. 1994). During cavitation, very fine bubbles are formed, which then grow and collapse. All these steps happen in a fraction of a second (micro to milliseconds). A large amount of energy per unit volume is released with the collapsing of bubbles, which results in cell disruption (Petrier et al. 1998).

The extraction efficiency of HPH is influenced by different process parameters, such as working pressure, process temperature, DCW concentration, the orifice and valve design, number of passes and the feed flow rate. The cell disruption efficiency is influenced also by cell wall rigidity (Engler and Robinson 1981; Naganuma et al. 1984; Spiden et al. 2013), which varies with the species and growth phase at which the cells are harvested. The extraction efficiency of HPH in a few cases is reported to be very high, as much as 80% to 90% (Balasundaram and Pandit 2001; Halim et al. 2012). The efficiency is directly proportional to the pressure (Follows et al. 1971; CR Engler 1985) and the number of passes. However, these results can be considered more theoretical. The increase in pressure and number of cycles result in an increase in the temperature of suspension by 2°C per pass for every 10,000 kPa of pressure applied (Becker et al. 1983). This is not suitable for thermolabile biomolecules, which demand instant cooling (energy-intensive operation) (Komaki et al. 1998; Janczyk et al. 2005). A few of the case studies on the extraction of biomolecules from microalgae using HPH is presented in Table 5.1.

HPH offers many advantages, such as ease of operation and scale-up and can be applied for high DCW feed (Middelberg 1995; Olmstead et al. 2013; Günerken et al. 2015). The non-selective intracellular compound release, generation of very fine cell debris, release of contaminants from the cytoplasm and cellular organelles along with target biomolecules (Balasundaram et al. 2009; Jamshad and Darby 2012) increase the cost of downstream processing. Large-scale homogenizers are expensive and have high maintenance costs. HPH for the extraction of thermolabile biomolecules require cooling, increasing the energy requirement for the process (Diels and Michiels 2006). As a result, this process is more suitable for extraction of lipids from microalgae when no thermolabile by-products are expected from the process (Eugene and Hackett 2000; Lee et al. 2010).

ii) High-Speed Homogenization

High-speed homogenizers (HSHs) are one of the simplest and most effective cell disruption methods. HSHs are also known as rotor-stator homogenizers and are already used for the extraction of different biomolecules (Khoo et al. 2011; González-Delgado and Kafarov 2012; Wang and Wang 2012; Rajesh Kumar Balasubramanian et al. 2013; Guedes et al. 2013). A few of the case studies on the extraction of biomolecules from microalgae using HSH is presented in Table 5.1. These are stirring-type devices consisting of a high-speed rotor enclosed within a stator. The mechanism of cell disruption is the combination of hydrodynamic cavitation resulting because of stirring and mechanical forces, such as turbulence or shear at the solid–liquid interphase. The efficiency of cell disruption is affected by different process parameters, such as the diameter and design of the rotor-stator probes, feed viscosity, flow rate and DCW concentration.

This method offers many advantages, such as short contact times, and can handle feeds with relatively high DCW concentrations (reduces downstream processing cost). In addition, the equipment can be easily adapted for batch and continuous operations (Maa and Hsu 1996). But this equipment is suitable only for feeds of viscosity less than 10 Pas and demand specially designed homogenization accessories to handle feeds of higher viscosity. One way of increasing the extraction efficiency is by combining this method with glass or plastic beads (Mogren et al. 1974). But this poses a problem of contamination of the product by small glass or plastic particles produced as a result of wear between the beads and the rotor-stator probe (Goldberg 2008; Charinpanitkul et al. 2008). The shear generated leads to an increase in local and bulk temperature, resulting in the denaturation of thermolabile biomolecules.

5.2.1.3 Ultrasound-Assisted Extraction

Ultrasonication (US) is another efficient mechanical cell disruption method based on high shear force created by high-frequency ultrasound. Cell disruption by US occurs because of the combined effect of cavitation, microscale eddies and free radical reaction. High-energy ultrasonic waves propagate through the medium and create tiny, unsteady cavitation bubbles around the cell. These bubbles then grow, and after reaching a certain volume fail to absorb any further energy and eventually collapse (implode), producing extreme shock waves. These shock waves travel with a velocity of over 400 km/h and exert pressure equivalent to 100 MPa, resulting in increased temperatures as high as 5000°C, which ultimately leads to cell disruption. The cavitation also results in highly reactive free radicals (H•, HO• and HOO•) around the bubbles produced because of the thermolysis of water. These free radicals react with cells, enhancing the cell disruption (Suslick and Price 1999; Guangming Zhang and Hua 2000; Brujan and Williams 2006; Lee et al. 2010).

Sonicators are of two types, horn and bath (Hosikian et al. 2010). The equipment for both laboratory and industrial scale, which can be operated on continuous mode, is available (Hosikian et al. 2010; Gerde et al. 2012). The efficiency of US to disrupt cells depend on different parameters, such as operation temperature (Condón et al. 2005; Zhang et al. 2014), viscosity of the cell suspension (Popinet and Zaleski 2002), DCW concentration, US time (Ellwart et al. 1988; Cravotto et al. 2008), US amplitude and output power (Kapturowska et al. 2007; Martín et al. 2013; Dan Liu et al. 2013; Li Zhang et al. 2014) and duty cycle (Borthwick et al. 2005; Cacciola et al. 2013; Ferraretto et al. 2013; Dan Liu et al. 2013).

Ultrasound-assisted extraction (UAE) is proven to be better than other extraction methods in many literature reports (Natarajan et al. 2014; Grimi et al. 2014; Keris-Sen et al. 2014; Zhang et al. 2014; Meng Wang et al. 2014; Safi et al. 2014a). A few of the case studies on the extraction of biomolecules from microalgae using US is presented in Table 5.1. Wang et al. compared the performances of high-frequency focused and low-frequency non-focused US for the extraction of lipid from *Nannochloropsis oculata* and *Scenedesmus dimorphus*. It was inferred that better cell disruption efficiencies were achieved when low- and high-frequency techniques were applied in series than the individual treatments under the same operating conditions (Meng Wang et al. 2014). In our work on C-PC extraction from *Arthrospira platensis* dry biomass, US in combination with other methods resulted in the highest extraction efficiency of 92% (Tavanandi et al. 2018a). Another accessory pigment, A-PC, in spite of having similar value and applications as C-PC has not received the required attention. About 50% to 60% of A-PC is left in spent biomass. US in combination with other methods was attempted, and extraction efficiency of ~92% could be achieved. Sequential extraction of A-PC after the release of C-PC was observed where ~40% of A-PC extracted was devoid of C-PC, clearly indicating that US was able to break the polypeptides that link C-PC with A-PC. The positive effect of US on the extraction of chlorophyll was clearly observed when the spent biomass left after extraction of C-PC and A-PC was utilized for recovery of chlorophylls. The chlorophyll yield of 9.49 mg/g, db achieved was significantly higher than the untreated sample, which resulted in a yield of 6.44 mg/g, db (Tavanandi and Raghavarao 2018). Similar results were observed in case of extraction of carotenoids from spent biomass as well (Tavanandi et al. 2019). The positive effect of cell disruption by UAE on the extraction of biomolecules is well studied (Simon and Helliwell 1998; Macías-Sánchez et al. 2009). The yield achieved by UAE was observed to be higher when compared to conventional methods like maceration (Cha et al. 2009) and grinding (Simon and Helliwell 1998). Attempts have been made to increase the efficiency of US by adding particles such as beads or quartz sand of a standard bead size and bead-to-biomass ratio (Wiltshire et al. 2000).

This technique is simple and can be easily integrated with other methods. The difficulty of separating out the fine cell debris is not there in the case of ultrasound-assisted methods as this results in only the permeabilization of the cells (Günerken et al. 2015). Higher yields can be achieved in short extraction times with relatively lower volumes of solvent, as US increases the penetration of solvent by a sonocapillary effect (Parniakov et al. 2014; Show et al. 2015; Tavanandi et al. 2018a). UAE is an environmentally friendly technique (Halim et al. 2012).

During US, the temperature of the sample increases significantly (50 to 90 °C) (Chandler et al. 2001; Taylor et al. 2001; Borthwick et al. 2005; Panyue Zhang et al. 2007), and this condition is not suitable for the extraction of heat-sensitive biomolecules (Sartory and Grobbelaar 1984; Suslick 1989; Borthwick et al. 2005; Schumann et al. 2005; Gogate and Pandit 2008). Hence, extraction is carried out under cold conditions by placing the samples in an ice bath (Piasecka et al. 2014), increasing the energy consumption, which adds to the cost. These factors are the bottlenecks for the industrial adoption of this method.

5.2.1.4 Microwave-Assisted Extraction

Microwaves have been used for the extraction of intracellular products (IL Choi et al. 2006), and a few of the case studies on the extraction of biomolecules from microalgae using microwave-assisted methods are presented in Table 5.1. It is applicable only when polar solvents such as water are used as a medium. Cell disruption occurs because of the water vapor produced within the cell due to intracellular heating caused when a polar material is exposed to microwaves. The intracellular heating is a result of frictional forces generated from inter- and intra-molecular movements (Rosenberg and Bogl 1987; Amarni and Kadi 2010). The recovery of intracellular biomolecules is because of the combined effect of pressure and local heat, along with cell membrane/wall damage induced by the microwaves (Rosenberg and Bogl 1987; Choi et al. 2006). Microwave-induced heating generates steam that ruptures the cell membrane/wall, releasing the intercellular contents (Mandal et al. 2007). The efficiency of this method to disrupt cells depends on DCW concentration, feed viscosity, energy input, solvent used, extraction temperature and time. Microwave treatment at 2.5 GHz is considered to be the most suitable frequency for cell disruption (Vasavada 1986).

The microwave-assisted extraction (MAE) method was reported to be superior to UAE in the case of extraction of C-PC from *A. platensis*. Both yield and purity of C-PC achieved by MAE were observed to be higher than UAE (Vali Aftari et al. 2015). MAE prior to Bligh–Dyer extraction was observed to be an efficient method for the recovery of oil from wet biomass of *C. vulgaris, Botryococcus* sp. and *Scenedesmus* sp. (Jae-Yon Lee et al. 2010) compared to other methods. Similarly, extraction efficiency of ~80% could be achieved for the recovery of oil from microwave pre-treated *Scenedesmus obliquus* biomass when hexane was used as the extraction solvent (Sundar Balasubramanian et al. 2011). The increase in microwave treatment time (up to 20 minutes) had a positive effect on the extraction of biomolecules, as a significant increase in the pore diameter and thickness was observed in the case of *Chlorella* sp. (Cheng et al. 2013). Microwave irradiation was applied for lipid extraction from *Nannochloropsis* sp. by different solvents and their mixtures (Teo and Idris 2014). The ratio of water present in the cell to that present in the surrounding medium is too small. The microwave-induced electroporation contribute up to 45% of the total yield when compared to microwave-induced temperature increase (Sundar Balasubramanian et al. 2011). Because microwave-induced heat contributes more to the yield, as it is mainly dependent on the water content, feeds with low DCW concentration are more suitable (Chuanbin et al. 1998; Amarni and Kadi 2010).

This method offers an advantage in terms of higher extraction efficiency with relatively low solvent and energy requirements (Amarni and Kadi 2010; Cheng et al. 2013). This method, besides being simple (Jae-Yon Lee et al. 2010), is very effective, robust and easy to scale up (Sundar Balasubramanian et al. 2011). One of the disadvantages is that it is applicable only for polar solvents. Because the method involves an increase in temperature, it is not suitable for volatile (Zheng et al. 2011) and/or thermolabile biomolecules. Because feeds with low DCW concentrations are more suitable, the extent to which the energy is absorbed by the medium is very high. This energy gets dissipated into heat (Lee et al. 2012), resulting in denaturation and functionality loss of bioactive compounds (Woo et al. 2000). Hence, this method is usually performed for a shorter duration prior to other methods to achieve higher efficiency.

5.2.1.5   Pulsed Electric Field–Assisted Extraction

Pulsed electric field (PEF) is a relatively recent technique for microalgal cell disruption (Saulis 2010). Electroporation of cell membranes is achieved by an external electric field, which induces a critical electrical potential across the cell membrane/wall. This potential results in electric field-induced tension and electromechanical compression, which creates permeabilization of the cell membrane (Zimmermann et al. 1985; Tsong 1990; Weaver and Chizmadzhev 1996; Ho and Mittal 1996; Barbosa-Canovas et al. 1999). Reports are available wherein it is shown that the permeabilization by the PEF method can be reversible or irreversible (Weaver et al. 1988; Rols et al. 1990; Tsong 1990), depending on the ratio of pore size and total surface area of the membrane/wall. PEF is applied for the extraction of different biomolecules from microalgae (Sheng et al. 2012; Mathilde Coustets et al. 2015; Luengo et al. 2015; Parniakov et al. 2015; Martínez et al. 2016), and a few of the case studies on the extraction of biomolecules from microalgae using the PEF-assisted method is presented in Table 5.1. The efficiency is affected by different process parameters, such as pulse waveform, applied energy, voltage, treatment time, etc. (Vorobiev and Lebovka 2009). Standardization of these parameters is very important for the economic feasibility of PEF (Mathilde Coustets et al. 2015; Martínez et al. 2016). It is reported that the size and number of the pores (Günerken et al. 2015) are directly proportional to the electric field strength, pulses and DCW concentration (Goettel et al. 2013), which in turn affect disruption efficiency.

PEF was applied for the extraction of biomolecules from *Auxenochlorella protothecoides,* and it was observed that PEF-treated samples could result in six times higher yields than the untreated biomass (Goettel et al. 2013). In the case of extraction of lipids from *Scenedesmus* sp., the yield achieved at standardized conditions is three times higher than the untreated samples (Lai et al. 2014). Similar results were observed during the extraction of lutein from *Chlorella vulgaris* (Luengo et al. 2015). PEF was successfully used also for the selective extraction of biomolecules (Zuckerman et al. 2002; Goettel et al. 2013; Grimi et al. 2014; Luengo et al. 2015; Parniakov et al. 2015).

The advantages of this method are (a) it offers an advantage of combining it with different methods and (b) it is easy to scale up. Because it is a cell permeabilization–based method, separation of cell debris is not difficult, as in the case of HPH and bead milling (Goettel et al. 2013; Martínez et al. 2016). This method can be used for the extraction of thermolabile biomolecules such as pigment, protein, and carbohydrates. The main disadvantage of this method is that the medium should be free from ions, limiting its application for processing marine microalgae, as it requires thorough washing and deionization. Further, the compounds released during extraction increase the conductivity of the medium, reducing the efficiency of PEF to disrupt cells. PEF is more effective for the release of thermolabile membrane proteins and enzymes (Ganeva et al. 2003) than for lipids (Shynkaryk et al. 2009; Vanthoor-Koopmans et al. 2013).

5.2.2 NON-MECHANICAL METHODS

Mechanical methods have been observed to result in very good yield but often with lower purity (Tavanandi et al. 2018a; Tavanandi et al. 2018b). This is because the biomass will be left completely broken (instead of permeabilization), resulting in the release of other contaminant proteins from the cytoplasm and other organelles of cells. The non-mechanical cell disruption methods have become a favorite procedure and are an easier alternative to physical disruption. Some examples of non-mechanical cell disruption methods are based on chemical treatments (like surfactants) (Pavlić et al. 2005; Ebenezer et al. 2012; Ulloa et al. 2012; Huang and Kim 2013) and enzymatic treatments (Fu et al. 2010; Jin et al. 2012; Cho et al. 2013). Standardization of primary extraction by non-mechanical methods for the extraction of C-PC from wet biomass of *Spirulina* sp. employing methods such as lysozyme enzymes (Boussiba and Richmond 1979) and enzymes from *Klebsiella pneumonia* bacteria (Zhu et al. 2007) was reported. These methods are perceived as more selective with respect to the target protein/biomolecules when compared to mechanical methods. Surfactants and the enzymatic methods offer the advantage of the selective extraction of target biomolecules because of their interaction with the lipid bilayer of the cell wall/membrane (Vogels and Kula 1992).

5.2.2.1 Enzyme-Assisted Extraction

One of the most popular non-mechanical cell disruption methods is enzyme-assisted extraction, which works on the principle of using digestive enzymes to decompose the microbial cell membrane/wall. Enzyme-assisted extraction is a promising, eco-friendly and non-destructive method that can prevent thermally sensitive biomolecules from degradation. The kind of cell walls and membranes vary with different cell types and strains, and hence the enzymes used vary from one microorganism to the other.

Lysozyme was used for the cell disruption of *C. vulgaris* during the extraction of lipids, and it was observed that yields were 7.46-fold more compared to untreated biomass (Zheng et al. 2011). Lysozyme was successfully used for the isolation of alkane hydroxylase from *Pseudomonas putida* (Chisti and Moo-Young 1986). Different methods of extraction of C-PC (sonication, enzymatic and maceration) were examined, where maximum concentration (281 mg/g) was achieved in the case of enzyme-assisted extraction at a lysozyme concentration of 1.5 mg/mL from wet biomass of *Spirulina maxima* (Antonio 2015). Extraction of phycocyanins from wet biomass of *Spirulina platensis* resulted in a purity of 0.9 with a lysozyme concentration of 100 ug/mL and 10 mm ethylenediaminetetraacetic acid (EDTA) (Boussiba and Richmond 1979). Although it has several advantages, there are disadvantages as well. The enzyme's high price, especially in purified form, limits its industrial application. Enzyme-assisted methods suffer from the drawback of low productivity (partly due to the high process time) and product inhibition (Harun and Danquah 2011). The enzyme (when used for cell disruption), itself being a protein, acts as a contaminant and complicates downstream processing. These disadvantages can be overcome by using immobilized enzymes (Crapisi et al. 1993). A few more case studies on the extraction of biomolecules from microalgae using the enzyme-assisted method is presented in Table 5.2.

TABLE 5.2

Non-mechanical Methods for the Extraction of Biomolecules From Microalgae

Species	Target biomolecule	Cell-disruption condition	Efficiency	Reference
Enzymatic-assisted extraction				
Haematococcus pluvialis	Astaxanthin	Enzyme: Protease K and driselase Enzyme concentration: 0.1% protease K and 0.5% driselase Incubation time: 1 h pH: 5.8 Extraction temperature: 30 °C	1.65-fold more extraction than untreated cells	(Mendes-Pinto et al. 2001)
Chlamydomonas reinhardtii UTEX 90	Dextrin	Enzyme: Protease Thermostable α-amylase Enzyme concentration: 0.005% Incubation time: 30 mins Extraction temperature: 90 °C	Yield: 25.21 g/l	(Seung Phill Choi et al. 2010)

Species	Target biomolecule	Cell-disruption condition	Efficiency	Reference
Chlorella vulgaris	Lipid	Enzyme: Cellulase Enzyme concentration: 5 mg/l Incubation time: 10 h pH: 4.8 Extraction temperature: 55 °C	8.1-fold more extraction than untreated cells	(Zheng et al. 2011)
R. toruloides	Lipid	Microwave pretreatment: 1 min, Enzyme: b-1,3-glycomannanase Incubation time: 2 h Extraction temperature: 37 °C DCW concentration: 12.80% Extraction solvent: Ethyl acetate	Recovery: 96.6%	(Jin et al. 2012).
Scenedesmus sp.	Lipid	Enzyme: Cellulase: pectinase: hemicellulase ratio of 1:1:1 Incubation time: 60 h Extraction temperature: 30 °C DCW concentration: 0.25% Extraction solvent: Hexane	Recovery: 86.1%	(Huo et al. 2015)
Nannochloropsis sp.	Lipid	Enzyme: cellulose, protease, pectinase, lysozyme Extraction solvent: Chloroform	Yield: 198 mg/g	(Wu et al. 2017)
Surfactant-assisted extraction				
Chlorella vulgaris	Lipid	Surfactant: Sodium dodecyl benzene sulfonate in presence of 2% H_2SO_4	Yield: 843.9 mg/g	(Park et al. 2014)

(Continued)

TABLE 5.2 (*Continued*)

Species	Target biomolecule	Cell-disruption condition	Efficiency	Reference
		Surfactant concentration: 0.2% Incubation temperature: 120 °C Incubation time: 1 h Extraction solvent: hexane/ methanol (7:3, v/v)		
Salt-based method				
Chlorella sp.	Lipid	Salt: 29 mM $Fe_2(SO_4)_3$ or 41 mM $FeCl_3$ with 2% H_2O_2 Incubation temperature: 120 °C Incubation time: 90 min DCW concentration: 2% Extraction solvent: hexane	Recovery: 89.5%–94.5%	(Dong-Yeon Kim et al. 2015)
Chlorella sp.	Lipid	Salt: 2 mM $K_2S_2O_8$ and 0.5% H_2O_2 Incubation temperature: 90 °C Incubation time: 60 min DCW concentration: 1.7% Extraction solvent: Chloroform	Recovery: 95%	(Yeong Hwan Seo et al. 2016b)
Autrantiochytrium	Lipid	Salt: Poly-dimethylamino methylstyrene (pDMAMS) membrane Incubation temperature: room temperature Incubation time: 10 min DCW concentration: 1.5% Extraction solvent: Hexane	Recovery: 25.6% yield compared to 0.77% yield in control	(Yoo et al. 2014)

Species	Target biomolecule	Cell-disruption condition	Efficiency	Reference
Nanoparticle-assisted extraction				
Chlorella vulgaris	Lipid	Nanoparticle: NiO (<50 nm) Incubation time: ~100 hours Extraction time: chloroform/ methanol (2:1, v/v)	Yield: 900 mg/g	(Huang and Kim 2016)
Chlorella sp.	Lipid	CTAB-decorated Fe_3O_4, hexane as extraction solvent	Yield: 460 mg/g	(Jung Yoon Seo et al. 2016a)
Ionic liquid–based extraction				
Spirulina platensis	Phycobiliproteins (C-PC, A-PC and phycoerythrin)	Ionic liquid: 2-hydroxy ethylammonium acetate (2-HEAA) and 2-hydroxy ethylammonium formate (2-HEAF) in 1:1 molar ratio pH: 6.50 L/S ratio: 7.93 mLg^{-1} Extraction time: 30 min	Allophycocyanin yield: 6.34 mg/g C-phycocyanin yield: 5.95 mg/g Phycoerythrin yield: 2.62 mg/g	(Rodrigues et al. 2018)
Haematococcus pluvialis	Astaxanthin	Ionic liquid: [Emim] $EtSO_4$ Incubation temperature: 25 °C Incubation time: 1 min Extraction solvent: Acetone/methanol (1:2, v/v)	Yield: 20 pg/g yield	(Praveenkumar et al. 2015)
Chlorella vulgaris	Lipid	Ionic liquid: [Bmim] CF_3SO_3/methanol mixture Incubation temperature: 65 °C Incubation time: 18 h Extraction solvent: methanol	Yield: 189.8 mg/g	(Young-Hoo Kim et al. 2012)

5.2.2.2 Surfactant-Assisted Extraction

Along with the cell wall/membrane composition and the location of the biomolecule of interest, some of the parameters of surfactants, such as type, concentration, hydrophilic-lipophilic balance (HLB) and critical micelle concentration (CMC) numbers of the surfactant, are also decided the efficiency of the permeabilization of cells (Galabova et al. 1996). Non-ionic surfactants possess uncharged and hydrophilic head-groups (Rocha 1999), which cleave lipid–lipid, lipid–protein associations but not protein–protein interactions (Bhairi and Mohan 2007). The majority of them do not alter the functionality of proteins and hence can be construed as mild surfactants and are more preferred for the isolation of membrane proteins. Mild surfactants like Tween 20, Tween 80 and Triton X-100 are used in low concentrations to extract biomolecules in the buffer. These surfactants exhibit very low CMC, providing a gentle environment, and do not affect protein while being effective in the solubilization of the membrane.

HLB is a very important parameter that determines the efficiency of surfactant to solubilize the membrane. HLB is an empirical number, varying from 0 to 20, which determines a surfactant's structural properties. The influence of HLB number on the extraction of biomolecules from fungi and bacteria by different surfactants was well studied (Bernot et al. 2005). The CMC is another important parameter that affects the extraction of biomolecules. It is the lowest concentration of surfactant required above which its monomers self-assemble to form non-covalent aggregates called micelles. Generally, the lower the CMC value of a given surfactant, the more hydrophobic it becomes and the greater its tendency to penetrate into the lipid bilayer even at low concentrations.

The major disadvantage with the surfactant-based extraction method is the variations in estimating the purity because of the presence of a phenyl ring that absorbs ultraviolet (UV) light and hence interferes with UV spectrophotometry. Protein determination at 280 nm in the presence of the surfactant would be imprecise, resulting in variation in estimating the purity. Separating these surfactants from the biomolecules is also difficult.

5.2.2.3 Salt-Based Extraction

Salt changes the osmotic pressure and acts on the cell wall membrane, leading to excessive shock, which results in disruption and cell death. In the area of biotechnical applications, the osmotic shock is used as a technology for lysis of cells, where they are exposed first to either low or high concentrations of salt, leading to osmotic pressure and cell wall degradation. This is due to the fact that water quickly flows from lower to higher salt concentrations. Thus, if cells are exposed first to a high salt concentration, water flows into the cell after exposure to a low salt concentration, resulting in a pressure increase in the cell, which then explodes (Stanbury et al. 2013). On the contrary, when the cells are exposed to a high salt concentration after exposure to a low concentration, water flows out of the cell, leading to cell wall disruption. Different salts such as $CaCl_2$ and $NaNO_3$ have been used for the extraction of biomolecules. Calcium chloride breaks down into ions in aqueous solution, such as a solution the cells are kept in, and the ions create a transient state of permeability in the cell membrane, allowing the small particles to pass through. Silveira et al. in their work achieved a maximum concentration and purity of 3.68 mg mL^{-1} and 0.46, respectively, at a solvent ratio of 1:12 and extraction time of 4 h at 25 °C (Silveira et al. 2007). A few more case studies on the extraction of biomolecules from microalgae using salts are presented in Table 5.2.

5.2.2.4 PEG-Based Extraction

Use of polyethylene glycol (PEG) solutions as a novel solvent has been increasing due to their wide applications in analytical chemistry. The Food and Drug Administration (FDA) has approved PEG for internal consumption because of its benign characteristics. PEGs of lower molecular weight are biodegradable, nonvolatile and have low flammability, unlike organic solvents. On the other hand, PEGs have been found to be stable to acid, base, high temperature, high oxidation and reduction systems. In addition, PEG is miscible with water and organic solvents. PEGs have been found to be more advantageous over ionic liquids in several aspects such as lower cost, low toxicity and non-halogenated. Thus, it is a preferred solvent for aqueous two-phase systems for sample preparation, mainly in separating analytes (Bulgariu and Bulgariu 2008; Griffin et al. 2006) and cloud point extraction (Liang et al. 2009; Zhilong Wang et al. 2008). The application of PEG aqueous solution as a green solvent in MAE was developed for the extraction of flavone and coumarin compounds from medicinal plants by (Zhou et al. 2011). In the extraction process, yield decreases with the increasing molecular weight of PEG, which could be due to the viscosity of PEG with increasing molecular weight.

5.2.2.5 Nanoparticle-Assisted Extraction

Nanoparticles have been used for the extraction of phycocyanins as an alternative approach to conventional methods of extraction. It is a greener, cost-effective and non-laborious method for extraction. The minute size of nanoparticles resulted in a large surface area, which in turn causes the cell wall to react strongly with the microorganisms (Hazani et al. 2013; Wong and Liu 2010). Nanoparticles break down cell wall molecules by forming pits or holes, releasing intracellular molecules (Sondi and Salopek-Sondi 2004; Li et al. 2008). The main composition of the cell wall is a cellulose and protein complex, which is the primary site for nanoparticle interaction (Dash et al. 2012).

Extraction of carbohydrate and oil by using nanoparticles from *C. vulgaris* has been reported (Razack et al. 2016). There was an increase in intracellular oil extraction (8.44% to 17.68%) with an increase in the concentration of silver nanoparticles (50 µg/g to 150 µg/g). Similar results were observed where the maximum yield (13.8%) of carbohydrate was found to be the same with the same concentration of silver nanoparticle. On the other hand, the yield was found to be constant with the concentration of silver nanoparticles above 150 µg/g. Rodea-Palomares et al. 2010 reported that that at high concentrations, nanoparticles break the cyanobacterium *Anabaena* cell wall in a shorter reaction time period (Rodea-Palomares et al. 2010). A few more case studies on the extraction of biomolecules from microalgae using the nanoparticle-assisted method is presented in Table 5.2

5.2.2.6   Ionic Liquid–Based Extraction

Ionic liquids (IL) are organic salts that remain in a liquid state at room temperature. They consist of a smaller organic or inorganic anion coupled with a large asymmetric organic cation (Praveenkumar et al. 2015; Orr and Rehmann 2016). They are also known as 'designer solvents' as the constituents of ILs, that is, cations and anions, can be selected and tailor-made to suit the solubility, hydrophobicity, polarity and conductivity of a specific biomolecule (Young et al. 2010; Grosso et al. 2015). A large number of possible combinations of anions and cations give an opportunity to produce innumerable varieties of ILs (Jampani et al. 2015). Their unique properties, such as high heat capacity, low vapor pressure, extremely low volatility, high chemical and thermal stability, nonflammability, low melting point and high solubility, give them an edge over conventional solvents (Bucar et al. 2013; Negi and Pandey 2015; Orr and Rehmann 2016). Literature reports are available on IL-based, aqueous, two-phase extraction systems where phases are formed by combining an IL solution with a salt solution, thus replacing the polymer component of the conventional polymer-salt two-phase system (A Tavanandi et al. 2015). Many reports have been published on the application of ILs for the purification of biomolecules (Han and Row 2010), antibiotics (Soto et al. 2005), alkaloids (Lu et al. 2008), phenolic compounds (Vidal et al. 2005; Fan et al. 2008) and amino acids (Smirnova et al. 2004). However, recently it was realized that these solvents can be applied for cell disruption as well (Negi and Pandey 2015). They can be designed to solubilize the cell membrane/wall of different microalgae. IL breaks the hydrogen bond network in the cell walls of microalgae, resulting in their solubilization (Kim et al. 2012; Negi and Pandey 2015). A few case studies on the extraction of biomolecules from microalgae using the IL-based method are presented in Table 5.2. ILs were employed for the extraction of natural colors such as phycocyanin and astaxanthin from *Spirulina* sp. (X Zhang et al. 2015) and *Haematococcus pluvialis* (Desai et al. 2016), respectively. Praveenkumar et al. compared IL-based cell disruption with French-pressure cell homogenization for the extraction of astaxanthins and observed that higher yields could be achieved by IL (1-ethyl-3-methylimidaxolium ethylsulfate) within 1 min of treatment time. These results were on par with that obtained by homogenization (Praveenkumar et al. 2015). Rodrigies et al. studied the efficiency of ILs for the extraction of phycobiliproteins (C-PC, A-PC and phycoerythrin) from *Spirulina platensis*. They observed that equimolar mixtures of 2-hydroxy ethylammonium acetate (2-HEAA) and 2-hydroxy ethylammonium formate (2-HEAF) were able to effectively solubilize the membrane at pH 6.50, L/S ratio of 7.93 mL·g^{-1} and extraction time of 30 min (Rodrigues et al. 2018). They also demonstrated the ease of combining IL-based cell disruption with US. There is a possibility of recovering ILs used for cell disruption, and in a study on cell disruption of *Synechocystis* sp., it was observed that the ILs recovered were equally efficient as fresh ILs (Fujita et al. 2013). The disadvantages of ILs are that they are very expensive. A few of ILs are toxic and not environmental friendly (Deetlefs and Seddon 2010).

5.3   COMBINATION OF METHODS

As discussed, the cell disruption efficiency is affected by the cell wall/membrane composition and location of target biomolecules in the cell. Many times, individual methods fail to offer the desired degree of cell disruption. Under such conditions, the synergistic combination of different methods are preferred (Anand et al. 2007). Most of the literature reported is a combination of mechanical and non-mechanical methods, for example, a combination of chemical and mechanical methods (Anand et al. 2007) or enzymatic hydrolysis with mechanical methods (Baldwin and Robinson 1990; Vogels and Kula 1992) and other relatively newer mechanical methods such as electropermeabilization (Ganeva et al. 2015) and US (Priego-Capote and de Castro 2007; Tavanandi et al. 2019). Baldwin et al. in their study on the disruption of commercially available pressed baker's yeast observed that mechanical disruption using a high-pressure homogenizer resulted in only 32% disruption at 95 MPa pressure after four passes, whereas combinations of enzymatic hydrolysis by Zymolyase and high-pressure homogenization resulted in 100% disruption after four passes at 95 MPa (Baldwin and Robinson 1990). Similarly, an improvement in extraction efficiency was observed when the combination of heat and/or enzymatic treatment with mechanical methods such as high-pressure homogenization and wet milling was employed for cell disruption of *Bacillus cereus* (Vogels and Kula 1992). Enzymatic hydrolysis was applied to microwave pre-treated samples. This resulted in lipid recovery of ~97% wherein a five-fold increase in yield was observed compared to enzymatic hydrolysis alone (Jin et al. 2012). The improved yield was attributed to the permeabilization of the cell wall after microwave pre-treatment. The damaged cell wall is expected to be more susceptible to enzyme hydrolysis, and the same observation has been made in other reports also (Okuda et al. 2008; Tan and Lee 2014). Ganeva et al. applied a combination of electropermeabilization and subsequent treatment with a lytic enzyme to achieve the efficient and selective recovery of proteins from yeast (Ganeva et al. 2015). The electropermeabilization facilitated the extraction of a portion of the proteins, and the subsequent addition of lyticase enzyme led to a protein yield of 70%.

A few reports on the combination of one mechanical method with other mechanical methods (Sudar et al. 2013; Luengo et al. 2014; Tavanandi et al. 2018a; Tavanandi et al. 2018b) and electrical methods with mechanical methods (Shynkaryk et al. 2009) are available. A few reports are available wherein combinations of non-mechanical with other non-mechanical primary extraction methods are attempted, for example, a combination of chemical and chemical or enzyme (Hernández et al. 2015). When a combination of two methods are not sufficient to achieve the desired cell disruption, multiple methods are also attempted (Xiao-Yong Liu et al. 2008).

5.4 CONCLUSIONS AND FUTURE PERSPECTIVES

Newer efficient cell disruption methods for microalgae need to be developed, keeping disruption and energy efficiencies in mind. Identifying the most suitable method for cell disruption to extract biomolecules from microalgae is an important step, as it determines the recovery and purity of the target biomolecule. Both mechanical and non-mechanical methods have their own advantages and disadvantages. The former is most preferred in industries when higher recovery is preferred, and the latter when higher purity is required. Mechanical methods like homogenization, maceration and bead milling result in fine cell debris, posing problems in downstream processing. A higher degree of disruption results also in the release of contaminant proteins, which complicates the purification process. On the other hand, non-mechanical methods cannot achieve higher yields. Relatively newer mechanical methods such as US, microwave and pulsed electric field assisted are able to achieve higher yields without compromising the purity of the target biomolecule, but they are energy intensive. As a result, cell disruption methods are carried out in combination to intensify the disruption process. There is a need to develop newer methods for analyzing cell disruption efficiencies, as the present methods based on absorbance, protein content estimation and particle size of the spent biomass are misleading and will not provide the actual results. A detailed study on cell wall/membrane structures of different microalgae is required in order to select the most suitable cell disruption method.

5.5   ACKNOWLEDGEMENTS

Ms. A. Chandralekha Devi and Mr. Hrishikesh A. Tavanandi thank the Council of Scientific and Industrial Research (CSIR), government of India, for providing the fellowship. The authors are thankful to Ms. Hamsavi GK and Ms. Prashi Verma for the kind help during the preparation of the chapter.

REFERENCES

Amarni, F., and H. Kadi. 2010. "Kinetics study of microwave-assisted solvent extraction of oil from olive cake using hexane: Comparison with the conventional extraction." *Innovative Food Science & Emerging Technologies* 11 (2):322–327.

Anand, H., B. Balasundaram, A. B. Pandit, and S. T. L. Harrison. 2007. "The effect of chemical pretreatment combined with mechanical disruption on the extent of disruption and release of intracellular protein from E. coli." *Biochem. Eng. J.* 35 (2):166–173.

Antonio, E. 2015. "C-phycocyanin from Arthrospira maxima LJGR1: Production, extraction and protection." *Journal of Advances in Biotechnology* 5 (2):659–666.

Balasubramanian, R. K., T. T. Y. Doan, and J. P. Obbard. 2013. "Factors affecting cellular lipid extraction from marine microalgae." *Chemical Engineering Journal* 215:929–936.

Balasubramanian, S., J. D. Allen, A. Kanitkar, and D. Boldor. 2011. "Oil extraction from Scenedesmus obliquus using a continuous microwave system—design, optimization, and quality characterization." *Bioresour. Technol.* 102 (3):3396–3403.

Balasundaram, B., and AB Pandit. 2001. "Selective release of invertase by hydrodynamic cavitation. Selective release of invertase by hydrodynamic cavitation." *Biochemical Engineering Journal* 8 (3):251–256.

Balasundaram, B., S. Harrison, and D. G. Bracewell. 2009. "Advances in product release strategies and impact on bioprocess design." *Trends in Biotechnology* 27 (8):477–485.

Baldwin, C., and C. W. Robinson. 1990. "Disruption of Saccharomyces cerevisiae using enzymatic lysis combined with high-pressure homogenization." *Biotechnol. Tech.* 4 (5):329–334.

Barbosa-Canovas, G. V., U. R. Pothakamury, M. M. Gongora-Nieto, and B. G. Swanson. 1999. *Preservation of Foods with Pulsed Electric Fields*. London: Elsevier.

Becker, T., J. R. Ogez, and S. E. Builder. 1983. "Downstream processing of proteins. Downstream processing of proteins." *Biotechnology Advances* 1 (2):247–261.

Bernot, R. J., E. E. Kennedy, and G. A. Lamberti. 2005. "Effects of ionic liquids on the survival, movement, and feeding behavior of the freshwater snail, Physa acuta." *Environ. Toxicol. Chem.* 24 (7):1759–1765.

Bhairi, S. M., and C. Mohan. 2007. Detergents: A guide to the properties and uses of detergents in biology and biochemistry. *San Diego, CA USA Calbiochem, EMD Biosciences* 200.

Bierau, H., Z. Zhang, and A. Lyddiatt. 1999. "Direct process integration of cell disruption and fluidised bed adsorption for the recovery of intracellular proteins." *Journal of Chemical Technology & Biotechnology: International Research in Process, Environmental & Clean Technology* 74 (3):208–212.

Borthwick, K. A. J., W. T. Coakley, M. B. McDonnell, H. Nowotny, E. Benes, and M. Gröschl. 2005. "Development of a novel compact sonicator for cell disruption. Development of a novel compact sonicator for cell disruption." *Journal of Microbiological Methods* 60 (2):207–216.

Boussiba, S., and A. E. Richmond. 1979. "Isolation and characterization of phycocyanins from the blue-green alga Spirulina platensis." *Arch. Microbiol.* 120 (2):155–159.

Brujan, EA., and PR Williams. 2006. "Cavitation phenomena in non-Newtonian liquids. Cavitation phenomena in non-Newtonian liquids." *Chem. Eng. Res. Des.* 84 (4):293–299.

Bucar, F., A. Wube, and M. Schmid. 2013. "Natural product isolation—how to get from biological material to pure compounds." *Natural Product Reports* 30 (4):525–545.

Bulgariu, L., and D. Bulgariu. 2008. "Extraction of metal ions in aqueous polyethylene glycol—inorganic salt two-phase systems in the presence of inorganic extractants: Correlation between extraction behaviour and stability constants of extracted species." *Journal of Chromatography A* 1196:117–124.

Bunge, F., M. Pietzsch, R. Müller, and C. Syldatk. 1992. "Mechanical disruption of Arthrobacter sp. DSM 3747 in stirred ball mills for the release of hydantoin-cleaving enzymes." *Chemical Engineering Science* 47 (1):225–232.

Cacciola, V., I. F. Batllò, P. Ferraretto, S. Vincenzi, and E. Celotti. 2013. "Study of the ultrasound effects on yeast lees lysis in winemaking." *European Food Research and Technology* 236 (2):311–317.

Cha, K. H., H. J. Lee, S. Y. Koo, D.-G. Song, D.-U. Lee, and C.-H. Pan. 2009. Optimization of pressurized liquid extraction of carotenoids and chlorophylls from Chlorella vulgaris." *J. Agric. Food Chem.* 58 (2):793–797.

Chandler, D. P., J. Brown, C. J. Bruckner-Lea, L. Olson, G. J. Posakony, J. R. Stults, N. B. Valentine, and L. J. Bond. 2001. "Continuous spore disruption using radially focused, high-frequency ultrasound." *Anal. Chem.* 73 (15):3784–3789.

Charinpanitkul, T., A. Soottitantawat, and W. Tanthapanichakoon. 2008. "A simple method for bakers' yeast cell disruption using a three-phase fluidized bed equipped with an agitator." *Bioresource Technology* 99 (18):8935–8939.

Cheng, J., J. Sun, Y. Huang, J. Feng, J. Zhou, and K. Cen. 2013. "Dynamic microstructures and fractal characterization of cell wall disruption for microwave irradiation-assisted lipid extraction from wet microalgae." *Bioresource Technology* 150:67–72.

Chethana, S., C. A. Nayak, M. C. Madhusudhan, and K. S. M. S. Raghavarao. 2015. "Single step aqueous two-phase extraction for downstream processing of C-phycocyanin from Spirulina platensis." *Journal of Food Science and Technology* 52 (4):2415–2421. doi: 10.1007/s13197-014-1287-9.

Chisti, Y., and M. Moo-Young. 1986. "Disruption of microbial cells for intracellular products. Disruption of microbial cells for intracellular products." *Enzyme Microb. Technol.* 8 (4):194–204.

Cho, H.-S., Y.-K. Oh, S.-C. Park, J.-W. Lee, and J.-Y. Park. 2013. "Effects of enzymatic hydrolysis on lipid extraction from Chlorella vulgaris." *Renewable Energy* 54:156–160.

Choi, I. L., S. J. Choi, J. K. Chun, and T. W. Moon. 2006. "Extraction yield of soluble protein and microstructure of soybean affected by microwave heating." *Journal of Food Processing and Preservation* 30 (4):407–419.

Choi, S. P., M. T. Nguyen, and S. J. Sim. 2010. "Enzymatic pretreatment of Chlamydomonas reinhardtii biomass for ethanol production." *Bioresource Technology* 101 (14):5330–5336.

Chuanbin, L., X. Jian, B. Fengwu, and S. Zhiguo. 1998. "Trehalose extraction from Saccharomyces cerevisiae after microwave treatment." *Biotechnology Techniques* 12 (12):941–943.

Clarke, A., T. Prescott, A. Khan, and A. G. Olabi. 2010. "Causes of breakage and disruption in a homogeniser. Causes of breakage and disruption in a homogeniser." *Applied Energy* 87 (12):3680–3690.

Condón, S., J. Raso, R. Pagán, G. Barbosa-Cánovas, M. Tapia, and M. Cano. 2005. "Microbial inactivation by ultrasound. Microbial inactivation by ultrasound." *Novel Food Processing Technologies*:423–442.

Coustets, M., N. Al-Karablieh, C. Thomsen, and J. Teissié. 2013. "Flow process for electro-extraction of total proteins from microalgae." *The Journal of Membrane Biology* 246 (10):751–760.

Coustets, M., V. Joubert-Durigneux, J. Hérault, B. Schoefs, V. Blanckaert, J.-P. Garnier, and J. Teissié. 2015. "Optimization of protein electroextraction from microalgae by a flow process." *Bioelectrochemistry* 103:74–81.

Crapisi, A., A. Lante, G. Pasini, and P. Spettoli. 1993. "Enhanced microbial cell lysis by the use of lysozyme immobilized on different carriers." *Process Biochem.* 28 (1):17–21.

Cravotto, G., L. Boffa, S. Mantegna, P. Perego, M. Avogadro, and P. Cintas. 2008. "Improved extraction of vegetable oils under high-intensity ultrasound and/or microwaves."*Ultrason. Sonochem.* 15 (5):898–902.

Currie, J. A., P. Dunnill, and M. D. Lilly. 1972. "Release of protein from Bakers' yeast (Saccharomyces cerevisiae) by disruption in an industrial agitator mill." *Biotechnology and Bioengineering* 14 (5):725–736.

Dash, A., A. P. Singh, B. R. Chaudhary, S. K. Singh, and D. Dash. 2012. "Effect of silver nanoparticles on growth of eukaryotic green algae." *Nano-micro Letters* 4 (3):158–165.

Deetlefs, M., and K. R. Seddon. 2010. "Assessing the greenness of some typical laboratory ionic liquid preparations." *Green Chemistry* 12 (1):17–30.

Desai, R. K., M. Streefland, R. H. Wijffels, and M. H. M. Eppink. 2016. "Novel astaxanthin extraction from Haematococcus pluvialis using cell permeabilising ionic liquids." *Green Chemistry* 18 (5):1261–1267.

Diels, A. M. J., and C. W. Michiels. 2006. "High-pressure homogenization as a non-thermal technique for the inactivation of microorganisms." *Critical Reviews in Microbiology* 32 (4):201–216.

Doucha, J., and K Lívanský. 2008. "Influence of processing parameters on disintegration of Chlorella cells in various types of homogenizers." *Applied Microbiology and Biotechnology* 81 (3):431.

Ebenezer, V., Y. V. Nancharaiah, and V. P. Venugopalan. 2012. "Chlorination-induced cellular damage and recovery in marine microalga, Chlorella salina." *Chemosphere* 89 (9):1042–1047.

Eing, C., M. Goettel, R. Straessner, C. Gusbeth, and W. Frey. 2013. "Pulsed electric field treatment of microalgae—benefits for microalgae biomass processing." *IEEE Transactions on Plasma Science* 41 (10):2901–2907.

Ellwart, J. W., H. Brettel, and L. O. Kober. 1988. "Cell membrane damage by ultrasound at different cell concentrations." *Ultrasound in Medicine and Biology* 14 (1):43–50.

Engler, C. R. 1985. Disruption of microbial cells. In *Comprehensive Biotechnology*, edited by M. Moo-Young, 305–324. Oxford: Pergamon Press.

Engler, C. R., and C. W. Robinson. 1981. "Disruption of Candida utilis cells in high pressure flow devices." *Biotechnology and Bioengineering* 23 (4):765–780.

Eugene, C Yi., and Murray Hackett. 2000. "Rapid isolation method for lipopolysaccharide and lipid A from gram-negative bacteria." *Analyst* 125 (4):651–656.

Fan, J., Y. Fan, Y. Pei, K. Wu, J. Wang, and M. Fan. 2008. "Solvent extraction of selected endocrine-disrupting phenols using ionic liquids." *Separation and Purification Technology* 61 (3):324–331.

Ferraretto, P., V. Cacciola, I. F. Batlló, and E. Celotti. 2013. "Ultrasounds application in winemaking: Grape maceration and yeast lysis." *Italian Journal of Food Science* 25 (2):160–168.

Follows, M., P. J. Hetherington, P. Dunnill, and M. D. Lilly. 1971. "Release of enzymes from bakers' yeast by disruption in an industrial homogenizer." *Biotechnol. Bioeng.* 13 (4):549–560. doi: 10.1002/bit.260130408.

Fonseca, L. P., and J. Cabral. 2002. "Penicillin acylase release from Escherichia coli cells by mechanical cell disruption and permeabilization." *J. Chem. Technol. Biotechnol.* 77 (2):159–167.

Fu, C.-C., T.-C. Hung, J.-Y. Chen, C.-H. Su, and W.-T. Wu. 2010. "Hydrolysis of microalgae cell walls for production of reducing sugar and lipid extraction." *Bioresour. Technol.* 101 (22):8750–8754.

Fujita, K., D. Kobayashi, N. Nakamura, and H. Ohno. 2013. "Direct dissolution of wet and saliferous marine microalgae by polar ionic liquids without heating." *Enzyme and Microbial Technology* 52 (3):199–202.

G. Brookman, and S. James. 1974. "Mechanism of cell disintegration in a high pressure homogenizer." *Biotechnology and Bioengineering* 16 (3):371–383.

Galabova, D., B. Tuleva, and D. Spasova. 1996. "Permeabilization of Yarrowia lipolytica cells by Triton X-100." *Enzyme Microb. Technol.* 18 (1):18–22.

Ganeva, V., B. Galutzov, and J. Teissié. 2003. "High yield electroextraction of proteins from yeast by a flow process." *Analytical Biochemistry* 315 (1):77–84.

Ganeva, V., D. Stefanova, B. Angelova, B. Galutzov, I. Velasco, and M. Arévalo-Rodríguez. 2015. "Electroinduced release of recombinant β-galactosidase from Saccharomyces cerevisiae." *J. Biotechnol.* 211:12–19.

Garrido, F., U. C. Banerjee, Y. Chisti, and M. Moo-Young. 1994. "Disruption of a recombinant yeast for the release of -galactosidase." *Bioseparation* 4:319–319.

Gerde, J. A., M. Montalbo-Lomboy, L. Yao, D. Grewell, and T. Wang. 2012. "Evaluation of microalgae cell disruption by ultrasonic treatment." *Bioresour. Technol.* 125:175–181.

Gerken, H. G., B. Donohoe, and E. P. Knoshaug. 2013. "Enzymatic cell wall degradation of Chlorellavulgaris and other microalgae for biofuels production." *Planta* 237 (1):239–253.

Gernaey, K. V., A. E. Cervera-Padrell, and J. M. Woodley. 2012. "A perspective on PSE in pharmaceutical process development and innovation." *Comput. Chem. Eng.* 42:15–29.

Gharibzahedi, S. M. T., S. M. Mousavi, M. Hamedi, K. Rezaei, and F. Khodaiyan. 2013. "Evaluation of physicochemical properties and antioxidant activities of Persian walnut oil obtained by several extraction methods." *Industrial Crops and Products* 45:133–140.

Goettel, M., C. Eing, C. Gusbeth, R. Straessner, and W. Frey. 2013. "Pulsed electric field assisted extraction of intracellular valuables from microalgae." *Algal Research* 2 (4):401–408.

Gogate, Parag R., and A. B. Pandit. 2008. "Application of cavitational reactors for cell disruption for recovery of intracellular enzymes." *Journal of Chemical Technology & Biotechnology: International Research in Process, Environmental & Clean Technology* 83 (8):1083–1093.

Goldberg, Stanley. 2008. "Mechanical/physical methods of cell disruption and tissue homogenization." In *2D Page: Sample Preparation and Fractionation*, 3–22. New York: Springer.

González-Delgado, AD., and Viatcheslav Kafarov. 2012. "Microalgae based biorefinery: Evaluation of several routes for joint production of biodiesel, chlorophylls, phycobiliproteins, crude oil and reducing sugars." *Chem Eng Trans* 29 (1):607–612.

Griffin, S. T., M. Dilip, S. K. Spear, J. G. Huddleston, and R. D. Rogers. 2006. "The opposite effect of temperature on polyethylene glycol-based aqueous biphasic systems versus aqueous biphasic extraction chromatographic resins." *Journal of Chromatography B* 844 (1):23–31.

Grimi, N., A. Dubois, L. Marchal, S. Jubeau, N. I. Lebovka, and E. Vorobiev. 2014. "Selective extraction from microalgae Nannochloropsis sp. using different methods of cell disruption." *Bioresource Technology* 153:254–259.

Grosso, C., P. Valentão, F. Ferreres, and P. Andrade. 2015. "Alternative and efficient extraction methods for marine-derived compounds." *Marine Drugs* 13 (5):3182–3230.

Guedes, A. C., H. M. Amaro, M. S. Gião, and F. X. Malcata. 2013. "Optimization of ABTS radical cation assay specifically for determination of antioxidant capacity of intracellular extracts of microalgae and cyanobacteria." *Food Chemistry* 138 (1):638–643.

Günerken, E., E. d'Hondt, M. H. M. Eppink, L. Garcia-Gonzalez, K. Elst, and R. H. Wijffels. 2015. "Cell disruption for microalgae biorefineries. Cell disruption for microalgae biorefineries." *Biotechnol. Adv.* 33 (2):243–260.

Halim, R., R. Harun, M. K. Danquah, and P. A. Webley. 2012. "Microalgal cell disruption for biofuel development." *Applied Energy* 91 (1):116–121.

Han, D., and K. H. Row. 2010. "Recent applications of ionic liquids in separation technology." *Molecules* 15 (4):2405–2426.

Harun, R., and M. K. Danquah. 2011. "Enzymatic hydrolysis of microalgal biomass for bioethanol production." *Chem. Eng. J.* 168 (3):1079–1084.

Hazani, A. A., M. M. Ibrahim, A. I. Shehata, Gehan A El-Gaaly, M. Daoud, D. Fouad, H. Rizwana, and N. Moubayed. 2013. "Ecotoxicity of Ag-nanoparticles on two microalgae, Chlorella vulgaris and Dunaliella tertiolecta." *Archives of Biological Sciences* 65 (4):1447–1457.

Hedenskog, G., L. Enebo, J. Vendlová, and B. Prokeš. 1969. "Investigation of some methods for increasing the digestibility in vitro of microalgae." *Biotechnology and Bioengineering* 11 (1):37–51.

Hernández, D., B. Riaño, M. Coca, and M. C. García-González. 2015. "Saccharification of carbohydrates in microalgal biomass by physical, chemical and enzymatic pre-treatments as a previous step for bioethanol production." *Chem. Eng. J.* 262:939–945.

Ho, S. Y., and GS Mittal. 1996. "Electroporation of cell membranes: A review. Electroporation of cell membranes: A review." *Critical Reviews in Biotechnology* 16 (4):349–362.

Hosikian, A., S. Lim, R. Halim, and M. K. Danquah. 2010. "Chlorophyll extraction from microalgae: A review on the process engineering aspects." *International Journal of Chemical Engineering* 2010:1–11.

Huang, W.-C., and J.-D. Kim. 2013. "Cationic surfactant-based method for simultaneous harvesting and cell disruption of a microalgal biomass." *Bioresour. Technol.* 149:579–581.

Huang, W.-C., and J.-D. Kim. 2016. "Nickel oxide nanoparticle-based method for simultaneous harvesting and disruption of microalgal cells." *Bioresource Technology* 218:1290–1293.

Huo, S., Z. Wang, F. Cui, B. Zou, P. Zhao, and Z. Yuan. 2015. "Enzyme-assisted extraction of oil from wet microalgae *Scenedesmus* sp. G4." *Energies* 8 (8):8165–8174.

Jahanshahi, M., Y. Sun, E. Santos, A. Pacek, T. T. Franco, A. Nienow, and A. Lyddiatt. 2002. "Operational intensification by direct product sequestration from cell disruptates: Application of a pellicular adsorbent in a mechanically integrated disruption-fluidised bed adsorption process." *Biotechnology and Bioengineering* 80 (2):201–212.

Jampani, C., H. A. Tavanandi, and K. S. M. S. Raghavarao. 2015. "Application of ionic liquids in separation and downstream processing of biomolecules." *Current Biochemical Engineering* 2 (2):135–147.

Jamshad, M., and R. A. J. Darby. 2012. "Disruption of yeast cells to isolate recombinant proteins." *Recombinant Protein Production in Yeast*, 237–246. Springer.

Janczyk, P., C. Wolf, and W. B. Souffrant. 2005. "Evaluation of nutritional value and safety of the green microalgae Chlorella vulgaris treated with novel processing methods." *Arch Zootech* 8:132–147.

Jin, G., F. Yang, C. Hu., H. Shen, and Z. K. Zhao. 2012. "Enzyme-assisted extraction of lipids directly from the culture of the oleaginous yeast Rhodosporidium toruloides." *Bioresour. Technol.* 111:378–382.

Kapturowska, A., I. Stolarzewicz, I. Chmielewska, and E. Bialecka-Florianczyk. 2007. "Ultrasounds-a tool to inactivate yeast and to extract intracellular protein." *Zywnosc Nauka Technologia Jakosc (Poland)*.

Kelly, W. J, and K. R. Muske. 2004. "Optimal operation of high-pressure homogenization for intracellular product recovery." *Bioprocess Biosystems Eng.* 27 (1):25–37.

Keris-Sen, U. D, U. Sen, G. Soydemir, and M. D. Gurol. 2014. "An investigation of ultrasound effect on microalgal cell integrity and lipid extraction efficiency." *Bioresour. Technol.* 152:407–413.

Khoo, H. H., P. N. Sharratt, P. Das, R. K. Balasubramanian, P. K. Naraharisetti, and S. Shaik. 2011. "Life cycle energy and CO_2 analysis of microalgae-to-biodiesel: Preliminary results and comparisons." *Bioresource Technology* 102 (10):5800–5807.

Kim, D.-Y., Y.-K. Oh, J.-Y. Park, B. Kim, S.-A. Choi, and J.-I. Han. 2015. "An integrated process for microalgae harvesting and cell disruption by the use of ferric ions." *Bioresource Technology* 191:469–474.

Kim, J., G. Yoo, H. Lee, J. Lim, K. Kim, C. W. Kim, M. S. Park, and J.-W. Yang. 2013. "Methods of downstream processing for the production of biodiesel from microalgae." *Biotechnology Advances* 31 (6):862–876.

Kim, Y.-H., Y.-K. Choi, J. Park, S. Lee, Y.-H. Yang, H. J. Kim, T.-J. Park, Y. H. Kim, and S. H. Lee. 2012. "Ionic liquid-mediated extraction of lipids from algal biomass." *Bioresource Technology* 109:312–315.

Komaki, H., M. Yamashita, Y. Niwa, Y. Tanaka, N. Kamiya, Y. Ando, and M. Furuse. 1998. "The effect of processing of Chlorella vulgaris: K-5 on in vitro and in vivo digestibility in rats." *Animal Feed Science and Technology* 70 (4):363–366.

Lai, Y. S., P. Parameswaran, A. Li, M. Baez, and B. E. Rittmann. 2014. "Effects of pulsed electric field treatment on enhancing lipid recovery from the microalga, Scenedesmus." *Bioresource Technology* 173:457–461.

Lee, A. K., D. M. Lewis, and P. J. Ashman. 2012. "Disruption of microalgal cells for the extraction of lipids for biofuels: Processes and specific energy requirements." *Biomass Bioenergy* 46:89–101.

Lee, J.-Y., C. Yoo, S. -Y Jun, C.-Y. Ahn, and H.-M. Oh. 2010. "Comparison of several methods for effective lipid extraction from microalgae." *Bioresour. Technol.* 101 (1):S7–577.

Li, Q., S. Mahendra, D. Y. Lyon, L. Brunet, M. V. Liga, D. Li, and P. J. J. Alvarez. 2008. "Antimicrobial nanomaterials for water disinfection and microbial control: Potential applications and implications." *Water Research* 42 (18):4591–4602.

Liang, R., Z. Wang, J.-H. Xu, W. Li, and H. Qi. 2009. "Novel polyethylene glycol induced cloud point system for extraction and back-extraction of organic compounds." *Separation and Purification Technology* 66 (2):248–256.

Liu, D., X.-A. Zeng, D.-W. Sun, and Z. Han. 2013. "Disruption and protein release by ultrasonication of yeast cells." *Innovative Food Science & Emerging Technologies* 18:132–137.

Liu, X.-Y., Q. Wang, S. W. Cui, and H.-Z. Liu. 2008. "A new isolation method of β-D-glucans from spent yeast Saccharomyces cerevisiae." *Food Hydrocolloids* 22 (2):239–247.

Lu, Y., W. Ma, R. Hu, X. Dai, and Y. Pan. 2008. "Ionic liquid-based microwave-assisted extraction of phenolic alkaloids from the medicinal plant Nelumbo nucifera Gaertn." *Journal of Chromatography A* 1208 (1–2):42–46.

Luengo, E., J. M. Martínez, A. Bordetas, I. Álvarez, and J. Raso. 2015. "Influence of the treatment medium temperature on lutein extraction assisted by pulsed electric fields from Chlorella vulgaris." *Innovative Food Science & Emerging Technologies* 29:15–22.

Luengo, E., S. Condón-Abanto, S. Condón, I. Álvarez, and J. Raso. 2014. "Improving the extraction of carotenoids from tomato waste by application of ultrasound under pressure." *Sep. Purif. Technol.* 136:130–136.

Maa, Y.-F., and C. Hsu. 1996. "Liquid-liquid emulsification by rotor/stator homogenization." *Journal of Controlled Release* 38 (2–3):219–228.

Macías-Sánchez, M. D., C. Mantell, M. Rodriguez, de la., E. M. D. L. Ossa, L. M. Lubián, and O. Montero. 2009. "Comparison of supercritical fluid and ultrasound-assisted extraction of carotenoids and chlorophyll a from Dunaliella salina." *Talanta* 77 (3):948–952.

MacNeill, C., J. R. Sneeringer, and A. Szatmary. 1985. "Optimization of cell disruption with bead mill. Optimization of cell disruption with bead mill." *Pharm Eng* 5:34–38.

Mandal, V., Y. Mohan, and S. Hemalatha. 2007. "Microwave assisted extraction—an innovative and promising extraction tool for medicinal plant research." *Pharmacognosy Reviews* 1 (1):7–18.

Maran, J. P., B. Priya, and C. V. Nivetha. 2015. "Optimization of ultrasound-assisted extraction of natural pigments from Bougainvillea glabra flowers." *Industrial Crops and Products* 63:182–189.

Martín, J. F. G., L. Guillemet, C. Feng, and D.-W. Sun. 2013. "Cell viability and proteins release during ultrasound-assisted yeast lysis of light lees in model wine." *Food Chemistry* 141 (2):934–939.

Martínez, J. M., E. Luengo, G. Saldaña, I. Álvarez, and J. Raso. 2016. "C-phycocyanin extraction assisted by pulsed electric field from Artrosphira platensis." *Food Res. Int.* 99 (3):1042–1047.

Mei, C. Y., T. B. Ti, M. N. Ibrahim, A. Ariff, and L. T. Chuan. 2005. "The disruption of Saccharomyces cerevisiae cells and release of glucose 6-phosphate dehydrogenase (G6PDH) in a horizontal dyno bead mill operated in continuous recycling mode." *Biotechnology and Bioprocess Engineering* 10 (3):284.

Melendres, A. V., H. Unno, N. Shiragami, and H. Honda. 1992. "A concept of critical velocity for cell disruption by bead mill." *Journal of Chemical Engineering of Japan* 25 (3):354–356.

Mendes-Pinto, M. M., M. F. J. Raposo, J. Bowen, A. J. Young, and R. Morais. 2001. "Evaluation of different cell disruption processes on encysted cells of Haematococcus pluvialis: Effects on astaxanthin recovery and implications for bio-availability." *J. Appl. Phycol.* 13 (1):19–24.

Middelberg, A. P. J. 1995. "Process-scale disruption of microorganisms. Process-scale disruption of microorganisms." *Biotechnology Advances* 13 (3):491–551.

Middelberg, APJ. 1993. "Extension of the wall-strength model for high-pressure homogenization to multiple passes." *Food and Bioproducts Processing: Transactions of the Institution of Chemical Engineers, Part C* 71 (3):215–219.

Miller, J., M. Rogowski, and W. Kelly. 2002. "Using a CFD model to understand the fluid dynamics promoting E. coli breakage in a high-pressure homogenizer." *Biotechnol. Prog.* 18 (5):1060–1067.

Miranda, J. R., P. C. Passarinho, and L. Gouveia. 2012. "Pre-treatment optimization of *Scenedesmus obliquus* microalga for bioethanol production." *Bioresource Technology* 104:342–348.

Mogren, H., M. Lindblom, and G. Hedenskog. 1974. "Mechanical disintegration of microorganisms in an industrial homogenizer." *Biotechnology and Bioengineering* 16 (2):261–274.

Naganuma, T., Y. Uzuka, and K. Tanaka. 1984. "Simple and small-scale breakdown of yeast. Simple and small-scale breakdown of yeast." *Analytical Biochemistry* 141 (1):74–78.

Natarajan, R., W. M. R. Ang, X. Chen, M. Voigtmann, and R. Lau. 2014. Lipid releasing characteristics of microalgae species through continuous ultrasonication." *Bioresour. Technol.* 158:7–11.

Negi, S., and A. K. Pandey. 2015. "Ionic liquid pretreatment." *Pretreatment of Biomass*, 137–155. Elsevier.

Okuda, K., K. Oka, A. Onda, K. Kajiyoshi, M. Hiraoka, and K. Yanagisawa. 2008. "Hydrothermal fractional pretreatment of sea algae and its enhanced enzymatic hydrolysis." *J. Chem. Technol. Biotechnol.* 83 (6):836–841.

Olmstead, I. L. D., S. E. Kentish, P. J. Scales, and G. J. O. Martin. 2013. "Low solvent, low temperature method for extracting biodiesel lipids from concentrated microalgal biomass." *Bioresource Technology* 148:615–619.

Orr, Valerie CA., and Lars Rehmann. 2016. "Ionic liquids for the fractionation of microalgae biomass. Ionic liquids for the fractionation of microalgae biomass." *Current Opinion in Green and Sustainable Chemistry* 2:22–27.

Ottman, J. 2011. *The New Rules of Green Marketing: Strategies, Tools, and Inspiration for Sustainable Branding.* San Francisco, CA: Berrett-Koehler Publishers.

Park, J.-Y., Y.-K. Oh, J.-S. Lee, K. Lee, M.-J. Jeong, and S.-A. Choi. 2014. "Acid-catalyzed hot-water extraction of lipids from Chlorella vulgaris." *Bioresource Technology* 153:408–412.

Parniakov, O., F. J. Barba, N. Grimi, L. Marchal, S. Jubeau, N. Lebovka, and E. Vorobiev. 2015. "Pulsed electric field and pH assisted selective extraction of intracellular components from microalgae Nannochloropsis." *Algal Research* 8:128–134.

Parniakov, O., F. J. Barba, N. Grimi, N. Lebovka, and E. Vorobiev. 2014. "Impact of pulsed electric fields and high voltage electrical discharges on extraction of high-added value compounds from papaya peels." *Food Research International* 65:337–343.

Pasquet, V., P. Morisset, S. Ihammouine, A. Chepied, L. Aumailley, J.-B. Berard, B. Serive, R. Kaas, I. Lanneluc, and V. Thiery. 2011. "Antiproliferative activity of violaxanthin isolated from bioguided fractionation of Dunaliella tertiolecta extracts." *Marine Drugs* 9 (5):819–831.

Patil, G., and K. S. M. S. Raghavarao. 2007. "Aqueous two phase extraction for purification of C-phycocyanin." *Biochem. Eng. J.* 34 (2):156–164. doi: 10.1016/j.bej.2006.11.026.

Patil, G., S. Chethana, A. S. Sridevi, and K. S. M. S. Raghavarao. 2006. "Method to obtain C-phycocyanin of high purity." *J. Chromatogr.* 1127 (1):76–81.

Pavlić, Ž., Ž. Vidaković-Cifrek, and D. Puntarić. 2005. "Toxicity of surfactants to green microalgae Pseudokirchneriella subcapitata and Scenedesmus subspicatus and to marine diatoms Phaeodactylum tricornutum and Skeletonema costatum." *Chemosphere* 61 (8):1061–1068.

Petrier, C., Y. Jiang, and M.-F. Lamy. 1998. "Ultrasound and environment: Sonochemical destruction of chloroaromatic derivatives." *Environmental Science & Technology* 32 (9):1316–1318.

Piasecka, A., I. Krzemińska, and J. Tys. 2014. "Physical methods of microalgal biomass pre-treatment." *International Agrophysics* 28 (3):341–348.

Popinet, S., and S. Zaleski. 2002. "Bubble collapse near a solid boundary: A numerical study of the influence of viscosity." *Journal of Fluid Mechanics* 464:137–163.

Postma, P. R., T. L. Miron, G. Olivieri, M. J. Barbosa, R. H. Wijffels, and M. H. M. Eppink. 2015. "Mild disintegration of the green microalgae Chlorella vulgaris using bead milling." *Bioresource Technology* 184:297–304.

Praveenkumar, R., K. Lee, J. Lee, and Y.-K. Oh. 2015. "Breaking dormancy: An energy-efficient means of recovering astaxanthin from microalgae." *Green Chemistry* 17 (2):1226–1234.

Priego-Capote, F., and Luque de Castro. 2007. "Ultrasound-assisted digestion: A useful alternative in sample preparation." *J. Biochem. Biophys. Methods* 70 (2):299–310.

Razack, S. A., S. Duraiarasan, and V. Mani. 2016. "Biosynthesis of silver nanoparticle and its application in cell wall disruption to release carbohydrate and lipid from C. vulgaris for biofuel production." *Biotechnology Reports* 11:70–76.

Ricci-Silva, M. E., M. Vitolo, and J. Abrahao-Neto. 2000. "Protein and glucose 6-phosphate dehydrogenase releasing from baker's yeast cells disrupted by a vertical bead mill." *Process Biochemistry* 35 (8):831–835.

Rocha, J. M. S. 1999. "Aplicações de agentes tensoativos em biotecnologia. Aplicações de agentes tensoativos em biotecnologia." *Bol Biotecnol* 64:5–11.

Rodea-Palomares, I., K. Boltes, F. Fernández-Pinas, F. Leganés, E. García-Calvo, J. Santiago, and R. Rosal. 2010. "Physicochemical characterization and ecotoxicological assessment of CeO2 nanoparticles using two aquatic microorganisms." *Toxicological Sciences* 119 (1):135–145.

Rodrigues, R. D. P., F. C. de Castro, R. S. de Santiago-Aguiar, and M. V. P. Rocha. 2018. "Ultrasound-assisted extraction of phycobiliproteins from Spirulina (Arthrospira) platensis using protic ionic liquids as solvent." *Algal Research* 31:454–462.

Rols, M. P., F. Dahhou, K. P. Mishra, and J. Teissié. 1990. "Control of electric field induced cell membrane permeabilization by membrane order." *Biochemistry* 29 (12):2960–2966.

Rosenberg, U., and W. Bogl. 1987. "Microwave thawing, drying, and baking in the food industry." *Food Technology* 41.

Safi, C., A. V. Ursu, C. Laroche, B. Zebib, O. Merah, P.-Y. Pontalier, and C. Vaca-Garcia. 2014a. "Aqueous extraction of proteins from microalgae: Effect of different cell disruption methods." *Algal Research* 3:61–65.

Safi, C., B. Zebib, O. Merah, P.-Y. Pontalier, and C. Vaca-Garcia. 2014b. "Morphology, composition, production, processing and applications of Chlorella vulgaris: A review." *Renewable and Sustainable Energy Reviews* 35:265–278.

Samarasinghe, N., S. Fernando, and W. B. Faulkner. 2012. "Effect of high pressure homogenization on aqueous phase solvent extraction of lipids from Nannochloris oculata microalgae." *Journal of Energy and Natural Resources* 1 (DOE-DANF-0003046-P4).

Sartory, D. P., and J. U. Grobbelaar. 1984. "Extraction of chlorophyll a from freshwater phytoplankton for spectrophotometric analysis." *Hydrobiologia* 114 (3):177–187.

Saulis, G. 2010. "Electroporation of cell membranes: The fundamental effects of pulsed electric fields in food processing." *Food Engineering Reviews* 2 (2):52–73.

Save, S. S., A. B. Pandit, and J. B. Joshi. 1994. "Microbial cell disruption: Role of cavitation. Microbial cell disruption: Role of cavitation." *The Chemical Engineering Journal and the Biochemical Engineering Journal* 55 (3):B67–B72.

Schuette, H., and M.-R. Kula. 1990. "Pilot-and process-scale techniques for cell disruption. Pilot-and process-scale techniques for cell disruption." *Biotechnology and applied biochemistry* 12 (6):599–620.

Schumann, R., N. Häubner, S. Klausch, and U. Karsten. 2005. "Chlorophyll extraction methods for the quantification of green microalgae colonizing building facades." *Int. Biodeterior. Biodegrad.* 55 (3):213–222.

Schwede, S., A. Kowalczyk, M. Gerber, and R. Span. 2011. "Influence of different cell disruption techniques on mono digestion of algal biomass." World Renewable Energy Congress-Sweden; 8–13 May; 2011; Linköping; Sweden.

Schwenzfeier, A., P. A. Wierenga, and H. Gruppen. 2011. "Isolation and characterization of soluble protein from the green microalgae Tetraselmis sp." *Bioresource Technology* 102 (19):9121–9127.

Seetharam, R., and S. K. Sharma. 1991. *Purification and Analysis of Recombinant Proteins.* New York: CRC Press.

Seo, J. Y., R. Praveenkumar, B. Kim, J.-C. Seo, J.-Y. Park, J.-G. Na, S. G. Jeon, S. B. Park, K. Lee, and Y.-K. Oh. 2016a. "Downstream integration of microalgae harvesting and cell disruption by means of cationic surfactant-decorated Fe 3 O 4 nanoparticles." *Green Chemistry* 18 (14):3981–3989.

Seo, Y. H., M. Sung, Y.-K. Oh, and J.-I. Han. 2016b. "Lipid extraction from microalgae cell using persulfate-based oxidation." *Bioresource Technology* 200:1073–1075.

Shen, Y., Z. Pei, W. Yuan, and E. Mao. 2009. "Effect of nitrogen and extraction method on algae lipid yield." *International Journal of Agricultural and Biological Engineering* 2 (1):51–57.

Shene, C., M. T. Monsalve, D. Vergara, M. E. Lienqueo, and M. Rubilar. 2016. "High pressure homogenization of Nannochloropsis oculata for the extraction of intracellular components: Effect of process conditions and culture age." *European Journal of Lipid Science and Technology* 118 (4):631–639.

Sheng, J., R. Vannela, and B. E. Rittmann. 2012. "Disruption of Synechocystis PCC 6803 for lipid extraction. Disruption of Synechocystis PCC 6803 for lipid extraction." *Water Sci. Technol.* 65 (3):567–573.

Show, K.-Y., D.-J. Lee, J.-H. Tay, T.-M. Lee, and J.-S. Chang. 2015. "Microalgal drying and cell disruption—recent advances." *Bioresour. Technol.* 184:258–266.

Shynkaryk, M. V., N. I. Lebovka, J.-L. Lanoisellé, M. Nonus, C. Bedel-Clotour, and E. Vorobiev. 2009. "Electrically-assisted extraction of bio-products using high pressure disruption of yeast cells (Saccharomyces cerevisiae)." *J. Food Eng.* 92 (2):189–195.

Silveira, S. T., J. F. de M. Burkert, J. A. V. Costa, C. A. V. Burkert, and S. J. Kalil. 2007. "Optimization of phycocyanin extraction from Spirulina platensis using factorial design." *Bioresour. Technol.* 98 (8):1629–1634.

Simon, D., and S. Helliwell. 1998. "Extraction and quantification of chlorophyll a from freshwater green algae." *Water Res.* 32 (7):2220–2223.

Smirnova, S. V., I. I. Torocheshnikova, A. A. Formanovsky, and I. V. Pletnev. 2004. "Solvent extraction of amino acids into a room temperature ionic liquid with dicyclohexano-18-crown-6." *Analytical and Bioanalytical Chemistry* 378 (5):1369–1375.

Sondi, I., and B. Salopek-Sondi. 2004. "Silver nanoparticles as antimicrobial agent: A case study on E. coli as a model for Gram-negative bacteria." *Journal of Colloid and Interface Science* 275 (1):177–182.

Soto, A., A. Arce, and M. K. Khoshkbarchi. 2005. "Partitioning of antibiotics in a two-liquid phase system formed by water and a room temperature ionic liquid." *Separation and Purification Technology* 44 (3):242–246.

Spiden, E. M., P. J. Scales, S. E. Kentish, and G. J. O. Martin. 2013. "Critical analysis of quantitative indicators of cell disruption applied to Saccharomyces cerevisiae processed with an industrial high pressure homogenizer." *Biochemical Engineering Journal* 70:120–126.

Stanbury, P. F., A. Whitaker, and S. J. Hall. 2013. *Principles of Fermentation Technology*: New York: Elsevier.

Sudar, M., D. Valinger, Z. Findrik, D. Vasić-Rački, and Ž. Kurtanjek. 2013. "Effect of different variables on the efficiency of the Baker's yeast cell disruption process to obtain alcohol dehydrogenase activity." *Appl. Biochem. Biotechnol.* 169 (3):1039–1055.

Suslick, K. S. 1989. "The chemical effects of ultrasound." *Scientific American* 260 (2):80–87.

Suslick, K. S., and G. J. Price. 1999. "Applications of ultrasound to materials chemistry. Applications of ultrasound to materials chemistry." *Annu. Rev. Mater. Sci.* 29 (1):295–326.

Tan, I. S., and K. T. Lee. 2014. "Enzymatic hydrolysis and fermentation of seaweed solid wastes for bioethanol production: An optimization study." *Energy* 78:53–62.

Tavanandi, H. A., D. Karley, R. Mittal, and K. V. Murthy. 2015. "Contactors for aqueous two-phase extraction: A review."*Current Biochemical Engineering* 2 (2):148–167.

Tavanandi, H. A., A. C. Devi, and K. S. M. S. Raghavarao. 2018a. "A newer approach for the primary extraction of Allophycocyanin with high purity and yield from dry biomass of Arthrospira platensis." *Separation and Purification Technology*. 204:162–174. doi: 10.1016/j.seppur.2018.04.057.

Tavanandi, H. A., R. Mittal, J. Chandrasekhar, and K. S. M. S. Raghavarao. 2018b. "Simple and efficient method for extraction of C-Phycocyanin from dry biomass of Arthospira platensis." *Algal Research* 31:239–251.

Tavanandi, H. A., and K. S. M. S. Raghavarao. 2019. "Recovery of chlorophylls from spent biomass of arthrospira platensis obtained after extraction of phycobiliproteins." *Bioresour. Technol.* 271:391–401.

Tavanandi, H. A., Vanjari, P., and K. Raghavarao. 2019. "Synergistic method for extraction of high purity Allophycocyanin from dry biomass of Arthrospira platensis and utilization of spent biomass for recovery of carotenoids." *Separation and Purification Technology*, 225: 97–111.

Taylor, M. T., P. Belgrader, B. J. Furman, F. Pourahmadi, G. T. A. Kovacs, and M. Allen Northrup. 2001. "Lysing bacterial spores by sonication through a flexible interface in a microfluidic system." *Analytical Chemistry* 73 (3):492–496.

Teo, C. L., and A. Idris. 2014. "Enhancing the various solvent extraction method via microwave irradiation for extraction of lipids from marine microalgae in biodiesel production." *Bioresource Technology* 171:477–481.

Tsong, T. Y. 1990. "On electroporation of cell membranes and some related phenomena." *Journal of Electroanalytical Chemistry and Interfacial Electrochemistry* 299 (3):271–295.

Ulloa, G., C. Coutens, M. Sánchez, J. Sineiro, J. Fábregas, F. J. Deive, A. Rodríguez, and M. J. Núñez. 2012. "On the double role of surfactants as microalga cell lysis agents and antioxidants extractants." *Green Chem.* 14 (4):1044–1051.

Vali Aftari, R., K. Rezaei, A. Mortazavi, and A. R. Bandani. 2015. The Optimized concentration and purity of Spirulina platensis C-phycocyanin: A comparative study on microwave-assisted and ultrasound-assisted extraction methods." *J. Food Process. Preserv.* 39 (6):3080–3091.

Van Gaver, D., and A. Huyghebaert. 1991. "Optimization of yeast cell disruption with a newly designed bead mill." *Enzyme and Microbial Technology* 13 (8):665–671.

Vanthoor-Koopmans, M., R. H Wijffels, M. J Barbosa, and M. H. M. Eppink. 2013. "Biorefinery of microalgae for food and fuel." *Bioresour. Technol.* 135:142–149.

Vasavada, P. C. 1986. "Effect of microwave energy on bacteria." *Journal of Microwave Power and Electromagnetic Energy* 21 (3):187–188.

Vidal, S. T. M., M. J. N. Correia, M. M. Marques, M. R. Ismael, and M. T. A. Reis. 2005. "Studies on the use of ionic liquids as potential extractants of phenolic compounds and metal ions." *Separation Science and Technology* 39 (9):2155–2169.

Vogels, G., and M.-R. Kula. 1992. "Combination of enzymatic and/or thermal pretreatment with mechanical cell disintegration." *Chem. Eng. Sci.* 47 (1):123–131.

Vorobiev, E., and N. Lebovka. 2009. "Pulsed-electric-fields-induced effects in plant tissues: Fundamental aspects and perspectives of applications." In *Electrotechnologies for Extraction from Food Plants and Biomaterials*, 39–81. New York: Springer.

Wang, G., and T. Wang. 2012. "Characterization of lipid components in two microalgae for biofuel application." *Journal of the American Oil Chemists' Society* 89 (1):135–143.

Wang, M., W. Yuan, X. Jiang, Y. Jing, and Z. Wang. 2014. "Disruption of microalgal cells using high-frequency focused ultrasound." *Bioresour. Technol.* 153:315–321.

Wang, Z., J. -H. Xu, W. Zhang, B. Zhuang, and H. Qi. 2008. "In situ extraction of polar product of whole cell microbial transformation with polyethylene glycol-induced cloud point system." *Biotechnology Progress* 24 (5):1090–1095.

Weaver, J. C., and Y. A. Chizmadzhev. 1996. "Theory of electroporation: A review. Theory of electroporation: A review." *Bioelectrochemistry and Bioenergetics* 41 (2):135–160.

Weaver, J. C., G. I. Harrison, J. G. Bliss, J. R. Mourant, and K. T. Powell. 1988. "Electroporation: High frequency of occurrence of a transient high-permeability state in erythrocytes and intact yeast." *FEBS Letters* 229 (1):30–34.

Wiltshire, K., M. Boersma, A. Möller, and H. Buhtz. 2000. "Extraction of pigments and fatty acids from the green alga Scenedesmus obliquus (Chlorophyceae)." *Aquatic Ecology* 34 (2):119–126.

Wong, K. K. Y., and X. Liu. 2010. "Silver nanoparticles—the real "silver bullet" in clinical medicine?" *MedChemComm* 1 (2):125–131.

Woo, I.-S., I.-K. Rhee, and H.-D. Park. 2000. "Differential damage in bacterial cells by microwave radiation on the basis of cell wall structure." *Applied and Environmental Microbiology* 66 (5):2243–2247.

Wu, C., Y. Xiao, W. Lin, J. Li., S. Zhang, J. Zhu, and J. Rong. 2017. "Aqueous enzymatic process for cell wall degradation and lipid extraction from Nannochloropsis sp." *Bioresource Technology* 223:312–316.

Yoo, G., Y. Yoo, J.-H. Kwon, C. Darpito, S. K. Mishra, K. Pak, M. S. Park, S. G. Im, and J.-W. Yang. 2014. "An effective, cost-efficient extraction method of biomass from wet microalgae with a functional polymeric membrane." *Green Chemistry* 16 (1):312–319.

Young, G., F. Nippgen, S. Titterbrandt, and M. J. Cooney. 2010. "Lipid extraction from biomass using co-solvent mixtures of ionic liquids and polar covalent molecules." *Separation and Purification Technology* 72 (1):118–121.

Zbinden, M. D. A., B. S. M. Sturm, R. D. Nord, W. J. Carey, D. Moore, H. Shinogle, and S. M. Stagg-Williams. 2013. "Pulsed electric field (PEF) as an intensification pretreatment for greener solvent lipid extraction from microalgae." *Biotechnology and Bioengineering* 110 (6):1605–1615.

Zhang, G., and I. Hua. 2000. "Cavitation chemistry of polychlorinated biphenyls: Decomposition mechanisms and rates." *Environmental Science & Technology* 34 (8):1529–1534.

Zhang, L., Y. Jin, Y. Xie, X. Wu, and T. Wu. 2014. "Releasing polysaccharide and protein from yeast cells by ultrasound: Selectivity and effects of processing parameters." *Ultrasonics Sonochemistry* 21 (2):576–581.

Zhang, P., G. Zhang, and W. Wang. 2007. "Ultrasonic treatment of biological sludge: Floc disintegration, cell lysis and inactivation." *Bioresource Technology* 98 (1):207–210.

Zhang, X., F. Zhang, G. Luo, S. Yang, and D. Wang. 2015. "Extraction and separation of phycocyanin from Spirulina using aqueous two-phase systems of ionic liquid and salt." *Journal of Food and Nutrition Research* 3 (1):15–19.

Zhang, X., S. Yan, R. D. Tyagi, P. Drogui, and R. Y. Surampalli. 2014. "Ultrasonication assisted lipid extraction from oleaginous microorganisms." *Bioresource Technology* 158:253–261.

Zheng, H., J. Yin, Z. Gao, H. Huang, X. Ji, and C. Dou. 2011. "Disruption of Chlorella vulgaris cells for the release of biodiesel-producing lipids: A comparison of grinding, ultrasonication, bead milling, enzymatic lysis, and microwaves." *Appl. Biochem. Biotechnol.* 164 (7):1215–1224.

Zhou, T., X. Xiao, G. Li, and Z.-w. Cai. 2011. "Study of polyethylene glycol as a green solvent in the microwave-assisted extraction of flavone and coumarin compounds from medicinal plants." *Journal of Chromatography A* 1218 (23):3608–3615.

Zhu, Y., X. B. Chen, K. B. Wang, Y. X. Li, K. Z. Bai, T. Y. Kuang, and H. B. Ji. 2007. "A simple method for extracting C-phycocyanin from Spirulina platensis using Klebsiella pneumoniae." *Appl. Microbiol. Biotechnol.* 74 (1):244–248.

Zimmermann, U., J. Vienken, J. Halfmann, and C. C. Emeis. 1985. "Electrofusion: A novel hybridization technique. Electrofusion: A novel hybridization technique." *Advances in Biotechnological Processes* 4:79.

Zuckerman, H., Y. E. Krasik, and J. Felsteiner. 2002. "Inactivation of microorganisms using pulsed high-current underwater discharges." *Innovative Food Science & Emerging Technologies* 3 (4):329–336.

6 A Sustainable Approach for Bioenergy and Biofuel Production from Microalgae

J. K. Bwapwa, T. Mutanda, A. Anandraj

CONTENTS

6.1   INTRODUCTION

The world demand for energy is growing every year, and the call to produce alternative fuels such as biodiesel, bio-jet fuel, bioethanol, and biohydrogen is increasing. However, petroleum-related products are still the most used for energy applications and transportation. According to the International Energy Agency (IEA), more than 85 million barrels of oil are consumed daily worldwide. The global growth rate regarding the primary energy demand is currently standing at 1.6% per year. Predictions from various models concur with the fact that there is a possibility of doubling the global demand for primary energy over the next 40 years. In addition, the IEA estimates that there is a strong need to introduce low carbon emission fuels into the global energy pool to stabilize atmospheric CO_2 levels. Because of the increasing focus on environmental issues and undeniable certainty of oil price volatility, many countries are considering the use of renewable energy more, especially biofuels, to prevent any energy crisis in the near future. It is well known that currently many countries are relying on fossil-based fuels for energy production. Unfortunately, fossil-based fuels, despite their competitive costs compared to biofuels, are considered the most polluting, with high volumes of greenhouse gases being emitted daily in the atmosphere. Greenhouse gas emissions are known to be the key origin of global warming and climate change. Hence, the consequences linked to climate change and global warming are immeasurable in today's environmental context. A need for a reliable source of energy that can generate less carbon emissions is needed to remediate the current environmental challenges. Biofuels can generate low carbon emissions below the environmental standards and be the alternative. Biofuels are also expected to be sustainable and have the same performance compared to fossil fuels. In some cases, they are used as a total replacement or are blended with their petroleum counterparts. This option has presented many advantages economically and technically. The global increasing demand for energy, as mentioned earlier, is becoming a challenge today because of the increasing need for products and chemicals used for the production of energy. Also, the limited reserve of fossil fuels become a cause for concern for future generations. A need for innovation to secure a better future for transportation fuels and energy production is very vital. The development of emerging technologies and sustaining innovations can be the entry points for groundbreaking solutions. There is a need for solutions that can defy the current definition of energy sources and the way energy should be utilized or deployed. In recent years, wind, solar and hydrogen energy have been developed, these energy forms are not practical in meeting energy and transportation needs due to size, scale, and costs application. However, biofuels such as bioethanol, biohydrogen, and biodiesel and bio-jet fuel that will not cause exacerbate the food crisis and will reduce environmental problems are among the most promising alternatives. In the past, there has been a certain evolution regarding the development stages of biofuels in terms of improving their effectiveness and sustainability. Biofuels have moved from the first generation to the fourth generation with different improvements and advancements introduced from one generation to the next.

The first-generation biofuels are generated from food crops by extracting the lipids or oil from seed rape; thereafter, transesterification is undertaken to produce biodiesel. From biodiesel, there is a prospect to produce green jet fuel via deoxygenation, decarboxylation, and isomerization (Bwapwa et al. 2017: 1345–1354). Regarding the production of bioethanol, this can be produced via fermentation using crops such as wheat or sugar as a feedstock. These biofuels have encountered some challenges, but the most contentious problem is the threat to the food industry to solve the issue related to food versus fuel (Naik et al. 2010: 578–597). The demand for first-generation biofuels caused the diversion of food crops from the global food market toward the fuel market, causing the prices of these food products to increase (Lee and Lavoie 2013: 6–11; Fairley 2011: 2–5, Saladini et al. 2016: 221–227). The second-generation biofuels were developed with the aim of overcoming the limitations encountered from the first generation. Non-food crops were chosen to produce biofuels using feedstocks such as wood, food crop wastes, or used edible oil and biomass (Stephens et al. 2008; Naik et al. 2010: 578–597). These biofuels are cost-competitive when compared to existing petroleum-based fuels. It has been shown from life cycle assessment that the net energy gains will be increased; this is another aspect that allows overcoming one of the main limitations from the first generation. Second-generation biofuels are made up from inedible resources, such as agriculture remainders, or non-crop plants, including switchgrass. It aimed to remove the pressure on resources that are edible to prevent stress on food supply (Stephens et al. 2008: 20–43; Naik et al. 2010: 578–597). Additionally, a smaller space is used for a higher output in terms of production; also, there is no requirement for a certain quality of lands for the cultivation of resources. However, large-scale production is not yet economically viable until new technology emerges. Technological advances will require time in order to use these biofuels. Consequently, second-generation biofuels are not likely to play an important role to supply fuels in the near and medium term, unless specific improvements and ground-breaking technologies are developed (Lee and Lavoie 2013: 6–11; Stephens et al. 2008: 20–43; Naik et al. 2010: 578–597).

The third generation of biofuels has aimed to improve the production of biomass. In this category, specially engineered organisms or crops with considerable energy content such as microalgae are highly explored (Behera et al. 2015: 90; Saladini et al. 2016: 221–227). Algae biomasses are cultivated at lower costs, and the energy content is very high with the capacity to be renewed in very short periods. Therefore, there is less stress on water resources for algae growth. Algae biotechnology is a highly explored option and could be one of the foremost players in the renewable energy market for biofuels such as biodiesel, biohydrogen, bioethanol, and bio-jet fuel (Behera et al. 2015: 90; Saladini et al. 2016: 221–227; Bwapwa et al. 2017: 1345–1354). Though algae-based biofuels present few challenges, such as high costs for harvesting and drying, algae require fewer inputs. It is the fastest-growing biomass and can be grown anywhere, and there is no strain on the food crisis, thereby solving most problems of traditional crop-based biomass (Behera et al. 2015: 90; Saladini et al. 2016: 221–227; Alam et al. 2015: 763–768; Wang et al. 2009: 1856–1868). The main benefit of using algae compared to other biofuels is that algae do not compete with agricultural or limited resources such as water and lands (Bajpai 2019: 7–10; Behera et al. 2015: 90; Saladini et al. 2016: 221–227). Microalgae can considerably decrease the impact of greenhouse gases (GHG) in the environment and recycle CO_2 emissions from industrial operations. This is a very important fact, because the growth of algae depends on the amount of carbon dioxide. Therefore, power plants can reduce their CO_2 emissions in the atmosphere by supplying CO_2 to algae producers. Consequently, algae processing plants will not only produce fuels but will also reduce the effects of climate change. It is essential to emphasize the fact that the literature on algae-based transportation biofuels is very scarce, and not many studies have been undertaken in this specific field (Bwapwa et al. 2017: 1345–1354). However, an additional advantage of microalgae-based biofuels is that fuel fractions such as biodiesel, biogasoline, and bio-jet fuel can also be produced using the same processes as their petroleum counterparts. Ongoing research for algae-based biofuels today seeks to improve technologies to enable the production of bio-jet fuel and other biofuels to the commercial level (Bajpai 2019: 7–10; Behera et al. 2015: 90; Saladini et al. 2016: 221–227). A number of projects on algae-based biofuels have been achieved currently in the laboratory and pilot tests, but not many of them have been implemented at a large scale for commercialization. The feasibility for cost-competitive products remains a challenge because of the absence of cost-competitive processes at large scale. For instance, in the area of aviation fuels, scientists have concluded from a few studies that the production of bio-jet fuel can technically be achievable (Bwapwa et al. 2017: 1345–1354). However, the challenging concern is to improve the operation costs and increase algae lipids content (Bwapwa et al. 2018; 522–535).

The fourth-generation biofuels have focused more on the production of a very clean source of energy such as biohydrogen from microalgae and the bioconversion of living organisms (microorganisms and plants) by means of techniques common to biotechnology (Kumar et al. 2018: 1–16; Ben-Iwo et al. 2016: 172–192). Also, the aim is not only to produce sustainable energy but also to find ways to capture and store carbon dioxide. The carbon dioxide will be used to feed the biomass in order to convert it into sustainable fuel. The sequestration of carbon dioxide is part of the process, and ways to achieve it are still being developed. This can be achieved by storage in gas fields or saline aquifers. The carbon capture can become negative rather than neutral. More carbon is locked than produced; therefore, less carbon emission will be recorded (Dutta et al. 2014: 114–22; Kumar et al. 2018: 1–16; Ben-Iwo et al. 2016: 172–192).

Future prospects will be focused more on the improvement of the third- and fourth-generation biofuels. Improvements in technology are needed to make the production more effective and cost-competitive compared to fossil fuels. The demand for fuels is high and will always increase. The economics of producing these biofuels is not yet viable compared to fossil fuels. To alleviate the concern regarding operation costs, an efficient process that provides high-quality biofuel at competitive costs needs to be developed. In general, future prospects are now involving the third- and fourth-generation biofuels, which are focusing on the identification of feedstocks with very high lipid content and the use of innovative ways to extract oil and produce low-cost, low-carbon-emission, and sustainable fuels. Also, there is much focus on improving the lipid content of feedstock, with a low lipid content presenting the high potential for biofuels. The objective is to develop ways to stimulate lipid increase using cost-competitive methods.

The main aim for future biofuel production will also be focused on how to get drop-in fuels comply with standards and ready to be commercialized. Further studies have to be completed for process modelling to optimize the production of biofuels and reduce costs. It is a fact that global consumption for biofuels and renewable energy will rise due to the number of policies advocating for environmentally friendly products. The World Energy Council has indicated that by 2050 biofuels will represent 30% of the total global energy demand. This shows that there is huge potential for biofuels to be used as an alternative once fossil fuels are depleted. As mentioned earlier, more innovative ways are needed to improve the production processes, to identify the most promising feedstocks that can generate more lipids, and to develop methods to stimulate lipid increase from the various feedstocks. More emphasis can be put on the third- and fourth-generation biofuels because they are not competing with the food industry and can generate low carbon fuels.

Energy production is among the key factors that are essential to economic and industrial development. However, the production of fossil fuels through traditional petrochemical processes is seen as the major cause of environmental pollution and climate change effects. Furthermore, reserves of world fossil fuel resources are very limited and may lead to scarcity in the near future. To reduce dependence on fossil fuels and produce cost-effective fuel, biofuels can be explored to meet the costs (capital and production) and environmental requirements.

6.2   ALGAE-TO-BIOFUELS: OPPORTUNITY AND CHALLENGES

Today sustainable microalgae species are considered the life force of the growing biofuels industry. In particular, the microalgae lipid content is the key aspect of producing advanced biofuels. This implies that species with high lipid content are expected to be the most effective in the production of biofuels. However, for the large-scale operation for microalgae cultivation, the dewatering is not cost-competitive because it is energy demanding and requires recurring maintenance. There are also biological challenges related to the nature of the species and chemical challenges related to oil extraction using solvents.

The latter is not the major concern because there are ways to overcome it by using various downstream processes. It is rare to find a large scale for algae-based biofuels. Many attempts have taken place in the past in terms of laboratory and pilot tests.

More studies on the development and investments in innovative technologies should be undertaken to find ways to assist in overcoming the challenge of escalating to commercial size. It is therefore required to intensify research and development for the improvement of algae-based biofuels. This will assist in the reduction of risk and uncertainty linked to the algae-to-biofuels process for its commercialization at a large scale. Regular reviews in terms of technology gaps and cross-cutting needs are essential for research and development to get a clear picture of the current state of technology for scaling up microalgae-based biofuels.

Figure 6.1 illustrates various steps, from bioprospecting to downstream processing, for sustainable biofuel production using microalgae. Many options can be explored and improved to produce microalgae-based fuel in a sustainable way. Most

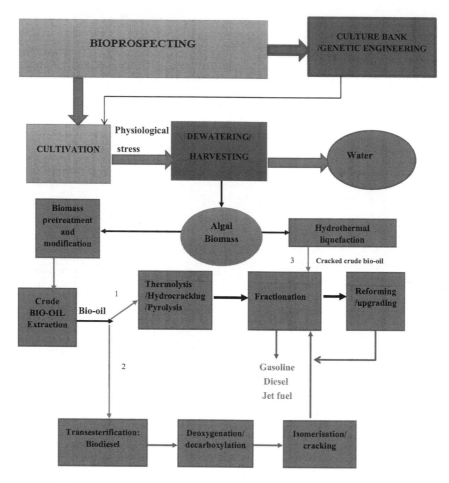

FIGURE 6.1   Sustainable biofuel production, from bioprospecting to green fuels, using microalgae.

of these processes are currently used in petrochemical processes to produce fossil-based fuels. It is therefore possible to follow the same methods and adapt these processes to microalgae crude bio-oil.

6.3   BIOPROSPECTING FOR OLEAGINOUS MICROALGAE

Successful downstream processing and subsequent industrial application of micro-algal biomass and metabolites hinge on a rigorous bioprospecting expedition. Microalgal bioprospecting is the searching, collection, isolation, and identification of indigenous microalgal strains with robust and unique properties such as hyper-lipid producers, as well as producers of novel metabolites and accessory pigments for the purpose of developing a commercial product (Montalvao et al. 2016: 399–406). Bioprospecting is a key driver and powerful tool for biodiversity conservation, as well as commercial utilization of algal bioproducts (Montalvao et al. 2016: 399–406). The best strategy for successful bioprospecting for hyper-lipid-producing microalgal strains has been previously reported (Mutanda et al. 2011: 57–70). According to Rizza et al. 2017, bioprospecting for indigenous microalgae is generally an initial step in the roadmap for the sustainable generation of biofuels and other useful commodities from microalgal biomass.

It is advantageous to consider indigenous microalgal strains because they are well adapted and naturally familiarized to the predominant environmental conditions (Rizza et al. 2017: 140–147). Briefly, the water samples are collected from local diverse aquatic habitats, such as lakes, dams, rivers, tributaries, ocean coastlines, hypersaline environments, brackish, and estuarine habitats. The collected water samples are processed immediately, whereby suitable artificial enrichment media are used for the growth of the microalgal strains that might be in the water samples. Thereafter, conventional microbiological and molecular biology methods are applied for the successful purification and identification of the microalgal strains. Bioprospecting for marine microalgae is technically difficult due to the diffuse location of the microalgal primary producers.

Temporal and spatial occurrence of algal blooms in marine and estuarine environments has been reported (Thornber et al. 2017: 82–96). Therefore, the collection of filter feeders is a promising approach for the collection of rare marine microalgal strains. The filter feeders are dissected and gut contents enriched in suitable media such as F/2 for the growth of marine microalgal strains. The bioprospecting approach for marine microalgal strains is reported to be a good strategy for the isolation of rare and robust microalgal strains from these aquatic habitats. However, the main drawback of the bioprospecting exercise is that some microalgal strains are fastidious and do not easily grow in artificial media. However, because most microalgal strains have been placed in culture collection centers, some researchers might prefer purchasing these strains directly from commercial vendors.

6.4 CRITERIA AND METHODOLOGY FOR SELECTING OLEAGINOUS MICROALGAE

The selection of the best-performing microalgal candidates requires some criteria to avoid overlooking potentially useful strains. Hence, it is desirable to devise a suitable selection strategy with well-defined criteria. The main question is: Are the microalgal strains producing the desired oil? If so, in what amount and under what growth conditions? Therefore, screening for the production of the desired oil is the first major step. Several techniques are applicable for the screening of lipid-producing microalgal strains, ranging from analytical techniques to fully automated high-throughput techniques.

The best method that is widely used for the screening of microalgal strains producing neutral lipids is the Nile red (9-diethylamino-5H-benzo (α) phenoxazine-5-one) staining technique employing the highly lipophilic dye Nile red (Chen et al. 2018: 71–81). Nile red staining is regularly used as an alternative method for microalgal neutral lipid detection due to its speed, simplicity, and low requirement in terms of microalgal biomass, as well as reduced number of samples and preparation time

(Halim and Webley 2015: 1–14). The basic principle of this method is that the Nile red dye binds to the lipid globules, forming a chromogenic complex that appears yellow in color for neutral lipids. The spectral properties of the dye are determined by the polarity of its surrounding medium. The Nile red dye emits intense fluorescence in hydrophobic organic solvents in contact with lipid globules, whereas in aqueous solutions, its fluorescence is totally quenched (Halim and Webley 2015: 1–14). Besides being quick and simple, the main advantage is that the method is not expensive because the equipment used is not expensive.

However, the Nile red method suffers from the main disadvantage of poor dye penetrance into cells with thick walls. Another drawback of this method is that the cells are killed and cannot be rejuvenated and cultured for further growth studies. Research has demonstrated that poor Nile red dye penetrance into the cells can be improved by adding 25% of dimethyl sulfoxide (DMSO) into the dye to increase membrane permeability. By adding glycerol or DMSO into the growing microalgae cultures, lipid staining efficiency is significantly improved, and it raises the fluorescence intensity of the stained cells (Chen et al. 2009: 41–47; Doan and Obbard 2011: 895–901; Cooper et al. 2010: 198–201). The Nile red technique has been used for the screening of microalgae producing neutral lipids such as *Chlorella, Scenedesmus, Chlamydomonas*, and *Nannochloropsis* (Chen et al. 2009: 41–47; 2018: 1–7).

A fairly new and improved staining method for the vital staining and monitoring of oil storage in live microalgal cells has been devised using a highly lipophilic fluorescent dye called BODIPY 505/515 (4,4-difluoro-1,3,5,7-tetramethyl-4-bora-3a,4adiaza-s-indacene) (Cooper et al. 2010:198–201). Due to its high oil/water partition coefficient, the dye can effortlessly cross cell membranes and organelle membranes, and the dye gathers in the lipidic intracellular sections by a diffusion-trap mechanism (Cooper et al. 2010:198–201). The main strength of this method is that the cells are not killed and can be further used for growth studies or other applications. Careful selection of the best microalgal strain is therefore informed from the presence of the desired lipid either by the Nile red or the BODIPY 505/515 staining method (Figure 6.2).

The sulfo-phospho-vanilin (SPV) method can also be used to quantitatively determine the lipid content of the whole-cell biomass without any physical extraction of the lipids (Byreddy et al. 2016: 28–32). The reaction mechanism as to how the pink color develops and the overall principle of the SPV method are not clearly understood. However, it is also desirable to physically extract lipids using the gravimetric method, solvent extraction, and perform simple and advanced chromatographic separation techniques such as thin layer chromatography (TLC) and gas chromatography and mass spectrometry (GC-MS) for the detection and identification, respectively, of lipids using suitable authentic pure standards. TLC is a simple chromatographic separation method whereby a mixture of lipids is separated according to their polarity and affinity for the stationary phase using a suitable mobile phase. TLC can quickly check for the presence of desired lipids by comparing the unknown lipids to the retardation factors (R_f) of authentic pure standards. However, TLC suffers from the main disadvantage of the co-migration of some lipids with the same molecular weight, it is only applicable to non-volatile

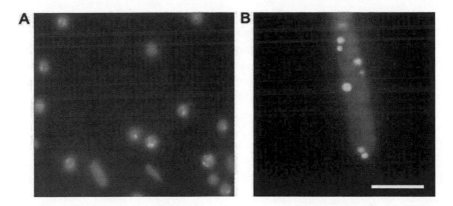

FIGURE 6.2 Micrographs of microalgal cells showing lipid globules stained with Nile red and BODIPY dyes. (A) Nile-red-stained *Chlorella* sp. observed at ×1000 using a fluorescence microscope at 490-nm excitation and 585-nm emission filters (Mutanda et al. 2011). (B) Oil-containing lipid bodies can be vitally stained and pictured in live oleaginous (oil-containing) microalgal cells using the green fluorescent dye, BODIPY 505/515. From the picture, vitally stained lipid bodies (green) are visibly differentiated from chloroplasts (red) in an *O. maius* Naegeli freshwater with algal cell in the form of a filament using an epifluorescence microscope named Zeiss Axioskop.

Source: Cooper et al. (2010:198–201)

compounds, it has limited resolution capability (separation numbers or peak capacities of 10 to 0), and it is currently not amenable to automation. The method of choice for lipid determination is GS-MS, whereby readily derivatized lipids are separated according to their mass-to-charge ratio. However, the GC-MS technique suffers from the main disadvantage of being too expensive (i.e. the equipment is expensive and beyond the reach of many research groups). Lastly, once desirable lipids are detected, identified, and quantified, the best microalgal candidate is carefully selected for further application.

6.5 IMPORTANT LIPIDS PRODUCED BY MICROALGAE

Microalgae and cyanobacteria produce diverse lipids under different growth conditions and are currently under extensive investigation for sustainable bioenergy and biochemical production. The quality and quantity of the lipids produced depend on the light intensity, nutrient composition, temperature, pH, salinity, and photoperiod. Because lipids are mainly storage compounds, their synthesis is mainly triggered by physiological stress in the growth environment. Nitrogen (N) and phosphorous (P) are the main macronutrients, and most microalgae are reported to produce various lipids under N- and P-limited growth conditions. The lipids produced by microalgae are saturated fatty acids, monounsaturated fatty acids, polyunsaturated fatty acids, and phospholipids *inter alia*. Therefore, the lipid content and fatty acid produced by microalgae are the key parameters for producing high-quality biodiesel fuel with the desired properties (Lin and Lin 2017: 399–403).

Microalgae species such as *Chlorella, Dunaliella, Nannochloris, Nannochloropsis, Neochloris, Porphyridium*, and *Scenedesmus* contain a high oil content of 20% to 50% of lipids by weight, making them appropriate for biofuel production (Piligaev et al. 2015: 368–376). Microalgae generate saturated fatty acids (SFAs) under certain growth conditions. SFAs are fatty acid chains that are predominantly made up of single bonds. These lipids have good oxidative stability due to their high melting points, which make them apt for the production of biofuels. SFAs play a key role in fuel properties, and the cetane number (CN) increases considerably in fuels with significant quantities of SFAs (Gopinath et al. 2009: 565–583). However, both saturation and unsaturation of fatty acid methyl esters (FAMEs) are expected to have an optimum balance for high biodiesel quality (Islam et al. 2013: 5676–5702). Microalgae produce monounsaturated fatty acids (MUFAs) in relatively low proportions compared to SFAs. MUFAs are fatty acids with one double bond in the fatty acid chain, with the remaining carbon atoms being single-bonded. The most important MUFAs are C16:1, C18:1, C20:1, C22:1, and C24:1.

The suitability of the microalgal lipids for biofuel production is due to the different fatty acids constituents, which vary as a result of structural features such as chain length, degree of saturation, branching of the carbon chain, positional isomers, configuration of double bonds, or other chemical groups (i.e., hydroxy, epoxy, cyclo, and keto) (Chen et al. 2018: 1–17).

To bioprospect and establish potential microalgae strains that are capable of producing high-quality biofuel, it is essential to screen for natural habitat strains with a high content of SFAs without additional requirements for costly supplements to hasten growth (e.g. vitamins) (Piligaev et al. 2015: 368–376). The most common SFA produced by microalgae such as *Chlorella vulgaris* and *Scenedesmus abundans* for biofuel production are mainly lauric acid (C12:0), myristic acid (C14:0), palmitic acid (C16:0), and stearic acid (C18:0). The preponderance and dominance of SFA—specifically palmitic acid and stearic acid—in microalgal lipids are desirable for the production of a high-quality biodiesel that satisfies international standards. Experiments to determine the biodiesel characteristics from the microalgal lipids extracted from *Isochrysis galbana* primarily comprised 35.34 wt.% myristic acid (C14:0) and 4.67 wt.% docosahexaenoic acid (DHA) in the biodiesel (Lin and Lin 2017: 399–403). Hence, the theoretical potential of microalgal biomass as a biofuel is ultimately determined by the acyl chains of the lipids, and therefore the lipid contents are quantified as the sum of their fatty acid constituents (Chen et al. 2018: 1–17; Wang et al. 2009: 1856–1868).

Freshwater and marine microalgae produce relatively different classes of lipids. The marine microalgae produce more long-chain lipids due to favorably low temperatures. In addition to other important metabolites produced by microalgae, highly polyunsaturated fatty acids (PUFAs) (e.g., eicosapentaenoic acid [EPA], arachidonic acid [AA], and DHA) are the most important (Spolaore et al. 2006: 87–96). Animals and higher plants lack the essential enzymes to synthesize PUFAs of more than 18 carbon chains, and therefore these must be obtained from food (Spolaore et al. 2006: 87–96). DHA is the unique algal PUFA that is commercially obtainable currently; hence, research on the screening of other microalgal strains is ongoing for the selection of PUFA producers.

6.6 PROCESSES FOR CONVERTING MICROALGAE TO BIOFUELS

6.6.1 CULTIVATION

The cultivation of microalgae is mainly undertaken in photobioreactors and open ponds as represented in Figures 6.3 and 6.4, using either photoautotrophic or heterotrophic methods.

Photoautotrophic cultivation. This method is achieved in conditions where microalgae cells are using light for their growth, aiming to generate more biomass. In this case, it is generally reported that capital costs needed to build closed photobioreactors are higher compared to the raceway ponds. Both photoautotrophic cultivation options have advantages and weaknesses, depending on the nature of the species/strain and environmental conditions.

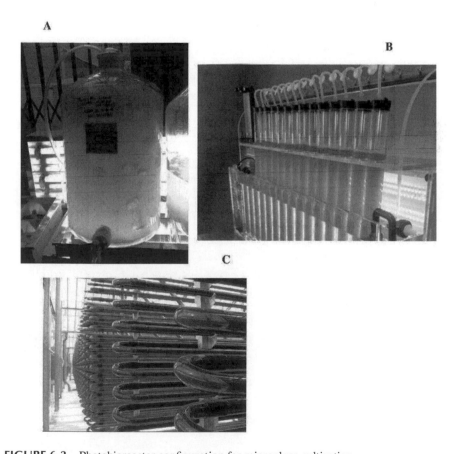

FIGURE 6.3 Photobioreactor configuration for microalgae cultivation.

Source: Algal Centre of Biotechnology, Mangosuthu University of Technology

FIGURE 6.4   Open pond configuration.

Usually, photobioreactors present weaknesses in terms of size escalation, the mixing, and the gas exchange for CO_2 and O_2. Therefore, this can cause a limitation in terms of biomass production. Although water loss is far less in photobioreactors than in open ponds because of the evaporation process, it appears that there are no benefits related to evaporative cooling. Consequently, the temperature should cautiously be monitored. In open ponds, the growing culture is exposed to regular changes with regard to temperature, humidity, and pH. These changes are recorded on a daily and seasonal basis. Therefore, in this case, the growing culture can directly be affected by environmental conditions. The species should be able to adapt to these changes, or the culture should be protected against these changes, which are external to the species' growing environment.

It is relatively challenging to undertake a regular sterilization for cultures growing in photobioreactors. However, periodic cleaning is required because of the formation of a biofilm in the photobioreactor tubes. Long-standing culture maintenance is reported to be higher compared to the one in open ponds. This is explained by the fact that in open ponds, the possibility of culture contamination is higher because of its direct exposure to the environment. Photobioreactors have the capacity to provide a greater surface-to-volume ratio. Therefore, they can support significant volumetric cell densities. This has the consequence of reducing water usage during cultivation and harvesting costs (Chisti 2007: 294–306). Light exposure in both photobioreactors and open ponds is a very important aspect of autotrophic growth.

Heterotrophic cultivation. For heterotrophic cultivation, microalgae cells are cultivated without a light source. Carbon sources such as sugars or CO_2 can be used to feed the growing culture, aiming to improve biomass productivity. This method is based on an established industrial fermentation technology. Fermentation is a mature technology widely used for the production of various products on a large scale. The heterotrophic cultivation method presents a number of advantages and challenges compared to the photoautotrophic method. First, it is affordable to maintain the optimum conditions required for an efficient and high biomass production rate. This has an advantage of reducing the extent and costs of the setup required for effective growth of microalgae cells or species (Xu et al. 2006: 499–507). Similarly, it is possible to prevent contamination during the cultivation period. Furthermore, there is a possibility to use cost-effective lignocellulosic sugars as food for the growing culture, for the sustainable growth of microalgae cells. The main challenges encountered by the heterotrophic cultivation method are related to the costs and accessibility of appropriate feedstocks such as lignocellulosic sugars. Also, there could be a competition for feedstocks with various biofuel technologies. A related approach suggested by many studies is mixotrophic cultivation. This approach combines both the photoautotrophic and heterotrophic ability of microalgae. Table 6.1 makes a comparison between these cultivation approaches.

TABLE 6.1
Comparative Features of Microalgal Cultivation Approaches

		Advantages	Weaknesses
Autotrophic cultivation	**Closed photobioreactors**	• Very little water loss compared to open ponds. • Long-term culture maintenance • Great surface-to-volume ratio supporting high volumetric cell densities.	• They have issues related to scalability. • They require a maintenance system for temperature because of the absence of the evaporative cooling system. • They may require sporadic cleaning because there is the formation of biofilm. • They require the greatest light exposure.
	Open ponds	• Cooling due to evaporation helps to maintain temperature. • Low capital costs.	• Show the possibility of maintaining optimal conditions for production and contamination prevention. • Prospect to use low-cost lignocellulosic sugars needed for cultivation. • High concentrations can be achieved.
Heterotrophic cultivation		• Optimal conditions are easily maintained for effective production and the hampering of contamination. • Provide an opportunity to use low-cost lignocellulosic sugars for growth. • High biomass concentration can be achieved.	• Cost-effective, and there is easy access for appropriate feedstocks such as lignocellulosic sugars. • There is a competition between feedstocks used for other biofuel technologies.

Source: Adapted from U.S. DOE (2010)

6.6.2 CHALLENGES RELATED TO MICROALGAE CULTIVATION

Water Management, Conservation, and Recycling

Microalgae have the ability to grow in water that is unsuitable for land crops, including saline water from the sea, municipal wastewater, brackish water, and water from aquifers. This is an important advantage that microalgae-based biofuels can have over many other biofuels. However, it is important to stress the fact that water management can cause many issues ranging from contamination to sustainability, to biomass productivity, to public support. Therefore, it is imperative to adequately address water management issues to avoid any consequences. If the cultivation systems are large in terms of their size, the demand for water to achieve a targeted biomass production will be massive. It is reported that for 1 1-ha pond that is 20 cm deep, more than 2 million liters of water will be required to fill the pond. It is reported that in desert areas, losses due to evaporation can surpass 0.5 cm on a daily basis (Weissman and Tillet 1989). This represents a daily water loss close to 50,000 liters from a 1-ha open pond. Even though the type of water being used for species cultivation can be saline, brackish, municipal wastewater, or any other low-quality water stream, it is important to stress the fact that the evaporated fraction is freshwater (U.S. DOE 2010). Consequently, with evaporation taking place, there will be salt concentrates, toxins, and other contaminants in the growing culture. The corrective measures should be the addition of high-quality freshwater to the growing culture. However, this option is perceived as not sustainable and costly. This is the situation where photobioreactors can be the acceptable option because no evaporation of water takes place. A water-conservation strategy must be put in place when using open ponds. This should be involved in the design of the open ponds systems and the site configuration. A recycling system is very important, but the volume of water to be recycled will depend on the strain/species, the quality of water used for cultivation, and the location (U.S. DOE 2010). However, it is also important to be aware of the fact that contaminants from concentrated salts or chemicals and biological inhibitors can be harmful to the growing culture if they are not properly removed during the recycling process. It is also a fact that the removal can be energy-intensive and costly. However, the water purification or recycling is still an important operation for the incoming or exiting water. Optimization of water use and an appropriate method for effective treatment are the key challenges of water management for sustainable cultivation of microalgae.

Scale-Up Challenges for Cultivation

The main challenge related to escalating from the laboratory to the large scale involves both technical and economic aspects. More focus should be on the innovative technologies that comprise both aspects involving technical advantages and cost-effectiveness. Small scales do not have major technical issues, and most of them are cost-effective. However, the need for high output in terms of biomass and crude bio-oil requires scaling up in order to make microalgae biofuels competitive. In this case, the use of municipal wastewater infrastructures and the domestic wastewater stream, mainly from households, which is rich in nutrients, can be an acceptable approach. However, the load of pathogens and contaminants could be harmful to the biomass (Hoffman et al. 2008: 70–81; Wilson et al. 2009: 1–42). More studies must be done in order to make this option more feasible.

Furthermore, there is a limited amount of knowledge regarding artificial pond ecology or pathology; therefore, further analyses or studies will be vital in terms of developing new approaches related to large-scale cultivation risk mitigation and remediation strategies from laboratory-scale units. Briefly, scaling-up challenges are also linked to culture stability, nutrient availability and water usage, conservation, management, and recycling. Consequently, the stability and success of large-scale cultures will depend on how the algae predators and pathogens can be prevented to be harmful to cultures—this is related to culture protection and maintenance. It will also depend on the presence of low-cost nutrients, the quality of water used for cultivation, and the knowledge of the operating mode of predators in order to prevent cell death during cultivation. This operating mode is not yet well understood, and further studies will be very beneficial (Becker 1994; Honda et al. 1999: 637–647; Cheng et al. 2004: 331–343; Brussaard 2004: 125–138; Hoffman et al. 2008: 70–81; Wilson et al. 2009: 1–42).

6.6.3 PHYSIOLOGICAL MODIFICATIONS

The main challenge hampering algae-based fuels from being competitive on the fuel market is the low oil content for many algae species. The composition of algae crude bio-oil is relatively similar in quality to that of fossil-based crude oil (Bwapwa et al. 2017: 1345–1354). Furthermore, the bio-oil from algae has the ability to generate low carbon emissions. The use of algae-based fuels, or any other biomass-based oil, can generate fuels with carbon emissions that can be 60% lower compared to that of fossil fuels. To improve algae bio-oil content, physiological modifications are used in order to get more oil out of algae species.

The physiological modifications can be achieved by using enzymes or processing by partial or total nutrient starvation. This process takes place after cultivation in most cases; however, it can also be undertaken during cultivation. Microalgae cells are placed in a cultivation environment made with either enzymes or without some of their main nutrients for a few days under illumination—the time frame for this technique depends on the species/strain type (Bwapwa et al. 2018: 522–535). During the time when microalgae cells are exposed to harsh conditions, their metabolism and genetics will be changing; this will consequently modify their physiology and their lipid content will be boosted. Physiological modification techniques still require more development to make them effective for any microalgae species. The enzyme-based techniques are known to be expensive, and a large amount of water is generated from the process. The starvation-based techniques are much more affordable, but the operating conditions depend on the nature and species or strain type. Once the species oil content is increased, conversion processes can be undertaken. Physiological modifications will not be needed if the species has a high oil content.

6.6.4 MICROALGAE HARVESTING METHODS

Harvesting of microalgae is an important cost component in the downstream processes. It is therefore imperative to analyse the various microalgae harvesting

TABLE 6.2
Strengths and Limitations of Harvesting Techniques

Method	Advantages	Limitations
Filtration	Affordable; a wide variety of filters and membranes available	Requires regular backwashing; it is time consuming. Highly reliant on algal species; best suited for large algal cells. Clogging or fouling is an issue.
Flotation	Cost-efficient and faster than sedimentation	Uses chemicals. Relies on suspended particles. Less reliable; it is used for specific algal species. High capital and operation costs.
Centrifugation	Quick, highly efficient, good recovery	Energy demanding and high capital costs.
Sedimentation	Low-cost technique, potentially can be used as a first stage for the reduction of energy input and costs of subsequent stages	Lower speed for separation; the output concentration might be low; may not be appropriate for some strains.
Microfiltration/ ultrafiltration	Can handle very small cells, highly performant with up to 98% dewatering Potentially good as pre-treatment method prior to centrifugation	High operating costs and membrane fouling. Current developments with antifouling membranes and low-pressure membranes can make this an efficient technique.

Source: Adapted from Mohn (1988); Molina Grima et al. (2003: 491–515); Shen et al. (2009: 1275–1287)

techniques that are available. This process is known as a dewatering stage in order to get the necessary biomass for downstream processes. However, dewatering is not a compulsory step for the hydrothermal liquefaction process that uses a wet slurry of microalgae. Some of the main methods generally used for the harvesting of microalgae are centrifugation, filtration, sedimentation, and flotation (Danquah et al. 2009: 1078–1083; Chen et al. 2011: 71–81). Table 6.2 highlights the strengths and limitations of these techniques.

6.6.5   BIO-OIL EXTRACTION AND PURIFICATION

Algal lipids are normally made up of mixtures of non-polar and polar components, including triglycerides, carotenoids, waxes, sterols, free fatty acids and xanthophylls, phospholipids, sphingolipids, and glycolipids (Becker 1994). To extract the maximum amount of bio-oil from microalgae cells, it is ideally recommended to have a solvent or solvent mixture that is sufficiently polar. However, the polarity should not be too high in order to ensure that the solvent readily dissolves non-polar lipids (Johnson 1983: 181–193). The process of extraction should be easy to carry out, and the recycling of solvent should be high in order to avoid adding fresh solvent continuously during the extraction. Furthermore, the solvent should have a low specific heat of evaporation and density. This aims to reduce the costs related to energy consumption. The solvent should also be environmentally friendly, renewable, and safe to handle. To collect the bio-oil, the solvent is mixed with dry or wet biomass and kept for a certain period of time in a decanting unit. A layer will be formed, separating two phases—the bio-oil will be at the top and the wet biomass will be at the bottom. A variety of methods are used to extract bio-oil from microalgae. Currently, Bligh and Dyer's method is the most used one, especially on the laboratory scale. It has been modified from time to time, depending on the species/strain. The extraction involves the mixture of chloroform and methanol in a volume ratio of 1:2 (Smedes and Askland 1999: 193–201; Bligh and Dyer 1959: 911–917). The Floch method is another one used for bio-oil extraction from samples with fairly high lipid content that provides more accurate results from the characterization. The Floch method uses a chloroform-to-methanol volume ratio of 2:1 for the extraction of microalgae bio-oil (Folch and Sloane-Stanley 1957: 497–509). This method provides convincing results in many studies, and it is considered more suitable for the extraction of total lipids, which are representative. The combination of chloroform and methanol requires careful handling in most cases because of the toxicity of the chloroform. This has raised the necessity of using solvents that are environmentally friendly. There are numerous types of solvents, but many of them have not been tested when it comes to the extraction of bio-oil from microalgae. However, at least one solvent— liquefied dimethyl ether (DME) has emerged and has produced satisfactory results. It is environmentally friendly, non-toxic, and has low energy demands. It has a boiling point of −24°C. It is partially miscible in water. This solvent was successfully used to extract oil from natural blue-green microalgae at moderate conditions: 20°C and 0.5MPa (Kanda and Li 2011: 1264–1266).

Once the bio-oil has been extracted, it must be purified to remove lipid fractions that are not suitable for biofuel production. This can be achieved via saponification. Non-saponifiable constituents, or the lipids that were not hydrolyzed during the saponification process, will remain non-polar and not soluble in water but soluble in hexane. Hexane is used to perform a post-saponification wash, and this allows the isolation of the unsaponifiable substances. These substances are made of carotenoids, sterols, and tocopherols. They are very useful in the nutraceutical industry (Ceron et al. 2008: 11761–11766). They can be isolated and processed separately to make a renewable biofuel, which is more cost-effective and economically viable. It is possible to skip the oil extraction process for catalytic processes such as gasification, pyrolysis, or hydrothermal liquefaction. In these processes, there is no need to produce biocrude oil; rather, but the whole algae biomass can be catalytically processed in order to produce biofuels such as green gasoline, diesel, or jet fuel. In many cases, bio-oil extraction has been completed on small scales in the laboratory.

6.6.6 HYDROTHERMAL LIQUEFACTION

In the course of this process, the reaction takes place between water and biomass at high temperatures (250 to 400°C) and pressures (50 to 200 bar) for a residence time of up to 30 minutes or under supercritical water conditions, which are defined as water kept in a liquid state above 100°C by applying pressure (Akhtar and Aishah 2011: 1615–1624; Balan et al. 2013: 732–759; Pandey et al. 2011; Patil et al. 2008: 1188–1195). In this process, wet biomass or a biomass slurry is fed to the reactor. There is no need for biomass harvesting after cultivation and bio-oil extraction. Consequently, the biomass is decomposed into reactive and small molecules with a higher energy density that is able to repolymerize in order to form the crude bio-oil (Zhang and Champagne 2010: 969–982). This implies that the molecules of bio-oil produced in this case are already cracked. There is no need for bio-oil cracking after the reaction. This technology is analogous to the natural geological processes taking place during the formation of petroleum oil and realized over significantly reduced time scales. Fractions made with CO_2, water, and biomass residue can be generated as by-products during the course of the process. While lipids are collected, the residual biomass is at the same time thermochemically processed to a crude bio-oil; the aqueous phase formed during the process is rich in nutrients, and it can be diluted and reused as microalgal growth media (Jena et al. 2011: 3380–3387; Levine et al. 2011; Biller et al. 2011a: 215–225; Biller et al. 2011b: 4841–4848).

The crude bio-oil output depends on the lipid content of the species/strain—a high lipid content in the species will obviously generate a high amount of crude bio-oil (Savage et al. 2009).

The advantage of this process is that the microalgae slurry can be directly processed and there is no need for biomass harvesting. Hydrothermal liquefaction of microalgae is a very promising technology. However, it is still a new approach, and a limited number of studies in this area have been published. More work was done in the private domain for commercialization purposes.

6.6.7 Pyrolysis

Pyrolysis is the thermal breakdown of organic substances achieved without oxygen. This decomposition can end up generating various products, depending on the operating conditions of the reaction. Low final temperatures (<450°C) with slow heating rates yield chars in most cases. Whereas at high temperatures (>800°C) with fast heating rates, gases are generally produced (U.S. DOE 2010). Fast pyrolysis is a variant that capitalizes on the oil production, and a yield of up to 80 wt% of dry feedstock is possible. During the process, the heating of finely ground feedstock is undertaken to a final temperature ranging between 400 and 600°C for a residence time of a few seconds (U.S. DOE 2010). Thereafter, there is rapid cooling (quenching) of the vapor produced to generate a bio-oil (Radlein and Quignard 2013; Güell et al. 2012). Pyrolysis for microalgae-based fuels is generally practiced in the liquid phase. In this case, it is performed on the microalgae crude bio-oil. The major advantage of pyrolysis over other conversion methods is its speed—the reaction can take place within a few seconds or minutes. A significant setback in using this technology for microalgae conversion to fuel is the moisture content of the biomass It is therefore imperative to perform substantial dehydration upstream in order to have an efficient process. It appears that pyrolysis will not be cost-competitive anytime soon unless a low-cost and innovative dewatering process is established. Furthermore, because pyrolysis is an established technology, it is expected that only significant developments or a breakthrough in conversion efficiency can take place.

Upgrading of Bio-oil

Bio-oil generated from pyrolysis requires catalytic upgrading. However, bio-oil produced from hydrothermal liquefaction does not require an upgrading process. Generally, the oxygen and water content is high in the bio-oil generated from pyrolysis. Also, the bio-oil heating value is low, and its thermal stability is relatively poor and exceedingly acidic. It is chemically unstable and corrosive and cannot be mixed with petroleum oil (Radlein and Quignard 2013; Güell et al. 2012; Snowden-Swan and Male 2012; Pandey et al. 2011). Two methods are used for the upgrading of bio-oil in order to remediate the weaknesses presented by pyrolyzed bio-oil from microalgae or any other biomass-related oil.

Two-Step Hydroprocessing

This approach involves two stages of hydrodeoxygenation (Radlein and Quignard 2013). Generally, the pyrolyzed bio-oil is hydrotreated at low temperatures (around 250°C) and high pressures lower than 2500 bars (Mawhood et al. 2014: 15). This is achieved in the presence of a catalyst before the hydrocracking at higher temperatures ranging between 350 and 400°C and the same high pressures (Brown et al. 2013: 463–469; Mawhood et al. 2014: 15). The generated product is made of H_2 with about 20 wt% liquid hydrocarbons.

Fluid Catalytic Cracking

The upgrading of bio-oil can be completed by a unique stage known as hydrotreatment. This is followed by fluid catalytic cracking, often using zeolite as a catalyst. Commodity chemicals are mainly produced during this process. It is less commonly used to produce transportation fuels (Zhang et al. 2013: 895–904).

6.6.8 GASIFICATION

Gasification of the microalgae biomass can provide a very flexible way to produce different liquid fuels. It usually operates at high temperatures in order to reduce impurities in the syngas and at high pressures without nitrogen dilution. This can primarily be achieved through the Fischer–Tropsch (FT) process or mixed alcohol synthesis of the resulting syngas. FT is a fairly established technology for which the syngas components (CO, CO_2, H_2O, H_2, and impurities) are purified and upgraded to suitable liquid fuels (Okabe et al. 2009: 171–176; Srinivas et al. 2007: 66–67; Balat 2006: 83–103). The conversion of bio-syngas presents numerous advantages compared to other methods. First, it is likely to produce an extensive variety of fuels with properties that comply with the relevant standards. Furthermore, bio-syngas is a useful feedstock for the production of several products, allowing the process to be flexible. Additionally, there is a possibility to use algae biomass in an existing petrochemical or thermochemical process plant. This is an advantage because the industrial processes in a petrochemical plant are efficient. It is therefore conceivable to feed a coal gasification plant with algae biomass in order to cut down on the costs related to the required capital investment and address the issue of availability.

6.6.9 TRANSESTERIFICATION

The transesterification reaction allows the conversion of triacylglycerols extracted from microalgae to FAMEs. This reaction takes place when triacylglycerols are reacted with an alcohol. In most cases, methanol or ethanol is used for transesterification. The process is done using either edible or non-edible vegetable oils. It can be completed via catalytic or non-catalytic reaction at a required temperature (or heat). This technology is relatively mature and uses an efficient conversion process (Hossain et al. 2008: 250–254). With regard to the base-catalyzed method, it is reported that transesterification of microalgal oil can be completed with ethanol/ methanol and sodium ethanolate as the catalyst (Zhou and Boocock 2006: 1047– 1052). The products from these reactions are generally separated by the addition of ether and saline water to the solution followed by thorough mixing. A vaporizer under high vacuum is used to separate biodiesel and ether. The route using the acid-catalyzed transesterification reaction is also an attractive option because acid catalysts such as H_2SO_4, HCl, and H_3PO_4 have low sensitivity in the presence of water and free fatty acids. Consequently, there is alleviation of saponification and emulsification, leading to the enhancement of product recovery (Ataya et al. 2008: 679–685; Wahlen et al. 2008: 4223–4228). Additionally, Lewis acid catalysts, such as $AlCl_3$ or $ZnCl_2$, are established as a feasible alternative in the conversion of triacylglycerols into FAMEs. Deoxygenation or decarboxylation can be undertaken after transesterification to help reduce the amount of oxygen and eliminate the esters, which are considered as undesirable if the final product is green gasoline, green diesel, or green jet fuel (Bwapwa et al. 2017: 1345–1354).

Enzymatic Conversion

The processes analysed in previous sections are all based on physico-chemical reactions, and they have a high conversion of bio-oil to methyl esters (US DOE 2010). However, these processes are energy intensive, the removal of glycerol is very challenging, they require a removal of alkaline catalysts from the product, and there is a generation of a complex alkaline effluent that will require appropriate treatment (U.S. DOE 2010). The enzymatic conversion, also known as a biochemical conversion, uses biocatalysts called lipases during the transesterification reaction for biodiesel production. These lipases are environmentally attractive compared to the conventional processes based on physico-chemical reactions (Svensson and Adlercreutz 2008: 1007–1013). Enzymatic methods are becoming attractive; however, they are not yet in operation on a large scale. This is due to the high costs of lipase and its short lifespan triggered by the negative effects from the excess of methanol and glycerol (U.S. DOE 2010). To establish a commercially viable enzymatic conversion, it is important to address all these aspects because they have a direct influence on the process's success. Therefore, a key area to be addressed is the

solvent and temperature tolerance of the enzymes to allow effective biocatalytic processing. It is reported that the presence of a solvent is substantial because it enhances
the degree of solubility of triacylglycerols (TAGs) during the extraction process. The
enzymes used in the conversion processes should operate effectively in the presence
of these solvents to allow cost-effective biofuel production (Fang et al. 2006: 510–
515). Furthermore, it is reported that enzymatic bioprospecting in extreme environments can generate new enzymes with required features that are more adequate for
industrial applications (Guncheva et al. 2008: 129–132). Another important aspect
relates to the development of enzymes that can break algal cell walls. More studies
are needed to determine which enzymes can produce reactions in different environments and on bio-oil feedstocks (Lopez-Hernandez et al. 2005: 365–372).

Supercritical Processing

This is a new process with the capacity of concurrently performing the extraction
and conversion of bio-oil into biofuels (Demirbas 2006: 933–940). Supercritical fluid
extraction of microalgal oil is faraway more effective compared to the basic solvent
separation methods, and it was shown to be highly effective in terms of extracting
various components present in microalgae (Mendes 2007: 189–213). The method is
a supercritical transesterification that can be also used for microalgal bio-oil. The
supercritical fluids used are selective, and they can provide high purity and high
product concentrations. Furthermore, no organic solvent residues are in the extract
or used biomass (Demirbas 2009: 163–168). Extraction is effective generally at less
than 50°C to ensure that great product stability and quality is achieved. Also, there
is no need for microalgal dewatering when using supercritical fluids; consequently,
the efficiency of the process is increased.

6.6.10 THERMAL AND CATALYTIC CRACKING

Microalgae bio-oil is made up of long-chain carbons that can reach up to C50 or
more depending on the species/strain (Bwapwa et al. 2018: 522–535). Different
biofuels such as green diesel, green gasoline, and green jet fuel have a specific
number of carbons when it comes to these chains. Cracking can assist in getting
the specific fraction for the required biofuel. Long carbon chains can be broken
down into small ones that can be relevant for green gasoline, green diesel, or
green jet fuel. Thermal cracking can be achieved in the temperature range between
250 and 370°C. This process can take longer in terms of time; some studies have
reported that thermal cracking can take up to 30 min (Bwapwa et al. 2018: 522–
535). To reduce the cracking time, catalytic cracking can be undertaken. In this
case, catalysts, together with the heat energy, play an important role in breaking
down long carbon chains into small and medium ones. Catalysts have the reputa-
tion of being very efficient and relatively cost-effective. However, they require
cleaning and elimination from the product stream—this consequently affects the
overall cost. The use of immobilized heterogeneous or homogeneous catalysts that
are affordable and effective can be the most likely answer to this problem (McNeff
et al. 2008: 39–48).

Fractionation of Cracked Bio-oil

The process is used to get "green" fuel fractions after fractional distillation, which is a separation method based on the difference of boiling points. This process is achieved with the help of a distillation unit that separates cracked bio-oil into various fuel fractions. During distillation, various fractions of fuels are collected at various boiling temperatures. The light distillate fractions such as naphtha and green gasoline are the first to be collected at temperatures right around (or below) 200°C. Thereafter, the middle distillates such as green jet fuel and a certain fraction of gasoline can be collected between 200 and 300°C, and the heavy distillate fractions such as green diesel are collected when the temperature reaches beyond 300°C. The fractionation process is necessary and cannot be avoided; it helps to get the needed fuel fraction at a relevant temperature. After fractionation, a process called catalytic reforming can be undertaken. During fractionation, below a temperature of 200°C, very light carbon fraction fuels called naphtha are collected. These low-carbon fractions can be converted into high-octane products via catalytic reforming. These are important high-quality blending stocks for green gasoline.

6.7 CONCLUSIONS

The algal biotechnology industry is developing at a fast pace. Therefore, it is incumbent upon researchers to develop technologies for sustainable biofuel production from microalgal biomass. The key procedure of microalgal bioprospecting is crucial in order to effectively utilize readily available bioresources, such as unique microalgal strains from diverse aquatic habitats. This will also facilitate the development of algal culture banks for other applications besides biofuel production. To date, the exploration and research on the novel microalgal metabolites and biopolymers such as phycobiliproteins, polysaccharides, lipids, and vitamins are gaining momentum. In addition to bioprospecting, it is critical to genetically modify the physiology of the existing microalgal strains to ensure the production of desired bioproducts, such as lipids, for sustainable biofuel production. Current techniques for microalgal biofuel production, such as hydrothermal liquefaction, cracking, pyrolysis, transesterification, gasification, and supercritical processing, are advancing. However, it is mandatory to develop cost-effective techniques for sustainable biofuel production to avert fuel shortages in the future. The tiny size of microalgal cells, as well as the occurrence of the microalgal entities in aqueous suspension, pose serious harvesting and dewatering challenges that must be addressed for successful commercial utilization of microalgal biomass. Current challenges in algal biotechnology are mainly the cultivation of microalgae at a large scale, dewatering, and metabolite extraction. In conclusion, the current developments in the upstream and downstream processing of microalgal biomass offer a bright outlook for the microalgal biofuel industry. In addition, the spin-off bioproducts generated from microalgal biomass will ultimately improve the techno-economic perspectives for the sustainable biofuel production for the future.

REFERENCES

Akhtar, J., and A. Nor, and S. Aishah. 2011 "A review on process conditions for optimum bio-oil yield in hydrothermal liquefaction of biomass." *Renewable and Sustainable Energy Reviews*, 15:1615–1624. doi: 10.1016/j.rser.2010.11.054.

Alam, F., S. Mobin, and H. Chowdhury. 2015. "Third generation biofuel from algae." *Procedia Engineering* 105:763–768. doi: 10.1016/j.proeng.2015.05.068.

Ataya, F., M. A. Dube, and M. Ternan. 2008. "Variables affecting the induction period during acid-catalysed transesterification of canola oil to FAME." *Energy Fuels* 22 (1):679–685. 10.1021/ef7005386.

Bajpai, P. 2019. "Fuel potential of third generation biofuels." In *Third Generation Biofuels*, edited by P. Bajpai, 7–10. Singapore: Springer. doi: 10.1007/978-981-13-2378-2_2.

Balan, V., D. Chiaramonti, and S. Kumar. 2013. "Review of US and EU initiatives toward development, demonstration, and commercialization of lignocellulosic biofuels." *Biofuels, Bioproducts and Biorefining* 7:732–759. doi: 10.1002/bbb.1436.

Balat, M. 2006. "Sustainable transportation fuels from biomass materials." *Energy, Education, Science and Technology* 17 (1–2):83–103.

Becker, E. W. 1994. *Microalgae: Biotechnology and Microbiology*. Cambridge: Cambridge University Press.

Behera, S., R. Singh, R. Arora, N. K. Sharma, M. Shukla, and S. Kumar. 2015. "Scope of algae as third generation biofuels." *Frontiers in Bioengineering and Biotechnology* 2:90. doi: 10.3389/fbioe.2014.00090.

Ben-Iwo, J., V. Manovic, and P. Longhurst. 2016. "Biomass resources and biofuels potential for the production of transportation fuels in Nigeria." *Renewable and Sustainable Energy Reviews* 63:172–192.

Biller, P., and A. P. Ross. 2011a: "Potential yields and properties of oil from the hydrothermal liquefaction of microalgae with different biochemical content." *Bioresource Technology* 102:215–225. doi: 10.1016/j.biortech.2010.06.028.

Biller, P., R Riley, and A. B. Ross. 2011b. "Catalytic hydrothermal processing of microalgae: Decomposition and upgrading of lipids." *Bioresource Technology* 102:4841–4848. doi: 10.1016/j.biortech.2010.12.113.

Bligh, E. G., and W. J. Dyer. 1959. "A rapid method of total lipid extraction and purification." *Canadian Journal of Biochemistry and Physiology* 37:911–917. doi: 10.1139/y59-099.

Brown, T. R., R. Thilakaratne, R. C. Brown, and G. P. Hu. 2013. "Techno-economic analysis of biomass to transportation fuels and electricity via fast pyrolysis and hydroprocessing." *Fuel* 106:463–469. doi: 10.1016/j.fuel.2012.11.029.

Brussaard, C. P. D. 2004. "Viral control of phytoplankton populations: A review." *The Journal of Eukaryotic Microbiology* 51:125–138. doi: 10.1111/j.1550-7408.2004.tb00537.x.

Bwapwa, J. K., A. Anandraj, and C. Trois. 2017. "Possibilities for conversion of microalgae oil into aviation fuel: A review." *Renewable and Sustainable Energy Reviews* 80:1345–1354. doi: 10.1016/j.rser.2017.05.224.

Bwapwa, J. K., A. Anandraj, and C. Trois. 2018. "Microalgae processing for jet fuel production." *Biofuels, Bioproducts and Biorefining* 12 (4):522–535. doi: 10.1002/bbb.1878.

Byreddy, A. R., A. Gupta, C. J. Barrow, and M. Puri. 2016. "A quick colorimetric method for total lipid quantification in microalgae." *Journal of Microbiological Methods* 125:28–32. doi: 10.1016/j.mimet.2016.04.002.

Ceron, C. M, I. I. Campos, J. F. Sanchez, F. G. Acien, E. Molina, and J. M. Fernandez-Sevilla. 2008. "Recovery of lutein from microalgae biomass: Development of a process for Scenedesmus almeriensis biomass." *Journal of Agricultural and Food Chemistry* 56:11761–11766. doi: 10.1021/jf8025875.

Chen, C.-Y., K. L. Yeh, R. Aisyah, D. J. Lee, J. S. Chang. 2011. "Cultivation, photobioreactor design and harvesting of microalgae for biodiesel production: A critical review." *Bioresource Technology* 102:71–81. doi: 10.1016/j.biortech.2010.06.159.

Chen, W., C. Zhang, L. Song, M. Sommerfield, Q. Hu. 2009. "A high throughput Nile red method for quantitative measurement of neutral lipids in microalgae." *Journal of Microbiological Methods* 77:41–47. doi: 10.1016/j.mimet.2009.01.001.

Chen, Z., L. Wang, S. Qiu, and S. Ge. 2018. "Determination of microalgal lipid content and fatty acid for biofuel production." *BioMed Research International*, 1–17 doi: 10.1155/2018/1503126.

Cheng, S. H., S. Aoki, M. Maeda, and A. Hino 2004. "Competition between the rotifer brachionus rotundiformis and the ciliate euplotes vannus fed on two different algae." *Aquaculture Engineering* 241 (1–4):331–343.

Chisti, Y. 2007. "Biodiesel from microalgae." *Biotechnology Advances* 25:294–306.

Cooper, M. S., W. R. Hardin, T. W. Petersen, and R. A. Cattolico. 2010. "Visualizing "green oil" in live algal cells." *Journal of Bioscience and Bioengineering* 109 (2):198–201. doi: 10.1016/j.jbiosc.2009.08.004.

Danquah, M. K., L. Ang, N. Uduman, N. Moheimani, and G. M. Fordea. 2009. "Dewatering of microalgal culture for biodiesel production: Exploring polymer flocculation and tangential flow filtration." *Journal of Chemical Technology and Biotechnology* 84:1078–1083. doi: 10.1002/jctb.2137.

Demirbas, A. 2006. "Oily products from mosses and algae via pyrolysis." *Energy Sources, Part A: Recovery, Utilization, and Environmental Effects* 28 (10):933–940. doi: 10.1080/009083190910389.

Demirbas, A. 2009. "Production of biodiesel from algae oils." *Energy Sources, Part A: Recovery, Utilization, and Environmental Effects* 31 (2):163–168. doi: 10.1080/15567030701521775.

Doan, T.-T. Y., and J. P. Obbard. 2011. "Improved Nile red staining of Nannochloropsis sp." *Journal of Applied Phycology* 23:895–901. doi: 10.1007/s10811-010-9608-5.

Dutta, K., A. Daverey, and J. G. Lin. 2014. "Evolution retrospective for alternative fuels: First to fourth generation." *Renewable Energy* 69:114–122. doi: 10.1016/j.renene.2014.02.044

Fairley, P. 2011. "Introduction: Next generation biofuels." *Nature* 474 (7352):2–5. doi: 10.1038/474S02a.

Fang, Y., Z. Lu, F. Lv, X. Bie, S. Liu, Z. Ding, and W. Xu. 2006. "A newly isolated organic solvent tolerant staphylococcus saprophyticus M36 produced organic solvent-stable lipase." *Current Microbiology* 53 (6):510–515.

Folch, J. M. L., and G. H. Sloane-Stanley 1957. "A simple method for the isolation and purification of total lipids from animal tissues." *Journal of Biological Chemistry* 226:497–509.

Gopinath, A., S. Puhan, and G. Nagarajan. 2009. "Relating the cetane number of biodiesel fuels to their fatty acid composition: A critical study." *Proceedings of the Institution of Mechanical Engineers, Part D: Journal of Automobile Engineering* 223 (4):565–583.

Güell, B. M., M. Bugge, R. S. Kempegowda, A. George, and S. M. Paap. 2012. Benchmark of conversion and production technologies for synthetic biofuels for aviation. Norway: SINTEF Energy Research.

Guncheva, M., D. Zhiryakova, N. Radchenkova, and M. Kambourova. 2008. "Acidolysis of tripalmitin with oleic acid catalyzed by a newly isolated thermostable lipase." *Journal of the American Oil Chemists' Society* 85 (2):129–132.

Halim, R., and P. A. Webley. 2015. "Nile red staining for oil determination in microalgal cells: A new insight through statistical modelling." *International Journal of Chemical Engineering* 2015:1–14 doi: 10.1155/2015/695061.

Hoffman, Y., C. Aflalo, A. Zarka, J. Gutman, T. Y. James, and S. Boussiba. 2008. "Isolation and characterization of a novel chytrid species (phylum Blastocladiomycota), parasitic on the green alga Haermatococcus." *Mycological Research* 112:70–81. doi: 10.1016/j.mycres.2007.09.002.

Honda, D., T. Yokochi, T. Nakahara, S. Raghukumar, A. Nakagiri, K. Schaumann, and T. Higashihara. 1999. "Molecular phylogeny of labyrinthulids and thraustochytrids based on the sequencing of 18S ribosomal RNA gene." *Journal of Eukaryotic Microbiology* 46 (6):637–647. doi: 10.1111/j.1550-7408.1999.tb05141.x.

Hossain, A. B. M. S., A. Salleh, A. N. Boyce, P. Chowdhury, and M. Naqiuddin. 2008. "Biodiesel fuel production from algae as renewable energy." *American Journal of Biochemistry and Biotechnology* 4 (3):250–254. doi: 10.3844/ajbbsp.2008.250.254.

Islam, M. A., M. Magnusson, R. J. Brown, G. A. Ayoko, N. Nabi, and K. Heimann. 2013. "Microalgal species selection for biodiesel production based on fuel properties derived from fatty acid profiles." *Energies* 6:5676–5702. doi: 10.3390/en6115676.

Jena, U., N. Vaidyanathan, S. Chinnasamy, and K. C. Das. 2011. "Evaluation of microalgae cultivation using recovered aqueous co-product from thermochemical liquefaction of algal biomass." *Bioresource Technology* 102 (3):3380–3387. doi: 10.1016/j.biortech.2010.09.111.

Johnson, L A. 1983. "Comparison of alternative solvents for oil extraction." *Journal of the American Oil Chemists' Society* 60 (2):18193. doi: 10.1007/BF02543490.

Kanda, H., and P. Li. 2011. "Simple extraction method of green crude from natural blue-green microalgae by dimethyl ether." *Fuel* 90 (3):1264–1266.

Kumar, A., S. Ogita, and Y. Y. Yau. (eds.) 2018. *Biofuels: Greenhouse Gas Mitigation and Global Warming: Next Generation Biofuels and Role of Biotechnology*, 1–16. Jaipur: Springer. doi: 10.1007/978-81-322-3763.

Lee, R. A., and J. M. Lavoie. 2013. "From first-to third-generation biofuels: Challenges of producing a commodity from a biomass of increasing complexity." *Animal Frontiers* 3 (2):6–11. doi: 10.2527/af.2013-0010.

Levine, R., P. Duan, T. Brown, and P. E. Savage. 2011. "Hydrothermal Liquefaction of Microalgae with Integrated Nutrient Recovery." 1st International Conference on Algal Biomass, Biofuels and Bioproducts. St. Louis.

Lin, C.-Y., and B.-Y. Lin. 2017. "Comparison of fatty acid compositions and fuel characteristics of biodiesels made from Isochrysis galbana lipids and from used cooking oil." *Journal of Marine Science and Technology* 25 (4):399–403. doi: 10.6119/JMST-017-0317-1.

Lopez-Hernandez, A., H. S. Garcia, and C. G. Hill, Jr. 2005. "Lipase-catalyzed transesterification of medium-chain triacylglycerols and a fully hydrogenated soybean oil." *Journal of Food Science* 70 (6): C365–372.

Mawhood R., A. R. Cobas, and R. Slade. 2014. Establishing a European renewable jet fuel supply chain: The technoeconomic potential of biomass conversion technologies, Final Report, Imperial College, London, November 2014, pp. 1–53.

McNeff, C. V., L. C. McNeff, B. Yan, D. T. Nowlan, M. Rasmussen, A. E. Gyberg, B. J. Krohn, R. L. Fedie, and T. R. Hoye. 2008. "A continuous catalytic system for biodiesel production." *Applied Catalysis, A: General* 343 (1–2):39–48. doi: 10.1016/j.apcata.2008.03.019.

Mendes, R. L. 2007. "Supercritical fluid extraction of active compounds from algae." In *Supercritical Fluid Extraction of Nutraceuticals and Bioactive Compounds*, edited by J. L. Martinez, 189–213. Boca Raton, FL: Taylor & Francis, Inc. doi: 10.1201/9781420006513.ch6.

Mohn, F. 1988. "Harvesting of micro-algal biomass." In *Micro-algal Biotechnology*, edited by L. J. Borowitzk and M. A. Borowitzka. Cambridge: Cambridge University Press.

Molina Grima, E., E.-H. Belarbi, F. G. Acien-Fernandez, A. Robles- Medina, C. Yusuf. 2003. "Recovery of microalgal biomass and metabolites: Process options and economics." *Biotechnol Adv* 20 (7–8):491–515. doi: 10.1016/S0734-9750(02)00050-2.

Montalvao, S., Z. Demirel, P. Devi, V. Lombardi, V. Hongisto, M. Perala, J. Hattara, E. Imamoglu, S. S. Tilvi, G. Turan, M. C. Dalay, and P. Tammela. 2016. "Bioprospecting for native microalgae as an alternative source of sugars for the production of bioethanol." *New Biotechnology* 33 (3):399–406.

Mutanda, T., D. Ramesh, S. Karthikeyan, S. Kumari, A. Anandraj, and F. Bux.2011. "Bioprospecting for hyper-lipid producing microalgal strains for sustainable biofuel production." *Bioresource Technology* 102 (1):57–70. doi: 10.1016/j.biortech.2010.06.077.

Naik, S. N., V. V. Goud, P. K. Rout, and A. K. Dalai. 2010. "Production of first- and second-generation biofuels: A comprehensive review." *Renewable and Sustainable Energy Reviews* 14 (2):578–597. doi: 10.1016/j.rser.2009.10.003.

Okabe, K., K. Murata, M. Nakanishi, T. Ogi, M. Nurunnabi, and Y. Liu. 2009. "Fischer-Tropsch Synthesis over Ru catalysts by using syngas derived from woody biomass." *Catalysis Letters* 128 (1–2):171–176. doi: 10.1007/s10562-008-9722-z.

Pandey, A., C. Larroche, S. Ricke, C.-G. Dussap, and E. Gnansounou. (eds.) 2011. *Biofuels—Alternative Feedstocks and Conversion Processes*. New York: Elsevier.

Patil, V., K. Q. Tran, and H. R. Giselroed. 2008. "Towards sustainable production of biofuels from microalgae." *International Journal of Molecular Sciences* 9 (7):1188–1195. doi: 10.3390/ijms9071188.

Piligaev, A. V., K. N. Sorokina, A. V. Bryanskaya, S. E. Peltek, L. A. Kolchanov, and V. N. Parmon. 2015. "Isolation of prospective microalgal strains with high saturated fatty acid content for biofuel production." *Algal Research* 12:368–376. doi: 10.1016/j.algal.2015.08.026.

Radlein, D., and A. Quignard. 2013. "A short historical review of fast pyrolysis of biomass." *Oil and Gas Science and Technology*. doi: 10.2516/ogst/2013162.

Rizza, L. S., M. E. S. Smachetti, M. Do Nascimento, G. L. Salerno, and L. Curatti. 2017. "Bioprospecting for native microalgae as an alternative source of sugars for the production of bioethanol." *Algal Research* 22:140–147. doi: 10.1016/j.algal.2016.12.021.

Saladini, F., N. Patrizi, F. M. Pulselli, N. Marchettini, and S. Bastianoni. 2016. "Guidelines for energy evaluation of first, second and third generation biofuels." *Renewable and Sustainable Energy Reviews* 66:221–227.

Savage, P. E., R. B. Levine, and C. M. Huelsman. 2009. "Hydrothermal Processing of Biomass." In Thermochemical Conversion of Biomass to Liquid Fuels and Chemicals, edited by M Crocker. Royal Society of Chemistry.

Shen, Y., W. Yuan, Z. J. Pei, Q. Wu, and E. Mao. 2009. "Microalgae mass production methods." *Transactions of the ASABE* 52 (4):1275–1287.

Smedes, F, and T. K. Askland. 1999. "Revisiting the development of the bligh and dyer total lipid determination method." *Marine Pollution Bulletin* 38:193–201. doi: 10.1016/ S0025-326X(98)00170-2.

Snowden-Swan, L. J., and J. L. Male. 2012. Summary of fast pyrolysis and upgrading GHG analyses (No. PNNL-22175). Pacific Northwest National Lab (PNNL), Richland, WA, USA.

Spolaore, P., C. Joannis-Cassan, E. Duran, and A. Isambert. 2006. "Commercial applications of microalgae." *Journal of Bioscience and Bioengineering* 101 (2):87–96. doi: 10.1263/ jbb.101.87.

Srinivas, S., R. K. Mahajani, and M. F. T. Sanjay. 2007. "Synthesis using bio-syngas and CO_2." *Energy for Sustainable Development* 11 (4):66–71. doi: 10.1016/S0973-0826(08)60411-1.

Stephens, E., U. C. Marx, J. H. Mussgnug, C. Posten, O. Kruse, and B. Hankamer. 2008. "Second generation biofuels: High-efficiency microalgae for biodiesel production." *Bioenergy Research* 1 (1):20–43.

Svensson, J., and P. Adlercreutz. 2008. "Identification of triacylglycerols in the enzymatic transesterification of rapeseed and butter oil." *European Journal of Lipid Science and Technology* 110 (11):1007–1013.

Thornber, C. S., M. Guidone, C. Deacutis, L. Green, C. N. Ramsayd, and M. Palmisciano. 2017. "Spatial and temporal variability in macroalgal blooms in a eutrophied coastal estuary." *Harmful Algae* 68:82–96. doi: 10.1016/j.hal.2017.07.011.

U.S. DOE. 2010. National Algal Biofuels Technology Roadmap. U.S. Department of Energy, Office of Energy Efficiency and Renewable Energy, Biomass Program. http://biomass. energy.gov

Wahlen, B. D., B. M. Barney, and L. C. Seefeldt. 2008. "Synthesis of biodiesel from mixed feedstocks and longer chain alcohols using an acid-catalysed method." *Energy and Fuels* 22 (6):4223–4228. doi: 10.1021/ef800279t.

Wang, Z., N. Ullrich, S. Joo, S. Waffenschmidt, and U. Goodenough. 2009. "Algal lipid bodies: Stress induction, purification, and biochemical characterization in wild-type and starchless Chlamydomonas reinhardtii." *Eukaryotic Cell* 8:1856–1868. doi: 10.1128/EC.00272-09.

Weissman, J. C., D. M. Tillett, and R. P. Goebel. 1989. *Design and operation of an outdoor microalgae test facility (No. SERI/STR-232-3569)*. Vacaville, CA: Microbial Products, Inc.

Wilson, W. H., J. L. Van Etten, and M. J. Allen. 2009. The Phycodnaviridae: The story of how tiny giants rule the world. Lesser Known Large dsDNA Viruses. Berlin: Springer-Verlag. 328:1–42.

Xu, H., X. Miao, and Q. Wu. 2006. "High quality biodiesel production from a microalga chlorella protothecoides by heterotrophic growth in fermenters." *Journal of Biotechnology* 126:499–507.

Zhang, L, C. Xu, and P. Champagne. 2010. "Overview of recent advances in thermochemical conversion of biomass." *Energy Conversion and Management* 51:969–982. doi: 10.1016/j.enconman.2009.11.038.

Zhang, Y., T. R. Brown, G. Hu, and R. C. Brown. 2013. "Techno-economic analysis of two bio-oil upgrading pathways." *Chemical Engineering Journal* 225:895–904. doi: 10.1016/j.cej.2013.01.030.

Zhou, W., and D. G. B. Boocock. 2006. "Phase distributions of alcohol, glycerol, and catalyst in the transesterification of soybean oil." *Journal of the American Oil Chemists' Society* 83 (12):1047–1052. doi: 10.1007/s11746-006-5161-4.

7 Sustainable Production of Bioproducts from Wastewater-Grown Microalgae

Ashiwin Vadiveloo, Emeka G. Nwoba,
Christiana Ogbonna, Preeti Mehta

CONTENTS

7.1   INTRODUCTION

Water is among the major necessities of life, as it accounts for up to 70% of the weight of all living cells including humans, plants and animals (Cooper and Hausman 2004). Water is not only consumed as a source of nourishment by humans but also extensively required for everyday activities such as cooking, cleaning and washing (Grant et al. 2012). The rapid growth in the global population, coupled with the overwhelming increase in urbanization and industrialization, has significantly enhanced the generation of wastewater, which has a negative effect on the natural ecosystem and human health (Rajasulochana and Preethy 2016, Van Drecht et al. 2009). It is estimated that approximately 1.1 billion people drink highly unsanitized water worldwide, while approximately 50% of the water utilized by humans is discharged as wastewater (Rajasulochana and Preethy 2016). Moreover, the discharge of sewage effluent rich in nitrogen and phosphorus into surface water bodies is expected to increase by a magnitude of up to four-fold in developing countries, such as those in Southeast Asia, before 2050 (Van Drecht et al. 2009). Apart from daily usage, huge proportions of wastewater are generated from various industrial and agricultural practices (Droste and Gehr 2018). Wastewater is generally used to describe any water body that is composed of a significant proportion of waste material and can also be termed as 'spent' or 'used water' containing various dissolved or suspended solids (Droste and Gehr 2018). Most wastewater is characterized as containing some form of chemical, physical or biological pollutants, which can significantly impact the quality of water for reuse or further application (Tchobanoglous, Burton, and Stensel 1991).

7.2   WASTEWATER

There is no clear-cut classification of wastewater, because there are always some overlap among the different classes. However, one can attempt to classify or group wastewater into categories based on some defined criteria, such as the source, composition, colour and stage of treatment. Wastewater can be broadly grouped into five prominent groups based on its source, namely domestic, industrial, commercial, storm and agricultural wastewater (Droste and Gehr 2018).

1. **Domestic wastewater:** Domestic wastewater includes spent water discarded from household activities and institutions, as well as water arising from rain splash and floods flowing within residential areas generated by humans from their day-to-day activities (Mara 2013). It includes all the used waters arising from toilets, kitchen, bathrooms and cleaning and washing equipment.
2. **Industrial wastewater:** This is wastewater that comes from commercial industrial production processes, and its composition varies greatly,

depending on the particular industry. However, most of this wastewater has been reported to contain contaminants such as oils, pharmaceuticals, various acidic or alkaline by-products and pesticides (Barakat 2011). Industrial wastewater may also contain toxic and harmful compounds such as heavy metals and xenobiotics

3. **Agricultural wastewater:** Wastewater generated from agricultural activities such as livestock production (pigs, poultry, cattle and sheep), aquaculture (fish farms, shrimps etc.), crop production, irrigation, washing and cooling farm machinery and washing farm implements (Lopez et al. 2006). This also includes water generated from food processing facilities such as abattoirs.

4. **Commercial wastewater:** This is wastewater generated from commercial establishments like markets, shopping centres, business areas, restaurants, café and banks. Wastewater from these places is similar in composition to residential/domestic wastewater, except that it may contain elevated concentrations of oils, fats and greases, which may vary depending on the nature of the business activities (Siegrist et al. 1985).

5. **Storm wastewater:** This is wastewater generated from floods, runoff and storms (Vasarevičius et al. 2010). This differs in composition from domestic, commercial and industrial wastewater. Storm water or flood water contains wastes it carries from the surrounding environments. When rain falls, the runoff mixes with whatever materials are on the surface of the surrounding environments It carries materials like dust particles, sand, gravel, pesticides, solid particles, oils, etc.

7.3   CONVENTIONAL WASTEWATER TREATMENT

Most wastewater, such as domestic wastewater, consists of approximately 99.9% water and 0.1% contaminants (both soluble and suspended particles) by weight (Abbas et al. 2007). Nevertheless, a significant challenge remains because this minute concentration of contaminants can be toxic and dangerous, and must therefore be removed before the water can be reused or released to the environment (Bolong et al. 2009). The extent of wastewater treatment depends on the intended final use of the treated effluent. Whereas some wastewater can be reused with little or no treatment (i.e., recycling certain food processing wastewater for irrigation), others do require extensive treatment procedures to remove organic and inorganic contaminants, such as microorganisms, nitrogen, phosphorus and heavy metals, before recycling the wastewater back to potable water (Droste and Gehr 2018). Elevated concentrations of inorganic nutrients such as nitrogen and phosphorus in discharged wastewater can result in the eutrophication of natural water bodies and ecosystem damage, and some microorganisms in wastewaters can be pathogenic and toxic to aquatic life and human health (Van Drecht et al. 2009, Khan and Ansari 2005).

Thus, there is great need for efficient and reliable wastewater treatment options. The main objectives of any wastewater treatment procedure must include (a) to recover fresh/clean water for further use, (b) to detoxify/decontaminate the water for environmental safety and (c) to maintain the hydrological cycle (Droste and Gehr

2018). The ultimate goal of any wastewater treatment system is the conservation of water and to provide an environment that is beneficial in terms of human health and socio-economic needs (Rawat et al. 2011).

Currently, wastewater treatment is typically carried out in wastewater treatment plants which employ a variety of physical, chemical and biological treatment methods to remove various contaminants from incoming effluent. The composition of wastewater determines the choice and severity of the different treatment options, which can be combined to make up specific stages that are commonly known as primary, secondary and tertiary wastewater treatment procedures (Droste and Gehr 2018, Rawat et al. 2011).

The primary treatment step involves the physical removal of solid matter from the wastewater using sediment tanks and filters (Droste and Gehr 2018). The settled solids are known as sludge and are removed for further disposal, while the liquid is channelled for subsequent treatment steps (Barakat 2011). The secondary treatment makes use of oxidation processes to reduce biological oxygen demand (BOD) through a combination of biofiltration, aeration and oxidation ponds (Tang and Ellis 1994). Indigenous heterotrophic microorganisms are used in the secondary stage to treat sewage effluent (Raouf, Homaidan, and Ibraheem 2012). Subsequently, tertiary treatment stages consist of the removal of ionic substances and nutrients such as nitrates and phosphates using either chemical or biological methods (Raouf, Homaidan, and Ibraheem 2012). Despite its relative efficiency, conventional chemical and physical tertiary treatment methods are expensive and energy intensive due to their complexity (Droste and Gehr 2018). The high volume of activated sludge generated through these methods are also expensive to treat or dispose of safely (Rawat et al. 2011). In some cases, the final treated wastewater is still restricted by the significant amount of impurities that can cause eutrophication when discharged into environmental water bodies (Grant et al. 2012). Additional biological tertiary treatment procedures have been proposed and are considered the most viable options for the further tertiary treatment of various types of wastewater due to its tolerance and cost-effectiveness (Grady Jr et al. 1999).

7.4 MICROALGAE AND PHYCOREMEDIATION

Microalgae are typically photoautotrophic microorganisms that can either be prokaryotic and filamentous cyanobacteria or eukaryotic unicellular algae. Microalgae are of substantial diversity, distributed globally and of great ecological importance, as they contribute to approximately 50% of global primary production and constitute the basis of most aquatic food chains (Masojídek and Prášil 2010). The rich biodiversity of microalgae in conjunction with their high growth rates, ability to grow on non-freshwater resources and diverse biochemical composition, make them excellent feedstock for the production of various bioproducts with potential applications in the food, energy, agricultural and health sectors (Stengel, Connan, and Popper 2011). Nevertheless, the true commercial potential of microalgae as a source of bioenergy and bioproducts is hampered by various challenges. Among the most prominent factors restricting the feasibility of microalgae cultivation for the production of low-cost commodities is the overall economics brought forward by elevated operating cost such as the requirement of water and fertilizers (i.e. carbon, phosphorus and nitrogen) (Spolaore, Duran, and Isambert 2006). Therefore, there is significant room for innovations that should not only allow for the optimization of microalgae production efficiency but also effectively reduce the associated cost. The cultivation of microalgae on wastewater is of great promise, as it not only reduces the water and nutrient (fertilizer) footprint of algal production but also allows for the simultaneous bioremediation of the waste stream (Cai, Park, and Li 2013). The use of wastewater for the cultivation of microalgae establishes a circular economic concept and promotes a sustainable approach for the production of bioproducts from microalgae, as illustrated in Figure 7.1 (Rawat et al. 2011).

Phycoremediation represents an innovative approach that involves the use of algae (micro- and macroalgae) for the simultaneous bioremediation and biotransformation of waste pollutants (i.e. nutrients and xenobiotics) from various types of wastewater, together with biomass propagation (Rawat et al. 2011, Olguín 2003, Renuka et al. 2015). The use of microalgae as a cost-effective, ecologically safer and environmentally friendly tertiary wastewater treatment option has been well established over the last 50 years (Oswald and Gotass 1957). Nevertheless, the dual potential of cultivating microalgae on wastewater for the production of various bioproducts has been only recently recognized and is currently gaining immense research interest, as it represents an efficient, cost-effective and sustainable method in which the algal biomass generated from treating the wastewater can be further used in various ways, such as feedstock for biodiesel, feed for animals, fertilizer and other value-added products (Harun et al. 2010a, Koutra et al. 2018; Rawat et al. 2011). Therefore, the aim of this chapter is to review the potential of wastewater-grown microalgae for the production of various bio-based products, while also providing in-depth analysis and discussion of current trends on the dual potential of microalgae for wastewater treatment and its biotechnological application as a potential source of bioproducts. The growth of microalgae in various types of wastewater, its bioremediation ability, potential generation of biomass and bioproducts and downstream processing of biomass for desired applications are discussed in this chapter.

7.4.1 Mechanism of Phycoremediation

Phycoremediation can be brought forward by several different inherent mechanisms and ability of microalgae, such as (a) bioconversion (b) bioabsorption and (c) bioadsorption for the removal of a wide range of contaminants.

(a) Biodetoxification/Conversion

The primary principle of phycoremediation involves the assumption that toxic contaminants in wastewater can de-toxified or converted to non-toxic forms through the assimilation of microalgae cells (Phang et al. 2015). For example, nitrogen and phosphorus in wastewater mostly occur in the form of ammonium (NH_4^+), nitrite (NO_2^-), nitrate (NO_3^-) and phosphate (PO_3^-) ions (Ayre 2013). In this case, biodetoxification involves conversion of these forms of nitrogen and phosphorus into biomass or non-toxic metabolites (Olguín 2003, Phang et al. 2015). For example, up to 88%, 75%, 64% and 51% removal efficiency of nitrite, nitrate, ammonia and phosphorus,

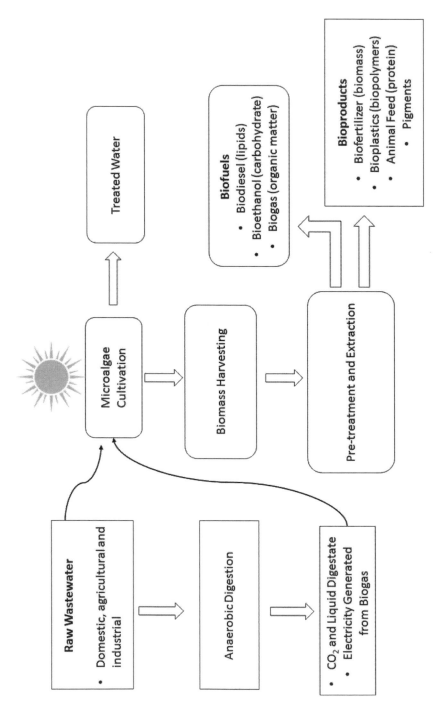

FIGURE 7.1   Flow chart representing the sustainable production of bioproducts from wastewater-grown microalgae.

respectively, was achieved for *Chlorella* culture grown in domestic sewage, industrial and aquaculture wastewater over a period of seven days (Kumar et al. 2015).

(b)   Bioabsorption/Bioaccumulation

Bioabsorption represents a pathway in which contaminants present in wastewater are directly absorbed and accumulated inside cells (Aksu 1998; Debelius et al. 2009). It has been reported that many species of microalgae are capable of accumulating various contaminants, especially heavy metals, into their cells through different metabolic pathways (Aksu 1998; Tüzün et al. 2005; Debelius et al. 2009). It has also been reported that *Phormidium ambiguum, Pseudochlorococcum typicum* and *Scenedesmus quadricauda* were able to tolerate and absorb mercury (Hg^{2+}), lead (Pb^{2+}) and cadmium (Cd^{2+}) ions in aqueous solutions as a single metal species at concentrations of 5 to 100 mg / L. Transmission electron microscopy confirmed that although there was adsorption of Cd, Hg and Pb (metals on cell surfaces), only Pb was accumulated (metals inside the cells) (Shanab et al. 2012). Some *Chlorella* species can accumulate heavy metals, including Ni, Mn, Fe, Cu, Zn, Cr, Mo, Al, Si, V, Ti and Sr, into their cells and are therefore good candidates for treating a mixture of domestic and industrial wastewater (Hammouda et al. 1995).

(c)   Bioadsorption

Bioadsorption is a process by which microalgae (either live cells or dead biomass) are used to adsorb contaminants on their cell surface without absorbing them (Aksu 1998). In some cases, adsorption of metals is the first step in bioaccumulation (Aksu 1998). Adsorption is a reaction between the cell wall and the metal ion, which occurs rapidly. This initial rapid process may be followed by a rather slow and prolonged process of bioabsorption and accumulation (Bilal et al. 2018). Wilke et al. (2006) studied the ability of 30 strains of algae to remove heavy metals from aqueous solution by biosorption. The different strains displayed different ranges of adsorption capacity for Pb, Ni, Cd and Zn. Their results showed that different types of microalgae have varied ability to adsorb heavy metals to their cell surfaces.

7.5   MICROALGAE USED FOR PHYCOREMEDIATION AND BIOPRODUCT PRODUCTION

In terms of potential application for phycoremediation and the production of bioproducts, the microalgae selected must have certain characteristics, such as high tolerance for the wastewater, extreme environmental conditions (i.e. temperature and light) and predators, while being non-pathogenic and having a favourable biochemical composition (Kiran et al. 2017). Phycoprospecting represents the identification and establishment of indigenous microalgae that are not only capable of growth in the wastewater but also have efficient waste mitigation ability (Phang et al. 2015). Phycoprospecting is typically investigated by culturing and growing microalgal cells in the direct presence of the targeted pollutants. Among the most common species of microalgae used for the bioremediation of various types of wastewater are green unicellular and eukaryotic microalgae such as *Scenedesmus* sp. and *Chlorella* sp.

(Ansari et al. 2017). In addition, culture of indigenous cyanobacteria such as *Spirulina* have been used to treat various types of wastewater such as primary, secondary and septic effluents, while also efficiently capturing and utilizing CO_2 from flue gas (Cheunbarn and Peerapornpisal 2010, Vonshak 1997). Both freshwater and marine microalgae have been reported to efficiently utilize contaminants in different types of wastewater for the production of biomass and lipids (Chinnasamy et al. 2010). The combined use of microalgal and bacterial consortia for wastewater treatment is also promising and an efficient method of wastewater treatment because both microorganisms can work in synergy (Chinnasamy et al. 2010; Phang et al. 2015).

A mixed consortium of microalgae cultures have been found capable of growth in high-strength wastewaters and shown to be more efficient for the assimilation of nutrients (Ayre 2013). For example, Pachés et al. (2018) compared the abilities of mono- and mixed cultures of four species of microalgae, namely *Chlamydomonas reinhardtii, Scenedesmus obliquus, Chlorella vulgaris* and *Monoraphidium braunii*, to remove nitrogen and phosphorus from effluent generated after anaerobic digestion. The results indicated that a mixed culture of the microalgae had a higher nutrient removal rate as a result of possible competition for nutrient among the species when compared to monocultures (Pachés et al. 2018).

The growth of microalgae in wastewater is also directly correlated to the composition and concentration of ions and contaminants in it (Kiran et al. 2017). Microalgae are able to effectively assimilate inorganic sources of nitrogen and phosphorus due to their high requirement for these nutrients, for protein and nucleic acid synthesis (Rawat et al. 2011; Oswald 2003). Nevertheless, the composition of the wastewater has been shown to significantly vary according to its origin and prior treatment steps employed (Christenson and Sims 2011). For example, the total nitrogen and phosphorus content of domestic wastewater can be between 10 and 100 mg.L^{-1}, whereas it can be above 1000 mg L^{-1} for effluents arising from agricultural facilities (Phang et al. 2015; de la Noüe et al. 1992). Among the most common nitrogen derivatives present in wastewaters is ammonia (NH_3), with values above 100 mg.L^{-1} being reported to be toxic to the microalgae, resulting in a decrease in lipid content. For instance, NH_3 concentrations equivalent to 17 and 143 mg.L^{-1} have been considered inhibitory and toxic to *Chlorella* sp., respectively (He et al. 2013). Nevertheless, recent studies have successfully grown some microalgae species such as *Chlorella, Scenedesmus* and pennate diatoms in high-strength NH_3 wastewaters with lipid contents >30% biomass dry weight (Nwoba, Ayre et al. 2016; Ayre, Moheimani, and Borowitzka 2017). The toxic effect of the high ammonia level on microalgae growth can be avoided through dilution using freshwater, and this approach is currently the most effective for the cultivation of algae in wastewaters. Dilution results in wastewaters that present a lower nitrogen content, enhancing the algal optimal biomass and lipid accumulation. Due to the high level of turbidity in wastewaters, dilution enhances light penetration and availability, which has a positive impact on lipid productivity and fatty acid composition (Cai, Park, and Li 2013). However, dilution of high-strength NH_3 digestate requires a large quantity of freshwater, which certainly raises the overall water footprint and cultivation cost; hence, it is an unsustainable strategy. It is economically attractive to dilute high-NH_3 wastewaters with other liquid wastes, considering that wastewaters can be acquired free of charge and are uncompetitive with natural resources

Growth factors such as light intensity and temperature have also been shown to significantly affect the growth and nutrient removal ability of microalgae cells grown in wastewater (Larsdotter 2006). Overall, an increase in light intensity below photoinhibition has been shown to increase microalgae growth and suppress the nitrification activities of bacteria, and an increase in temperature up to 32°C at constant irradiance has been shown to favour bacterial growth and nitrate and nitrite accumulation in the medium (Gonzalez-Camejo et al. 2018).

7.6 MICROALGAE CULTIVATION SYSTEMS

7.6.1 Open Systems

High-rate algae ponds (HRAP) represent the most practical and preferred system for the cultivation of microalgae in wastewater (Phang et al. 2015; Nurdogan and Oswald 1995). Microalgae grown in HRAP utilize energy from sunlight for either photoautotrophic or mixotrophic growth, depending on the composition of the wastewater (Nurdogan and Oswald 1995). HRAP-grown cultures are mixed using paddle wheels and can be operated in either batch mode or semi-continuous and continuous mode, which is more practical and effective (Phang et al. 2015; Nurdogan and Oswald 1995). Continuous cultures involve wastewater being fed into the pond at a predetermined rate, and the treated water flows out of the pond adjusted to the same rate as the inflow. The hydraulic retention rate is adjusted, depending on the concentration of the pollutants in the water and the standing biomass concentration (Nurdogan and Oswald 1995, Borowitzka 2005). The culture depth is also an important parameter and depends on the composition of the wastewater and the solar light intensity in the region (Borowitzka 2005). HRAP is a valuable system that can be readily integrated as a tertiary treatment option for conventional wastewater treatment facilities before discharge (Phang et al. 2015). Although open ponds remain the most economically viable culture system, there is a problem with the variability in the culture conditions, which can affect the growth and pollutant removal efficiencies of the system.

7.6.2 Closed Photobioreactors

Although various types of closed photobioreactors, such as tubular and panel photobioreactors, have been developed, and some of them have been successfully used for the commercial production of specialized products, their application in wastewater treatment is mainly limited by the costs (construction and operation) associated with the design and scability (Borowitzka 1996).

7.6.3 Immobilized Cultures

Wastewater treatment using immobilized microalgae is a promising technology that involves the physical restriction of cells inside an external carrier (De-Bashan and Bashan 2010). Various carriers, such as synthetic and natural fibres, multiple types of beads and synthetic and biopolymers, have been employed to immobilize algae (De-Bashan and Bashan 2010). In order to succeed with wastewater-grown cultures, these matrixes must sustain and promote the growth of living cells without any negative inhibition, while allowing the flow of the waste medium for nutrient uptake by the cells (Zamani et al. 2012; De-Bashan and Bashan 2010). The immobilization of *Chlorella* sp. and *Scenedesmus* sp. cells in alginate has been shown to be successful in removing up to 80% of ammonia from various wastewaters (Shi et al. 2007; Cai et al. 2013). High cell density can also be achieved when cells are immobilized, while it allows for the easy separation of cells from the treated effluent and recycling of the cells (De-Bashan and Bashan 2010). In the case of mixed cultures, the cells can be immobilized separately and then the immobilized cells mixed together, or the different species can be immobilized together. Monocultures of microalgae, mixed cultures of microalgae or even mixtures of algae and bacteria can be co-immobilized for efficient treatment of wastewater. For example, Shen et al. (2017) co-immobilized *Pseudomonas putida* and *Chlorella vulgaris* for the treatment of municipal wastewater under batch and continuous cultures and found their ability to uptake ammonium and phosphate and the chemical oxygen demand (COD) was more efficient than when immobilized individually. Shaker et al. (2015) worked on the use of immobilized *Chlamydomonas* sp. and *Chlorella vulgaris* in calcium alginate gel beads and reported the ability of *C. vulgaris* to remove 72% and 99% of nitrate and orthophosphate, respectively, from the urban wastewater. The major benefit of using immobilization-based methods for microalgae cultivated in wastewaters is its tolerance for wastewater's characteristics, such as high salinity, metal toxicity and pH, which can significantly inhibit the functioning of external flocculants (De-Bashan and Bashan 2010; Pahazri et al. 2016). The biomass obtained from these matrixes can be directly subjected to thermal or fermentative downstream processes for the production of biofuel. Immobilization procedures also allow for the treatment of wastewaters without affecting the quality of the biomass generated, while also offering greater operational control, being non-toxic and flexible (De-Bashan and Bashan 2010, Pahazri et al. 2016).

7.7 BIOPRODUCTS SOURCED FROM WASTEWATER-GROWN MICROALGAE

Microalgae are an exciting bioresource that can accumulate a variety of different biochemicals such as lipids, vitamins, proteins, carbohydrates, pigments and high-value bioactive compounds in their biomass (Batista et al. 2017, Adarme-Vega, Thomas-Hall, and Schenk 2014). Due to the ever-increasing demand for food; energy; feed and bio-based materials, including chemicals for applications in the pharmaceutical, nutraceutical and cosmetic industries, microalgae represent a sustainable source of multiple novel bio-based ingredients, as illustrated in Table 7.1.

7.7.1 ALGAL BIOFUEL PRODUCTION

Currently, up to 90% of the world's energy requirements are obtained from fossil fuels (Chen et al. 2015). Nevertheless, fossil fuel dependence is burdened by multifaceted challenges, ranging from sustainability, to emission of greenhouse gases (GHGs), to global warming, to energy insecurity issues (Demirbas 2010). The production of biofuel from first- and second-generation feedstocks such as food crops

TABLE 7.1
Potential Bio-products that Can be Sourced From Microalgae Biomass and Their Applications

Product	Examples	Application	Microalgal sources	Reference
Lipids	Hydrocarbons (Botryococcene), Triglycerides, Polyunsaturated fatty acids (eicosapentaenoic acid, docosahexaenoic acid, ϒ-linolenic acid, arachidonic acid)	Biofuels, nutritional/food supplements, feed additives, infant formula.	*Botryococcus braunii, Chlorella vulgaris, Nannochloropsis salina, Phaeodactylum tricornutum, Scenedesmus* sp., *Arthrospira* sp., *Schizochytrium mangrovei.*	(Borowitzka 2013; Kim et al. 2015; Yildiz-Ozturk and Yesil-Celiktas 2017)
Pigments	Chlorophyll a, chlorophyll b, C-phycocyanin, β-carotene, phycoerythrin, lutein, astaxanthin, allophycocyanin	Natural food colourant, pigmentation, cosmetics, food additive, feed additive, pharmaceutical, fluorescent agent	*Chlorella vulgaris, Chlorella sorokiniana, Dunalella salina, Haematococcus pluvialis, Arthrospira platensis, Aphanizomenon flos-aquae, Porphyridium cruentum.*	(Borowitzka 2013; Markou and Nerantzis 2013; Zuliani et al. 2016)
Minerals and vitamins	Magnesium, selenium, calcium, phosphate, iodine, vitamins B1, B2, B6, B12, C, E, folic acid, biotin	Food supplement, feed supplements, biofertilizers	*Euglena gracilis, Dunaliella tertiolecta, Isochrysis galbana, Arthrospira platensis, Nannochloropsis oculate, Tetraselmis suecica*	(Markou and Nerantzis 2013; Borowitzka 2013)
Phytosterols	Crinosterol, stigmasterol, fucosterol, dinosterol, isofucosterol	Food additives, pharmaceuticals	*Phaeodactylum tricornutum, Nannochloropsis gaditana, Isochrysis galbana, Dunaliella tertiolecta, Tetraselmis suecica, Pavlova lutheri*	(Borowitzka 2013, Luo, Su, and Zhang 2015)

(Continued)

TABLE 7.1 (*Continued*)

Product	Examples	Application	Microalgal sources	Reference
Polymers	Polyhydroxy alkanoates, exopolysaccharides	Industrial and medical uses	*Chlorella vulgaris, Scenedesmus acuminatus, Arthrospira platensis, Synechocystis* sp., *Porphydrium* sp., *Nostoc muscorum*	(Markou and Nerantzis 2013, Kovalcik et al. 2017, Borowitzka 2013)
Amino acids	Leucine, glycine, aspartic acid, glutamine, alanine, lysine, arginine, valine, cysteine, asparagines	Food supplements, feed additive, nutraceuticals	*Euglena gracilis, Anthrospira platensis, Nannochloropsis oculata, Porphyridium cruentum, Rhodomonas salina*	(Villarruel-López, Ascencio, and Nuño 2017; Borowitzka 2013)
Phenols	Gallic, ρ-coumaric, ferulic, caffeic, salicylic acids	Cosmetics	*Phaeodactylum tricornutum, Dunaliella salina, Nannochloropsis salina, Desmodesmus communis, Chlorella vulgaris*	(Borowitzka 2013; Safafar et al. 2015)
Phycotoxins	Okadaic acid, ciguatoxin, domoic acid, gambieric acid, gonyautoxins, yessotoxins	Health and pharmaceutical applications	*Nitzchia purgens, Gambierdiscus toxicus, Alexandrium lusitanicum, Amphidinium* sp., *Dinophysis* sp.	(de Jesus Raposo, de Morais, and de Morais 2013, Borowitzka 2013)
Mycosporine-like amino acids	Sporopollenin, Scytonemin, Shinorine	Sunscreen, UV-screening agent	*Aphanizomenon flos-aquae, Chlorella fusca, Chlorella minutissima, Scytonema* sp.	(Chu 2012)

(e.g., corn, sugar cane, beets, oil palms), non-food crops (Jatropha, Pongamia), lignocellulosic, agricultural and forest residues is currently unfeasible due to unsustainability issues and competition for arable land use (Sims et al. 2010). On the other hand, microalgae have been gaining immense interest for the production of bioenergy due to their innate ability in efficiently harvest light energy (e.g., from the sun) through photosynthesis and transforming it into a biomass that is rich in biomolecules such as lipids and oils. Depending on the biomolecular history of the biomass and downstream processing, several feasible energy options such as biodiesel (produced by transesterification reaction), biobutanol (fermentation), bioethanol (fermentation), biomethane (anaerobic digestion) and biohydrogen (photobiological reaction) can be derived from microalgae (Chaudry, Bahri, and Moheimani 2015). Microalgae biofuels have great potential as a sustainable and carbon-neutral source of bioenergy that includes biodiesel, biogas, bio-oil, bioethanol and biohydrogen (Ishika et al. 2017). The high lipid content (up to 60% dry weight), improved photosynthetic conversion competence (10 to 50 times higher than terrestrial plants), rapid reproduction cycles, capacity for considerable CO_2 sequestration, ability to be cultivated using seawater and on non-arable land, widespread availability and ability to recycle nutrients in wastewater and flue gas have significantly propelled the use of microalgae as a source of bioenergy over other conventional crops (Ishika et al. 2017). In addition, it has been well established and recommended that biofuel production from microalgae, in conjunction with wastewater treatment, represents the most viable commercial bioenergy production system in the near future (Van Harmelen and Oonk 2006, Brennan and Owende 2010).

7.7.2 BIODIESEL

Microalgae represent a feasible source of renewable oil for the production of biodiesel due to their fast growth rates and innate ability to store large amounts of lipids and fatty acids (i.e. fatty acid methyl esters, or FAMEs) which are similar to that of animal fat and vegetable oil in terms of their biomass (Koutra et al. 2018). The quality (physical and chemical attributes) of the FAMEs is a direct function of the fatty acid chain length, degree of unsaturation and branching (Yu et al. 2017, Shin et al. 2015). Aliphatic fatty acids with chain lengths between 16 and 18 carbon atoms and that have high degrees of saturation (e.g. palmitic acid, C16:0) and mono-unsaturation (e.g. oleic acid, C18:) are preferred for the production of high-quality biodiesel (Shin et al. 2015). It should be noted that algal-derived biodiesel is normally characterized by a high level of linolenic acid (18:3(9, 12, 15)), an n-3 fatty acid, which is highly susceptible to oxidation (Chisti 2007). A high proportion of linolenic acid is undesirable, and the C18:3 of biodiesel must not surpass 12% of total fatty acids to prevent oxidation and comply with the European Biodiesel Standards EN14214 (Shin et al. 2015). Several studies that have investigated the potential of microalgae for biodiesel production have primarily focused on lipid contents and the fatty acid profile of the biomass. Only limited studies have evaluated key variables that affect final biodiesel quality, such as the fuel quality during storage (kinematic viscosity), ignition time (cetane number), free fatty acids (high value affects quantity and quality) and percentage of unsaturated lipids (iodine value) (Cai et al. 2013; Kim et al. 2015). In general, the lipid content and fatty acid composition of microalgae cultivated in wastewaters is directly influenced by (a) growth conditions (pH, temperature, CO_2 addition, size of inoculum);(b) availability of nitrogen;(c) species of microalgae;(d) algal–bacterial interactions;(e) source and characteristics of wastewater; and (f) growth phase (exponential and stationary) and mode (batch, fed-batch, continuous and semi-continuous). The variation in lipid content is highly speciesspecific and can range from 4.5% to 80% of the dry biomass (Hu et al. 2008). Under nutrient-depleted conditions (nitrogen starvation), the lipid content of microalgae has been shown to increase substantially in all taxa, except for cyanobacteria (Hu et al. 2008). The cell growth phase influences the lipid content and composition of any microalgae species. Microalgae typically have lower lipid yields during the exponential phase, followed by an increase in the late exponential phase that either stabilizes or increases further during the stationary phase (Hu et al. 2008; Xu et al. 2008). Nutrient stress often results in growth inhibition and shutdown of the photosynthetic pathway, thereby channeling the energy to triacylglycerols (TAGs) production (Guschina and Harwood 2009, 2006, Ajjawi et al. 2017).

7.7.3 Fatty Acid Profile

The most common fatty acids distributed in microalgae are palmitic (C16:0), stearic (C18:0), oleic (C18:1), linoleic (C18:2) and linolenic (C18:3) acids (Knothe 2009). Some fatty acids, such as eicosapentaenoic acid (EPA) (C20:5) and docosahexaenoic acid (DHA) (C22:6), are also present in small quantities in most microalgae (Table 7.2). However, some microalgae species such as diatoms and eustigmatophytes produce substantial amounts of EPA and DHA based on cultivation conditions (Huerlimann et al. 2010), whereas dinoflagellates and haptophytes naturally make EPA and DHA, with DHA being the most dominant (Brown et al. 1996). For biodiesel application, the smaller the number of double bonds (the higher degree of saturation), the more suitable the fatty acids would be to guarantee the oxidative stability of the biodiesel product. Furthermore, the content of free fatty acids affects the oxidative stability of biodiesel, and a ratio of 5:4:1 for C16:1, C18:1 and C14:0 have been found to be optimal (Schenk et al. 2008). Saturated fatty acids

TABLE 7.2

The Fatty Acid Composition and Biodiesel Properties of Wastewater-grown Microalgae

Wastewater	Microalgal species	Fatty acid composition						Biodiesel property		Reference
		16:0	16:1	18:0	18:1	18:2	18:3 n-3	Kinetic viscosity	Cetane number	
Food wastewater	*Scenedesmus bijuga*	24.7	7.2	2.0	35.8	14.3	15.6	/	48–54	(Shin et al. 2015)
Municipal wastewater	*Chlorella sorokiniana*	2.2	/	/	0.4	2.3	1.5	3.7	61.9	(Bohutskyi et al. 2016)
Synthetic wastewater	*Scenedesmus* sp.	25.5	1.9	19.5	10.2	17.5	/	/	58.2	(Kim et al. 2015)
Kitchen wastewater	*Scenedesmus* SDEC-8	68.3	/	/	16.4	10.0	5.3	4.9	59.4	29(Yu et al. 2017)
Piggery wastewater	*Neochloris Oleoabundans*	23.9	0.6	3.1	39.7	24.5	4.3	/	/	(Olguín et al. 2015)

(SFAs) are necessary for determining the properties of the biodiesel, as a significant amount of SFAs increases the cetane number (Islam et al. 2013, Gopinath, Puhan, and Nagarajan 2009). In contrast, the degree of unsaturation of fatty acids also affects biodiesel properties, such as density and kinenatic viscosity, with fatty acids C18:3 and ≥4 double bonds limited to 12% and 1%, respectively, in biodiesel (Table 7.2) (Barabás and Todoruţ 2011; Pratas et al. 2010; Moser 2014). Hence, a high-quality biodiesel has an optimal blend of saturated and unsaturated fatty acids.

7.7.4 BIOMETHANE (BIOGAS)

The anaerobic digestion (AD) of the entire algal biomass or remnants after biomolecule extraction to produce biogas is also an interesting option for wastewater-grown algae (Zamalloa et al. 2011; Ward et al. 2014). Biogas is a mixture of both methane (55% to 75%) and carbon dioxide (25% to 45%) that is produced during anaerobic digestion by microorganisms (Zamalloa et al. 2011). Under anaerobic conditions, the microalgal biomass can be naturally degraded to produce methane at an emission volume of over 6.20 (±1.34) × 10^8 tons.yr^{-1} (Kwietniewska and Tys 2014). Therefore, microalgal biomass has been gaining attention as a feedstock for the generation of various types of biogas such as biomethane, biohydrogen and biohythane (methane in 5% to 25% hydrogen). Methane can be used to produce electricity or as a source of fuel. The AD of spent microalgal biomass used for oil extraction and hydrogen production has been demonstrated and has been found to be favourable due to the absence of lignin (Shuba and Kifle 2018). The remaining biomass in the form of sludge after AD can be further exploited as a possible biofertilizer (Ward, Lewis, and Green 2014). The economic feasibility and effectiveness of microalgal-based biogas production is dependent on a range of factors, including AD conditions, biomass pretreatment, microalgae strain and culture methods. Adopting a closed-loop approach where various types of anaerobically digested effluent (i.e. domestic, agricultural and industrial) are utilized as a culture media for microalgae cultivation, with the biomass produced subsequently exploited as AD feedstock, could drive microalgal biogas production close to economic feasibility. Several microalgae species, including *Chlorella, Chroococcus* and *Tetraselmis*, have been investigated under wastewater conditions for biogas production (Prajapati et al. 2014). However, the carbon–nitrogen ratio, which is a function of the wastewater source, was seen to influence the final composition of the biomass produced, which subsequently affected the volumes of biomethane generated (Ward et al. 2014). Low carbon–nitrogen ratios of wastewater can enhance ammonia build-up, as well as drive up pH (Ward et al. 2014). This can have a negative effect on the biogas production process, such as inhibiting methanogenesis. Given that the biogas produced from AD has 40% CO_2 and other impurities such as H_2S, N_2 and H_2, it is therefore ideal to remove CO_2 from biogas to improve its calorific value, a process known as biogas upgrading (Ward et al. 2014). The use of microalgal cultures for biogas upgrading through a complete removal of CO_2 is due to the latter's inherent ability to photosynthetically sequester CO_2 in large amounts (Ward et al. 2014).

7.7.5 BIOETHANOL

Algal biomass containing carbohydrate and protein can be used as feedstock for the production of ethanol through fermentation by other microorganisms (i.e. bacteria, yeast and fungi) (Harun et al. 2010; Ho et al. 2013). The fact that the carbohydrate content of the microalgal biomass can be readily harnessed, coupled with the low requirement of agricultural land and high affinity for CO_2 sequestration, makes these organisms promising candidates for bioethanol production (Jambo et al. 2016). Bioethanol from wastewater-grown microalgae biomass can be produced through either a biochemical process (fermentation) or a thermo-chemical process (gasification) (Harun et al. 2010; Ho et al. 2013). Compared with the traditional carbohydrate sources such as corn, sugar cane/beets and lignocellulosic materials, which require arable land to grow on, microalgal biomass are favoured due to the lack of recalcitrant lignin and their considerable amount of cellulose and starch, which can be readily degraded to reducing sugars (Odjadjare et al. 2017). Generally, the carbohydrate content of microalgae is relatively low, 5% to 23% dry weight of their biomass; however, with an appropriate cultivation medium and algal strains, the accumulated yield may be enhanced (Koutra et al. 2018; Vadiveloo et al. 2015).

Although reports on the bioethanol production from algal biomass cultivated in wastewaters is non-existent, investigation of the effect of different wastewaters on carbohydrate production is emphasized. The carbohydrate component of the biomass produced from wastewaters is either similar to or sometimes higher than values reported in culture grown on artificial media (Ledda et al. 2016; Koutra et al. 2018). Under limited nitrogen and phosphorus conditions, microalgae tend to accumulate carbohydrates, just as it has been observed with lipids (Michelon et al. 2016). Harvesting methods such as tannin-assisted coagulation-flocculation has been found to influence carbohydrate content positively compared to centrifugation, whereas the culture system appears to have no considerable effect on carbohydrate production (Michelon et al. 2016). Given this discussion, algal biofuels are currently unfeasible, but with technology maturation, coupled with a biorefinery approach, the microalgal biomass can be wholly used, resulting in the techno-economic feasibility of microalgal technology.

7.7.6 PIGMENTS

Pigments are a vital component of microalgae cells, as they are crucial for photosynthetic activity, metabolic regulation and the appearance of microalgae (Koutra et al. 2018). The major pigments found in microalgae are chlorophylls, carotenoids, and phycobilins (Nwoba et al. 2018, Borowitzka 2013). The pigment productivity of microalgae is generally lower than other macromolecular components of the biomass (i.e. lipids, carbohydrates, proteins). Nevertheless, microalgal pigments still offer immense economic attractiveness and profitability due to their high nutritional and market values. Algal pigments are currently used as organic colourants in foods, fish, animal (product colouration) and nutritional supplements (Borowitzka 2013). They are composed of several bioactive substances that can provide antibiotic, antitumour, anti-inflammatory, neuroprotective and anti-aging activities (Nwoba et al.

2018). They are also used as natural source of antioxidants, nutraceuticals, pharmaceuticals and cosmeceuticals (Markou and Nerantzis 2013). The use of wastewaters as a source of nutrients for the production of pigments is promising due to the lower cost associated with it and its potential to increase pigment productivity. The accumulation of chlorophyll *a*, α/β- carotene, astaxanthin, lutein and phycocyanin can be enhanced under wastewater conditions, principally due to the turbidity of the wastewaters, which limit light penetration (Nwoba, Moheimani et al. 2016). Microalgae have been shown to significantly increase pigment content under light-limiting conditions as a measure to efficiently absorb more available light (Vadiveloo et al. 2015). Moreover, pigments such as chlorophyll have also been found to be enhanced in wastewaters with high nitrogen concentration, whereas only β-carotene has been shown to increase under lower nitrogen concentrations (Michelon et al. 2016).

7.7.7 SOURCE OF PROTEINS

Microalgae have also been gaining interest as a rich and novel source of protein to overcome the challenges associated with the rise in the global population and the food crisis brought about by it. The use of wastewater-grown microalgal biomass as an alternative for animal feed additives represents a sustainable solution and has been recently evaluated (Batista et al. 2013; Van der Spiegel et al. 2013). Some microalgae strains, especially *Chlorella* sp., cultivated in wastewaters have a high protein content up to 46% of ash-free dry weight (AFDW) (Nwoba, Ayre et al. 2016; Moheimani et al. 2018). The protein content and amino acids composition of microalgae grown in wastewater have been shown to be similar to that of conventional protein sources, such as soybeans and eggs, and their overall sustainability is expected to improve with an increase in biomass productivity and energy efficiency (Moheimani et al. 2018). Despite its promise as a source of protein, microalgae biomass sourced from wastewater is still restricted by safety, public health, hygiene and quality concerns of potential contaminants, such as the presence of heavy metals, toxins, allergens, pathogens and herbicides, as well as the overall biomass digestibility and proportion of nucleic acids.

For successful application as a sustainable feed source, the algal biomass sourced from wastewaters must meet regulatory permissible thresholds of the previously mentioned contaminants. In terms of potential application as a nutrition source for humans, the biomass sourced must adhere to more stringent regulations and be subjected to rigorous authorization procedures before being introduced to the market.

7.7.8 POLYHYDROXYALKANOATES

Polyhydroxyalkanoates (PHAs) represent a group of naturally produced polymers with high biodegradability and biocompatibility features that are typically biosynthesized by heterotrophic bacteria (Troschl, Meixner, and Drosg 2017, Liu, Pohnert, and Wei 2016). These compounds are an attractive alternative to commercial plastics, such as polyethylene and polypropylene. Polyhydroxybutyrate (PHB) has been shortlisted as the most viable and commercially available biopolymer to replace conventional plastics that can be accumulated intracellularly by microbial cells (Troschl et al. 2017). Production of PHB can exceed 80% of the dry weight of the microalgal biomass under nutrient-limited and carbon-surplus conditions (Koutra et al. 2018). Several cyanobacteria, such as *Arthrospira (Spirulina) platensis, Synechocystis* sp. And *Nostocmuscorum*, can accumulate PHB (ranging between 29% and 85% dry weight, depending on the species and growth medium) in their biomass when cultivated mixotrophically and subjected to nutrient starvation (Troschl et al. 2017; Koutra et al. 2018). Cultivation of PHB-producing cyanobacteria in wastewaters could be a sustainable pathway for sourcing bioplastics. For instance, cultivation of *Synechocystis* sp. in wastewater for PHB production resulted in a yield comparable to bacterial species with similar thermal and rheological characteristics comparable to commercial PHB (Kovalcik et al. 2017). Nevertheless, there is a need for both laboratory and pilot-scale research studies to examine growth conditions that are optimal for simultaneous PHB and biomass production using wastewaters as low-cost media for cyanobacterial cultivation.

7.7.9 EXOPOLYSACCHARIDES

Exopolysaccharides (EPS) are a class of high-molecular-weight biopolymers synthesized and excreted into the cultivation medium by microbial cells during their growth. EPS have received increased attention for their medical, biotechnological and industrial applications. They are widely applied in the food industry as thickeners and gelling agents to improve food quality and texture (Liu et al. 2016). Microalgal EPS are also promising due to their antioxidant, antibacterial, anti-tumorigenic, and immunomodulatory properties (Delattre et al. 2016). They have also received interest for application as a flocculant in biomass dewatering and purification of effluent (Liu et al. 2016). Several microalgae species, including *Chlorella vulgaris, Porphyridium cruentum, Lyngbya majuscula,* and *Micractinium*, have been investigated for EPSs production (Liu et al. 2016; Wang et al. 2014). Despite the biotechnological significance of EPSs, very few studies (e.g., (Wang et al. 2014)) have evaluated EPS biosynthesis by wastewater-grown microalgae

7.7.10 BIOFERTILIZERS

The increase in food demand due to population growth necessitates a larger output from agricultural practices and food production systems, which in turn would require a larger volume of fertilizers such as nitrogen (N) and phosphorus (P). Conventional fertilizer production consumes a high amount of energy and is derived from fossil fuels, which is unsustainable and detrimental to the environment. Microalgae biomass cultured from wastewater has the potential to be used as a soil fertilizer for crop production due to its high content in N and P. It has also been suggested that the microalgae biomass may be utilized for the biological control of pathogens in certain crops due to their anti-fungal, antibacterial, and anti-viral properties (Kulik 1995, Prasanna et al. 2015). Microalgal biomass also appears to have comparable nutrient characteristics to agro-based chemicals with environmentally friendly and slow-releasing nutrient properties, while also exhibiting potential in bio-based crop production for improved growth, germination and floral traits (Garcia-Gonzalez and Sommerfeld 2016). However, further studies specifically focusing on the role of microalgae as a source of N, its pathogen load and its potential environmental effects (such as the transformation of organic N in the soil) are needed to determine the economic rates of the microalgae biomass as a bio-fertilizer.

7.8 DOWNSTREAM PROCESSING OF WASTEWATER-GROWN MICROALGAE

7.8.1 Harvesting

After the successful cultivation of microalgae in wastewater, the biomass produced must be subjected to a series of downstream processes in order to obtain the desired bioproducts. These downstream processes are typically made up of three distinct procedures: (a) harvesting of algal biomass, (b) pre-treatment and cell lysis, and (c) extraction and purification of the targeted bio-products (Kim et al. 2013; Grima et al. 2004).

Efficient and cost-favourable downstream processes that involve the separation of biomass from the aqueous media and the extraction of the desired cellular components after cultivation represent major bottlenecks in any microalgae production system (Amaro et al. 2011; Udom et al. 2013; Danquah et al. 2009). On its own, harvesting can account for up to 30% of the total production cost of microalgae (Mata et al. 2010; Kim et al. 2013). Therefore, the selection of the most appropriate and efficient harvesting procedure based on the characteristics of the selected microalgae and the desired end product is vital to enhance the feasibility and economics of the entire process (Christenson and Sims 2011, Danquah et al. 2009).

As previously highlighted, microalgae that are able to grow efficiently on wastewater are typically unicellular, small in size (between 4 and 20 μm in diameter), have low specific gravity and have low cellular concentration, which poses a significant challenge in the efficient recovery of biomass and the cost associated with it (Henderson et al. 2008b; Yuan et al. 2012; Grima et al. 2003). In addition, most microalgae cells are generally composed of a negatively charged surface area that restricts their natural settlement through gravity (Udom et al. 2013; Milledge and Heaven 2013). The inherent physicochemical properties of wastewaters, such as its pH and ionic strength, can also affect the efficacy of harvesting procedures (Milledge and Heaven 2013; Pahazri et al. 2016). Common harvesting techniques that are commercially employed for the recovery of microalgae biomass are generally based on water purification methods that either involve filtration, sedimentation, floatation, flocculation, centrifugation or a combination of them (Rawat et al. 2011; Christenson and Sims 2011). Table 7.3 illustrates the characteristics and performance of several conventional harvesting techniques that have been employed for common microalgae capable of efficient growth in wastewater, such as *Chlorella* sp. and *Scenedesmus* sp.

Centrifugation is commonly employed for harvesting the microalgae biomass as it represents a rapid, efficient (up to 90% recovery) and reliable method (Mutanda et al. 2011; Pahazri et al. 2016). Centrifugation uses the principle of gravitational force to increase sedimentation rates of even small microalgae cells (Heasman et al. 2000, Pahazri et al. 2016). Nonetheless, the high energy requirement and elevated

TABLE 7.3

Conventional Harvesting Procedures and Their Effectiveness on Common Wastewater-grown Microalgae

Microalgae	Harvesting Method	Efficiency	Conditions and Performance	Reference
Chlorella vulgaris	Gravity sedimentation	60% biomass recovered	No significant increase in biomass recovery after 1 hour of settling time	(Ras et al. 2011)
Chlorella vulgaris	Chemical flocculation	95% biomass recovered	Used sodium hydroxide to increase culture pH between 11 and 12	(Yahi, Elmaleh, and Coma 1994)
Chlorella vulgaris	Chemical flocculation	85% to 95% biomass recovered	Used aluminium sulphate at a concentration above 25mmol L^{-1}	(Jiang, Graham, and Harward 1993)
Chlorella minutissima	Chemical flocculation and sedimentation	60% biomass recovered	Used aluminium-, ferric- and zinc-based chloride and sulfate salts; aluminium salts were most efficient	(Papazi, Makridis, and Divanach 2010)
Chlorella vulgaris	Chemical flocculation and centrifugation	16% dry weight recovered		(Xu, Brilman et al. 2011)
Scenedesmus sp.	Bioflocculation and chemical flocculation	95% of biomass recovered	8.5mM CaCl2 and 0.2mM FeCl3 together with 1% of *Paenibacillus polymyxa* AM49 was used as flocculants	(Kim et al. 2011)
Chlorella vulgaris	Submerged microfiltration and centrifugation	22% w/v biomass recovered	The harvesting efficiency of submerged filtrations were more than 82%	(Bilad et al. 2012)
Chlorella vulgaris	Chemical flocculation and flotation	94%–99% cell removal efficiency	Aluminium sulphate was used as a flocculant to neutralize cells followed by dissolved air flotation	(Henderson, Parsons, and Jefferson 2010)

Source: Modified from Pragya, Pandey, and Sahoo (2013)

cost hamper the economic feasibility of centrifugation on a large scale (Pittman, Dean, and Osundeko 2011, Brennan and Owende 2010).

Sedimentation through gravity deposition represents a simple and straightforward method for concentrating biomass ($\approx$ 1.5% solid) that is also highly energy effective (Milledge and Heaven 2013; Udom et al. 2013). However, this method is relatively slow, resulting in the potential deterioration of biomass, it has low reliability and it is not suitable for motile microalgae species (Pahazri et al. 2016).

Filtration is also widely employed for the solid–liquid separation of microalgae cultures (Grima et al. 2003). Vacuum filtration is the preferred method for the separation of large algae cells (above 70µm), whereas the membrane-based microfiltration and ultrafiltration techniques are typically used to concentrate smaller cells (Grima et al. 2003; Mata et al. 2010, Milledge and Heaven 2013). Despite its high efficiency, filtration methods are greatly limited by the high cost and energy requirement due to the frequent need to replace clogged-up filters (biofouling) and the need for external pumps (Grima et al. 2003; Mata et al. 2010, Milledge and Heaven 2013).

Flotation systems are commonly referred to as 'inverted sedimentation' and employ the use of air bubbles produced by air diffusers or through external pressurization that destabilizes algae cells and causes them to float and form dense layers, which can be later skimmed off (Pahazri et al. 2016; Singh et al. 2011). Recent studies have shown flotation procedures to be favourable for harvesting small unicellular microalgae cells over other mechanical methods due to the gas vehicles that allow for the microalgae to be in a vertical position and move upward rather than downward (Hanotu et al. 2012). Dissolved air flotation (DAF) is gaining promise as an application in wastewater treatment plants and large-scale algae cultivation due to its economic benefits but is generally preceded by additional coagulation or flocculation steps (Teixeira and Rosa 2006, Christenson and Sims 2011). Further improvement on the efficiency of DAF systems can be achieved through the modification of the inherent negative charge of air bubbles (Henderson et al. 2008a, 2009). This can be performed through the use of cationic surfactants or other chemicals that can bring about a net positive charge (Henderson et al. 2008a, 2009).

Flocculation employs the adhesion of solute particles to form flocs after some form of collision (Park, Craggs, and Shilton 2011, Mata, Martins, and Caetano 2010). It can naturally occur in some mono- and mixed microalgae cultures known as autoflocculation (Schenk et al. 2008; Uduman et al. 2010). Flocculation is often employed as a pre-treatment step to concentrate the biomass and the particle size in conjunction with other mechanical harvesting procedures (Brennan and Owende 2010). External flocculants or chemicals can be also added to algae cultures to further induce and accelerate flocculation through changes in the physicochemical properties of cultures, such as an increase in the culture pH (Grima et al. 2003; Shen et al. 2009). The inherent negative surface charge of microalgae cells typically prevents their aggregation in aqueous solutions (Harun et al. 2010b). Destabilization and neutralization of these charges through the external addition of positively charged coagulants such as inorganic polyvalent cations (i.e. $FeCl_3$ and $Al_2(SO_4)_3$) and organic cationic polymers (i.e. chitosan and cationic starch) have been shown to promote flocculation (Papazi et al. 2010, Uduman et al. 2010). Nevertheless, the use of chemical flocculants such aluminium and sulfate are not cost-favourable and may negatively affect further downstream processes for the production of bioproducts from algae (Grima et al. 2003; Papazi et al. 2010; Schenk et al. 2008). Some microalgae have been shown to naturally flocculate due to the production of EPS (Bhaskar and Bhosle 2005; Ndikubwimana et al. 2014). Auto-flocculation has been observed in wastewater-grown microalgae due to natural changes in pH induced by microalgal–bacterial interactions (Christenson and Sims 2011). The presence of bacteria in the wastewater also enhances the bioflocculation of microalgae due to the production of bacterial EPS (Ndikubwimana et al. 2014).

Therefore, better understanding and controlled optimization of natural flocculation processes such as auto-flocculation and bioflocculation could provide valuable insight into reducing the overall cost of harvesting microalgae, while also eliminating the need for harmful chemicals (Christenson and Sims 2011; Pahazri et al. 2016).

The ineffectiveness and cost limitations of current harvesting methods have hampered the commercial exploitation of microalgae cultivation and have necessitated the need for new and innovative technologies (Matos et al. 2013). Among them is electrocoagulation, which is commonly used in wastewater treatment plants for the removal of difficult pollutants (Uduman et al. 2011). Electrocoagulation is an eco-friendly, effective and low-energy method but may pose a risk to the harvested biomass due to high concentrations of electrode metal ions (Matos et al. 2013; Uduman et al. 2011). Magnetic separation is another option for harvesting biomass that uses magnetic nanoparticles and the application of a strong magnetic field (Hu et al. 2014; Liu, Li, and Zhang 2009; Xu, Guo et al. 2011). The nanoparticles are coated with natural flocculants, such as chitosan or silica, that can adhere to algal cells and improve the harvesting procedure (Hu et al. 2014; Liu, Li, and Zhang 2009; Xu, Guo et al. 2011). Ultrasound represents a non-toxic, energy-saving and efficient technique for harvesting microalgae biomass based on its ability to disrupt cells and promote agglomeration, allowing for improved settlement of cells through gravity (Bosma et al. 2003).

Overall, with regard to harvesting technologies applicable for wastewater-grown microalgae, there is still significant room for improvement in existing and future methods alike to address the technical needs of such an integrated system. Further improvement of harvesting methods should consider the following aspects: (a) the characteristics of the microalgae species and wastewater used,(b) Its compatibility with further downstream processes for the extraction of desired bioproducts and (c) scalability and techno-economics for commercial-scale application (Christenson and Sims 2011, Kim et al. 2013; Udom et al. 2013).

7.8.2 EXTRACTION

Existing methods that are currently used for the extraction of molecules from microalgae are highly specific to the corresponding end product of interest, and the advantages and disadvantages of some of these methods have been well discussed in Show et al. (2015) and Wu et al. (2014). In general, technologies employed for the conversion of algal biomass to targeted endproducts either involve (a) the conversion of the whole biomass, (b) the extraction of specific biomolecules (i.e. lipid, protein and carbohydrate) or (c) the processing of any leftover biomass after extracting the desired biomolecules (Wu et al. 2014; Brennan and Owende 2010).

The efficient extraction (quality and quantity) of the desired biomolecules is vital for the successful exploitation of bioproducts from wastewater-grown microalgae.

Extraction of biomolecules from microalgae can be achieved through a variety of different chemical (i.e. acid and base), biological (i.e. enzymes) and mechanical (i.e. ultrasound, bead milling and microwave) methods or through a combination of them (Kim et al. 2013). Nonetheless, the extraction methods applied must be highly selective to the targeted product, rapid, effective and commercially viable (Lee et al. 2010). Extraction performed using organic solvents has been shown to be efficient in the recovery of specific biomolecules from microalgae such as lipids, proteins and pigments (Lee et al. 2010; Sathish and Sims 2012). Some solvents are able to disrupt algal cell walls and efficiently extract lipophilic molecules (Kim et al. 2013; Sathish and Sims 2012; Sarada et al. 2006). Solvent containing the extracts is then subjected to subsequent distillation steps to separate the extracted compounds from the solvent (Mercer and Armenta 2011). For example, hexane is commonly used for the extraction of oil from algae due to its high extraction efficiency and low cost (Jackson et al. 2017). Nevertheless, chemical solvents used for such extraction must be cost friendly, non-toxic, volatile and environmentally friendly.

Simple oil presses have been used to extract lipids from microalgae that have been subjected to prior pre-treatment steps, such as drying (Mercer and Armenta 2011). This straightforward technique utilizes pressure on ruptured cells to excrete oil from within the cells and can remove up to 75% of oil without the need for any specialized equipment (Mercer and Armenta 2011, Show et al. 2015). However, this technology is hampered by its slow extraction time requirement (Show et al. 2015).

Supercritical fluid extraction (SFE) represents another highly efficient extraction method that utilizes high pressure and temperatures to break algal cells and extract compounds (Mercer and Armenta 2011). SFE is currently gaining promise due to its short extraction time, the absence of toxic/harmful chemical solvents and high extraction yield but is still somewhat limited by its overall elevated cost (Mercer and Armenta 2011). CO_2 is typically employed as a extraction fluid for SFE, as it is cheap, easily available, inert and non-combustible (Cooney et al. 2009).

Another viable tool for the disruption of microalgae cells and extraction of intracellular compounds is ultrasound, which uses high-intensity waves to create small cavitation bubbles around cells that result in the disruption of the cell wall (Show et al. 2015). Ultrasound has been shown to have high extraction efficiency and does not alter the quality of the product during extraction (Lee et al. 2010).

After extraction, crude extracts containing the desired biomolecules are subsequently filtered and purified using different chromatographic methods. Pigments such as astaxanthin are generally purified using supercritical fluid chromatography, whereas fatty acids and proteins have been isolated using reverse-phase chromatography and ion exchange chromatography, respectively (Lim et al. 2002; Medina et al. 1995; Román et al. 2002)

7.9 CURRENT STATUS OF BIOPRODUCTS SOURCED FROM WASTEWATER-GROWN MICROALGAE

It has been well established that the integration of microalgae cultivation with wastewater treatment serves multiple benefits and can remarkably reduce the production cost of bioproducts sourced from microalgae (Chen, Zhao, and Qi 2015; Christenson and Sims 2011; Pahazri et al. 2016; Rawat et al. 2011). When employed as a tertiary treatment option, microalgae represent a significantly cheaper and much more efficient alternative to current nutrient and metal recovery processes (Cai, Park, and Li 2013; Christenson and Sims 2011; Pahazri et al. 2016). Microalgae-based wastewater treatment methods also eliminate the need for additional sludge treatment procedures after conventional wastewater treatment approaches (Koutra et al. 2018). In addition, the cultivation of microalgae on nutrient-rich wastewater also significantly reduces or can potentially eliminate the cost of chemical fertilizers, and revenues generated from wastewater treatment can offset production costs, representing a shift from a linear to a circular economy (Cai, Park, and Li 2013; Christenson and Sims 2011; Pahazri et al. 2016). The potential for integrating microalgae cultivation with wastewater treatment has been clearly acknowledged and recommended by the U.S. Department of Energy as a feasible pathway to offset the production cost of biofuel from microalgae (U.S. DOE 2010).

Nevertheless, in order to succeed with the production of bioproducts from wastewater-grown microalgae on a commercial scale, the following criteria and parameters must first be addressed: (a) high biomass productivity when cultivated in the choice of wastewater, (b) high tolerance to the wastewater and resilient to seasonal outdoor growth conditions and (c) high yield and productivity of desired bioproduct.

Currently, the commercial cultivation of microalgae for the production of bioenergy and bioproducts is restricted by overall low biomass productivity, resulting in higher production costs and expensive downstream processing methods such as harvesting (Amaro, Guedes, and Malcata 2011; Medina et al. 1998; Rawat et al. 2011). Lower productivity necessitates a larger cultivation area, more equipment and higher energy consumption (Wu et al. 2014). Various growth factors, such as nutrient supply, availability and supply of photosynthetically active radiation (PAR), biotic contamination and other varying environmental parameters (i.e. temperature, rain and pH), also significantly affect the productivity of large-scale algal cultivation systems (Chen et al. 2015; Wu et al. 2014; Christenson and Sims 2011). High tolerance of microalgae strains to the composition of wastewater increase the long-term health and viability of cultures, while also reducing the risk of severe contamination by other microorganisms (Chen et al. 2015).

A more pressing challenge hampering the production of bioproducts from wastewater-grown microalgae is the need for more efficient and cost-effective harvesting and downstream processing procedures that are highly compatible with the desired end product (Grima et al. 2003; Uduman et al. 2010). When compared to the conventional production of bioproducts from microalgae grown in a synthetic culture medium, certain risks and precautions must first be taken into account before utilizing the wastewater-grown biomass (Pahazri et al. 2016). Among them are the composition and quality of the final biomass and its potential application (Christenson and Sims 2011; Pahazri et al. 2016). The nature and composition of the wastewater can also pose a risk to downstream processes and can affect the quality of the final algal biomass. Wastewater rich in metal ions requires the addition of higher doses of chemical coagulant to improve harvesting efficiency, which can be detrimental to the biomass quality (Pahazri et al. 2016). In such cases, natural flocculation methods represent the best option (in terms of quality and quantity) for harvesting the microalgae biomass (Pahazri et al. 2016).

The primary concern of any wastewater-grown algal biomass is the presence of pathogenic microorganisms (i.e. bacteria, fungi and viruses) (Pahazri et al. 2016). Biotic contamination may pose a risk to the application of microalgae as a source of animal feed or biofertilizer while not affecting its application as a feedstock for bioenergy production (Moheimani et al. 2018; Pahazri et al. 2016). Therefore, quality assessment studies must first be performed on such a biomass to identify any potential risk associated with it and to determine its suitability for the final application (Moheimani et al. 2018).

The distribution of heavy metals in wastewater is also a concern for algal biomass application in the feed industry (Christenson and Sims 2011). The accumulation of heavy metals in the algae biomass can deteriorate the health of animals and humans at high concentrations (Cai, Park, and Li 2013; Christenson and Sims 2011). As such, wastewater rich in heavy metals must first be subjected to preliminary treatment for removal prior to the cultivation of microalgae (Barakat 2011). Overall, due to some of the limitations highlighted earlier, it is widely accepted that wastewater-grown microalgae would represent a more reliable feedstock for bioenergy production rather than agricultural use (Chen, Zhao, and Qi 2015; Rawat et al. 2011).

Nevertheless, characterizing both the initial wastewater and final biomass produced in terms of pathogen concentrations and potential environmental and health risks would most certainly determine the exact final application of the microalgae biomass (Moheimani et al. 2018).

7.10   FUTURE DIRECTIONS AND RECOMMENDATIONS

There is still considerable room for R&D innovations to be undertaken for the commercial and economic realization of bioproducts from wastewater-grown microalgae. Future research should look into the optimization of growth parameters for enhancing the yield of targeted bioproducts from wastewater-grown microalgae while also developing cost-effective and efficient downstream processes (Chen et al. 2015).

Among the most important parameters for the successful production of targeted bioproducts from microalgae is the choice of microalgae strain and species, which affects the productivity and yield of the endproduct obtained from the system (Chen et al. 2015; Wu et al. 2014). Based on available literature, it is suggested that locally isolated mixed algae species would represent the most efficient and viable option (Chen et al. 2015; Wu et al. 2014). When compared to monocultures, native microalgae consortiums have been shown to be more resilient, have higher productivity and are less susceptible to invasion and infection in wastewater under local environmental conditions compared to other outsourced species (Ptacnik et al. 2008; Reich et al. 2001; Ayre 2013; Chen et al. 2015). Therefore, it is vital that studies looking into the phycoremediation of wastewater and the production of bioproducts should first be devoted to isolating microalgal species with superior performance and that demonstrate ability to maintain stable biomass/lipid production against the biotic inhibitory factors within the targeted wastewater (Chen et al. 2015). Higher tolerance of the algal species to the targeted wastewater would certainly allow for the long-term stability and successful proliferation of cultures (Wu et al. 2014). It will also be interesting to develop algae–bacteria or algae–fungi consortia, depending on the nature of the wastewater. Algal–bacterial synergy ameliorates the harmful consequences of high oxygen concentration in the cultivation systems, leading to increased lipid production. Several microalgae (*Chlorella vulgaris, C. sorokiniana, Parachlorella kessleri, Scenedesmus dimorphus,* etc.) have shown high lipid productivity in non-sterile bacteria-dominated wastewater (Nwoba, Ayre et al. 2016; Erkelens et al. 2014). The utilization of microalgae consortia for wastewater treatment has also been highlighted by several reports, and algal consortia thrive well in a high-pH medium and demonstrate increased lipid productivity (Nwoba, Ayre et al. 2016).

In addition, optimization of cultivation conditions will most certainly result in improved biomass productivity and reduce the need for input resources and energy consumption required for the growth and downstream processing of microalgae grown in wastewater (Chen et al. 2015). The optimal values of these parameters depend on the type of microalgae, the type of wastewater and the target metabolites. Heterotrophic photobioreactors also have some potential in the case of microalgae with heterotrophic metabolism, and especially when heterotrophic microorganisms are combined with microalgae with heterotrophic metabolism. Immobilization techniques should also be further investigated for microalgae-based wastewater treatment. Depending on the nature of the wastewater and desired metabolites, research on selection of the carriers, immobilization methods and stability of the immobilized cells should be optimized.

Two-stage cultivation systems have also been proposed to improve the productivity of biomolecules derived from wastewater-grown microalgae (Courchesne et al. 2009). Such systems would initially compromise a growth and cell proliferation phase followed by a secondary stage involving and focusing on the production of the desired bioproduct (Chen et al. 2015; Courchesne et al. 2009). For example, nutrient-replete growth stages have been initially employed for cell proliferation and biomass accumulation, followed by a nutrient-deplete phase to enhance the production of lipids in microalgae (Tang et al. 2011; Chen et al. 2015). Such two-stage systems would allow for the optimized production of biomass and lipid, but maybe cost-ineffective due to the costs associated with additional harvesting procedures after each growth stage (Tang et al. 2011; Chen et al. 2015).

Despite its enormous potential, there are still limited studies evaluating the production of bioproducts from wastewater-grown microalgae at a pilot scale. Thus, there is great need for collaborative efforts from large-scale microalgae producers and the wastewater industry to evaluate the long-term feasibility of this innovative technology at a pilot scale (Christenson and Sims 2011). Such efforts will certainly confirm and provide information regarding the scalability of robust microalgae strains that are capable of growth in wastewater and are able to produce desired endproducts. The existing infrastructure available in wastewater treatment facilities can be optimized for the integration of microalgae cultivation, thus reducing capital cost and scalability issues while providing a platform for evaluating the viability of the system (Christenson and Sims 2011).

The implementation of the biorefinery concept for wastewater-grown microalgae is expected to generate additional revenue and significantly offset production cost while enhancing its feasibility at a commercial scale (Vanthoor-Koopmans et al.

2013; Olguín 2012). For example, the residual biomass left over after the extraction of targeted biomolecules (i.e. protein and lipid) can be anaerobically digested to produce biogas for electricity generation (Vanthoor-Koopmans et al. 2013; Olguín 2012). In terms of downstream processing, inherent physiological characteristics of microalgae consortium grown in wastewater and their preferred aggregation mechanism should be explored and optimized further to significantly reduce costs associated with harvesting (Grima et al. 2003).

In conclusion, it is without a doubt that a better understanding and further optimization of crucial parameters affecting the production and harvesting of microalgae, with the final aim of increasing productivity and reducing production costs, will be instrumental for the successful production of bioproducts from wastewater-grown microalgae. The potential environmental and economic incentives seem sufficient to warrant further investigation and potential commercialization.

REFERENCES

Abbas, S. T., M. Sarfraz, S. M. Mehdi, and G. Hassan. 2007. "Trace elements accumulation in soil and rice plants irrigated with the contaminated water." *Soil and Tillage Research* 94 (2):503–509.

Adarme-Vega, T. Catalina, S. R. Thomas-Hall, and P. M. Schenk. 2014. "Towards sustainable sources for omega-3 fatty acids production." *Current Opinion in Biotechnology* 26:14–18.

Ajjawi, I., J. Verruto, M. Aqui, L. B. Soriaga, J. Coppersmith, K. Kwok, L. Peach, E. Orchard, R. Kalb, and W. Xu. 2017. "Lipid production in *Nannochloropsis gaditana* is doubled by decreasing expression of a single transcriptional regulator." *Nature Biotechnology* 35 (7):647.

Aksu, Z. 1998. "Biosorption of heavy metals by microalgae in batch and continuous systems." In *Wastewater Treatment with Algae*, 37–53. New York: Springer.

Amaro, H. M., A. C. Guedes, and F. X. Malcata. 2011. "Advances and perspectives in using microalgae to produce biodiesel." *Applied Energy* 88 (10):3402–3410.

Ansari, F. A., P. Singh, A. Guldhe, and F. Bux. 2017. "Microalgal cultivation using aquaculture wastewater: Integrated biomass generation and nutrient remediation." *Algal Research* 21:169–177.

Ayre, J. 2013. "Microalgae culture to treat piggery anaerobic digestion effluent." Honours thesis, Murdoch University.

Ayre, J. M., N. R. Moheimani, and M. A. Borowitzka. 2017. "Growth of microalgae on undiluted anaerobic digestate of piggery effluent with high ammonium concentrations." *Algal Research* 24:218–226.

Barabás, I., and I.-A. Todoruț. 2011. "Biodiesel quality, standards and properties." In *Biodiesel-Quality, Emissions* and By-Products, edited by G. Montero and M. Stoytcheva, 3–28. Rijeka, Croacia: InTech.

Barakat, M. A. 2011. "New trends in removing heavy metals from industrial wastewater." *Arabian Journal of Chemistry* 4 (4):361–377.

Batista, A. P., A. Niccolai, P. Fradinho, S. Fragoso, I. Bursic, L. Rodolfi, N. Biondi, M. R. Tredici, I. Sousa, and A. Raymundo. 2017. "Microalgae biomass as an alternative ingredient in cookies: Sensory, physical and chemical properties, antioxidant activity and in vitro digestibility." *Algal Research* 26:161–171.

Batista, A. P., L. Gouveia, N. M. Bandarra, J. M. Franco, and A. Raymundo. 2013. "Comparison of microalgal biomass profiles as novel functional ingredient for food products." *Algal Research* 2 (2):164–173.

Bhaskar, P. V., and Narayan B Bhosle. 2005. "Microbial extracellular polymeric substances in marine biogeochemical processes." *Current Science*:45–53.

Bilad, M. R., D. Vandamme, I. Foubert, K. Muylaert, and I. F. J. Vankelecom. 2012. "Harvesting microalgal biomass using submerged microfiltration membranes." *Bioresource Technology* 111:343–352.

Bilal, M., T. Rasheed, J. Sosa-Hernández, A. Raza, F. Nabeel, and H. Iqbal. 2018. "Biosorption: An interplay between marine algae and potentially toxic elements—a review." *Marine Drugs* 16 (2):65.

Bohutskyi, P., D. C. Kligerman, N. Byers, L. K. Nasr, C. Cua, S. Chow, C. Su, Y. Tang, M. J. Betenbaugh, and E. J. Bouwer. 2016. "Effects of inoculum size, light intensity, and dose of anaerobic digestion centrate on growth and productivity of Chlorella and Scenedesmus microalgae and their poly-culture in primary and secondary wastewater." *Algal Research* 19:278–290.

Bolong, N., A. F. Ismail, M. R. Salim, and T. Matsuura. 2009. "A review of the effects of emerging contaminants in wastewater and options for their removal." *Desalination* 239 (1–3):229–246.

Borowitzka, M. A. 1996. "Closed algal photobioreactors: Design considerations for large-scale systems." *Journal of Marine Biotechnology* 4:185–191.

Borowitzka, M. A. 2005. "Culturing microalgae in outdoor ponds." In *Algal Culturing Techniques*, edited by A. Anderson, 205–217. London: Academic Press.

Borowitzka, M. A. 2013. "High-value products from microalgae—their development and commercialisation." *Journal of Applied Phycology* 25 (3):743–756.

Bosma, R., W. A. van Spronsen, J. Tramper, and R. H. Wijffels. 2003. "Ultrasound, a new separation technique to harvest microalgae." *Journal of Applied Phycology* 15 (2–3):143–153.

Brennan, L., and P. Owende. 2010. "Biofuels from microalgae—a review of technologies for production, processing, and extractions of biofuels and co-products." *Renewable and Sustainable Energy Reviews* 14 (2):557–577.

Brown, M. R., G. A. Dunstan, S. Norwood, and K. A. Miller. 1996. "Effects of harvest stage and light on the biochemical composition of the diatom Thalassiosira pseudonana." *Journal of Phycology* 32 (1):64–73.

Cai, T., S. Y. Park, and Y. Li. 2013. "Nutrient recovery from wastewater streams by microalgae: Status and prospects." *Renewable and Sustainable Energy Reviews* 19:360–369.

Cai, T., X. Ge, S. Y. Park, and Y. Li. 2013. "Comparison of Synechocystis sp. PCC6803 and Nannochloropsis salina for lipid production using artificial seawater and nutrients from anaerobic digestion effluent." *Bioresource Technology* 144:255–260.

Chaudry, S., P. A. Bahri, and N. R. Moheimani. 2015. "Pathways of processing of wet microalgae for liquid fuel production: A critical review." *Renewable and Sustainable Energy Reviews* 52:1240–1250.

Chen, G., L. Zhao, and Y. Qi. 2015. "Enhancing the productivity of microalgae cultivated in wastewater toward biofuel production: A critical review." *Applied Energy* 137:282–291.

Cheunbarn, S., and Y. Peerapornpisal. 2010. "Cultivation of Spirulina platensis using anaerobically swine wastewater treatment effluent." *Int. J. Agric. Biol* 12 (4):586–590.

Chinnasamy, S., A. Bhatnagar, R. W. Hunt, and K. C. Das. 2010. "Microalgae cultivation in a wastewater dominated by carpet mill effluents for biofuel applications." *Bioresource Technology* 101 (9):3097–3105.

Chisti, Y. 2007. "Biodiesel from microalgae." *Biotechnology Advances* 25 (3):294–306.

Christenson, L., and R. Sims. 2011. "Production and harvesting of microalgae for wastewater treatment, biofuels, and bioproducts." *Biotechnology Advances* 29 (6):686–702.

Chu, Wan-Loy. 2012. "Biotechnological applications of microalgae." *IeJSME* 6 (1):S24-37.

Cooney, M., G. Young, and N. Nagle. 2009. "Extraction of bio-oils from microalgae." *Separation & Purification Reviews* 38 (4):291–325.

Cooper, G. M., and R. E. Hausman. 2004. *The Cell: Molecular Approach*. Medicinska naklada.

Courchesne, N., M. Dorval, A. Parisien, B. Wang, and C. Q. Lan. 2009. "Enhancement of lipid production using biochemical, genetic and transcription factor engineering approaches." *Journal of Biotechnology* 141 (1–2):31–41.

Danquah, M. K., B. Gladman, N. Moheimani, and G. M. Forde. 2009. "Microalgal growth characteristics and subsequent influence on dewatering efficiency." *Chemical Engineering Journal* 151 (1–3):73–78.

de Jesus R., M. Filomena, R. M. S. Costa de Morais, and A. M. M. Bernardo de Morais. 2013. "Health applications of bioactive compounds from marine microalgae." *Life Sciences* 93 (15):479–486.

de la Noüe, J., G. Laliberté, and D. Proulx. 1992. "Algae and waste water." *Journal of Applied Phycology* 4 (3):247–254.

De-Bashan, L. E, and Y. Bashan. 2010. "Immobilized microalgae for removing pollutants: Review of practical aspects." *Bioresource Technology* 101 (6):1611–1627.

Debelius, B., J. M. Forja, Á. DelValls, and L. M. Lubián. 2009. "Toxicity and bioaccumulation of copper and lead in five marine microalgae." *Ecotoxicology and Environmental Safety* 72 (5):1503–1513.

Delattre, C., G. Pierre, C. Laroche, and P. Michaud. 2016. "Production, extraction and characterization of microalgal and cyanobacterial exopolysaccharides." *Biotechnology Advances* 34 (7):1159–1179.

Demirbas, A. 2010. "Use of algae as biofuel sources." *Energy Conversion and Management* 51 (12):2738–2749.

Droste, R. L, and R. L Gehr. 2018. Theory and Practice of Water and Wastewater *Treatment*. Hoboken, NJ: John Wiley & Sons.

Erkelens, M., A. J. Ward, A. S. Ball, and D. M. Lewis. 2014. "Microalgae digestate effluent as a growth medium for Tetraselmis sp. in the production of biofuels." *Bioresource Technology* 167:81–86.

Garcia-Gonzalez, J., and M. Sommerfeld. 2016. "Biofertilizer and biostimulant properties of the microalga Acutodesmus dimorphus." *Journal of Applied Phycology* 28 (2):1051–1061.

Gonzalez-Camejo, J., R. Barat, M. Pachés, M. Murgui, A. Seco, and J. Ferrer. 2018. "Wastewater nutrient removal in a mixed microalgae—bacteria culture: Effect of light and temperature on the microalgae—bacteria competition." *Environmental Technology* 39 (4):503–515.

Gopinath, A., S. Puhan, and G. Nagarajan. 2009. "Relating the cetane number of biodiesel fuels to their fatty acid composition: A critical study." *Proceedings of the Institution of Mechanical Engineers, Part D: Journal of Automobile Engineering* 223 (4):565–583.

Grady Jr., C. P. Leslie, G. T. Daigger, N. G. Love, and C. D. M. Filipe. 1999. *Biological Wastewater Treatment*. New York: Marcel Dekker.

Grant, S. B., Jean-Daniel Saphores, David L Feldman, Andrew J Hamilton, Tim D Fletcher, Perran LM Cook, Michael Stewardson, Brett F Sanders, Lisa A Levin, and Richard F Ambrose. 2012. "Taking the "waste" out of "wastewater" for human water security and ecosystem sustainability." *Science* 337 (6095):681–686.

Grima, E. M., E.-H. Belarbi, F. G. Acién Fernández, A. R. Medina, and Y. Chisti. 2003. "Recovery of microalgal biomass and metabolites: Process options and economics." *Biotechnology Advances* 20 (7):491–515.

Grima, E. M., F. G. Acién Fernández, and A. R. Medina. 2004. "10 Downstream Processing of Cell-mass and Products." *Handbook of Microalgal Culture: Biotechnology and Applied Phycology*:215.

Guschina, I. A, and J. L. Harwood. 2006. "Lipids and lipid metabolism in eukaryotic algae." *Progress in Lipid Research* 45 (2):160–186.

Guschina, I. A., and J. L. Harwood. 2009. "Algal lipids and effect of the environment on their biochemistry." In *Lipids in Aquatic Ecosystems*, 1–24. New York: Springer.

Hammouda, O., A. Gaber, and N. Abdelraouf. 1995. "Microalgae and wastewater treatment." *Ecotoxicology and Environmental Safety* 31 (3):205–210.

Hanotu, J., H. C. H. Bandulasena, and W. B. Zimmerman. 2012. "Microflotation performance for algal separation." *Biotechnology and Bioengineering* 109 (7):1663–1673.

Harun, R., M. K. Danquah, and G. M. Forde. 2010. "Microalgal biomass as a fermentation feedstock for bioethanol production." *Journal of Chemical Technology & Biotechnology* 85 (2):199–203.

Harun, R., M. Singh, G. M. Forde, and M. K. Danquah. 2010a. "Bioprocess engineering of microalgae to produce a variety of consumer products." *Renew Sust Energ Rev* 14:1037–1047.

Harun, R., M. Singh, G. M. Forde, and M. K. Danquah. 2010b. "Bioprocess engineering of microalgae to produce a variety of consumer products." *Renewable and Sustainable Energy Reviews* 14 (3):1037–1047.

He, P. J., B. Mao, C. M. Shen, L. M. Shao, D. J. Lee, and J. S. Chang. 2013. "Cultivation of Chlorella vulgaris on wastewater containing high levels of ammonia for biodiesel production." *Bioresource Technology* 129:177–181.

Heasman, M., J. Diemar, W. O'connor, T. Sushames, and L. Foulkes. 2000. "Development of extended shelf-life microalgae concentrate diets harvested by centrifugation for bivalve molluscs—a summary." *Aquaculture Research* 31 (8–9):637–659.

Henderson, R. K., S. A. Parsons, and B. Jefferson. 2008a. "Surfactants as bubble surface modifiers in the flotation of algae: Dissolved air flotation that utilizes a chemically modified bubble surface." *Environmental science & technology* 42 (13):4883–4888.

Henderson, R. K., S. A. Parsons, and B. Jefferson. 2009. "The potential for using bubble modification chemicals in dissolved air flotation for algae removal." *Separation Science and Technology* 44 (9):1923–1940.

Henderson, R. K., S. A. Parsons, and B. Jefferson. 2010. "The impact of differing cell and algogenic organic matter (AOM) characteristics on the coagulation and flotation of algae." *Water Research* 44 (12):3617–3624.

Henderson, R., S. A. Parsons, and B. Jefferson. 2008b. "The impact of algal properties and pre-oxidation on solid–liquid separation of algae." *Water research* 42 (8–9):1827–1845.

Ho, S.-H., S.-W. Huang, C.-Y. Chen, T. Hasunuma, A. Kondo, and J. -Shu Chang. 2013. "Bioethanol production using carbohydrate-rich microalgae biomass as feedstock." *Bioresource Technology* 135:191–198.

Hu, Q., M. Sommerfeld, E. Jarvis, M. Ghirardi, M. Posewitz, M. Seibert, and Al. Darzins. 2008. "Microalgal triacylglycerols as feedstocks for biofuel production: Perspectives and advances." *The Plant Journal* 54 (4):621–639.

Hu, Y.-R., C. Guo, L. Xu, F. Wang, S.-K. Wang, Z. Hu, and C.-Z. Liu. 2014. "A magnetic separator for efficient microalgae harvesting." *Bioresource Technology* 158:388–391.

Huerlimann, R., R. De Nys, and K. Heimann. 2010. "Growth, lipid content, productivity, and fatty acid composition of tropical microalgae for scale-up production." *Biotechnology and Bioengineering* 107 (2):245–257.

Ishika, T., N. R. Moheimani, and P. A. Bahri. 2017. "Sustainable saline microalgae co-cultivation for biofuel production: A critical review." *Renewable and Sustainable Energy Reviews* 78:356–368.

Islam, M. A., M. Magnusson, R. J. Brown, G. A. Ayoko, Md N. Nabi, and K. Heimann. 2013. "Microalgal species selection for biodiesel production based on fuel properties derived from fatty acid profiles." *Energies* 6 (11):5676–5702.

Jackson, B. A., P. A. Bahri, and N. R. Moheimani. 2017. "Repetitive non-destructive milking of hydrocarbons from Botryococcus braunii." *Renewable and Sustainable Energy Reviews* 79:1229–1240.

Jambo, S. A., R. Abdulla, S. H. M. Azhar, H. Marbawi, J. A. Gansau, and P. Ravindra. 2016. "A review on third generation bioethanol feedstock." *Renewable and Sustainable Energy Reviews* 65:756–769.

Jiang, J.-Q., N. J. D. Graham, and C. Harward. 1993. "Comparison of polyferric sulphate with other coagulants for the removal of algae and algae-derived organic matter." *Water Science and Technology* 27 (11):221–230.

Khan, F. A., and A. A. Ansari. 2005. "Eutrophication: An ecological vision." *The Botanical Review* 71 (4):449–482.

Kim, D.-G., H.-J. La, C.-Y. Ahn, Y.-H. Park, and H.-M. Oh. 2011. "Harvest of Scenedesmus sp. with bioflocculant and reuse of culture medium for subsequent high-density cultures." *Bioresource Technology* 102 (3):3163–3168.

Kim, G.-Y., Y.-M. Yun, H.-S. Shin, H.-S. Kim, and J.-I. Han. 2015. "Scenedesmus-based treatment of nitrogen and phosphorus from effluent of anaerobic digester and bio-oil production." *Bioresource Technology* 196:235–240.

Kim, J., G. Yoo, H. Lee, J. Lim, K. Kim, C. W. Kim, M. S. Park, and J.-W. Yang. 2013. "Methods of downstream processing for the production of biodiesel from microalgae." *Biotechnology Advances* 31 (6):862–876.

Kiran, B., K. Pathak, R. Kumar, and D. Deshmukh. 2017. "Phycoremediation: An Eco-friendly Approach to Solve Water Pollution Problems." *Microbial Applications* 1:3–28. Springer.

Knothe, G. 2009. "Improving biodiesel fuel properties by modifying fatty ester composition." *Energy & Environmental Science* 2 (7):759–766.

Koutra, E., C. N. Economou, P. Tsafrakidou, and M. Kornaros. 2018. "Bio-based products from microalgae cultivated in digestates." *Trends in Biotechnology* 36 (8): 819–833.

Kovalcik, A., K. Meixner, M. Mihalic, W. Zeilinger, I. Fritz, W. Fuchs, P. Kucharczyk, F. Stelzer, and B. Drosg. 2017. "Characterization of polyhydroxyalkanoates produced by Synechocystis salina from digestate supernatant." *International Journal of Biological Macromolecules* 102:497–504.

Kulik, M. M. 1995. "The potential for using cyanobacteria (blue-green algae) and algae in the biological control of plant pathogenic bacteria and fungi." *European Journal of Plant Pathology* 101 (6):585–599.

Kumar, D., S. P. Santhanam, T. Jayalakshmi, R. Nandakumar, S. Ananth, A. Shenbaga Devi, and B. Balaji Prasath. 2015. "Excessive nutrients and heavy metals removal from diverse wastewaters using marine microalga Chlorella marina (Butcher)." *Indian Journal of Geo-Marine Sciences*, 44 (2015).

Kwietniewska, E., and J. Tys. 2014. "Process characteristics, inhibition factors and methane yields of anaerobic digestion process, with particular focus on microalgal biomass fermentation." *Renewable and Sustainable Energy Reviews* 34:491–500.

Larsdotter, K. 2006. "Wastewater treatment with microalgae-a literature review." *Vatten* 62 (1):31.

Ledda, C., A. Schievano, B. Scaglia, M. Rossoni, F. G. Acién Fernández, and F. Adani. 2016. "Integration of microalgae production with anaerobic digestion of dairy cattle manure: An overall mass and energy balance of the process." *Journal of Cleaner Production* 112:103–112.

Lee, J.-Y., C. Yoo, S.-Y. Jun, C.-Y. Ahn, and H.-M. Oh. 2010. "Comparison of several methods for effective lipid extraction from microalgae." *Bioresource Technology* 101 (1):S75-S77.

Lim, G.-B., S.-Y. Lee, E.-K. Lee, S.-J. Haam, and W.-S. Kim. 2002. "Separation of astaxanthin from red yeast Phaffia rhodozyma by supercritical carbon dioxide extraction." *Biochemical Engineering Journal* 11 (2–3):181–187.

Liu, D., F. Li, and B. Zhang. 2009. "Removal of algal blooms in freshwater using magnetic polymer." *Water Science and Technology* 59 (6):1085–1091.

Liu, L., G. Pohnert, and D. Wei. 2016. "Extracellular metabolites from industrial microalgae and their biotechnological potential." *Marine Drugs* 14 (10):191.

Lopez, A., A. Pollice, A. Lonigro, S. Masi, A. M. Palese, G. L. Cirelli, A. Toscano, and R. Passino. 2006. "Agricultural wastewater reuse in southern Italy." *Desalination* 187 (1–3):323–334.

Luo, X., P. Su, and W. Zhang. 2015. "Advances in microalgae-derived phytosterols for functional food and pharmaceutical applications." *Marine Drugs* 13 (7):4231–4254.

Mara, D. 2013. *Domestic Wastewater Treatment in Developing Countries.* London: Routledge.

Markou, G., and E. Nerantziss. 2013. "Microalgae for high-value compounds and biofuels production: A review with focus on cultivation under stress conditions." *Biotechnology Advances* 31 (8):1532–1542.

Masojídek, J., and O. Prášil. 2010. "The development of microalgal biotechnology in the Czech Republic." *Journal of Industrial Microbiology & Biotechnology* 37 (12):1307–1317.

Mata, T. M, A. A. Martins, and N. S. Caetano. 2010. "Microalgae for biodiesel production and other applications: A review." *Renewable and Sustainable Energy Reviews* 14 (1):217–232.

Matos, C. T., M. Santos, B. P. Nobre, and L. Gouveia. 2013. "Nannochloropsis sp. biomass recovery by electro-coagulation for biodiesel and pigment production." *Bioresource Technology* 134:219–226.

Medina, A. R., A. Giménez Giménez, F. García Camacho, J. A. Sánchez Pérez, E. Molina Grima, and A. Contreras Gómez. 1995. "Concentration and purification of stearidonic, eicosapentaenoic, and docosahexaenoic acids from cod liver oil and the marine microalgaIsochrysis galbana." *Journal of the American Oil Chemists' Society* 72 (5):575–583.

Medina, A. R., E. M. Grima, A. G. Giménez, and M. J. I. González. 1998. "Downstream processing of algal polyunsaturated fatty acids." *Biotechnology Advances* 16 (3):517–580.

Mercer, P., and R. E. Armenta. 2011. "Developments in oil extraction from microalgae." *European Journal of Lipid Science and Technology* 113 (5):539–547.

Michelon, W., M. L. B. Da Silva, M. P. Mezzari, M. Pirolli, J. M. Prandini, and H. M. Soares. 2016. "Effects of nitrogen and phosphorus on biochemical composition of microalgae polyculture harvested from phycoremediation of piggery wastewater digestate." *Applied Biochemistry and Biotechnology* 178 (7):1407–1419.

Milledge, J. J., and S. Heaven. 2013. "A review of the harvesting of micro-algae for biofuel production." *Reviews in Environmental Science and Bio/Technology* 12 (2):165–178.

Moheimani, N. R., A. Vadiveloo, J. M. Ayre, and J. R. Pluske. 2018. "Nutritional profile and in vitro digestibility of microalgae grown in anaerobically digested piggery effluent." *Algal Research* 35:362–369.

Moser, B. R. 2014. "Impact of fatty ester composition on low temperature properties of biodiesel–petroleum diesel blends." *Fuel* 115:500–506.

Mutanda, T., D. Ramesh, S. Karthikeyan, S. Kumari, A. Anandraj, and F. Bux. 2011. "Bioprospecting for hyper-lipid producing microalgal strains for sustainable biofuel production." *Bioresource Technology* 102 (1):57–70.

Ndikubwimana, T., X. Zeng, Y. Liu, J.-S. Chang, and Y. Lu. 2014. "Harvesting of microalgae Desmodesmus sp. F51 by bioflocculation with bacterial bioflocculant." *Algal Research* 6:186–193.

Nurdogan, Y., and W. J. Oswald. 1995. "Enhanced nutrient removal in high-rate ponds." *Water Science and Technology* 31 (12):33–43.

Nwoba, E. G, D. A. Parlevliet, D. W. Laird, K. Alameh, and N. R. Moheimani. 2018. "Sustainable phycocyanin production from Arthrospira platensis using solar-control thin film coated photobioreactor." *Biochemical Engineering Journal* 141:232–238.

Nwoba, E. G., J. M. Ayre, N. R. Moheimani, B. E. Ubi, and J. C. Ogbonna. 2016. "Growth comparison of microalgae in tubular photobioreactor and open pond for treating anaerobic digestion piggery effluent." *Algal Research* 17:268–276.

Nwoba, E. G., N. R. Moheimani, B. E. Ubi, J. Ch. Ogbonna, A. Vadiveloo, J. R. Pluske, and J. M. Huisman. 2016. "Macroalgae culture to treat anaerobic digestion piggery effluent (ADPE)." *Bioresource Technology* 227:15–23.

Odjadjare, E. C., T. Mutanda, and A. O. Olaniran. 2017. "Potential biotechnological application of microalgae: A critical review." *Critical Reviews in Biotechnology* 37 (1):37–52.

Olguín, E. J. 2003. "Phycoremediation: Key issues for cost-effective nutrient removal processes." *Biotechnology Advances* 22 (1–2):81–91.

Olguín, E. J. 2012. "Dual purpose microalgae—bacteria-based systems that treat wastewater and produce biodiesel and chemical products within a Biorefinery." *Biotechnology Advances* 30 (5):1031–1046.

Olguín, E. J., O. S. Castillo, A. Mendoza, K. Tapia, R. E. González-Portela, and V. J. Hernández-Landa. 2015. "Dual purpose system that treats anaerobic effluents from pig waste and produce Neochloris oleoabundans as lipid rich biomass." *New Biotechnology* 32 (3):387–395.

Oswald, W. J. 2003. "My sixty years in applied algology." *Journal of Applied Phycology* 15 (2–3):99–106.

Oswald, W., and H. Gotass. 1957. "Photosynthesis in sewage treatment." *Trans. Amer. Soc. Civil Engrs. (United States)* 122.

Pachés, M., R. Martínez-Guijarro, J. González-Camejo, A. Seco, and R. Barat. 2018. "Selecting the most suitable microalgae species to treat the effluent from an anaerobic membrane bioreactor." *Environmental Technology*:1–10.

Pahazri, N. F., R. M. S. R. Mohamed, A. A. Al-Gheethi, and A. H. M. Kassim. 2016. "Production and harvesting of microalgae biomass from wastewater: A critical review." *Environmental Technology Reviews* 5 (1):39–56.

Papazi, A., P. Makridis, and P. Divanach. 2010. "Harvesting Chlorella minutissima using cell coagulants." *Journal of Applied Phycology* 22 (3):349–355.

Park, J. B. K., R. J. Craggs, and A. N. Shilton. 2011. "Wastewater treatment high rate algal ponds for biofuel production." *Bioresource Technology* 102 (1):35–42.

Phang, S.-M., W.-L. Chu, and R. Rabiei. 2015. "Phycoremediation." In *The Algae World*, edited by Dinabandhu Sahoo and Joseph Seckbach, 357–389. Dordrecht: Springer Netherlands.

Pittman, J. K., A. P. Dean, and O. Osundeko. 2011. "The potential of sustainable algal biofuel production using wastewater resources." *Bioresource Technology* 102 (1):17–25.

Pragya, N., K. K. Pandey, and P. K. Sahoo. 2013. "A review on harvesting, oil extraction and biofuels production technologies from microalgae." *Renewable and Sustainable Energy Reviews* 24:159–171.

Prajapati, S. K., P. Kumar, A. Malik, and V. Kumar Vijay. 2014. "Bioconversion of algae to methane and subsequent utilization of digestate for algae cultivation: A closed loop bioenergy generation process." *Bioresource Technology* 158:174–180.

Prasanna, R., S. Babu, N. Bidyarani, A. Kumar, S. Triveni, D. Monga, A. K. Mukherjee, S. Kranthi, N. Gokte-Narkhedkar, and A. Adak. 2015. "Prospecting cyanobacteria-fortified composts as plant growth promoting and biocontrol agents in cotton." *Experimental Agriculture* 51 (1):42–65.

Pratas, M. J., S. Freitas, M. B. Oliveira, S. C. Monteiro, A. S. Lima, and J. A. P. Coutinho. 2010. "Densities and viscosities of fatty acid methyl and ethyl esters." *Journal of Chemical & Engineering Data* 55 (9):3983–3990.

Ptacnik, R., A. G. Solimini, T. Andersen, T. Tamminen, P. Brettum, L. Lepistö, E. Willén, and S. Rekolainen. 2008. "Diversity predicts stability and resource use efficiency in natural phytoplankton communities." *Proceedings of the National Academy of Sciences* 105 (13):5134–5138.

Rajasulochana, P., and V. Preethy. 2016. "Comparison on efficiency of various techniques in treatment of waste and sewage water—A comprehensive review." *Resource-Efficient Technologies* 2 (4):175–184. doi: 10.1016/j.reffit.2016.09.004.

Raouf, N. A., A. A. Al Homaidan, and I. B. M. Ibraheem. 2012. "Microalgae and waste-water treatment." *Saudi Journal of Biological Sciences* 19 (3):257–275. doi: 10.1016/j.sjbs.2012.04.005.

Ras, M., L. Lardon, S. Bruno, N. Bernet, and J.-P. Steyer. 2011. "Experimental study on a coupled process of production and anaerobic digestion of Chlorella vulgaris." *Bioresource Technology* 102 (1):200–206.

Rawat, I., R. R. Kumar, T. Mutanda, and F. Bux. 2011. "Dual role of microalgae: Phycoremediation of domestic wastewater and biomass production for sustainable biofuels production." *Applied Energy* 88 (10):3411–3424.

Reich, P. B., J. Knops, D. Tilman, J. Craine, D. Ellsworth, M. Tjoelker, T. Lee, D. Wedin, S. Naeem, and D. Bahauddin. 2001. "Plant diversity enhances ecosystem responses to elevated CO 2 and nitrogen deposition." *Nature* 410 (6830):809.

Renuka, N., A. Sood, R. Prasanna, and A. S. Ahluwalia. 2015. "Phycoremediation of waste-waters: A synergistic approach using microalgae for bioremediation and biomass generation." *International Journal of Environmental Science and Technology* 12 (4):1443–1460.

Román, R. B., J. M. Alvarez-Pez, F. G. Acién Fernández, and E. Molina Grima. 2002. "Recovery of pure B-phycoerythrin from the microalga Porphyridium cruentum." *Journal of Biotechnology* 93 (1):73–85.

Safafar, H., J. Van Wagenen, P. Møller, and C. Jacobsen. 2015. "Carotenoids, phenolic compounds and tocopherols contribute to the antioxidative properties of some microalgae species grown on industrial wastewater." *Marine Drugs* 13 (12):7339–7356.

Sarada, R., R. Vidhyavathi, D. Usha, and G. A. Ravishankar. 2006. "An efficient method for extraction of astaxanthin from green alga Haematococcus pluvialis." *Journal of Agricultural and Food Chemistry* 54 (20):7585–7588.

Sathish, A., and R. C. Sims. 2012. "Biodiesel from mixed culture algae via a wet lipid extraction procedure." *Bioresource aechnology* 118:643–647.

Schenk, P. M., S. R. Thomas-Hall, E. Stephens, U. C. Marx, J. H. Mussgnug, C. Posten, O. Kruse, and B. Hankamer. 2008. "Second generation biofuels: High-efficiency microalgae for biodiesel production." *Bioenergy Research* 1 (1):20–43.

Shaker, S., A. Nemati, N. Montazeri-Najafabady, M. A. Mobasher, M. H. Morowvat, and Y. Ghasemi. 2015. "Treating urban wastewater: Nutrient removal by using immobilized green algae in batch cultures." *International Journal of Phytoremediation* 17 (12):1177–1182.

Shanab, S., A. Essa, and E. Shalaby. 2012. "Bioremoval capacity of three heavy metals by some microalgae species (Egyptian Isolates)." *Plant Signaling & Behavior* 7 (3):392–399.

Shen, Y., J. Gao, and L. Li. 2017. "Municipal wastewater treatment via co-immobilized microalgal-bacterial symbiosis: Microorganism growth and nutrients removal." *Bioresource Technology* 243:905–913.

Shen, Y., W. Yuan, Z. J. Pei, Q. Wu, and E. Mao. 2009. "Microalgae mass production methods." *Transactions of the ASABE* 52 (4):1275–1287.

Shi, J., B. Podola, and M. Melkonian. 2007. "Removal of nitrogen and phosphorus from wastewater using microalgae immobilized on twin layers: An experimental study." *Journal of Applied Phycology* 19 (5):417–423.

Shin, D. Y., H. U. K. Cho, J. C. Utomo, Y.-N. Choi, X. Xu, and J. Moon Park. 2015. "Biodiesel production from Scenedesmus bijuga grown in anaerobically digested food wastewater effluent." *Bioresource Technology* 184:215–221.

Show, K.-Y., D.-J. Lee, J.-H. Tay, T.-M. Lee, and J.-S. Chang. 2015. "Microalgal drying and cell disruption—recent advances." *Bioresource Technology* 184:258–266.

Shuba, E. S., and D. Kifle. 2018. "Microalgae to biofuels: 'Promising' alternative and renewable energy, review." *Renewable and Sustainable Energy Reviews* 81:743–755.

Siegrist, R. L., D. L. Anderson, and J. C. Converse. 1985. "Commercial Wastewater Onsite Treatment and Disposal." *Proc. 4th Natl. Symp. on Individual and Small Community Sewage Treatment.* American Society of Agricultural Engineers, ASAE Publ. 7-85:210–219.

Sims, R. E. H., W. Mabee, J. N. Saddler, and M. Taylor. 2010. "An overview of second generation biofuel technologies." *Bioresource Technology* 101 (6):1570–1580.

Singh, A., P. S. Nigam, and J. D. Murphy. 2011. "Mechanism and challenges in commercialisation of algal biofuels." *Bioresource Technology* 102 (1):26–34.

Spolaore, P., E. Duran, and A. Isambert. 2006. "Commercial applications of microalgae." *Journal of Bioscience and Bioengineering* 101 (2):87–96. doi: 10.1263/Jbb.101.87.

Stengel, D. B., S. Connan, and Z. A. Popper. 2011. "Algal chemodiversity and bioactivity: Sources of natural variability and implications for commercial application." *Biotechnology Advances* 29 (5):483–501.

Tang, H., M. Chen, M. E. D. Garcia, N. Abunasser, K. Y. Simon Ng, and S. O. Salley. 2011. "Culture of microalgae Chlorella minutissima for biodiesel feedstock production." *Biotechnology and Bioengineering* 108 (10):2280–2287.

Tang, S. L., and K. V. Ellis. 1994. "Wastewater treatment optimization model for developing world. II: Model testing." *Journal of Environmental Engineering* 120 (3):610–624.

Tchobanoglous, G., F. L. Burton, and H. D. Stensel. 1991. "Wastewater engineering." *Management* 7:1–4.

Teixeira, M. R., and M. J. Rosa. 2006. "Comparing dissolved air flotation and conventional sedimentation to remove cyanobacterial cells of Microcystis aeruginosa: part I: The key operating conditions." *Separation and Purification Technology* 52 (1):84–94.

Troschl, C., K. Meixner, and B. Drosg. 2017. "Cyanobacterial PHA production—Review of recent advances and a summary of three years' working experience running a pilot plant." *Bioengineering* 4 (2):26.

Tüzün, I., G. Bayramoğlu, E. Yalçın, G. Başaran, G. Celik, and M. Yakup Arıca. 2005. "Equilibrium and kinetic studies on biosorption of Hg (II), Cd (II) and Pb (II) ions onto microalgae Chlamydomonas reinhardtii." *Journal of Environmental Management* 77 (2):85–92.

Udom, I., B. H. Zaribaf, T. Halfhide, B. Gillie, O. Dalrymple, Q. Zhang, and S. J. Ergas. 2013. "Harvesting microalgae grown on wastewater." *Bioresource Technology* 139:101–106.

Uduman, N., V. Bourniquel, M. K. Danquah, and A. F. A. Hoadley. 2011. "A parametric study of electrocoagulation as a recovery process of marine microalgae for biodiesel production." *Chemical Engineering Journal* 174 (1):249–257.

Uduman, N., Y. Qi, M. K. Danquah, G. M. Forde, and A. Hoadley. 2010. "Dewatering of microalgal cultures: A major bottleneck to algae-based fuels." *Journal of Renewable and Sustainable Energy* 2 (1):012701.

Vadiveloo, A., N. R. Moheimani, J. J. Cosgrove, P. A. Bahri, and D. Parlevliet. 2015. "Effect of different light spectra on the growth and productivity of acclimated Nannochloropsis sp. (Eustigmatophyceae)." *Algal Research* 8:121–127.

Van der Spiegel, M., M. Y. Noordam, and H. J. Van der Fels-Klerx. 2013. "Safety of novel protein sources (insects, microalgae, seaweed, duckweed, and rapeseed) and legislative aspects for their application in food and feed production." *Comprehensive Reviews in Food Science and Food Safety* 12 (6):662–678.

Van Drecht, G., A. F. Bouwman, J. Harrison, and J. M. Knoop. 2009. "Global nitrogen and phosphate in urban wastewater for the period 1970 to 2050." *Global Biogeochemical Cycles* 23 (4).

Van Harmelen, T., and H. Oonk. 2006. "Microalgae biofixation processes: Applications and potential contributions to greenhouse gas mitigation options." *TNO Built Environment and Geosciences, Apeldoorn, The Netherlands* 56.

Vanthoor-Koopmans, M., R. H. Wijffels, M. J. Barbosa, and M. H. M. Eppink. 2013. "Biorefinery of microalgae for food and fuel." *Bioresource Technology* 135:142–149.

Vasarevičius, S., A. Mineikaite, and P. Vaitiekūnas. 2010. "Investigation into heavy metals in storm wastewater from Vilnius Žirmūnai district and pollutantits spread model in Neris river." *Journal of Environmental Engineering and Landscape Management* 18 (3):242–249.

Villarruel-López, A., F. Ascencio, and K. Nuño. 2017. "Microalgae, a potential natural functional food source—a review." *Polish Journal of Food and Nutrition Sciences* 67 (4):251–264.

Vonshak, A. 1997. *Spirulina Platensis Arthrospira: Physiology, Cell-Biology and Biotechnology*. London: Taylor & Francis.

Wang, M., W. C. Kuo-Dahab, S. Dolan, and C. Park. 2014. "Kinetics of nutrient removal and expression of extracellular polymeric substances of the microalgae, Chlorella sp. and Micractinium sp., in wastewater treatment." *Bioresource Technology* 154:131–137.

Ward, A. J., D. M. Lewis, and F. B. Green. 2014. "Anaerobic digestion of algae biomass: A review." *Algal Research* 5:204–214.

Wilke, A., R. Buchholz, and G. Bunke. 2006. "Selective biosorption of heavy metals by algae." *Environmental Biotechnology* 2:47–56.

Wu, Y.-H., H.-Y. Hu, Y. Yu, T.-Y. Zhang, S.-F. Zhu, L.-L. Zhuang, X. Zhang, and Y. Lu. 2014. "Microalgal species for sustainable biomass/lipid production using wastewater as resource: A review." *Renewable and Sustainable Energy Reviews* 33:675–688.

Xu, L., C. Guo, F. Wang, S. Zheng, and C.-Z. Liu. 2011. "A simple and rapid harvesting method for microalgae by in situ magnetic separation." *Bioresource Technology* 102 (21):10047–10051.

Xu, L., D. W. F. Wim Brilman, J. A. M. Withag, G. Brem, and S. Kersten. 2011. "Assessment of a dry and a wet route for the production of biofuels from microalgae: Energy balance analysis." *Bioresource Technology* 102 (8):5113–5122.

Xu, Z., X. Yan, L. Pei, Q. Luo, and J. Xu. 2008. "Changes in fatty acids and sterols during batch growth of Pavlova viridis in photobioreactor." *Journal of Applied Phycology* 20 (3):237–243.

Yahi, H., S. Elmaleh, and J. Coma. 1994. "Algal flocculation-sedimentation by pH increase in a continuous reactor." *Water Science and Technology* 30 (8):259–267.

Yildiz-Ozturk, E., and O. Yesil-Celiktas. 2017. "Supercritical CO_2 extraction of hydrocarbons from Botryococcus braunii as a promising bioresource." *The Journal of Supercritical Fluids* 130:261–266.

Yu, Z., M. Song, H. Pei, F. Han, L. Jiang, and Q. Hou. 2017. "The growth characteristics and biodiesel production of ten algae strains cultivated in anaerobically digested effluent from kitchen waste." *Algal Research* 24:265–275.

Yuan, X., M. Wang, C. Park, A. K. Sahu, and S. J. Ergas. 2012. "Microalgae growth using high-strength wastewater followed by anaerobic co-digestion." *Water Environment Research* 84 (5):396–404.

Zamalloa, C., E. Vulsteke, J. Albrecht, and W. Verstraete. 2011. "The techno-economic potential of renewable energy through the anaerobic digestion of microalgae." *Bioresource Technology* 102 (2):1149–1158.

Zamani, N., M. Noshadi, S. Amin, A. Niazi, and Y. Ghasemi. 2012. "Effect of alginate structure and microalgae immobilization method on orthophosphate removal from wastewater." *Journal of Applied Phycology* 24 (4):649–656.

Zuliani, L., N. Frison, A. Jelic, F. Fatone, D. Bolzonella, and M. Ballottari. 2016. "Microalgae cultivation on anaerobic digestate of municipal wastewater, sewage sludge and agrowaste." *International Journal of Molecular Sciences* 17 (10):1692.

8 Integrated Approach for the Sustainable Extraction of Carbohydrates and Proteins from Microalgae

Sambit Sarkar, Mriganka Sekhar Manna, Sunil Kumar Maity, Tridib Kumar Bhowmik, Kalyan Gayen

CONTENTS

8.1 INTRODUCTION

Microalgae became the focus of scientific research primarily due to its high growth rate through photosynthesis and high lipid reserve suitable for biodiesel production (Ghosh et al. 2017; Mondal et al. 2016). Interest has also been growing in using microalgae for the production of several valuable biomolecules such as carbohydrate, proteins, lipids, and pigments (Khanra et al. 2018). However, the availability of efficient techniques for the separation of these biomolecules from the biomass and their purification has been the primary constraint in terms of the commercialization (Ghosh et al. 2016). The economic feasibility of the large-scale production of microalgae for low-value bulk products such as proteins (for food/feed), carbohydrates, or biofuels has yet to be achieved (Williams and Laurens 2010). Photoautotrophic microalgae are currently being explored for the industrial-scale production of several high-value products like pigments (astaxanthin, β-carotene) and water-soluble proteins such as phycobiliproteins (Williams and Laurens 2010; Vermuë et al. 2018). Astaxanthin and β-carotene are commercially produced from *Haematococcus pluvialis* and *Dunaliella salina*, respectively, and the cyanobacterium *Spirulina platensis* is currently used to obtain phycobiliproteins (Williams and Laurens 2010). In the prevailing trend of downstream processing of microalgal biomass, only one specific substance of interest (e.g. astaxanthin, β-carotene) is targeted for the extraction, whereas other potential components (e.g. proteins, carbohydrates) are either discarded or not adequately valorized (Vermuë et al. 2018). Looking at the preliminary results of the extraction of valuable products from microalgae, it seems that the valorization of all fractions should be emphasized along with the reduction of cost for the simultaneous extraction of products. Parallel payback from other aspects, like wastewater treatment and carbon dioxide sequestration, that are associated with the production of microalgal biomass should be incorporated in the integrated extraction process for economic sustainability (Lorente et al. 2018). This concept of complete fractionation and valorization of the microalgal biomass is referred to as a microalgae biorefinery (Vermuë et al. 2018; Vanthoor-Koopmans et al. 2013). This concept is analogous to the petroleum refinery. The microalgae biorefinery with multiple end products should be implemented for the production of commodities in an economically feasible manner (Vermuë et al. 2018). Proteins and carbohydrates are generally underpriced in a microalgae biorefinery in comparison with various pigments (e.g., chlorophyll and β-carotene) and lipids. However, the proportions of these components are substantially high in the microalgal biomass. Different microalgae and cyanobacterial strains contain 6% to 70% proteins and 30% to 60% carbohydrate on a dry mass basis (Chen et al. 2013). These fractions have substantial market values and their recovery will enhance the profit ratio in commercial production of microalgal products. Dibenedetto et al. estimated the value of the protein fraction used as a food and feed supplement at 5 and 0.75 €/kg, respectively, and the value of sugars used in chemical industries was estimated at 1 €/kg (Dibenedetto, Colucci, and Aresta 2016).

Microalgal proteins, having all essential amino acids, possess a great potential to be an alternative source of proteins (Waghmare et al. 2016). Microalgae proteins are considered a functional food due to their positive health effects. A number of studies have revealed that *Chlorella* used as a dietary supplement may help in the reduction of high blood pressure, serum cholesterol, and glucose levels. It also accelerates wound healing and enhances immune functions (Jong-Yuh and Mei-Fen 2005; Halperin et al. 2003; Merchant and Andre 2001). Microalgal proteins containing functional properties have the potential to be used as an emulsifying agent. Protein hydrolysates with peptides of molecular size greater than 10 to 20 kDa are favoured for their emulsifying and foaming capability with better stability (Schröder et al. 2017). Proteins, protein hydrolysates, or peptides are the sources of the majority of the functional components possessing health-promoting effects (Phong et al. 2018). Consequently, a great deal of attention has been paid to the proteins and peptides from microalgae that possess biofunctional properties such as antioxidants, antihypertensive, antitumor, anticoagulative, and immunostimulant activities (Samarakoon and Jeon 2012). The isolated or pure or partially purified bioactive peptides from protein hydrolysates are injected into the human bloodstream, and they deliver beneficial biological activity (Ovando et al. 2018). Several beneficial properties of peptides, such as antioxidant (Ovando et al. 2018; Kim, Mujtaba, and Lee 2016), antihypertensive (Kim, Mujtaba, and Lee 2016), immune-modulatory (Morris et al. 2008), anticancer (Sheih, Wu, and Fang 2009), hepatic protective (Sheih, Wu, and Fang 2009), and anticoagulant activities (Athukorala et al. 2007) have been reported.

Depending on the culture condition and the nutrient status of the culture medium, microalgae can accumulate a large amount of structural and storage carbohydrates (Williams and Laurens 2010; Laurens et al. 2015). However, this was often regarded as disadvantageous under the strain improvement scheme for biofuel production (as the carbon flux was shifted from lipid to carbohydrate accumulation). Nevertheless, the utilization of microalgal carbohydrates in the fermentation medium has recently been increased for the production of biofuels (e.g., ethanol, butanol) (Laurens et al. 2015). The acid-catalyzed hydrolysis of carbohydrates derived from *Chlorococcum* sp. and *Chlorella* sp. biomass has been explored to obtain sugars for a fermentation broth (Doan et al. 2012). Cellulose from microalgae is also used for the production of bioplastic (Mihranyan 2011). Beside cellulose, other types of polysaccharides are also derived from microalgae for valuable products, such as alginates, agar, and carrageenan (Khanra et al. 2018). Two different red algae species, namely *Gracilaria* and *Gelidium*, are the prevalent sources of agar (Bixler and Porse 2011) that are used as a gelling agent in scientific laboratories and as a thickening agent in bakeries (Khanra et al. 2018). Carrageenan extracted from various red algae strains, such as *Chondrus* sp., *Gigartina* sp. *Eucheuma* sp., *Hypnea* sp., and *Furcel-laran* sp., is commercially employed in meat, dairy, desserts, and beverages industries (Khanra et al. 2018). Fucoidan, found in the cell wall of brown algae—*Phaeophyceae* sp. (Hahn et al. 2011)—is another polysaccharide with several bioactive properties such as anticoagulant, antitumor, antivirus, and antioxidant activities (Li et al. 2008).

Fucoidan thus finds potential application in the pharmaceutical industry. An integrated approach for the extraction of proteins and carbohydrates is a sustainable idea for the biorefining of the microalgal biomass. These biomolecules can subsequently be used for manufacturing different value-added products. The extraction of these two metabolites from the same biomass in both industrial and laboratory-scale applications has two major bottlenecks: (i) Both of these two metabolites are water soluble. The separation of these two metabolites is thus challenging. (ii) The selection of a suitable method for the disruption of the cell wall is another challenge.

The integrated extraction strategy contains three general stages, namely cell disruption, extraction, and product separation/fractionation. Both the proteins and carbohydrates can be simultaneously extracted into the aqueous medium via the disruption of the cell wall. These proteins and carbohydrates then become available for further separation and purification processes. The efficiency of the disruption method has a huge impact on the subsequent steps in the integrated extraction process. The efficient methods for cell wall disruption at the industrial-scale include acid hydrolysis at low pH, steam explosion, and microwave treatment at high temperatures. These disruption methods are very useful for the extraction of carbohydrates from a lignocellulosic biomass (Haldar, Sen, and Gayen 2016). However, these methods of cell wall disruption yield a lower amount of protein due to the tendency of proteins to denature under such severe conditions. The temperature should be maintained below 35 °C during the cell wall disruption to inhibit protein aggregation and maintain its solubility in the aqueous media (Postma et al. 2015). Following cell wall disruption, the proteins and carbohydrates are separated from the aqueous solution by fractionation. The proteins and polysaccharides tend to get precipitated upon the addition of trichloroacetic acid (TCA) and cold ethanol, respectively. The proteins precipitated by the application of TCA, however, are prone to denaturization. The proteins can be selectively separated using chromatographic separation techniques, for example, ion-exchange chromatography (Schwenzfeier, Wierenga, and Gruppen 2011; Schwenzfeier et al. 2014; Schwenzfeier et al. 2013) and size-exclusion chromatography (Cuellar-Bermudez et al. 2015). The chromatographic techniques, however, are highly expensive. The fractionation by aqueous two-phase separation (ATPS), three-phase partitioning (TPP), and membrane separation are alternative and effective methods for the simultaneous extraction of carbohydrates and proteins from the microalgal biomass. The possible scale-up of these integrated processes to the industrial level is an added advantage. The chapter is thus intended to provide an outline of the technological advancements in the simultaneous recovery of proteins and carbohydrates from the microalgal biomass in an integrated approach.

8.2   CELL WALL DISRUPTION OF MICROALGAE

The cell wall of microalgae is more rigid and thicker compared to other microorganisms and some higher plants (Phong et al. 2018). The tensile strength of the microalgal cell wall is reported to be as high as 9.5 MPa, which is approximately three-fold greater compared to that of *Daucus carota* (Lee, Lewis, and Ashman 2012). The microalgal cell wall is a three-layered structure and usually composed of polysaccharides such as cellulose, pectin, mannose, and xylan; minerals like calcium or silicates; and proteins in the form of glycoproteins in different proportions

(Phong et al. 2018). A fairly large proportion of cellulose in the cell wall provides the structural stability and rigidity for most of the microalgae species. The presence of covalent and hydrogen bonds and van der Waals force also contribute to the strength of the cell wall (Lee, Lewis, and Ashman 2012). The strong cell wall provides great resistance against the release of intracellular components from the cell (Vermuë et al. 2018). Understanding the chemical composition and the structure is essential to develop a suitable method for the selective disruption of the microalgal cell wall (i.e., breaking a portion of the cell wall instead of complete disintegration) (Günerken et al. 2015). The disruption of the cell wall under mild conditions (shear strength and heat) is obligatory for preserving the inherent structure of the individual intracellular components (Schwenzfeier, Wierenga, and Gruppen 2011; Schwenzfeier et al. 2014; Schwenzfeier et al. 2013; Cuellar-Bermudez et al. 2015; Lee, Lewis, and Ashman 2012; Günerken et al. 2015). In particular, the functionality of water-soluble, high-value proteins should be preserved due to their functional properties like foaming and emulsifying (Vermuë et al. 2018). Severe treatment conditions in (bio)chemical processes such as low/high pH, high temperatures and pressure, and shear stress can have a negative impact on these functionalities (Schwenzfeier, Wierenga, and Gruppen 2011). Exposing microalgal cells to such harsh conditions might also damage the structure and subsequently lead to the loss of specific functionality and activity of proteins (Postma et al. 2015).

Based on the nature of the force applied, disruption methods can be classified as physical (e.g., drying, sonication, and pulsed electric field), mechanical (e.g., bead milling, homogenization), and chemical/biological (e.g., acid, base, and enzymes) (Soto-Sierra, Stoykova, and Nikolov 2018). The selection of the disruption method depends mainly on the structure of the cell wall, operational cost, and targeted end products and their applications (Phong et al. 2018).

Cell disruption by mechanical methods, such as bead milling, does not use harsh conditions such as high temperature and extreme pH (Vermuë et al. 2018). Consequently, this method has proved to be suitable for the extraction of highly sensitive biomolecules such as proteins (Schwenzfeier, Wierenga, and Gruppen 2011). In spite of the mildness, the mechanical methods are highly energy intensive (Coons et al. 2014) and therefore they are not suitable for industrial applications. For cell wall disruption of *Chlorella vulgaris*, for example, the optimization of the operating condition has been shown to reduce the energy consumption of bead milling to 0.5 kWh (Postma et al. 2017) from 2.8 to 10 kWh per kg dry weight (Doucha and Lívanský 2008). The bead milling completely disintegrates the cells into ultra-small fragments (Postma et al. 2017). This is still one of the major drawbacks. Consequently, a heterogeneous and emulsified mixture of different molecules, like proteins, sugars, pigments, and lipids, is formed. The phase separation and the isolation of specific molecules from the mixture becomes a difficult task (Vermuë et al. 2018).

Ultrasonication is another physical method for gently rupturing the microalgal cell wall (Vanthoor-Koopmans et al. 2013). This method has been used in a number of studies for the extraction of various components, such as proteins, pigments, and lipids, from microalgae. Sonication alone, however, is not sufficient for the complete release of intracellular components due to the high resistance afforded by the double protective layers (cell wall and cell membrane) of microalgal cells (Soto-Sierra, Stoykova, and Nikolov 2018; Coons et al. 2014; Postma et al. 2017; Doucha and Lívanský 2008; Sierra, Dixon, and Wilken 2017). Additional treatments such as mixing with high shear, enzymatic, or chemical treatment were shown to release additional soluble protein from *Nannochloropsis* sp., *Chlamydomonas reinhardtii*, and *S. platensis* (Sierra, Dixon, and Wilken 2017; Gerde et al. 2013; Lupatini et al. 2017). Energy consumption in sonication is relatively lower compared to mechanical processes such as bead beating and homogenization. The energy consumption can be further reduced by using high-frequency ultrasound (Günerken et al. 2015). Coupling chemical treatment with ultrasonication is often considered a low-energy-consumption process for this purpose.

The pulsed electric field method utilizes low-shear and low-temperature conditions for cell disruption. Thus, this method is proven to be an alternative mild process of physical cell disruption. It is best suited for the selective release of water-soluble intracellular components (e.g., proteins). The release of entire proteins from a mutant strain of *C. reinhardtii* with no cell wall has been explored ('t Lam et al. 2017). The method creates an opening for hydrophilic protein release, while the pigments still remain inside the cell. It does not break the cell into fragments, which has a severe effect on the fractionations of different components ('t Lam et al. 2017). Nevertheless, this method is inefficient for breaking a thick cell wall. This is due to the fact that only 25% to 39% carbohydrates and 3% to 5% proteins could be released from the cells of *Chlorella vulgaris* by pulsed electric field treatment (Postma et al. 2016). The two-stage strategy comprising enzymatic treatment followed by pulsed electric field treatment has been proposed to overcome the drawbacks of this cell disruption method. A two-fold increase in protein yield was observed for the pulsed electric field treatment of the protease pre-treated cells of *C. reinhardtii* as compared to pulsed electric field treatment alone ('t Lam et al. 2017).

Enzymatic treatment is the mildest among the chemical and biochemical methods that disrupt the cell wall by hydrolyzing the polysaccharide. Enzymatic hydrolysis has successfully been utilized for cell wall disruption of various microalgal species (e.g., *Scenedesmus sp., Chlorella vulgaris*) using cellulase or a combination of cellulase with other enzymes like lysozyme. Enzymes are specific to the characteristics of the cell wall structure. Large-scale implementation of the enzymatic disruption method might be hindered by high processing time and cost. This problem, however, can be diminished by either combining it with other methods or immobilizing the enzyme(s). Immobilized cellulose enzyme on electrospun polyacrylonitrile nanofibrous membrane was able to hydrolyze ~62% of polysaccharides from the cell wall of *Chlorella* sp. (Fu et al. 2010). Interestingly, the enzyme remained useful even after five passes (Fu et al. 2010). The reusability and ease of separation of the immobilized enzyme are the benefits of this technique.

8.3   SELECTIVE SEQUENTIAL EXTRACTION

8.3.1   Two-Step Sequence for the Extraction of Proteins and Carbohydrates

The success of the sequential extraction of metabolites by using conventional extraction methods from the microalgal biomass is highly dependent on the sequence of the extraction. The yield of a particular metabolite is high when it is primarily extracted from the initial biomass. But during the secondary extraction from spent biomass, the yield drastically reduces. The yield of proteins from the sun-dried biomass of *Scenedesmus obliquus* has been found to be decreased by ~11% when extracted from carbohydrate-extracted biomass. On the other hand, the extraction yield of carbohydrate decreased by 41% when extracted from the protein-extracted biomass (Ansari et al. 2017). A decrease of 19.4% yield has been observed during the extraction of protein from lipid-extracted cells of *Navicula* sp. (Patterson and Gatlin 2013). The loss of yield may be due to product degradation and loss of biomass at each step. So, this loss should be calculated for each metabolite in each sequence. Therefore, the most economical sequence should be selected.

Ansari et al. investigated the different sequences of extraction of different metabolites from *Scenedesmus obliquus*. The residual biomass from the primary step was vacuum filtered, dried, and then used for the subsequent steps. Proteins were extracted from dried microalgal cells mixed with water by means of grinding and centrifugation and then re-extracted with 0.1N NaOH with 0.5% β-mercaptoethanol (v/v). Carbohydrates were extracted from the remaining biomass by mixing 2% H_2SO_4 (v/v) followed by autoclaving at 121 °C and at a pressure of 15 psi (Ansari et al. 2017). Protein and carbohydrate yields were found to be ~28% and ~10%, respectively, when proteins were extracted first, followed by the carbohydrates. Figure 8.1 depicts a flow diagram of the different steps of the sequential extraction process.

The yield was ~25% for proteins and ~16% for carbohydrates when carbohydrate extraction was prioritized over protein extraction. The percentage loss of carbohydrate is more than that of protein when extracted from the residual biomass. But the remaining biomass for subsequent steps is much lower when carbohydrates were extracted first (65%) than when proteins were extracted first (80%). A low residual biomass signifies the low yield of all the metabolites in subsequent steps (Ansari et al. 2017). Interestingly, Munoz et al. observed the accumulation of carbohydrates in the spent biomass of *B. braunii* after the extraction of proteins and lipids (Muñoz et al. 2015). The recent findings (Gerde et al. 2013; Lupatini et al. 2017; 't Lam et al. 2017; Postma et al. 2016; Fu et al. 2010; Ansari et al. 2017; Patterson and Gatlin 2013; Muñoz et al. 2015; Goettel et al. 2013) further suggested that protein extraction should be prioritized in the biorefinery process (Ansari et al. 2017).

8.3.2 COMBINED/PARALLEL ALGAL PROCESSING

Hydrolysis of microalgal carbohydrates by dilute acid has been established as an effective method for better utilization of the biomass. It releases fermentable sugar and makes the proteins and lipid fractions available for extraction. The sugar-rich liquor can be separated out from the residual biomass by means of solid–liquid separation (SLS) (Laurens et al. 2015), or the entire sugar-rich mixture can be used as the media for the fermentation (Zhao et al. 2014). The approach that uses an entire mixture of sugar and residual biomass to produce fuel by means of fermentation is called combined algal processing (CAP). On the other hand, the parallel algal processing (PAP) uses a separation of sugar mixture from the residual biomass. Methods like ATPS or TPP can be employed for the fractionation of lipid and protein from the

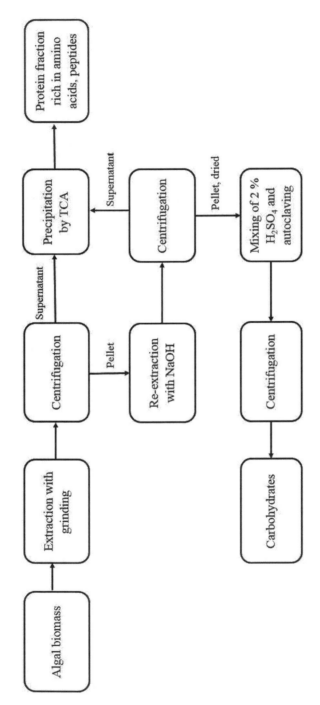

FIGURE 8.1   Sequence for the selective extraction of proteins and carbohydrate from a biomass.

residual biomass, followed by additional cell wall disruption processes like ultra-sonication or homogenization. These two methods are helpful in the separation of protein from lipid (Mulchandani, Kar, and Singhal 2015).

Dong et al. achieved 80% total sugar yield using a wet algal biomass of *Scenedesmus acutus* by extracting with 2% aqueous H_2SO_4 solution for 15 min at 155 °C. Moreover, they achieved an ethanol yield of 31 gasoline gallon equivalent (GGE) per ton of biomass by using the entire mixture for fermentation (Dong et al. 2016). Laurens et al. achieved an ethanol yield of 20 GGE/ton biomass by using the sugar solution obtained by similar acid pre-treatment, but after separating the solids from it by SLS (Laurens et al. 2015). The loss of sugar in the solid cake was found to be 37% in PAP process, and the entire fermentable sugar is utilized in the CAP process. They employed hexane to extract the lipid from the leftover biomass, which left proteins as residue. It was reported that the lipid extracted by a suitable solvent like hexane had undergone no degradation and could be directly converted to biofuel (Dong et al. 2016). The recovered protein fraction can be used in food and feed production and in anaerobic digestion for biogas production (Dong et al. 2016). The amino acids can be used for the production of branched-chained higher alcohols (Atsumi, Hanai, and Liao 2008).

8.4 CO-EXTRACTION OF PROTEINS AND CARBOHYDRATES BY PRESSURIZED LIQUID EXTRACTION

An extraction technique that employs liquids at high temperature and pressures (below the critical limit) is called pressurized liquid extraction (PLE). Other terminologies, such as accelerated solvent extraction (ASE), pressurized fluid extraction (PFE), subcritical water extraction (SWE) when using water as the fluid, or pressurized hot-water extraction (PHWE), are also used (Herrero et al. 2015). PLE is characterized by the requirement of the comparatively low volume of solvents than conventional extraction techniques (Hu et al. 2011). The higher temperature improves the solubility of the solute and reduces the viscosity of the solvent, which enhances the solvent penetration and hence the rate of mass transfer (Herrero et al. 2015). Moreover, the dielectric constant (ε) of water, which signifies the strength of the solvent, decreases significantly, up to 30 at a high temperature close to 250 °C. The value of the dielectric constant of the water (ε) is approximately 80 at ambient temperature. The performance of water thus becomes similar to an organic solvent like methanol, ethanol, acetone, etc., in terms of polarity due to the reduction in the dielectric constant (Herrero, Cifuentes, and Ibañez 2006). The temperature and pressure, however, should be optimized for the solutes (proteins and carbohydrate) to keep water in the liquid state for the highest extraction yield.

Microalgal carbohydrates are extracted by the conventional methods at low temperatures. However, none of these low-temperature methods are efficient to completely extract carbohydrate from the biomass due to their complex structure. Hence, the techniques, that employ higher temperature and pressure, are recommended for enhancing the yield of carbohydrate.

Carbohydrates are highly soluble in water at temperatures above 100 °C to 150 °C. The highest extraction yield is thus exhibited near 200 °C under the influence of a reduced dielectric constant. Most of the studies on the extraction of carbohydrates by PLE indicate that temperature, extraction time, biomass loading, and particle size are the important factors that influence the extraction (Zakaria et al. 2017; Awaluddin et al. 2016). The degradation of polysaccharide at higher temperature often results in a lower yield and the effect of the reduced dielectric constant may be surpassed (Gallego et al. 2018). The antiviral activity of carbohydrates extracted by the SWE method from the microalgae *Haematococcus pluvialis* and *Dunaliella salina*, has been investigated to be greater than that by carbohydrates extracted by water at ambient temperature. Glucose was the main component of *D. salina* (94.34%), whereas mannose (main component), along with glucose and galactose, were found in high quantities from *H. pluvialis* (Santoyo et al. 2012). However, very little information is available about selective extraction of protein by PLE.

Proteins have been co-extracted along with carbohydrate by the PLE method using either water, ethanol, or their mixture as a solvent. It has been reported that ~31% (wt. dry-biomass) proteins along with 14% (wt. dry-biomass) carbohydrates can be extracted at 277 °C from the wet biomass of *Chlorella* sp. using SWE at 5 min extraction time with 5% biomass concentration and 90 μm particle size. The yield of proteins decreases above 270 to 300 °C due to the thermal denaturation of proteins through the destruction of their tertiary structure (Awaluddin et al. 2016). Rapid vibration of proteins at high thermal energy leads to the breakage of hydrogen bonds among functional groups and non-polar hydrophobic interactions. Extract from the biomass of *Isocrysis galbana* obtained by PLE using ethanol at 80 °C and 100 bar was found to contain ~8% to 9% (w/w of total content) of both proteins and sugars. Whereas the yield of the proteins and sugars in the subsequent step of SWE (80 °C, 100 bar) was ~18 % and ~10%, respectively, using the residual biomass of the ethanol PLE step (Gilbert-López et al. 2015). The antioxidant activity of the ethanolic extract was found to be twice that of the water extract. Therefore, it is evident that the types of proteins and sugars present in ethanol and water extracts are different. Ethanol is known to have the property of precipitating polysaccharides, so ethanol extract may contain monosaccharides and oligosaccharides, whereas water extract would preferentially contain polysaccharides in greater proportion (Gilbert-López et al. 2015). However, a very low amount of proteins (~1%) and carbohydrates (~2.8%) were estimated in the extract of *Scenedesmus obliquus* obtained by PLE using water under the operating conditions of 100 bar pressure, 50 °C temperature, and 45 min of extraction time in another biorefinery downstream process design (Gilbert-López et al. 2017). In this process, the previous step of extraction involved gas-expanded ethanol (75% ethanol, 25% scCO$_2$) at 50° C for 150 min under 70 bar pressure. The ethanol extract contained ~2% and 4% of proteins and sugars, respectively. These biorefinery processes also contained a supercritical CO$_2$ extraction step as an initial step for the extraction of hydrophobic molecules such as pigments, lipids, and free fatty acids. Therefore, the extraction of hydrophilic compounds was not prioritized in the process. The low yield of both the proteins and sugars may be due to their retention in the residue. The PLE technique for co-extraction may not be fully developed, and further optimization of the process may enhance the yield. However, the additional separation is required to obtain a separate fraction of proteins and sugars.

Separation of proteins may be achieved by either shifting the pH beyond the iso-electric point, or chromatographic separation methods, or solvent-based separation methods.

8.5 INTEGRATION OF FRACTIONATION WITH CO-EXTRACTION

Following co-extraction, fractionation is a prerequisite to obtain a separate fraction of proteins and carbohydrates (pure or partially pure) from the aqueous mixture. The extract from the selective sequential extraction also possesses some amount of impurity, as both the proteins and sugars are water-soluble. And the fractionation incurs an additional cost. Nevertheless, liquid–liquid extraction processes viz. TPP, ATPS, or mechanical processes viz. ultrafiltration and dia-filtration lead to substantially pure proteins and sugars in small- to large-scale operation.

8.5.1 THREE-PHASE PARTITIONING

TPP employs water-miscible aliphatic alcohol (mostly *t*-butanol) and an aqueous antichaotropic salt (most commonly ammonium sulphate) to form three distinct phases with different biomolecules in each phase. Despite the miscibility of alcohols in water, they can form two distinct phases when mixed with the aqueous solutions of antichaotropic salts. The salty mixture remains at the bottom, making the alcohols float above it (Chew, Ling, and Show 2019).

TPP was proposed by Tan and Lovrein in 1972 for the first time as an alternative to the conventional extraction and separation techniques. A solvent separation process with toxic and volatile organic solvents is not recommended because it is not environmentally friendly. It is also unfavourable due to the longer time and higher energy requirements (Dennison and Lovrien 1997). The crude protein mixture containing impurities is mixed with an inorganic salt, followed by the addition of alcohol in a specific amount in TPP. In the presence of the antichaotropic salt, the alcohol shoves the protein out of the salt solution. A bond is formed between the hydrophobic portions of the mixture, which reduces the density of the proteins. The proteins, as a result, float above the denser aqueous-salt phase that contains hydrophilic components (impurities). Proteins are later recovered in purified form in a third phase created between the two phases. The hydrophobic and hydrophilic contaminants are found in the alcohol and the aqueous-salt phase, respectively (Yan et al. 2017; Pike and Dennison 1989).

The TPP separation process is believed to involve a combination of diverse operating phenomena *viz.* salting out, kosmotropy, co-solvent precipitation, conformation tightening, isoionic precipitation, protein hydration shifts, and osmolytic electrostatic forces (Dennison and Lovrien 1997; Yan et al. 2017). The behaviours unveiled by the proteins subjected to the different parameters depend on the source of the biomass, charge, molecular weight, hydrophobicity, isoelectric point (pI) of the proteins, and temperature of the solvent (Pike and Dennison 1989). Moreover, the dehydration properties of ammonium sulfate (the most commonly used salt in TPP) are also believed to contribute to the protein concentration in the interphase, and *t*-butanol (the most commonly used alcohol) facilitates the protein's buoyancy (Yan et al. 2017; Pike and Dennison 1989; Paule et al. 2004). It has been reported that in this process, proteins are actually precipitated as protein-*t*-butanol co-precipitates whose buoyancies can be amplified by increasing the amount of *t*-butanol (Dennison and Lovrien 1997).

The TPP technique possesses the advantages of having simple operating parameters and ability to recover solute rapidly. It has also the potential to be scaled up to the industrial level. Tertiary butanol has a unique capacity of forming three phases with water in the presence of inorganic salt. Alcohols other than *t*-butanol do not exhibit the properties to the same extent as shown by *t*-butanol under room temperature. Ethanol and isopropanol were reported to provide two-fold less yield than *t*-butanol for the extraction of fat from kokum (Vidhate and Singhal 2013). The process can be applied for the extraction and purification of a wide range of biological compounds such as proteins, enzymes, enzyme inhibitors, carbohydrates, lipids, etc. Besides providing improved concentration, TPP is able to purify products at an equivalent level to chromatographic purification (Saxena, Iyer, and Ananthanarayan 2007). It has the advantage of being applicable at room temperature. The time requirement is small (less than one hour), and extracting chemicals are recyclable. Tertiary butanol is chemically less hazardous due to the high boiling point compared to other common solvents such as the hexane or chloroform used in oil extraction (Panadare and Rathod 2017). It has a melting point of 25 °C (Reeve, Erikson, and Aluotto 1979), which facilitates its recovery as a solid crystal from the final extract below 25 °C (Mulchandani, Kar, and Singhal 2015). Various strategies for the fractionation of algal biomass into carbohydrates and proteins have been demonstrated using the capability of fractionation of TPP in three distinct phases (a general approach for extraction through TPP is depicted in Figure 8.2). Following are two strategies using TPP.

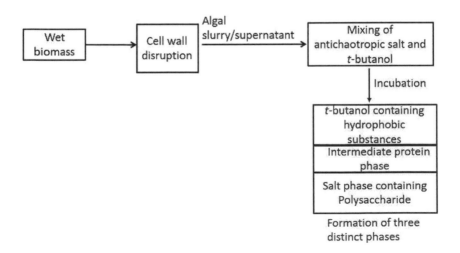

FIGURE 8.2   General steps for the application of TPP on microalgae for carbohydrate and protein extraction.

8.5.1.1   Hot Water and Enzymatic Treatment–Assisted TPP

The extract from the algal biomass may contain soluble proteins and polysaccharides if pre-treatment of the biomass is mild (e.g., mild acid treatment, hot water digestion, or enzymatic treatment). Proteins can be easily separated out from carbohydrates from the mixture by employing the TPP method. The residual biomass contains the insoluble proteins along with the pigments and lipids. Zhao et al. demonstrated that 80.24% lipid, 90.63% polysaccharide, and 78.87% protein extraction could be achieved by using TPP (Zhao et al. 2018). Hot water and enzymatic hydrolysis were used in combination as a cell wall disruption method. The best combinations of parameters for cell disruption were 4 hours of enzymatic hydrolysis, 12 g/l of enzyme concentration, 2.5 hours of time, and temperature of 90 °C for the water bath. The optimization was done based on the yield of protein and polysaccharide, consumption of energy, and overall cost. The enzyme concentration was reported to be the most influential factor for the disruption of the cell wall. The hydrolyzed biomass was centrifuged, and the supernatant was used for the separation of soluble proteins and polysaccharides using the TPP method. $(NH_4)_2SO_4$ (40 wt.%) was used as the inorganic salt and a 1:2 ratio of the sample solution to t-butanol was found to be optimum for the concentration of protein and preservation of polysaccharide as well at room temperature. On the other hand, the residual biomass was used to separate the lipids and chlorophylls using a mixture of ethanol or n-hexane and sulfuric acid solution. The insoluble proteins were concentrated in the interphase between hexane and ethanol in this step (Zhao et al. 2018).

8.5.1.2   High-Pressure Homogenization–Assisted Three-Phase Partition

All the intracellular components are exposed to the solvent with the adoption of more severe cell wall disruption methods. A more energy-intensive process, namely high-pressure homogenization or bead milling, may be required for the complete disruption of the cell wall. Mulchandani et al. (Mulchandani, Kar, and Singhal 2015) investigated the efficacy of TPP for the fractionation of intracellular components of a microalgal biomass and observed that the proteins were concentrated in the middle phase during the extraction of lipid in t-butanol. This observation eventually explored the potential of TPP in the fractionation of biomolecules from the microalgae. Disruption of the cells of *Chlorella saccharophila* was accomplished by high-pressure homogenization (10 passes at 800 bar). They employed TPP using 30% (w/v) ammonium sulfate, 1:0.75 slurry:t-butanol ratio, and 60 min reaction time at 35 °C. The yield of proteins and lipids were found to be ~12% (w/w) and ~19% (w/w), respectively which corresponds to a recovery of ~22% and ~90% respectively. Nearly 78% of proteins were lost in this process, and the amount may be retained in the cellular debris, as their primary objective was the extraction of lipid (Mulchandani, Kar, and Singhal 2015). High-pressure homogenization has been proved to be an effective cell wall disruption technique for microalgae. So, it can be said that the polysaccharides (the main constituents of the cell wall) were separated in the bottom aqueous phase. However, the constituent of the bottom phase was not reported in this study. The recovery of protein can be enhanced, and the separation of carbohydrates and protein from algal biomass might be achieved through the proper optimization of the entire methods.

8.5.2 Aqueous Two-Phase Separation

An array of biomaterials—for example, proteins, nucleic acids, enzymes, pigments, and phytochemicals—have been separated by liquid–liquid extraction in the field of industrial biotechnology (Raja et al. 2011). Since its inception in 1950s by Albertsson (Albertsson 1958), ATPS has emerged as an auspicious bioseparation tool. Remarkable advancement towards sustainable downstream processing has been denoted by this simple yet useful liquid–liquid extraction technique. It is mainly due to several advantages, for instance, simplicity, efficiency, rapid separation, flexibility, economy, and biocompatibility (Phong et al. 2018). A couple of immiscible phases are formed when two water-miscible substances are mixed in a concentration more than their critical limits (Asenjo and Andrews 2012). An ATPS mixture typically comprises either different polymers or a mixture of low-molecular-weight alcohol/polymer and kosmotropic salt (Phong et al. 2018) or two salts (Desai et al. 2014). The most commonly used polymers are ethylene oxide–propylene oxide co-polymer (EOPO), polyethylene glycol (PEG), polyacrylates, and dextran (Phong et al. 2018).

ATPS has been widely reported for the separation of intracellular compounds from either a cellular biomass (Asenjo and Andrews 2012) or a mixture of soluble substances (Gu 2014; Shahbaz Mohammadi, Omidinia, and Taherkhani 2008). ATPS is regarded as more versatile than conventional solvent extraction methods. ATPS is more energy-efficient and cost-effective due to the reduction of intermediate stages in separation/purification processes for multiple components (Raja et al. 2011). The modest operational process, quick separation, great ability in terms of distinction, low cost, and low power intake are the advantages of this technique (Raja et al. 2011; Albertsson 1958; Phong et al. 2018; Asenjo and Andrews 2012; Desai et al. 2014; Gu 2014; Shahbaz Mohammadi, Omidinia, and Taherkhani 2008; Goja et al. 2013; Soares et al. 2015). It can provide a harmonious environment for the segregation of biomolecules due to the existence of polar solvents in all phases (Agasøster 1998). The rate of solute transfer from one phase to other is very high due to the enhanced interfacial contact area of the dispersed phases, which is due to very low interfacial tension of this system (Peters 1987). The partition profile of the substance is determined by various physicochemical interactions such as van der Waals force, electrostatic interaction, hydrogen bond, hydrophobicity, steric effects, specific affinity, and conformational effects between the substance and the phase-forming materials (Phong et al. 2018; Asenjo and Andrews 2012; Desai et al. 2014; Gu 2014; Shahbaz Mohammadi, Omidinia, and Taherkhani 2008; Goja et al. 2013; Soares et al. 2015; Agasøster 1998; Peters 1987). The partitioning activity of a substance is also influenced by the following factors: size and molecular weight of the polymer, nature of ions, nature and proportion of phase materials, presence of neutral salts, pH, and temperature of system.

8.5.2.1 Ionic Liquid–Based ATPS

The use of ionic liquid (IL) as an extractant is an advanced conventional aqueous biphasic systems. It provides exclusive design flexibility of the processes, along with many other benefits, while keeping them green in nature (Suarez Ruiz et al. 2018). It offers the possibility to tailor the extraction process by selecting the appropriate cations and anions, depending on the target biomolecules (Gutowski et al. 2003). Moreover, IL enhances extraction efficiency in terms of yield and selectivity by extending the polarity range further than polymer-based conventional biphasic systems.

Du et al. first demonstrated the extraction of proteins from biological fluids using 1-butyl-3-methylimidazolium chloride (BmimCl) and K_2HPO_4 without altering their natural characteristics (Du, Yu, and Wang 2007). Subsequent studies used imidazolium-based ILs for the extraction of proteins (Pei et al. 2009; Freire et al. 2012). Some of the ILs are associated with a strong alkaline or acidic character (Lou et al. 2006). In recent studies, the use of biodegradable and environmentally friendly IL has been explored in IL-based ATPS. The buffering capacities of ILs have also been explored to ensure a mild medium to proteins for their integrity in the medium (Taha et al. 2014). Different phosphonium- and ammonium-based ILs are widely used for the extraction of proteins (Taha et al. 2014; Pereira et al. 2015), with promising results. Ammonium-based ILs in combination with inorganic salts have proven potential for the biocompatible extraction of catalytically active enzymes (Dreyer and Kragl 2008). More recently, the stability of RuBisCO, bovin serum albumin (BSA), and IgG1 in aqueous solutions of two ILs, namely Iolilyte 221PG and Cyphos 108, has been investigated. This study indicates that a high concentration of ILs affects the stability of proteins (Desai et al. 2014). Cholinium-based ILs with buffering characteristics were proposed for extraction purposes due to their inherent properties such as low toxicity, excellent biodegradability, and lower cost in comparison with other types of ILs. Few studies demonstrated that this type of IL can retain the structure of proteins and functionality of enzymes (Li et al. 2012; Lee et al. 2015; Song et al. 2015).

Ruiz et al. conducted comparative studies to investigate the efficiency of three biocompatible ATPSs viz. PEG 400 (potassium citrate), iolilyte 221PG13 (potassium citrate), and PEG 400 (cholinium dihydrogen phosphate) for the extraction of ribulose-1,5-biphosphate carboxylase/oxygenase (RuBisCO, a protein predominantly present in microalgae) from the mixture. Iolilyte 221PG-citrate has been found to be the most efficient. 80% to 100% extraction efficiency has been achieved by optimizing the factors like tie-line length (TLL) and pH. However, PEG-based ATPSs have been found to be more proficient for the sake of the integrity of extracted RuBisCO (Suarez Ruiz et al. 2018).

8.5.2.2 Application of ATPSs on Microalgae
The concept of extractive disruption of the thick cell wall of microalgae through ATPS is being embraced currently to simplify the overall downstream processing, which is otherwise time consuming and challenging (Phong et al. 2018). Integration of the cell disruption and ATPS was claimed to recover ~84% of total proteins from microalgae (Phong et al. 2017). An integrated method permits efficient downstream processing in a shorter time and with minimum deterioration of products. Consequently, both the yield and quality of end products can be enhanced (Rito-Palomares and Lyddiatt 2002; Buyel, Twyman, and Fischer 2015). An ATPS system composed of PEG and cholinium dihydrogen phosphate is an efficient medium to obtain pigment-free proteins using the cell-disrupted biomass of *Neochloris oleoabundans*. Ninety-three

percent of proteins in their intact form could be extracted in the interphase between the two aqueous phases (Suarez Ruiz et al. 2018).

8.5.2.2.1. Extraction of Proteins and Carbohydrates from Microalgae Using IL-ATPS

Proteins and carbohydrates can be separated using IL-based ATPSs, despite their polar nature. The contributing factors for the high yield are as follows: electrochemical interactions between ions of phase-forming materials with a charged groups of the biomolecules, a salting out effect, π-π interactions between the IL cation of the aromatic ring and aromatic amino acids of the proteins, and the hydrophilicity/hydrophobicity of the target molecules. Pei et al. is one of the forerunners in the application of IL-ATPS for the selective fractionation of proteins from the mixture of proteins and carbohydrates. They found that 80% to 100% of the proteins (bovine serum albumin) drifted to the top phase, whereas the carbohydrates accumulated in the bottom phase. These findings paved the way for use of ATPS for fractionation of proteins and carbohydrates (Pei et al. 2010). Table 8.1 lists the application of ATPS for the extraction of carbohydrate and proteins.

Garcia et al. used an ATPS consisting of iolilyte 221PG, citrate, and water for the simultaneous extraction and separation of proteins and carbohydrates from a crude microalgal extract of *Neochloris oleoabundans* and *Tetraselmis suecica*. The extract was obtained after bead milling followed by centrifugation and filtration. The resulting crude proteins (CP) extract contained ~45% and ~50% (dry weight of the CP) proteins for *N. oleoabundans* and *T. Suecica*, respectively. The total carbohydrates present in CP were ~40% (dry weight of the CP), and the fraction of free glucose to total carbohydrates in the CP was found to be ~22% and ~38% for *N. oleoabundans* and *T. Suecica*, respectively. Negatively charged proteins are accumulated in the

---

TABLE 8.1
Application of ATPS in the Extraction of Carbohydrate and Proteins From Mixtures

Source	Type of ATPS	Yield of products, %	Ref.
Mixture of BSA and different carbohydrates	$[C_4Clim][N(CN)_2]/$ K_2HPO_4	Proteins: 80%–100% Carbohydrates- 100% (approx.)	(Pei et al. 2010)
Leaves of aloevera	$[C_4Clim][BF_4]/NaH_2PO_4$	Proteins: ~96% Carbohydrates: ~93%	(Tan, Z.-J. et al. 2012)
Cordyceps sinensis	$[C_4Clim]Cl/K_3PO_4$	Proteins: ~88% Polysaccharides: ~89%	[Yan et al. 2014)
Crude protein extract of *Neochloris oleoabundans* and *Tetraselmis suecica*	iolilyte 221PG-citrate	Proteins: 75%(% dw) Sugar: 25% of total carbohydrate	(Garcia et al. 2018)
Aqueous extract of *I. galbana*	$[C_8mim]Cl—K_3PO_4$	Proteins: 100% Polysaccharides: 75%	(Santos 2018)

IL-rich phase due to the electrostatic interactions between the cations present in IL. The negatively charged anions of citrate further influenced the migration of proteins into the top IL phase. On the other hand, sugars, being hydrophilic in nature, are concentrated in the bottom aqueous phase. Up to 75% of proteins (% dry weight of CP) and 25% of free sugar to total carbohydrate in the CP extract were reported to be separated in this investigation (Garcia et al. 2018).

Santos et al. (Santos et al. 2018) demonstrated that not only can proteins be separated from polysaccharides using IL-based ATPS but also polysaccharides with two different compositions can be fractioned. Proteins and carbohydrates were extracted from the lyophilized biomass of *I. galbana* at 65 °C for 3 hours following the degreasing steps with different solvents (chloroform/methanol, ethanol, and acetone). The supernatant obtained from the aqueous extract (centrifuged at 15,000 rpm for 25 min at 2 °C) was subjected to fractionation using an IL-ATPS system. They primarily tested the ability of different salts like tripotassium phosphate (K_3PO_4), dipotassium hydrogen phosphate (K_2HPO_4), and potassium phosphate buffer (consisting of a mixture of K_2HPO_4 and KH_2PO_4) to form two aqueous phases with an array of ILs. Finally, a system composed of 1-methyl-3-octylimidazolium chloride [C_8mim] Cl (15 wt%) and K3PO4 (20 wt%) was selected and used with aqueous extract of *Isocrasis galbana* (65 wt%) to achieve extraction efficiency of ~75% for carbohydrates and 100% for proteins. Interestingly, the presence of arabinose-rich polysaccharides (~16% w/w) was observed in the top phase along with the proteins. So, ethanol was used to precipitate the polysaccharides from this mixture for further fractionation of proteins and arabinans. Polysaccharides recovered in the bottom phase contained glucose (56 mol%), mannose (20 mol%), and uronic acids (12 mol%) as major sugars. The composition of the polysaccharides recovered in the top phase was arabinose (65 mol%), uronic acids (20 mol%), and xylose (16 mol%) primarily. Both the salts and ILs were recovered in dialysis with the membrane of 12 to 14 kDa cut-off.

8.5.3 Membrane Separation of the Cell Wall–Disrupted Microalgal Biomass

Unidirectional and selective transfer of target solute happens between two phases using a membrane. This separation is based on the size of the molecule and affinity of the membrane for the solute. The driving force for the membrane separation process is the difference in concentration, pressure, electrical, or chemical potential. Ultrafiltration (UF) and microfiltration (MF) are based on the molecular sieving mechanism. Membranes and the process can also be classified according to the configuration and fabrication constituents. Membrane technology has been widely used across various industries. But the use of this technology in the biorefinery is still limited (Gerardo, Oatley-Radcliffe, and Lovitt 2014). Protein- and carbohydrate-rich fractions can be obtained by an appropriate membrane separation process, while the bulk of the lipids remain in the biomass (Gerardo, Oatley-Radcliffe, and Lovitt 2014). The UF membranes with the molecular weight cut-off range of 1 to 100 kDa may play an important role in the fractionation of the microalgae metabolites (Gerardo, Oatley-Radcliffe, and Lovitt 2014).

Gerardo et al. described the separation scheme of proteins and carbohydrates based on size and charge exclusion using membrane separation processes (Gerardo,

Oatley-Radcliffe, and Lovitt 2014). They compared the microalgal biorefinery inclusive of membrane-based separation with membrane fractionation of milk in the dairy industry. Pilot-scale micro filtration (MF), ultrafiltration (UF), and nano filtration (NF) were used for the separation of casein micelles from milk serum proteins (Nelson and Barbano 2005), lactose (Vyas and Tong 2003), and minerals (Vyas and Tong 2003; Morr and Brandon 2008) from milk. Components of the milk, namely fat globules, casein micelles, serum proteins, lactose, salts, and minerals, were separated using a MF to reverse osmosis (RO) membrane based on the size of the components (Pouliot 2008; Brans et al. 2004). They described that either disrupted microalgae or whole cells may be separated by MF membranes, whereas the individual components are likely to be separated by UF–NF membranes. However, the fractionation of microalgae products using membranes has yet to be verified at the industrial level (Gerardo, Oatley-Radcliffe, and Lovitt 2014).

The application of membranes for the downstream processing of microalgal metabolites is very limited. A few studies have been reported for the purification of a single component, such as the polysaccharides from *Porphyridium cruentum* (Patel et al. 2013; Marcati et al. 2014), *Spirulina platensis*, and *Chlorella pyrenoidosa* (Pugh et al. 2001) and to concentrate proteins from *Chlorella vulgaris* and *Haematococcus pluvialis* in the retentate (Ba et al. 2016; Ursu et al. 2014). The role of EPS of *Chlorella* sp. and *Porphyridium purpureum* in the fouling of UF membranes has also been examined (Morineau-Thomas, Jaouen, and Legentilhomme 2002). Recently, different strategies for the separation of protein and carbohydrate using membranes were accomplished from a pre-treated biomass of different microalgal species (see Table 8.2).

TABLE 8.2

Application of the Membrane Process in the Extraction of Carbohydrate and Protein

Name of the source microalgae	Technique applied	Membrane cut-off	Product separated	Ref.
Nannochloropsis gaditana	Enzymatic cell wall disruption- Ultrafiltration/ diafiltration	300 kDa–1000 KDa	Proteins	(Safi et al. 2017)
Tetraselmis suecica	High-pressure homogenization: two stage ultrafiltration	100 kDa and 10k Da	Proteins and polysaccharides	(Safi et al. 2014)
Nannochloropsis gaditana, Chlorella sorokiniana, and Dunaliella tertiolecta	Steam explosion: dynamic filtration	5000 Da	Sugar and protein	(Lorente et al. 2018)
Nanochloropsis gaditana	Steam explosion: dynamic tangential cross-flow filtration	5000 Da and 100,000 Da	Monosaccharide	(Lorente et al. 2017)

8.5.3.1 Enzymatic Hydrolysis–Assisted Ultrafiltration Coupled with Diafiltration

Safi et al. investigated the combination of UF and diafiltration of *Nannochloropsis gaditana* biomass after enzymatic treatment and high-pressure homogenization (HPH) for the extraction of soluble proteins free from chlorophyll (Safi et al. 2017). Membranes with a molecular weight cut-off between 300 kDa and 1000 kDa were used with a constant transmembrane pressure of 2.07 bar in these experiments. The UF was done in the first step with supernatant from pre-treated algae, which showed complete retention of polysaccharides in the retentate. Diafiltration of retentate was done at the second step to obtain more proteins. It was found that ~50% of the proteins were released to the aqueous phase after the homogenization (one passage at 1500 bar), while treatment with alcalase (5% v/w) for 4 hours at 50 °C released ~36% (w/w) of total proteins into the supernatant. The alcalase-treated samples provided more permeate flux due to the fact that alcalase hydrolyzes the proteins to reduce the molecular weight, which results in the reduction of fouling of the membrane. The combination of enzymatic hydrolysis–UF/DF resulted in a larger overall yield of water-soluble proteins (~25%) in permeate, compared to the combination of HPH–UF/DF (~17%). However, the flow rate of the permeate was found to be the highest for 300 kD and lowest for 1000 kD, and that didn't increase with increasing molecular weight cut-off (Safi et al. 2017). This result can be attributed to the absorptive fouling of polysaccharides (Susanto, Franzka, and Ulbricht 2007). Additionally, it can be concluded that the adsorption can only happen on the surface of the membrane for 300 kD, whereas the membrane pores can be clogged due to the penetration of some retained molecules in the case of 1000 kDa (De la Torre et al. 2009; Susanto et al. 2008).

8.5.3.2 High-Pressure Homogenization–Assisted Two-Stage Ultrafiltration

Molecules of different sizes and weights can be retained by UF membranes of different pore sizes. UF can be employed at multiple stages to yield different products as retentate at each stage, and the permeate could be used for the subsequent stage of UF. Safi et al. demonstrated a two-stage ultrafiltration strategy for the separation of proteins and polysaccharides from *Tetraselmis suecica* after cell wall disruption with HPH (Safi et al. 2014). Two consecutive membranes of 100 kDa and 10 kDa cut-off were used at 2.07 bar of transmembrane pressure with 30 min operational time. The membrane of 100 kDa cut-off retained all starch and pigments at the first stage, whereas the 10 kDa membrane retained proteins from the remaining mixture, giving the sugars in permeate at the second stage. The protein concentration of permeate from the first stage of UF increased from 50% to 80% with the increasing pressure of homogenization between 200 bar and 1,000 bar. The permeation rate of sugar through the 10 kDa membrane was 90% when the cells were treated at 600 bar pressure, and this amount corresponds to 65% of total sugar present in the supernatant (Safi et al. 2014). The molecular weights of proteins from *T. suecica* are between 15 and 50 kDa (Schwenzfeier, Wierenga, and Gruppen 2011). Hence, the retention of proteins at 10 kDa was justified for this species. The use of diafiltration after each step for increasing the yield and separation efficiency is a subject of evaluation. Depending on the cell wall and biomass composition, this strategy can be applied to other micro-algal species in conjunction with various other cell disruption methods, although the molecular weight cut-off of the membrane should be optimized according to the species (Safi et al. 2014). The molecular weight of proteins from *Chlorella vulgaris* is typically within the range of 12 to 120 kDa, whereas that from *Haematococcus pluvialis* is between 10 and 100 kDa. On account of the strategy discussed earlier, it can be concluded that the two-stage sequential UF process is a potential green process for the fractionation of carbohydrates and proteins from microalgal biomass.

8.5.3.3   Steam Explosion–Assisted Dynamic Filtration

The steam explosion is one of the pre-treatment processes of lignocellulosic materials. It offers both the mechanical and chemical disruption in unison. The technique has recently been proved to be effective in the disruption of the microalgal cell wall (Nurra et al. 2014), as polysaccharides are the main constituent of this organism. Steam explosion hydrolyzes the polysaccharides to monosaccharides (Lorente et al. 2017) that remains in the aqueous phase after solid–liquid separation, and the monosaccharides can eventually be concentrated using NF (Gerardo, Oatley-Radcliffe, and Lovitt 2014; Nelson and Barbano 2005; Vyas and Tong 2003; Morr and Brandon 2008; Pouliot 2008; Brans et al. 2004; Patel et al. 2013; Marcati et al. 2014). The dynamic tangential filtration has been introduced to avoid membrane fouling (Nurra et al. 2014a; Rios et al. 2011; Ríos et al. 2012).

Lorente et al. used three different strains of microalgae (*Nannochloropsis gaditana, Chlorella sorokiniana,* and *Dunaliella tertiolecta*) with different cell wall characteristics for fractionation of the sugars, proteins, and lipids via dynamic filtration using the polyethylene membrane (PE5) with a molecular weight cut-off of 5 kDa and membrane area of 0.0446 m². The 5 bar trans-membrane pressure and 55.4 Hz vibrational frequency were employed. Each sample was treated with 5% (w/w) sulphuric acid for 2 hours at room temperature followed by a steam explosion at 150 °C and 4.7 bar for 5 min. This acid–catalyzed steam explosion process hydrolyzed polysaccharides to sugars. The proteins were hydrolyzed partially during this process. Sugars were then separated from proteins by membrane filtration. Complete retention of proteins and lipids was observed in the retentate with sugars in the permeate stream (Lorente et al. 2018).

Similar results were found with *Nanochloropsis gaditana* with similar pre-treatment and filtration using PE5 (5 kDa) and PEV400 (100 kDa) membranes operating at 5 bar pressure and 55 Hz frequency. The permeability of 6 L/h/m²/bar was obtained. The complete retention of lipid was reported. For the steam-exploded sample, the concentration of sugars was the same in both retentate and permeate streams. The results implied that both of the membranes (PE5 and PV400) were unable to retain the sugars. Therefore, MF can provide separation of purified sugars from a microalgal biomass with a very large permeate rate (Lorente et al. 2017).

Finally, it can be concluded that the combination of steam explosion and dynamic filtration is a very good technique for the separation of fermentable sugar and partially hydrolyzed proteins (peptides). This strategy might also be viable for commercialization, as the irreversible membrane fouling factor can be reduced to two- to seven-fold by dynamic filtration (Lorente et al. 2017).

8.6 CONCLUSION AND FUTURE PERSPECTIVES

The integrated systems for the extraction of carbohydrates and proteins from micro-algae are not fully developed yet. On the other hand, sequential extraction approaches are time consuming and involve several steps and solvents. This type of extraction system is thus not economically sustainable. Although the membrane separation, ATPS, and TPP help to obtain carbohydrates and proteins in high quality, the yield is not often very high. This is due to the fact that, in addition to carbohydrates and proteins, these investigations were aimed towards lipids and other biomolecules such as pigments. Membrane separation is the greenest route, as it doesn't require any additional chemicals other than water, but this process has its own restrictions. However, this method has the potential to become cost-effective in the near future. Eventually, this method can be used for the extraction of carbohydrate and protein, as well as in the development of a complete biorefinery. Aqueous biphasic systems are also efficient in terms of greenness and product quality. The involvement of polymers and ILs, however, have made these systems economically unfavourable until now. Recovery of these phase-forming materials is difficult. Biphasic systems with simple phase-forming materials or with the capability to recover and reuse the phase-forming materials are necessary in the future. However, the most crucial step for any of these methods is cell disruption. Choosing the proper cell disruption technique is essential, as this step determines the energy consumption and efficiency of the overall process. Selective release of materials from the disrupted cell is helpful in separation and product recovery. The pulsed electric field is a promising technology in the selective release of proteins from the cell. However, this method should be combined with other techniques for high-efficiency cell wall disruption and release of polysaccharides from the cell wall. In this regard, a strain-specific combination is required in future investigations. Some prominent evolving technologies, like explosive decompression and cationic polymer-coated membrane treatment, require a substantial amount of research to achieve efficiency comparable to conventional cell wall disruption techniques. Integration of cell wall disruption with extraction can be achieved by using ILs where no separate cell disruption step is required due to the disruptive nature of ILs. In the future, suitable IL-based biphasic systems should be designed for the extraction of proteins and carbohydrates. In the end, future research should be directed towards designing a process with the fewest possible number of steps and minimum amount of chemicals, along with end-to-end integration of the process.

ACKNOWLEDGEMENT

The authors thankfully acknowledge the financial support from Department of Biotechnology, government of India.

REFERENCES

Agasøster, T. 1998. "Aqueous two-phase partitioning sample preparation prior to liquid chromatography of hydrophilic drugs in blood." *Journal of Chromatography B: Biomedical Sciences and Applications* 716 (1–2): 293–298.

Albertsson, P. A. 1958. "Partition of proteins in liquid polymer-polymer two-phase systems." *Nature* 182 (4637):709–711.

Ansari, F. A., et al. 2017. "Exploration of microalgae biorefinery by optimizing sequential extraction of major metabolites from *scenedesmus obliquus*." *Industrial & Engineering Chemistry Research* 56 (12):3407–3412.

Asenjo, J. A., and B. A. Andrews. 2012. "Aqueous two-phase systems for protein separation: Phase separation and applications." *Journal of Chromatography A* 1238:1–10.

Athukorala, Y., et al. 2007. "Anticoagulant activity of marine green and brown algae collected from Jeju Island in Korea." *Bioresource Technology* 98 (9):1711–1716.

Atsumi, S., T. Hanai, and J. C. Liao. 2008. "Non-fermentative pathways for synthesis of branched-chain higher alcohols as biofuels." *Nature* 451:86.

Awaluddin, S., et al. 2016. "Subcritical water technology for enhanced extraction of biochemical compounds from Chlorella vulgaris." *BioMed Research International* 2016:10, Article ID 5816974.

Ba, F., et al. 2016. "Haematococcus pluvialis soluble proteins: Extraction, characterization, concentration/fractionation and emulsifying properties." *Bioresour Technol* 200:147–152.

Bixler, H. J., and H. Porse. 2011. "A decade of change in the seaweed hydrocolloids industry." *Journal of Applied Phycology* 23 (3):321–335.

Brans, G., et al. 2004. "Membrane fractionation of milk: State of the art and challenges." *Journal of Membrane Science* 243 (1): 263–272.

Buyel, J., R. Twyman, and R. Fischer. 2015. "Extraction and downstream processing of plant-derived recombinant proteins." *Biotechnology Advances* 33 (6):902–913.

Chen, C.-Y., et al. 2013. "Microalgae-based carbohydrates for biofuel production." *Biochemical Engineering Journal* 78:1–10.

Chew, K. W., T. C. Ling, and P. L. Show. 2019. "Recent developments and applications of three-phase partitioning for the recovery of proteins." *Separation & Purification Reviews* 48 (1):52–64.

Coons, J. E., et al. 2014. "Getting to low-cost algal biofuels: A monograph on conventional and cutting-edge harvesting and extraction technologies." *Algal Research* 6:250–270.

Cuellar-Bermudez, S. P., et al. 2015. "Extraction and purification of high-value metabolites from microalgae: Essential lipids, astaxanthin and phycobiliproteins." *Microbial Biotechnology* 8 (2):190–209.

De la Torre, T., et al. 2009. "Filtration charaterization methods in MBR systems: A practical comparison." *Desalination and Water Treatment* 9 (1–3):15–21.

Dennison, C., and R. Lovrien. 1997. "Three phase partitioning: Concentration and purification of proteins." *Protein Expression and Purification* 11 (2):149–161.

Desai, R. K., et al. 2014. "Extraction and stability of selected proteins in ionic liquid based aqueous two phase systems." *Green Chemistry* 16 (5):2670–2679.

Dibenedetto, A., A. Colucci, and M. Aresta. 2016. "The need to implement an efficient biomass fractionation and full utilization based on the concept of "biorefinery" for a viable economic utilization of microalgae." *Environmental Science and Pollution Research* 23 (22):22274–22283.

Doan, Q. C., et al. 2012. "Microalgal biomass for bioethanol fermentation: Implications for hypersaline systems with an industrial focus." *Biomass and Bioenergy* 46:79–88.

Dong, T., et al. 2016. "Combined algal processing: A novel integrated biorefinery process to produce algal biofuels and bioproducts." *Algal Research* 19:316–323.

Doucha, J., and K. Lívanský. 2008. "Influence of processing parameters on disintegration of Chlorella cells in various types of homogenizers." *Applied Microbiology and Biotechnology* 81 (3):431.

Dreyer, S., and U. Kragl. 2008. "Ionic liquids for aqueous two-phase extraction and stabilization of enzymes." *Biotechnology and Bioengineering* 99 (6):1416–1424.

Du, Z., Y. L. Yu, and J. H. Wang. 2007. "Extraction of proteins from biological fluids by use of an ionic liquid/aqueous two-phase system." *Chemistry—A European Journal* 13 (7):2130–2137.

Freire, M. G., et al. 2012. "Aqueous biphasic systems: A boost brought about by using ionic liquids." *Chemical Society Reviews* 41 (14):4966–4995.

Fu, C.-C., et al. 2010. "Hydrolysis of microalgae cell walls for production of reducing sugar and lipid extraction." *Bioresource Technology* 101 (22):8750–8754.

Gallego, R., et al. 2018. "Green extraction of bioactive compounds from microalgae." *Journal of Analysis and Testing* 2 (2):109–123.

Garcia, E. S., et al. 2018. "Fractionation of proteins and carbohydrates from crude microalgae extracts using an ionic liquid based-aqueous two phase system." *Separation and Purification Technology* 204:56–65.

Gerardo, M. L., D. L. "Oatley-Radcliffe, and R. W. Lovitt. 2014. Integration of membrane technology in microalgae biorefineries." *Journal of Membrane Science* 464:86–99.

Gerde, J. A., et al. 2013. "Optimizing protein isolation from defatted and non-defatted *Nannochloropsis* microalgae biomass." *Algal Research* 2 (2):145–153.

Ghosh, A., et al. 2016. "Progress toward isolation of strains and genetically engineered strains of microalgae for production of biofuel and other value added chemicals: A review." *Energy Conversion and Management* 113:104–118.

Ghosh, A., et al. 2017. "Effect of macronutrient supplements on growth and biochemical compositions in photoautotrophic cultivation of isolated Asterarcys sp.(BTA9034)." *Energy Conversion and Management* 149:39–51.

Gilbert-López, B., et al. 2015. "Downstream processing of Isochrysis galbana: A step towards microalgal biorefinery." *Green Chemistry* 17 (9):4599–4609.

Gilbert-López, B., et al. 2017. "Green compressed fluid technologies for downstream processing of *Scenedesmus obliquus* in a biorefinery approach." *Algal Research* 24:111–121.

Goettel, M., et al. 2013. "Pulsed electric field assisted extraction of intracellular valuables from microalgae." *Algal Research* 2 (4):401–408.

Goja, A. M., et al. 2013. "Aqueous two-phase extraction advances for bioseparation." *J. Bioprocess. Biotechnol* 4 (1):1–8.

Gu, Z. 2014. "Recovery of recombinant proteins from plants using aqueous two-phase partitioning systems: An outline." In *Protein Downstream Processing*, 77–87. Totowa, NJ: Humana Press.

Günerken, E., et al. 2015. "Cell disruption for microalgae biorefineries." *Biotechnology Advances* 33 (2):243–260.

Gutowski, K. E., et al. 2003. "Controlling the aqueous miscibility of ionic liquids: Aqueous biphasic systems of water-miscible ionic liquids and water-structuring salts for recycle, metathesis, and separations." *Journal of the American Chemical Society* 125 (22):6632–6633.

Hahn, T., et al. 2011. "Extraction of lignocellulose and algae for the production of bulk and fine chemicals." *Industrial Scale Natural Products Extraction*: 221–245.

Haldar, D., D. Sen, and K. Gayen. 2016. "A review on the production of fermentable sugars from lignocellulosic biomass through conventional and enzymatic route—A comparison." *International Journal of Green Energy* 13 (12):1232–1253.

Halperin, S. A., et al. 2003. "Safety and immunoenhancing effect of a Chlorella-derived dietary supplement in healthy adults undergoing influenza vaccination: Randomized, double-blind, placebo-controlled trial." *Canadian Medical Association Journal* 169 (2):111–117.

Herrero, M., A. Cifuentes, and E. Ibañez. 2006. "Sub-and supercritical fluid extraction of functional ingredients from different natural sources: Plants, food-by-products, algae and microalgae: A review." *Food Chemistry* 98 (1):136–148.

Herrero, M., et al. 2015. "Plants, seaweeds, microalgae and food by-products as natural sources of functional ingredients obtained using pressurized liquid extraction and supercritical fluid extraction." *TrAC Trends in Analytical Chemistry* 71:26–38.

Hu, J., et al. 2011. "Pressurized liquid extraction of ginger (Zingiber officinale Roscoe) with bioethanol: An efficient and sustainable approach." *Journal of Chromatography A* 1218 (34):5765–5773.

Jong-Yuh, C., and S. Mei-Fen. 2005. "Potential hypoglycemic effects of Chlorella in strepto-zotocin-induced diabetic mice." *Life Sciences* 77 (9):980–990.

Khanra, S., et al. 2018. "Downstream processing of microalgae for pigments, protein and carbohydrate in industrial application: A review." *Food and Bioproducts Processing* 110:60–84.

Kim, G., G. Mujtaba, and K. Lee. 2016. "Effects of nitrogen sources on cell growth and biochemical composition of marine chlorophyte Tetraselmis sp. for lipid production." *Algae* 31 (3):257–266.

Laurens, L., et al. 2015. "Acid-catalyzed algal biomass pretreatment for integrated lipid and carbohydrate-based biofuels production." *Green Chemistry* 17 (2):1145–1158.

Lee, A. K., D. M. Lewis, and P. J. Ashman. 2012. "Disruption of microalgal cells for the extraction of lipids for biofuels: Processes and specific energy requirements." *Biomass and Bioenergy* 46:89–101.

Lee, S. Y., et al. 2015. "Evaluating self-buffering ionic liquids for biotechnological applica-tions." *ACS Sustainable Chemistry & Engineering* 3 (12):3420–3428.

Li, B., et al. 2008. "Fucoidan: Structure and bioactivity." *Molecules* 13 (8):1671–1695.

Li, Z., et al. 2012. "Design of environmentally friendly ionic liquid aqueous two-phase sys-tems for the efficient and high activity extraction of proteins." *Green Chemistry* 14 (10):2941–2950.

Lorente, E., et al. 2017. "Microalgae fractionation using steam explosion, dynamic and tan-gential cross-flow membrane filtration." *Bioresour Technol* 237:3–10.

Lorente, E., et al. 2018. "Steam explosion and vibrating membrane filtration to improve the processing cost of microalgae cell disruption and fractionation." *Processes* 6 (4):28.

Lou, W.-Y., et al. 2006. "Impact of ionic liquids on papain: An investigation of structure—function relationships." *Green Chemistry* 8 (6):509–512.

Lupatini, A. L., et al. 2017. "Protein and carbohydrate extraction from S. platensis biomass by ultrasound and mechanical agitation." *Food Research International* 99:1028–1035.

Marcati, A., et al. 2014. "Extraction and fractionation of polysaccharides and B-phycoerythrin from the microalga Porphyridium cruentum by membrane technology." *Algal Research* 5:258–263.

Merchant, R. E., and C. A. 2001. "Andre. A review of recent clinical trials of the nutritional supplement Chlorella pyrenoidosa in the treatment of fibromyalgia, hypertension, and ulcerative colitis." *Alternative Therapies in Health and Medicine* 7 (3):79–92.

Mihranyan, A. 2011. "Cellulose from cladophorales green algae: From environmental problem to high-tech composite materials." *Journal of Applied Polymer Science* 119 (4):2449–2460.

Mondal, M., et al. 2016. "Mixotrophic cultivation of Chlorella sp. BTA 9031 and Chlamydomonas sp. BTA 9032 isolated from coal field using various carbon sources for biodiesel production." *Energy Conversion and Management* 124:297–304.

Morineau-Thomas, O., P. Jaouen, and P. Legentilhomme. 2002. "The role of exopolysac-charides in fouling phenomenon during ultrafiltration of microalgae (Chlorellasp. and Porphyridium purpureum): Advantage of a swirling decaying flow." *Bioprocess Biosyst Eng* 25 (1):35–42.

Morr, C. V., and S. C. Brandon. 2008. "Membrane fractionation processes for removing 90% to 95% of the lactose and sodium from skim milk and for preparing lactose and sodium-reduced skim milk." *J Food Sci* 73 (9):C639–C647.

Morris, H. J., et al. 2008. "Utilisation of Chlorellavulgaris cell biomass for the production of enzymatic protein hydrolysates." *Bioresource Technology* 99 (16):7723–7729.

Mulchandani, K., J. R. Kar, and R. S. Singhal. 2015. "Extraction of lipids from chlorella saccharophila using high-pressure homogenization followed by three phase partitioning." *Appl Biochem Biotechnol* 176 (6):613–626.

Muñoz, R., et al. 2015. "Preliminary biorefinery process proposal for protein and biofuels recovery from microalgae." *Fuel* 150:425–433.

Nelson, B. K., and D. M. Barbano. 2005. "A microfiltration process to maximize removal of serum proteins from skim milk before cheese making*." *Journal of Dairy Science* 88 (5):1891–1900.

Nurra, C., et al. 2014a. "Biorefinery concept in a microalgae pilot plant. Culturing, dynamic filtration and steam explosion fractionation." *Bioresour Technol* 163:136–142.

Nurra, C., et al. 2014b. "Vibrating membrane filtration as improved technology for microalgae dewatering." *Bioresource Technology* 157:247–253.

Ovando, C. A., et al. 2018. "Functional properties and health benefits of bioactive peptides derived from Spirulina: A review." *Food Reviews International* 34 (1):34–51.

Panadare, D., and V. Rathod. 2017. "Three phase partitioning for extraction of oil: A review." *Trends in Food Science & Technology* 68:145–151.

Patel, A. K., et al. 2013. "Separation and fractionation of exopolysaccharides from Porphyridium cruentum." *Bioresource Technology* 145:345–350.

Patterson, D., and D. M. Gatlin. 2013. "Evaluation of whole and lipid-extracted algae meals in the diets of juvenile red drum (Sciaenops ocellatus)." *Aquaculture*: 92–98, 416–417.

Paule, B., et al. 2004. "Three-phase partitioning as an efficient method for extraction/concentration of immunoreactive excreted—secreted proteins of Corynebacterium pseudotuberculosis." *Protein Expression and Purification* 34 (2): 311–316.

Pei, Y., et al. 2009. "Ionic liquid-based aqueous two-phase extraction of selected proteins." *Separation and Purification Technology* 64 (3):288–295.

Pei, Y., et al. 2010. "Selective separation of protein and saccharides by ionic liquids aqueous two-phase systems." *Science China Chemistry* 53 (7):1554–1560.

Pereira, M. M., et al. 2015. "Enhanced extraction of bovine serum albumin with aqueous biphasic systems of phosphonium-and ammonium-based ionic liquids." *Journal of Biotechnology* 206:17–25.

Peters, T. J. 1987. "Partition of cell particles and macromolecules: Separation and purification of biomolecules, cell organelles, membranes and cells in aqueous polymer two phase systems and their use in biochemical analysis and biotechnology. P-A. Albertsson. Third Edition, 1986, John Wiley and Sons, Chichester, £61.35 pages 346." *Cell Biochemistry and Function*, 5 (3):233–234.

Phong, W. N., et al. 2017. "Extractive disruption process integration using ultrasonication and an aqueous two-phase system for protein recovery from Chlorella sorokiniana." *Engineering in Life Sciences* 17 (4):357–369.

Phong, W. N., et al. 2018. "Mild cell disruption methods for bio-functional proteins recovery from microalgae—Recent developments and future perspectives." *Algal Research* 31:506–516.

Phong, W. N., et al. 2018. "Recovery of biotechnological products using aqueous two phase systems." *Journal of Bioscience and Bioengineering* 126 (3):273–281.

Pike, R. N., and C. Dennison. 1989. "Protein fractionation by three phase partitioning (TPP) in aqueous/t-butanol mixtures." *Biotechnology and Bioengineering* 33 (2):221–228.

Postma, P., et al. 2015. "Mild disintegration of the green microalgae Chlorella vulgaris using bead milling." *Bioresource Technology* 184:297–304.

Postma, P., et al. 2016. "Selective extraction of intracellular components from the microalga Chlorella vulgaris by combined pulsed electric field—temperature treatment." *Bioresource Technology* 203:80–88.

Postma, P., et al. 2017. "Energy efficient bead milling of microalgae: Effect of bead size on disintegration and release of proteins and carbohydrates." *Bioresource Technology* 224:670–679.

Pouliot, Y. 2008. "Membrane processes in dairy technology—From a simple idea to worldwide panacea." *International Dairy Journal* 18 (7):735–740.

Pugh, N., et al. 2001. "Isolation of three high molecular weight polysaccharide preparations with potent immunostimulatory activity from Spirulina platensis, aphanizomenon flosaquae and Chlorella pyrenoidosa." *Planta Med* 67 (8):737–742.

Raja, S., et al. 2011. "Aqueous two phase systems for the recovery of biomolecules—a review." *Science and Technology* 1 (1):7–16.

Reeve, W., C. M. Erikson, and P. F. Aluotto. 1979. "A new method for the determination of the relative acidities of alcohols in alcoholic solutions. The nucleophilicities and competitive reactivities of alkoxides and phenoxides." *Canadian Journal of Chemistry* 57 (20):2747–2754.

Rios, S. D., et al. 2011. "Dynamic microfiltration in microalgae harvesting for biodiesel production." *Industrial & Engineering Chemistry Research* 50 (4):2455–2460.

Ríos, S. D., et al. 2012. "Antifouling microfiltration strategies to harvest microalgae for biofuel." *Bioresource Technology* 119:406–418.

Rito-Palomares, M., and A. Lyddiatt. 2002. "Process integration using aqueous two-phase partition for the recovery of intracellular proteins." *Chemical Engineering Journal* 87 (3):313–319.

Safi, C., et al. 2014. "A two-stage ultrafiltration process for separating multiple components of Tetraselmis suecica after cell disruption." *Journal of Applied Phycology* 26 (6):2379–2387.

Safi, C., et al. 2017. "Biorefinery of microalgal soluble proteins by sequential processing and membrane filtration." *Bioresour Technol* 225:151–158.

Samarakoon, K., and Y.-J. Jeon. 2012. "Bio-functionalities of proteins derived from marine algae—A review." *Food Research International* 48 (2):948–960.

Santos, J. H. P. M., et al. 2018. "Fractionation of isochrysis galbana proteins, arabinans, and glucans using ionic-liquid-based aqueous biphasic systems." *ACS Sustainable Chemistry & Engineering* 6 (11):14042–14053.

Santoyo, S., et al. 2012. "Antiviral compounds obtained from microalgae commonly used as carotenoid sources." *Journal of Applied Phycology* 24 (4):731–741.

Saxena, L., B. K. Iyer, and L. Ananthanarayan. 2007. "Three phase partitioning as a novel method for purification of ragi (Eleusine coracana) bifunctional amylase/protease inhibitor." *Process Biochemistry* 42 (3):491–495.

Schröder, A., et al. 2017. "Interfacial properties of whey protein and whey protein hydrolysates and their influence on O/W emulsion stability." *Food Hydrocolloids* 73:129–140.

Schwenzfeier, A., et al. 2013. "Emulsion properties of algae soluble protein isolate from Tetraselmis sp." *Food Hydrocolloids* 30 (1):258–263.

Schwenzfeier, A., et al. 2014. "Effect of charged polysaccharides on the techno-functional properties of fractions obtained from algae soluble protein isolate." *Food Hydrocolloids* 35:9–18.

Schwenzfeier, A., P. A. Wierenga, and H. Gruppen. 2011. "Isolation and characterization of soluble protein from the green microalgae Tetraselmis sp." *Bioresource Technology*, 102 (19):9121–9127.

Shahbaz Mohammadi, H., E. Omidinia, and H. Taherkhani. 2008. "Rapid one-step separation and purification of recombinant phenylalanine dehydrogenase in aqueous two-phase systems." *Iranian Biomedical Journal* 12 (2):115–122.

Sheih, I.-C., T.-K. Wu, and T. J. Fang. 2009. "Antioxidant properties of a new antioxidative peptide from algae protein waste hydrolysate in different oxidation systems." *Bioresource Technology* 100 (13):3419–3425.

Sierra, L. S., C. K. Dixon, and L. R. Wilken. 2017. "Enzymatic cell disruption of the microalgae Chlamydomonas reinhardtii for lipid and protein extraction." *Algal Research* 25:149–159.

Soares, R. R., et al. 2015. "Partitioning in aqueous two-phase systems: Analysis of strengths, weaknesses, opportunities and threats." *Biotechnology Journal* 10 (8):1158–1169.

Song, C. P., et al. 2015. "Green, aqueous two-phase systems based on cholinium aminoate ionic liquids with tunable hydrophobicity and charge density." *ACS Sustainable Chemistry & Engineering* 3 (12):3291–3298.

Soto-Sierra, L., P. Stoykova, and Z. L. Nikolov. 2018. "Extraction and fractionation of microalgae-based protein products." *Algal Research* 36:175–192.

Suarez Ruiz, C. A., et al. 2018. "Rubisco separation using biocompatible aqueous two-phase systems." *Separation and Purification Technology* 196:254–261.

Suarez Ruiz, C. A., et al. 2018. "Selective and mild fractionation of microalgal proteins and pigments using aqueous two-phase systems." *J Chem Technol Biotechnol* 93 (9):2774–2783.

Susanto, H., et al. 2008. "Ultrafiltration of polysaccharide—protein mixtures: Elucidation of fouling mechanisms and fouling control by membrane surface modification." *Separation and Purification Technology* 63 (3):558–565.

Susanto, H., S. Franzka, and M. Ulbricht. 2007. "Dextran fouling of polyethersulfone ultrafiltration membranes—Causes, extent and consequences." *Journal of Membrane Science* 296 (1):147–155.

Taha, M., et al. 2014. "Good's buffers as a basis for developing self-buffering and biocompatible ionic liquids for biological research." *Green Chemistry* 16 (6):3149–3159.

Tan, Z.-J., et al. 2012. "Simultaneous extraction and purification of aloe polysaccharides and proteins using ionic liquid based aqueous two-phase system coupled with dialysis membrane." *Desalination* 286:389–393.

't Lam, G. P., et al. 2017. "Mild and selective protein release of cell wall deficient microalgae with pulsed electric field." *ACS Sustainable Chemistry & Engineering* 5 (7):6046–6053.

Ursu, A.-V., et al. 2014. "Extraction, fractionation and functional properties of proteins from the microalgae Chlorella vulgaris." *Bioresource Technology* 157:134–139.

Vanthoor-Koopmans, M., et al. 2013. "Biorefinery of microalgae for food and fuel." *Bioresource Technology* 135:142–149.

Vermuë, M., et al. 2018. "Multi-product microalgae biorefineries: From concept towards reality." *Trends in Biotechnology* 36 (2):216–227.

Vidhate, G. S., and R. S. Singhal. 2013. "Extraction of cocoa butter alternative from kokum (Garcinia indica) kernel by three phase partitioning." *Journal of Food Engineering* 117 (4):464–466.

Vyas, H. K., and P. S. Tong. 2003. "Process for calcium retention during skim milk ultrafiltration." *Journal of Dairy Science* 86 (9):2761–2766.

Waghmare, A. G., et al. 2016. "Concentration and characterization of microalgae proteins from Chlorella pyrenoidosa." *Bioresources and Bioprocessing* 3 (1):16.

Williams, P. J. l. B., and L. M. Laurens. 2010. "Microalgae as biodiesel & biomass feedstocks: Review & analysis of the biochemistry, energetics & economics." *Energy & Environmental Science* 3 (5):554–590.

Yan, J. K., et al. 2017. "Three-phase partitioning as an elegant and versatile platform applied to nonchromatographic bioseparation processes." *Crit Rev Food Sci Nutr*: 1–16.

Yan, J.-K., et al. 2014. "Facile and effective separation of polysaccharides and proteins from Cordyceps sinensis mycelia by ionic liquid aqueous two-phase system." *Separation and Purification Technology* 135:278–284.

Zakaria, S., et al. 2017. *Extraction of antioxidants from Chlorella sp. using subcritical water treatment.* In IOP Conference Series: Materials Science and Engineering. Vol. 206. No. 1. IOP Publishing.

Zhao, B., et al. 2014. "Efficient anaerobic digestion of whole microalgae and lipid-extracted microalgae residues for methane energy production." *Bioresource Technology* 161:423–430.

Zhao, W., et al. 2018. "A mild extraction and separation procedure of polysaccharide, lipid, chlorophyll and protein from Chlorella spp." *Renewable Energy* 118:701–708.

Section 3

Optimized Downstream
Processing of Microalgae

9 Current Issues with a Sustainable Approach in the Industrial-Scale Operations of Microalgae

Avinash Sinha, Amol Pandharbale,
Jigar Rameshbhai Patel

CONTENTS

9.1   INTRODUCTION

Overexploitation of natural fossil fuel resources over the last several decades has led to unprecedented levels of pollution, so much so that it has caught global attention like never before. Large -scale climate initiatives are currently in play to cap the emission levels and global rise in temperature. The world is scrambling to develop long-term, reliable, robust, mass scale and sustainable renewable technologies that can be an economically viable alternative to conventional fossil fuels. The biomass industry is emerging to provide biofuels, biochemical, bioproducts and biopower products, giving this world a glimmer of hope to reduce its dependence on fossil fuels, providing a potentially viable and sustainable substitute.

The main impetus for the commercial use of microalgae came into being because of the adoption and acceptability of algae as a viable feedstock for third-generation biofuels. First-generation biofuels competed directly with food crops, whereas second-generation biofuels did not compete for food crops, but did directly compete with arable land meant for food crops, and thus both these generations conflict with food supplies (Behera et al. 2015; Alam et al. 2012). Microalgae, being sustainable and renewable, can act as an economically viable source for not only biofuels (solid: bio-char; liquid: bioethanol, biodiesel, vegetable oil; gaseous: biohydrogen and bio-syngas) (Fon Sing et al. 2013), but a variety of products, including but not limited to direct use (human food, animal fodder/feed (de Cruz, Lubrano, and Gatlin 2018), food and health supplements) and bioproducts (polyunsaturated fatty acids [PUFAs], antioxidants, colouring agents, vitamins, anticancer drugs, antimicrobial drugs) (Khan, Shin, and Kim 2018; Brennan and Owende 2010; Borowitzka 2013).

Although algae is uniquely positioned to replace both the first- and second-generation biofuels (Behera et al. 2015) and act as an viable feedstock for various processes, in contrast to cellulosic biofuels, algae operations cannot derive learnings from direct agricultural and process engineering lineages, as there is no legacy knowledge or practices established for cultivating, protecting, harvesting, drying and processing algae at mass industrial scales. Thus, strategic investments in both research and development and critical infrastructure are required to support algae-based commercialization activities.

In spite of decades of extensive research activity around the globe, there has been limited but noticeable commercial success in the global-scale deployment of algae-based biofuels or bioproducts (Paul Abishek, Patel, and Prem Rajan 2014; Silva 2015; Usher et al. 2014; Khan, Shin, and Kim 2018; Kumar, Singh, and Sharma 2017). Grassroots-level research is needed to develop technologies and optimize processes and systems to reduce the level of risk and uncertainty associated with the commercialization of the algae-to-biofuels/biochemical/bioproducts process. By engaging in an in-depth assessment of the progress made in developing algal products and biofuels and identifying the current technology gaps and cross-disciplinary needs, this chapter provides a review of the current state of the art of industrial-scale operations of microalgae and identifies major challenges that must be overcome by sustained developmental focus to make the greatest impact.

9.2   SETTING THE AGENDA AND THE SCOPE OF DISCUSSION

A typical algae production facility is shown in Figure 9.1. The layout depicted in the figure is merely illustrative in nature and should not be construed as the only possible layout or the most common layout. Many of the unit operations depicted in the figure are commonly used in algae facilities or industries; however, with an extremely large number of alternative unit operations available and several different combinations possible based on the end-product specifications, tens of different layouts are possible, if not hundreds. In any algae production facility, the major unit operations that are followed are:

1. **Strain selection:** Strain selection is not a straightforward process. It is an iterative process that must be carried out scientifically and rationally, because it is the backbone of the entire process. If the strain is not selected judiciously, it can break the process, and all other processes in the overall architecture can only do so much.
2. **Inoculum development**: This usually starts from a preserved (lyophilized or as cryo-stock) production strain, which is first inoculated in flasks in controlled-environment shakers and then progresses towards larger sterile/ semi-sterile photo-bioreactors (PBRs).
3. **Pre-production pond or PBR/baby ponds (raceways) or PBR**: From here the culture can either first go to an array of baby ponds, if the inoculum volume is not too high, or directly to pre-production ponds (or a PBR as the case may be) if there is sufficient inoculum volume.
4. **Cultivation in production ponds (raceways)/PBR**: This is the largest operation in the production facility, where large-scale cultivation of algae will take place to meet the algae production target. It is one of the most important steps and directly determines if one will meet the production targets or not and has a significant effect on the process economics. It is also one of the most researched areas due to its complexity, as will be illustrated later.
5. **Crop protection**: When the algae culture is growing, especially in fully open raceways/ponds, they are exposed to the complete biome of the local geography. There are a number of ways in which undesired biotic invasion

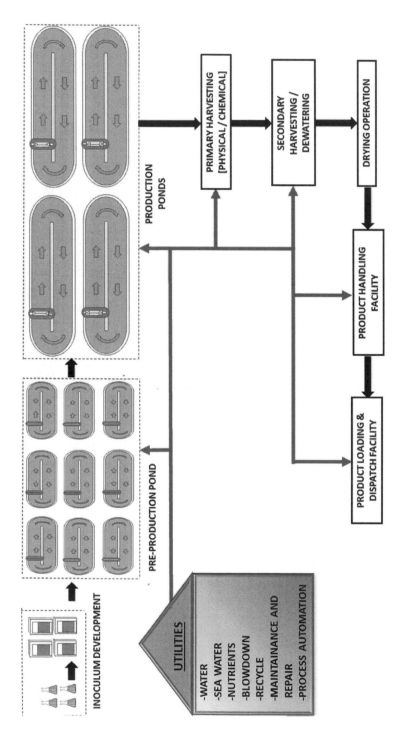

FIGURE 9.1   Layout of unit operations in a typical algae production facility.

Source: Mariam AI, Abdel, and Amal (2015); Ullah et al. (2014); Alam et al. (2012)

can take place, resulting in either an undesired algae strain growing and competing with the desired algae strain for nutrition or a grazer growth that can consume the desired algae itself. To protect the desired algae, it is generally necessary to either continuously add an appropriately designed chemical or provide an intermittent treatment to keep the contamination at bay (Karuppasamy et al. 2018). Treatments may not be limited to chemical means only; other suitable techniques can also be applied that are most suitable for the given biome.

6. **Harvesting**: Once the cells reach the desired volumetric concentration, algae and water will need to be separated (i.e., water will need to be removed) to reduce the algae to the desired concentration—this process is referred to as harvesting. Harvesting is typically divided into two parts: primary and secondary:

    a. *Primary harvesting*: Primary harvesting typically concentrates algae from 0.02% to 0.1% to 1% to 6%. This can follow a chemical route (for example: coagulation, followed by flocculation, followed by settling) or a physical route (such as filtration or centrifugation).

    b. *Secondary harvesting/dewatering*: This step concentrates algae from 1% to 6% to anywhere between 10% and 30%, depending on the process downstream. This can be a purely physical process or a chemical step before the final physical dewatering.

7. **Drying:** If the requirement is to concentrate the algae further to near-complete dryness (moisture <10%) then the algae slurry/cake coming out of the secondary harvesting unit is sent to a dryer. There are more than 20 types of dryers, and the right choice depends on the feed concentration, desired final moisture content, final product specs, acceptable heat contact time, acceptable temperature exposure and final acceptable algae nutritional level.

8. **Oil Extraction:** If crude bio-oil (CBO) is not the final product, then generally lipids/oil is extracted either after drying the biomass or wet milling is done on the slurry to get lipids and lipid-extracted algae (LEA). Lipids can be converted to bio-diesel, and the LEA can be used for a variety of co-product applications. If there is an extremely high-value compound in the algae, some kind of solvent extraction is conducted to selectively extract it (for example, astaxanthin).

9. **Hydrothermal liquefaction (HTL):** If CBO is the final desired product, then the algae slurry can bypass the drying step and go directly from primary or secondary harvesting to the HTL reactor.

10. **Utilities**: There are several utilities that a typical large algae production facility has, such as demineralized (DM) water, sea water, nutrient stock and formulation unit, blowdown unit, water recycle unit, effluent treatment plant (ETP), maintenance and repair, process automation, CO_2 bullet, etc.

Each of the unit operations listed is in itself a big and intense area of research, and any combination of these processes or their sub-types or alternatives can be used for different products and co-product production.

Thus, further in-depth discussion in this chapter will be restricted to the sustainability of strain selection, cultivation and harvesting. These three steps ensure that we have the right strain, the required amount and concentration of biomass for any further downstream processing that is desired as per the final required product. Also, except for the strain chosen, the steps of algae cultivation and harvesting must be completed, irrespective of the final end use for the algae. Second, the majority of the discussion will be limited to the perspective of producing CBO from algae (i.e., biofuel). This is because hundreds of different products and co-products can be looked at, but the details are beyond the scope of this chapter.

9.3 FACTORS TO CONSIDER IN ALGAL STRAIN SELECTION FOR THE COMMERCIALIZATION OF ALGAE

9.3.1 BACKGROUND

There are some hurdles in the successful commercialization of algal-based processes for biofuels and co-products:

- few commercial plants in operation
- unavailability of reliable legacy data
- low productivity
- contamination by invasive species
- unsustainable supply of macro-nutrients such as nitrogen and phosphorous
- costly availability and transfer of carbon dioxide
- availability of fresh water
- high energy inputs required for water pumping
- discharge of spent medium
- cost of constant mixing energy
- unfavourable climatic and environmental factors
- limited availability of land
- costly and inefficient harvesting, dewatering and drying, product extraction, thermochemical conversion, hydro-treatment, and fuel upgrading.

To overcome these hurdles will require considerable, long-term and sustained investment in end-to-end fundamental research and technology development if algae-based fuel or co-products are to become commercially attractive at a meaningful scale. One of the factors that has an important role to play in almost each of the hurdles outlined is strain selection. The physiological, physiochemical, physical and morphological nature of the cell—such as unicellular versus multicellular nature, cellular dimension and shape, preference for fresh water or sea water or brackish water, nutritional requirement, survival capability, preference of ecology and environment, lipid accumulation capability, growth rate, susceptibility to contamination or competition, amenability to harvesting, dewatering and drying—have a significant bearing on how difficult or easy a particular hurdle is to overcome. This is because some of these characteristics of algae have inherent advantages or disadvantages for a given unit operation in algae processing under a given set of conditions. Thus, it is clear that a scientific approach to selecting a strain that has some inherent advantages (which it would have acquired over millions of years of evolution) for a given unit of operation will definitely simplify some of the process-related challenges, so that time and effort can be more suitably allocated in solving the challenges that nature has not already overcome for us.

Studies have been conducted to systematically asses the effect of site selection, biomass-to-biofuels conversion technology and strain selection on the economic viability of the algae biofuels industry (Venteris et al. 2014). Their modelling-based analysis shows that strain selection and the biomass-to-biofuels conversion technology in different permutations and combinations can have a variation on economic impact up to 10 million USD/year/UF (unit farm). This is a huge impact. According to Duong et al., another important factor very closely related to strain selection is strain isolation (Duong et al. 2012). They present a framework for the most optimal strain isolation and selection procedure, which would help to identify strains that are most suitable for the commercial algae biofuels industry, as well as the multiproduct biorefinery concept. Depending on the end use, isolation and selection criteria will differ, but the logical approach should remain similar.

9.3.2 Strain Collection, Isolation and Selection

9.3.2.1 Strain Collection Criterion

When collecting a strain, one must be extremely thoughtful about the intended end use. This is the single most important aspect. Once the site selection for the production facility is made, the algae strain must be collected from the local environment (Lim et al. 2012) because a strain thriving in the local ecology probably has evolved a competitive advantage in that ecology and may have a higher probability and mechanisms to ensure its survival over other organisms and against the elements of the environment (Duong et al. 2012; Thomas, Tornabene, and Weissman 1984; Tadros 1985). This collection criterion can be further refined once the end use of the organism is clearly known. Let's suppose that a high-lipid-yielding algal strain is desired; then it would be prudent to isolate strains from local aquatic ecologies (Cobos et al. 2017) that see frequent changes in environmental conditions and stress cycles because lipid accumulation is known to be triggered under such conditions (Duong et al. 2012). Examples of such locations could include lagoons, estuaries and tidal rock pools (Duong et al. 2012; Sánchez Roque et al. 2018; Thangavel et al. 2018). Consider another scenario: let's say the end use of the algae being isolated from nature is for bioremediation of a toxic compound or metal. In such a scenario, it may be a good idea to look for areas that have an excess of this compound or metal, either because of natural occurrence (Massimi and Kirkwood 2016) or because of some accidental spill or industrial effluent, or it is a mining site for that metal. Organisms surviving in such regions may have developed a specialized survival metabolism, which gives them a survival advantage (Massimi and Kirkwood 2016) and may be more amenable for end use and further manipulation and modification for a higher degree of genetic or metabolic customization.

9.3.2.2   Strain Isolation

Conventionally, techniques such as single-cell isolation and medium nutrient selection pressure-based methods have been used for strain isolation. Single-cell isolation requires painstakingly working under a microscope with a pipette to separate a single cell or colonies to have pure strains. Nutrient selection pressure-based methods utilize enrichment of the growth medium with specific nutrients that encourage or favour growth of a certain species over others. Recently, automated methods for single-cell isolation, such as automated cell sorting (Reckermann 2000; Davey and Kell 1996), based on autofluorescence or other suitable markers have been gaining prominence but are still not very prevalent. Using only the described techniques can only take algal research so far in terms of both the speed of research and isolation of the right strain. This is because the algae industry is nascent, and there is a lack of legacy data on strain capability, biochemistry and genomic information. To develop an effective isolation program, efforts would be required to combine the search for an organism in a larger taxonomic group of microalgae across a wide ecological and geographical distribution, combining it with reliable data on species-specific oil content and other co-product and growth-stimulating environmental conditions, linking it further with curated phylogenetic information and finally cross-referencing it against genomic data. This approach might eventually help in identifying molecular or phenotypic markers that help in identifying the potentially promising isolates (Duong et al. 2012).

9.3.2.3   Analytical Method for Qualitative and Quantitative Assessment of the Trait of Interest

All the effort put into strain collection and isolation may not, after all, yield desired results if the right kind of analytical method is not available to make an accurate assessment of the trait for which the microalgae is being collected and isolated for. Depending on the desired trait, whether is it lipid, bioactive pigment or protein, etc., the appropriate analytical technique must be in place for both qualitative and quantitative estimate of the desired trait so that the organism can be properly assessed. Qualitative or semi-quantitative estimation techniques would be initially sufficient for quickly ranking the different species or strains for the desired trait, but eventually when the final selection is to be made in the context of meeting the production targets, quantitative assessment techniques are a must. Typically, microalgae grow as very dilute suspensions, and when the initial screening is being carried out, sample volumes are very low—in the range of tens of millilitres. Thus, the estimation method being deployed must not only be rapid, reproducible, robust and free from the matrix effect but should not require large sample volumes.

9.3.2.4   Strain Selection and Screening

Once the different algal species/strains have been collected and the pure colonies isolated and quantified for the desired metabolic product (for example, lipid or pigment or other high-value compounds), the next logical step is the selection of the most appropriate production strain. This final selection should never be based simply on the titre of the desired metabolic product measured in cultures grown at small scale under controlled conditions, but of course, it would be the starting point for a more involved selection process. Once the initial selection of a list of species or strains is made based on the desired metabolic product, one must quickly move to optimize conditions that are suitable for each selected species to maximize growth and productivity (Rodolfi et al. 2009). This is important because the strains that might appear close in terms of performance for a desired trait might appear widely separated once they are exposed to their respective optimal conditions (Duong et al. 2012). But one must remain very careful to limit the access to only such optimal process parameters that can be applied under outdoor conditions within reasonable bounds of technical feasibility and economic viability (Duong et al. 2012). This will help in further narrowing down the list to candidate strains that can be taken outdoors. Then the selection process should quickly move towards medium-scale cultivation vessels (pond/raceway or PBR) under outdoor conditions. This is an extremely important step because some of the environmental parameters cannot be replicated in their entirety under indoor conditions, such as sunlight illumination, elements of the weather, temperature (Li and Qin 2005), humidity, dust, rain, biotic competition or stress, changes in water quality and intra-day and inter-day fluctuations across the listed parameters. Under these outdoor conditions, strains should not only be assessed based on their product titers but should also be ranked, taking into consideration factors such as productivity, susceptibility to environmental biotic competition, ability to cope up with abiotic stresses and the general robustness of the strain to the outdoor biome (Rodolfi et al. 2009; Li and Qin 2005; Thomas, Tornabene, and Weissman 1984; Al-Hasan et al. 1990; Pulz 2001). This is because some of the considerations other than product titer values may offer a substantial benefit or disadvantage to the overall process, which might offset the benefit of the high-product titer alone. Once this characterization is complete, the final decision of strain selection should take place. The final set of strains should be tested outdoors for an extended period to evaluate robustness.

9.3.3 Relative Advantages or Disadvantages of the Different Physiological, Physiochemical, Physical and Morphological Natures of the Algae Cell

The different physiological, physiochemical, physical and morphological natures of the algae cell can influence the different unit operations. There has not been much focus in exploiting these correlations for strain selection in the industrial-scale operation of algae, and thus it would be prudent to give due attention to strain selection in this context.

9.3.3.1 Small Spherical Cells versus Larger Filamentous Cells

1. **Small spherical morphology may be advantageous because:**
   a. They lend themselves more easily to new evolving techniques such as ultrasonic acoustic wave focusing–based harvesting methods (Kurokawa et al. 2016; Rajasekhar et al. 2012).
   b. Hydrogen (Proton) nuclear magnetic resonance (H-NMR)–based quantitative assays is rapidly evolving as a preferred non-destructive and rapid analytical methods for determination of triacylglycerols (TAG). This

method works well with small and spherical algae, which are better at bending the respective electromagnetic fields (Prabakaran and Ravindran 2011).

c. Small cell size means larger surface area per unit cell, which would favour improved adhesion rates between the cell and the bubble in dissolved air-flotation-based harvesting methods. Improved adhesion would improve algae capture rate per unit energy expenditure, improving both the process throughput and process economy (Ndiaye, Gadoin, and Gentric 2018; Palaniandy et al. 2017; Zhang et al. 2014; Han et al. 2002).

d. A smaller cell is always more efficient in the transfer of gases and liquids to meet its metabolic requirement. Thus, smaller algae cells may have enhanced ability to transport materials (including waste products), minerals, light transfer and dissolved nutrients across the cell membrane (Huang et al. 2017; Ndiaye, Gadoin, and Gentric 2018; Farajzadeh, Zitha, and Bruining 2009).

e. Evolutionarily speaking, smaller cells have a higher likelihood of survival under non-conducive conditions, as they seem to be must efficient in the transport of mass, and this becomes a significant advantage when resources are scarce (Huang et al. 2017; Ndiaye, Gadoin, and Gentric 2018; Farajzadeh, Zitha, and Bruining 2009).

f. Small spherical solid bodies do not directly contribute to increases in hydrodynamic viscosity, which could negatively affect heat, mass and momentum transfer operations (Gudin and Chaumont 1991).

g. Unlike filamentous cells, spherical cells do not entangle with paddle wheels or rotors used to agitate algae in open raceway ponds (Gudin and Chaumont 1991).

2. Small spherical morphology may be disadvantageous because:

a. Algae cells have negatively charged surfaces. The smaller the size, the higher the charge density. Higher surface charge density enables algae cells to repel each other and thus maintain high suspension stability. All things being equal, algae with a relatively smaller size will form a more stable suspension and thus require a higher dosage of the coagulating chemical to bring about the destabilization of algae suspension for a given degree of algae water separation in a chemical harvesting approach (Henderson, Parsons, and Jefferson 2008a; Henderson, Parsons, and Jefferson 2010; Ghernaout, Ghernaout, and Saiba 2010; Pestana et al. 2015; Ghernaout et al. 2015; Ghernaout and Ghernaout 2012). Higher dosages of harvesting chemicals such as poly-electrolytes or metal salts or cationic polymers would not only mean higher process cost but also additional chemical contamination of the harvested biomass. Higher chemical input on some occasions might lead to non-compliance with product specifications or would necessitate additional steps for removal of the undesired chemical species—in either case, this will add to the process cost and complexity.

b. The most economical method of cell concentration is gravity settling; however, the smaller the cells, the lower the settling velocity and the higher the residence times, which could make gravity settling an unviable option, as the holding tank size will increase considerably as will the settling time (Peperzak et al. 2003).

c. Acoustic wave focusing–based harvesting technology relies on the fast settling of algae post-focusing (Bosma et al. 2003; Zhang, Zhang, and Fan 2009); however, smaller cells are focused to smaller flocs and have a slower settling velocity.

d. Smaller algae cells have a higher potential for causing fouling of membranes used in filtration-based harvesting technologies. As the cell size becomes smaller and starts approaching membrane pore size, the potential to clog the pores and cause fouling increases. De-fouling by chemicals is time consuming, costly and reduces the membrane's operational life (Uduman et al. 2010; Mariam Al, Abdel, and Amal 2015).

e. Smaller algae cells become an easy target for contaminating organisms (grazers, amoeba, fungi, rotifers, bacteria, undesirable autotrophs). Many of these contaminating organisms swallow whole cells, and if the cells are smaller, the contaminating organism can easily engulf the cells in large numbers, thus leading to a rapid decline of the desired phototroph (Wang et al. 2013). Once the contamination takes over the raceway pond, toxins and growth inhibitors are released, and there is competition for nutrients and light (Valério, Chaves, and Tenreiro 2010; Katırcıoğlu, Akın, and Atıcı 2004). The higher susceptibility for contamination would result in the deployment of additional resources for contamination monitoring, detection and control, fluctuating productivity, occasional washout resulting in shutdown, unnecessary downtime and need for decontamination and clean-up—each of these aspects will add cost to the process.

9.3.3.2 Eukaryotic vs. Prokaryotic

1. **Compared to prokaryotic cyanobacteria (blue-green algae—BGA), many of the eukaryotic green algal strains might offer the following advantages:**

   a. Some filamentous BGA (like wild-type *Anabaena* sp.) are known to secrete toxins (including neurotoxins) as defence mechanism, while green unicellular algae are generally regarded as much safer (Valério, Chaves, and Tenreiro 2010; Katırcıoğlu, Akın, and Atıcı 2004). The presence of toxic metabolic products, even if released as a defence mechanism, can render the biomass product unfit for any human- (nutraceutical, cosmetic, essential oils, food supplements) or animal-related usage (animal feed).

   b. Except for a few molecules, such as the photosynthetic pigments phycobilisomes (for example, phycocyanin, phycoerythrin, etc.) that BGA and red algae like *Gualdieria* sp. make, the eukaryotic green algae have inherently higher productivity kinetics for both high-value chemicals (HVC) and low-value commodity (LVC) biofuel precursors. Examples of HVC include astaxanthin, omega-3 free fatty acids, lutein, etc., to name a few, and lipid precursors, and low-heating-value commodity biofuels [i.e., monoacylglycerols (MAG), triacylglycerols (TAG), diacylglycerols (DAG)] are examples of LVC (Masojídek, Torzillo, and Koblížek 2013; Andersen 2005).

c. Many BGA species are known to secrete high-molecular-weight polymeric compounds such as extracellular polysaccharides (ECP) (Tamaru et al. 2005; Yang and Kong 2012; Myklestad 1995), which not only tends to increase the viscosity, resulting in higher cost of mixing and inadequate mass transfer; it also represents channelling of metabolic flux away from the desired products into undesirable metabolites, resulting in a loss of carbon and energy.

2. Compared to prokaryotic cyanobacteria (BGA), many of the eukaryotic green algal strains might offer the following disadvantages:

a. Eukaryotes generally have a slower growth rate and hence lower productivities than BGA (Nielsen 2006). Typically, marine or fresh water algae blooms are associated with BGA. Some research groups have also suggested that the bloom-causing ability of BGA is aided by their ability to migrate across the water depth and thus prevent sedimentation in warmer and highly non-homogeneous waters, and their ability to avoid becoming prey to zooplankton, especially when warming reduces zooplankton body size (LÜRLING et al. 2013). This means that green algae have a lower throughput, and hence for the same amount of biomass requirement, a larger volume of culture will have to be cultivated, and that would mean larger cultivation acreage and hence higher production costs.

b. Green algae have much more complex genomes, transcriptomes (Cooper and Hausman 2007; Tirichine and Bowler 2011), cell walls (Popper, Ralet, and Domozych 2014), nuclei, mitochondria and chloroplasts than BGA and thus are difficult systems for genetic modification in terms of low copy numbers, expression of recombinant DNA and directed secretion of desired products into the extracellular space. Such factors limit the usage of green algae as bio-factories for the production of relatively pure, volatile, high heating-value hydrocarbons as secreted products, and thus necessitate the conversion of biomass to hydrocarbons using hydrothermal liquefaction (Biller 2018; Liu, Huang, and Chen 2011).

9.3.3.3   Sea Water vs. Fresh Water

1. Compared to the cultivation of marine algae, the cultivation of fresh water algae is advantageous for the following reasons or in the following scenarios:

a. The air–gas bubbles delivered by an orifice sparger coalesce together to form larger bubbles much more easily in fresh water than in sea water.

Larger bubbles are able to transfer adequate momentum from buoyancy forces for the agitation of algal cells and O2 removal (Han et al. 2002; Bondelind, Sasic, and Bergdahl 2013). The transition point of bubbles from coalescence to non-coalescence is 8 to 10 g/L of dissolved salts, and sea water is much higher, in the range of 35 to 60 g/L, where bubbles formed will not recombine with adjacent bubbles (Monahan 2002, 1969; Monahan 1971).

b. Gases in general, including CO_2, have higher solubility in fresh water than in sea water. The solubility of gases decreased in water with increasing salinity (Pérez-Salado Kamps et al. 2006; Holzammer et al. 2016). Lower solubility means higher probability of loss of CO_2 to the atmosphere, and this wastage will add to the process cost.

c. Fresh water neither has high concentration of salts nor other stronger surfactants that are typically found in sea water, thus presenting a higher gas–liquid mass transfer coefficient than sea water (Ruen-ngam et al. 2008; Sardeing, Painmanakul, and Hébrard 2006; Hanwright et al. 2005).

d. Because of the high salt content, sea water or saltwater promotes cell desiccation, which is not observed in the case of fresh water algae strains. As a response to this stress, in saline water, many algae strains are known to excrete ECP as a coping mechanism (De Philippis and Vincenzini 1998; Tamaru et al. 2005; Caiola, Billi, and Friedmann 1996; Holzinger and Karsten 2013). These secretions have a downside on several fronts: ECP are polymeric in nature and hence increase the viscosity, which negatively affects the mass transfer coefficient, it makes algae water separation (harvesting) significantly more difficult, and promotes susceptibility to bacterial contamination because extracellular polysaccharides (ECP) acts as a source of carbon and represents an unnecessary drain of carbon and energy flux to undesired product (Usher et al. 2014; Farajzadeh, Zitha, and Bruining 2009; Hanwright et al. 2005; Masojídek and Torzillo 2008).

e. Fresh water has extremely low salts or total dissolved solids compared to sea water and is thus not corrosive, and specialized material of construction (MOC) is not required for algae handling. When handling marine algae, corrosion is a big issue (Fink 1960; Melchers and Jeffrey 2005; Xiangyu et al. 2018), and thus extremely inert material needs to be used as the MOC for algae handling equipment.

f. Most specialty chemical suppliers manufacture coagulants, flocculants and dewatering chemicals that are more suited for a fresh water matrix rather than sea water matrix, and thus several chemicals are available on the market to choose from to make a tailor-made recipe for solid–liquid separation.

2. Compared to cultivation of marine algae, cultivation of fresh water algae is disadvantageous for the following reasons:

a. The CO_2 gas bubbles delivered by an orifice sparger or other appropriate micro-diffusers into the algae cultivation vessel for the supply of carbon and pH control coalesce together to form larger bubbles much more easily in fresh water than in sea water (Ruen-ngam et al. 2008; Ndiaye, Gadoin, and Gentric 2018). Because of these large bubbles, fresh water suffers from high bubble rise velocity, lower gas hold-up, lower surface tension and lower surface area per unit volume, and thus an inferior gas–liquid mass transfer coefficient in comparison to sparging CO_2 in a sea water matrix (Monahan 2002, 1969; Monahan 1971; Ndiaye, Gadoin, and Gentric 2018).

b. Harvesting techniques such as dissolved air flotation (DAF) or froth flotation rely on the formation of micro-size bubbles with a high specific area (Han 2002; Palaniandy et al. 2017). However, because in fresh water bubbles formed coalesce together to form larger bubbles, there is significant loss of efficiency of cell removal for the harvesting methods in question (Bondelind, Sasic, and Bergdahl 2013).

c. The concentration of total dissolved solids (TDS) in a solution or suspension is a measure of the ionic strength (Montes, A. Rotz, and Chaoui 2009). The higher the TDS concentration, the higher the ionic strength. The higher the ionic strength, the lower the dissociation constant of ammonium in water solution, resulting in lower ammonia volatilization (Arogo, W. Westerman, and S. Liang 2003). Fresh water, in absence of high TDS, will result in a higher loss of ammoniacal-N to the atmosphere, resulting in additional cost. Adsorption of ammoniacal-N on TDS decreases dissociation of ammonium ions and reduces the effective concentration of ammoniacal-N in the solution, thus resulting in decreased potential for ammonia volatilization. In the absence of TDS in fresh water, the potential for volatilization will become significantly elevated.

d. Only 3% of all water present on the earth's surface is fresh water; the rest of it is salt water (Venteris et al. 2013; Cosgrove and Loucks 2015). With rapid human population growth and mindless overexploitation, the world currently has an acute drinking water shortage. In this scenario, the use of a fresh water-based algae cultivation system will further strain already heavily burdened fresh water resources.

e. Non-halotolerant species cannot propagate in a saline environment; however, they can contaminate fresh water algae cultures because of the low ionic strength, which cannot offer salt stress inhibition to the growth of non-halotolerant contaminating species (Hellebust 1985). Thus, fresh water-based cultivation systems will always be more susceptible to additional contamination, which would have been impossible with salt water-based cultivation systems.

From this discussion on the advantages and disadvantages of different scenarios, cultivation systems, model organism systems and cultivation medium, it seems prudent that the algae research community and industry focus their attention and efforts on developing understanding and technologies for unicellular, small, spherical algae or filamentous marine green algae or BGA in open pond or closed vessel (PBR-based cultivation systems) to strategically select and rightly position the appropriate algae strain for developing a sustainable algae production program.

9.4 CULTIVATION OF ALGAE

The very first step on which the successful outcome of any sustainable algae-based product or fuel business depends is the ability to successfully and economically grow algae, which is referred to as cultivation. Cultivation of any algae requires some basic inputs: water (fresh or sea or of different grades of salinities like brackish, estuary, creek), sunlight, CO_2, air, optimal temperature, micro-nutrients and macro-nutrients such as nitrogen and phosphorus and a vessel to hold water where algae grows as a suspension. The discussion in this section will be limited to unicellular, freely suspended algae, because various species of algae also grow as clumps or as an attached growth or film growth and are beyond the scope of this chapter. Although cultivation requirements look simple, it is plagued with difficulties, making algae cultivation one of the most challenging and costly components of the algae industry. Various estimates show the cultivation cost component to be in the range of 30% to 60% of the total cost of producing algal biofuels (Acién et al. 2012; Norsker et al. 2011; Griffin et al. 2013; Chisti 2008), and this is a very significant number.

The parameter that gives the ultimate verdict on the success or failure of any cultivation process is productivity, expressed as mass of algae grown per unit cultivation volume or area per unit time [Mass / (Volume or Area * Time)]. If productivity is expressed per unit volume, it is called volumetric productivity (typical units are g/m^3/day) and if expressed per unit area, it is called areal productivity (typical units are g/m^2.day). Productivity is basically a change in concentration per unit time—it would mean more algae has grown in the same unit of time. The objective of most cultivation facilities is to maximize productivity; however, it is not easy to bring about high productivity. In general, most algae strains that have been deployed in algae production systems are wild-type strains which have evolved mechanisms and metabolic systems over millions of years to survive and thrive in their respective environments. However, we are exploiting them for human benefit and expect them to grow at a rate that satisfies the human need for energy; unfortunately, this is not always possible. There two ways to get more productivity out of algae: first by optimizing the factors that govern the growth; although this approach can be highly effective, it has its limits, such as the upper metabolic limit of the species. If even after this level of optimization, desired productivities are not achieved, the next logical step is genetic modification to either enhance the capacity of the native trait or to code new traits into the wild-type strain to achieve desired growth rates or product production. Whereas the latter approach is beyond the scope of this chapter, the former is not.

The factors that can be optimized to enhance productivity are water quality, salinity, salts, sunlight, CO_2, air, temperature, micro-nutrients, macro-nutrients and optimal strain selection. The importance of strain selection for designing an optimal algae production facility was discussed in the last section, but optimization of other factors can make a substantial improvement in algal productivity. Cultivation vessels, which contain the bulk volume of the growth medium, are the most essential part of the optimization process. Cultivation vessel design enables control of almost all of the abiotic factors listed, and it can also be designed to control biotic factors such as contamination or competing organisms. Cultivation vessels are of two basic types: PBRs and open ponds or raceways. All other formats or configurations are manifestations of these two basic types.

9.4.1 PHOTO-BIOREACTORS

9.4.1.1 Factors Governing PBR Technology and the Shortcomings

PBRs are either partially contained or fully contained vessels that can be designed to provide a 100% sterile environment, whereas ponds or raceways are fully open and can be of the order of 1,000 to even 10,000 times larger than PBRs. The semi-closed or fully closed nature of the PBR (Cristóbal García Cañedo and Lizárraga 2016) allows precise control of operating variables, such as evaporation losses, CO_2, O_2, air, light exposure (at least in indoor settings), temperature, nutrient addition and contamination control (Wen et al. 2016; Chen et al. 2011; Cristóbal García Cañedo and Lizárraga 2016; Kunjapur and Eldridge 2010). Several types of commercial PBRs are available (Figure 9.2) (Huang et al. 2017; Posten 2009), and these can be classified as sea-weed type (Chetsumon et al. 1998), conical (Contreras et al. 1998; Watanabe and Saiki 1997), bubble column (Degen et al. 2001; Oncel and Sukan 2008), airlift reactors (Ranjbar et al. 2008; Rubio et al. 1999; Ugwu, Ogbonna, and Tanaka 2002; Harker, Tsavalos, and Young 1996; Kaewpintong et al. 2007), tubular (Richmond et al. 1993; Hall et al. 2003), torus (Pruvost, Pottier, and Legrand 2006), annular (Zittelli, Rodolfi, and Tredici 2003), stirred tank (Zhang 2013; Ogbonna,

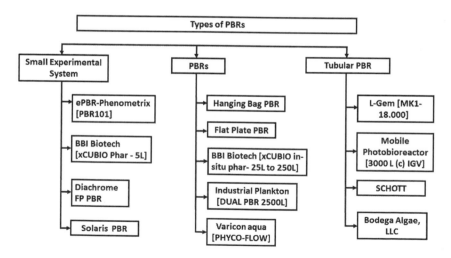

FIGURE 9.2 Types of PBRs.

Source: (Oncel and Sukan 2008; Gupta, Lee, and Choi 2015; Cristóbal García Cañedo and Lizárraga 2016; Xu et al. 2009; Degen et al. 2001)

Soejima, and Tanaka 1999), flat plate (Hu, Guterman, and Richmond 1996) and possibly some more.

Despite the several models and configurations of PBRs available, only a very few of them (tubular PBR, plastic bag PBR, column airlift PBR, flat panel air-lift PBR) are actually suitable for large-scale mass cultivation of algae. The primary reason for this scenario is that less research effort has gone into the development of PBRs for mass cultivation of algae. Most designs available today are based on semi-empirical data, at best, and a fundamental understanding is highly underdeveloped, leading to non-ideal coupling of operating variables such as mass transfer, heat transfer, light distribution, hydrodynamics and cell growth and pH control, resulting in sub-optimal performance of PBRs (Degen et al. 2001; Sforza et al. 2012). A suitable MOC for PBRs is also lacking that offers properties like good clarity, light weight, longevity, low scaling and resistance to abrasion at low cost. Various fittings, spargers and sensors used in PBRs are also very expensive. All these factors combined have resulted in a current generation of PBRs that require heavy capital investment, have significant operating costs, have short life spans and have sub-optimal productivities.

Another factor to consider is use of sea water vs. fresh water as a cultivation medium. Most research groups involved in PBR development have mainly used fresh water algae as the model organism (Wen et al. 2016; Ting et al. 2017), but as the algae-based biofuels industry is maturing, it is becoming more and more apparent that marine algae with sea water-based cultivation is the only viable and sustainable approach. Only a handful of researchers are actually working towards the development of PBR technology suitable for sea water-based media (Narala et al. 2016). Thus, the current generation of already sub-optimal and expensive PBRs designed for fresh water will further face the unique challenges of handling sea water. Typically, sea water handling would result in a higher degree of scaling, material aberration and corrosion, and thus there would be a much higher maintenance cost. There would also be challenges associated with gas sparging and salinity increases due to evaporation losses—and much more so in open pond systems.

9.4.1.2   Algae Productivity in PBRs

The productivities required to make an algae biofuels project practical should be in the range of 40 to 60 g/m^2.day (Benemann and Oswald 1996; Davis et al. 2016; Sun et al. 2011); however, current productivities in almost all commercial systems hover at a much lower range, between 12 and 40 g/m^2.day. For example, (Wen et al. 2016) have reported a maximum productivity in the range of 17 to 20 g/m^2.day, and that too, in fresh water, whereas (Ting et al. 2017) have reported productivities of 34.6 g/m^2.day in fresh water. (Narala et al. 2016), on the other hand, have reported productivities in the range of 8.2 to 14.4 g/m^2.day in a sea water matrix. Clearly these productivities are far lower than what is commercially viable.

9.4.1.3   Parameters Affecting Algae Productivity in PBRs

The primary reason why higher productivities have not yet been achieved in commercial outdoor PBRs is because there is very limited knowledge about the effect of individual operating parameters on algal growth and their interplay.

1. **Light:** Because there is light in abundance, it might appear that light is the least of the worries as far as optimization is concerned; however, effective light utilization per unit PBR surface or per unit culture volume held within these surfaces is a major technical challenge (Kunjapur and Eldridge 2010). This is because if all the cost that goes into construction (capital cost = CAPEX) and operation (operating cost = OPEX) is gathered into a single cost and normalized with respect to the PBR surface area or the PBR volume, then each such unit becomes a cost centre (Huang et al. 2017). And the return of investment is the productivity one gets out of each of these cost centres, meaning every unit area or unit volume of PBR costs a lot (CAPEX + OPEX), and each such unit must be able to convert the maximum amount of photon (sunlight) and carbon into energy to get a good return on investment (Cristóbal García Cañedo and Lizárraga 2016). Just because light is free, does not mean it can be freely converted to usable organic energy (biomass), because to harvest light effectively, one will need to create the PBR surface or volume available for conversion, and the cost per unit area or volume of PBR creation is pretty high at the moment. When sunlight (or artificial light in the case of indoor PBRs) falls on the PBR, it is distributed, and this has a finite effect on cell growth (Gupta, Lee, and Choi 2015). There are gaps in the knowledge about the effect of light distribution, polychromatic radiation transfer, photo-limitation, photo-inhibition and light and dark cycles (L-D cycles) on cell growth and productivity, and unless there are understood, real optimization is not really possible (Sforza et al. 2012; Xu et al. 2009).

2. **Environmental conditions:** Various environmental parameters like nutrients, pH, temperature, dissolved CO_2, shear force and air sparging have a profound effect on the performance and hence the productivity of algae (Degen et al. 2001; Gupta, Lee, and Choi 2015). It is extremely necessary to understand that it is not just the individual effect of these parameters but their interdependence that is even more important and complex. Algae typically require nitrogen (N), phosphorus (P), potassium (K) and some micronutrients for growth, which is very similar to what plants require (Zhang and Chen 2015; Csavina et al. 2011; Jia et al. 2014). However, many sources of these nutrients are not truly sustainable and represent a considerable cost. Thus, media optimization remains a very important aspect of algae cultivation. Recycling the medium to utilize unused nutrients post-harvesting is also something that still needs to be developed and integrated in large-scale cultivation systems (Wang, Yabar, and Higano 2013; Farooq et al. 2015; Stephenson et al. 2010). As far as algae is concerned, CO_2 is one of most important nutrients after photons. Typically, CO_2 needs to be sparged to make it available to algae; however, CO_2 does not remain only dissolved in water—it exists in an equilibrium with carbonate and bicarbonate, and this equilibrium is governed by the pH of the medium. pH, in turn, is affected by the rate of CO_2 sparging, cell metabolism and alkalinity. Typically, CO_2 is sparged along with air, whose primary function is to mix the culture (mass transfer) and provide flow patterns (Gupta, Lee, and Choi 2015; Azov

1982; Singh and Singh 2014; Minillo, Godoy, and Fonseca 2013; Moreira et al. 2016; Valiorgue et al. 2014). Flow patterns, in turn, determine the light exposure of cells and the frequency of the L/D cycle experienced by the cells and thus the light utilization, light inhibition and photosynthetic efficiency. The method and intensity of sparging determine the gas–liquid mass transfer, degree of shear force experienced by the cells, degree of gassing out and the overall mixing patterns (Khoo, Lam, and Lee 2016; Huang et al. 2017; Valiorgue et al. 2014). Clearly, the environmental factors that govern algae cultivation and ultimately its productivity are intricately linked with each other. Their mechanistic understanding of these complex interactions is lacking, and unless significant fundamental research is undertaken, it will be difficult to arrive at a rational reactor design that is sustainable in the long run.

Thus, a lot needs to be done to understand the dynamics of these operating parameters and their interdependence in order to extend the mechanism-based growth kinetic model to a more holistic level, coupling it with hydrodynamics, mass and heat transfer for a rationally optimized PBR design and operation. Once this is achieved, a sustainable commercial PBR can possibly be envisioned that has efficient temperature control; provides excellent mass transfer, high illuminated area/volume ratio, effective light utilization and high efficiency carbon utilization; does not expose cells to high shear; is easy to clean and maintain; is robust in the face of environmental elements; and is low in CAPEX and OPEX. Several research groups and specialized start-ups across the world have been focusing on developing a fundamental understanding of physical phenomena that govern operational parameters of PBRs, and the current need is for the large algae-based corporations to collaborate effectively with these research groups to integrate this understanding and accelerate technology development and leapfrog the laboratory to field deployment.

9.4.2 Open Ponds or Raceways

Closed PBRs have several advantages when compared to open ponds, such as:

- significantly higher volumetric productivities [PBR: 0.2 to 3.8 g/L.day; raceway: 0.12 to 0.48 g/L.day (Ketheesan and Nirmalakhandan 2012; Brennan and Owende 2010)]
- higher photon conversion efficiency [PBR: ~5% or higher; raceway: ~1.5% (Schlagermann et al. 2012)]
- higher biomass concentration [PBR: 2–9 g/L; raceway: 0.25–1 g/L (Schlagermann et al. 2012)]
- possibility of using different strains in the same setup (Jiménez et al. 2003)
- lower land area requirement (Alabi, Tampier, and Bibeau 2009)
- very high surface-to-volume ratio (Alabi, Tampier, and Bibeau 2009)
- lower cost for harvesting due to higher volumetric biomass concentration
- higher gas contact or retention time, better transfer efficiency
- comparatively efficient light utilization, mixing, L-D cycle management

- more precise control of operational variables
- potential to be operated in sterile mode or semi-sterile mode—thus less susceptibility biotic contamination or competition.

However, there are several disadvantages too. For example, the gas exchange requirement prohibits PBR scale-up above 100 m² (Klein-Marcuschamer et al. 2013); operation and clean-up is expensive, cumbersome and time consuming; the capital expenditure is very high (some estimate it to be 100 times higher than raceways on a per unit area basis); and MOC longevity under direct sunlight and resistance to abrasion during cleaning have not been established. Although resistant to contamination, once contaminated, decontamination is extremely difficult and costly (Klein-Marcuschamer et al. 2013). One of the very few commercial facilities operating a commercial-scale PBR facility (tubular PBR) is Algatech, based out of Israel, who produces astaxanthin from *Haematococcus pluvialis*. Thus, except where the end product is a very high-value chemical or non-sterile operation cannot be tolerated, ponds or raceways have remained the mainstay for large-scale commercial cultivation of algae.

The term high-rate algal ponds (HRAPs) or raceway ponds (Oswald and Golueke 1960) was first suggested by Oswald for the raceway-type design (open and shallow configuration) of a pond with a large-scale recirculation system; however, the earliest known example of mass production of algae for wastewater treatment was started in the 1950s in Japan, and they cultivated *Chlorella* sp. Commercial wastewater treatment plants and algae cultivation facilities started using the raceway type of pond design from this point onwards (de Godos et al. 2014; Chisti 2013; Terry and Raymond 1985; Pulz 2001; Spolaore et al. 2006). Several types of open pond–based cultivation systems have been in existence; the majority of them fall into four categories: tanks, shallow big ponds, circular ponds and raceway ponds or HRAPs (Borowitzka 1999). The raceway configuration consists of a cylindrical closed loop flow channel with oval ends, with typical depths ranging from 20 cm to 50 cm (more often 20 to 30 cm) (Chisti 2007). It also consists of one or two sets of paddle wheels, which keep the algae culture in a continuous gentle motion. The raceway is divided in the centre by a small dividing wall, which is just under a foot over the water surface. The raceway has a flat bottom and a vertical periphery wall, which again rises just under a foot above the soil level. These raceways can be as small as 0.5 to 1 m² in area to as large as a couple of acres (Chisti 2007).

Raceway ponds offer several significant benefits over PBR technology, such as:

- lower capital expenditure (Borowitzka 1999)
- less energy required for mixing (raceway: 0.5 to 4 W/m³; PBR: 2.5 to 15 W/m³, assuming power input = 50 to 150 W/m³ and volume-to-surface ratio of PBR = in the range 0.05 to 0.1 m) (Mendoza et al. 2013; Jorquera et al. 2010)
- lower energy (power and head) demand [raceway: ~ 57 MJ/m².year; PBR: ~ 207 MJ/m².year (Schlagermann et al. 2012)]
- lower lipid production cost [raceway: ~ 12.73 \$/gallon; PBR: ~ 31.61 \$/gallon (Richardson, Johnson, and Outlaw 2012)]

- higher scale-up capacities [raceway: up to 10,000 m²; PBR: up to 100 m² (Klein-Marcuschamer et al. 2013; Alabi, Tampier, and Bibeau 2009)]
- approximately 95% of large algae production facilities globally use raceways compared to the mere 5% domination by PBR (Mendoza et al. 2013)
- have lower global warming potential [GWP] than fossil fuel, whereas the GWP of PBR was higher than fossil fuel (Adesanya et al. 2014; Stephenson et al. 2010)
- are easier to clean, maintain and operate (PBRs, on the other hand, are costly to maintain and are laborious and time consuming to clean) (Klein-Marcuschamer et al. 2013)
- fabrication, fitting and investment cost could be 100 times lower than PBR [(Klein-Marcuschamer et al. 2013)] and
- there is much lower build-up of O_2 than with PBRs.

Open raceways are fully open to the atmosphere, and thus it is very difficult to control the same operating variables as a PBR, and they are susceptible to biotic contamination (Wang et al. 2013). They, of course, have lower volumetric concentration (Schlagermann et al. 2012), volumetric and aerial productivities due to a much lower surface-to-volume ratio than PBRs. Just like PBRs, the productivities and performance of a raceway are dependent on several factors, including but not limited to choice of geographical location of the plant or its co-location with other major raw material sources, depth of the pond, power consumption, mixing, CO_2 delivery and utilization in the pond, light, salinity, evaporation losses, oxygen accumulation, competition with non-desired algae strains, effect of biotic contamination from grazers, etc. There is a severe lack of understanding of all these operation parameters in the literature, and even more so for their combined effect on algal growth and productivity. As a result of insufficient fundamental knowledge, most plants involved with large-scale algae cultivation, have not been able to maintain an operating volumetric concentration of more than 0.5 g/L under a semi-turbidostat mode of operation. Although there have been some occasional reports of certain research groups operating algae raceways in semi-turbidostat mode with volumetric concentrations as high as 1 g/L (Schlagermann et al. 2012), none of them have demonstrated it at any significant large scale for a sustained period.

9.4.3 PBRs vs Raceways

Given the several relative merits and demerits of PBRs and raceways, their complex dependence on dozens of operating parameters (some controllable, some not) and their interplay, their dependence of the local weather and biome, complex fabrication issues in case of PBRs, possibility of contamination with non-desired species or crop failure due to biotic contamination and lack of a fundamental understanding of different biotic and abiotic factors affecting algae cultivation and productivity, the sustainability question is more or less unanswered as of today.

Richardson et al., conducted an in-depth financial assessment of two competing cultivation systems, PBR vs. raceway, and their contributions to algae biofuel economic viability (Richardson et al. 2014). The outcome of their analysis suggested neither cultivation systems are currently profitable at the present cost. Analysis also showed that PBR provides for a much greater net cash income (NCI: the gross cash income less all cash expenses, such as for feed, seed, fertilizer, property taxes, interest on debt, wagers, contract labour and rent to non-operator landlords) and lower cost per gallon of CBO when compared to open raceways. The authors suggest that raceways have more inherent risk in terms of inconsistent performance because of their open nature, and as a result algae plants with raceways will have to invest in larger raceway acreage (to mitigate productivity fluctuation risks), resulting in higher OPEX for water handling, harvesting and everything downstream. These larger facilities will have higher maintenance and labour costs compared to a PBR facility of similar algae production capacity. As per the authors' analysis for both PBRs and raceways, the cultivation vessel OPEX and interest costs make up for the highest fraction the total operating cost of the algae biofuels project. These costs are so high at the present that only with a dramatic change in the cultivation vessel architecture with next-generation low-cost study of MOC and an order of magnitude increase in algae productivities, combined with substantial reduction in OPEX and CAPEX, will algae biofuels will be commercially viable.

Narala et al. has made a serious attempt to compare three types of cultivation systems: PBR, open raceways and a hydrid system (combining algae cultivation with stimulating lipid accumulation) (Narala et al. 2016). The outcome of their study suggests that neither PBR nor open raceway, but rather the hybrid cultivation system, is a much more promising cultivation system for the mass production of lipid-rich algae from a technical standpoint (the techno-economic analysis is pending). The hybrid system's ability to separate algae growth and lipid induction phases enables the system to run the two phases simultaneously, while at the same time reducing the probability of contamination.

Borowitzka et al. has studied the commercial production of microalgae using four different cultivation systems: ponds, tanks, tubes and fermenters (Borowitzka 1999). Closed cultivation vessels (PBRs), according to them, have the most promising future because many of the new algae and algae products must be grown in an environment that is free from contaminating microbes, toxins and metal residues, and closed systems represent the best bet for this. That said, the current CAPEX- and OPEX-intensive PBRs are a bottleneck that needs to be overcome by developing technology that addresses the interplay of light availability, mixing requirement and ultra-low-cost MOC for PBRs.

Jorquera et al. conducted a comparative energy life-cycle analysis of microalgal biomass production in open ponds and photobioreactors [tubular and flat-plate PBRs] (Jorquera et al. 2010). The outcome of their study suggested that net energy ratio [NER: the ratio of the total energy produced (in this case, the energy value of the lipid and lipid-free biomass) to total energy incurred (all inclusive) to produce biomass or lipid or both] for tubular PBR <1, strongly suggesting that this option is economically unviable. The study further suggested the NER for flat PBR and raceways is >1 and could be further enhanced if the lipid content of the biomass was taken up to 60% (m/m). Some strains have 60% or even higher lipid content; however, they generally lack other desirable trails like growth, robustness and sturdiness, to name a few. So generally, it is extremely difficult to find algae strains that score high in each desirable trait. Additionally, the authors did not consider the cost of algae–water separation, that is, harvesting and oil extraction or hydrothermal liquefaction (or equivalent technology) in the NER calculation, which could be the reason why NER values for flat-panel PBRs and raceways appear >1. The authors further conclude that none of the systems compared in the study could compete with conventional non-renewable crude at the present cost.

Richardson et al., in their paper (Richardson, Johnson, and Outlaw 2012), have used a Monte Carlo financial feasibility model recursively over several years to estimate the algae production cost and the probability of commercial success for large-scale algae production facilities operating either PBRs or raceways. Modelling results show that open raceways have much higher financial feasibility than PBRs. For the base case scenario, for both PBRs and open raceways, the probability of economic success was zero. Modelling further showed that the only way to stack the odds for a 95% chance of economic success for the PBR was to reduce the CAPEX by ≥80% and OPEX by ≥90%; similar values for the raceway case stood at CAPEX reduction by ≥60% and OPEX by ≥90%. Thus, the raceway assured a more favourable outcome by a small reduction in CAPEX, if not OPEX.

Roselet et al. conducted a comparison of the performance of fully open raceways and semi-enclosed raceways constructed inside a greenhouse for marine algae to determine the best option for the cultivation of algae in massive algae production facilities in sub-tropical and temperate latitudes (Roselet et al. 2013). The outcome of their study shows that during winters, when temperatures drop, indoor raceways show better productivity than outdoor ones, because the temperature inside the greenhouse is 2 to 4 °C higher than outside. There are also advantages in cultivating inside during the rainy season with lower evaporation losses during summers. However, the authors do not present a detailed economic analysis of the trade-off between the higher production under the semi-controlled (greenhouse) environment and the cost penalty of creating the indoor infrastructure. Considerable work is required to arrive at a logical conclusion.

Thus, clearly, several attempts have been made to establish the most sustainable approach to microalgae cultivation; however, the outcomes are conflicting, and there is no clear path forward. Various research groups have shown very different results, and their opinions on the most optimistic path towards commercially viable algae cultivation or production technology development diverge widely. Also, almost no study available in the published literature is wholistic in its approach and considers all possible scenarios with balanced assumption when required. Thus, significant time, attention, and R&D effort will have to be dedicated for a significant number of years to develop the fundamental knowledge and validation data to reach a position where disruptive cultivation technology can be eventually developed at fraction of the present OPEX and CAPEX to make industrial-scale algae cultivation sustainable. Another fact that comes out from the published literature is that, although almost no group considers all possible scenarios, each one has in-depth data and details for at least a few options. It would be prudent to overlap these data sets to develop a more comprehensive and holistic data set to provide non-intuitive insights. Thus, information sharing becomes a critical catalyst here.

9.5 HARVESTING

9.5.1 CHALLENGE AND THE SUSTAINABILITY ISSUE

Throughout the history of algae production, parallels have been drawn between crop farming and algae cultivation. Thus, just like when the crop is mature and it's time to remove the produce from the farm, this is called crop harvesting, similarly, when the algae culture in a raceway or a PBR has achieved the desired cell concentration or cell density, it's time to separate the algae from the water, a process commonly referred to and accepted as algae harvesting. Unlike vegetable or grain or fruit farming, where the produce occurs as discrete entities, which, when plucked, can be directly sold and used, in case of unicellular autotrophic algae, this is not the case. Algae does not grow as discrete bodies that can simply be plucked or harvested from the raceway or PBR, as it grows as a dilute, uniform, stable suspension of ultra-low or low cell concentration. Thus, when one refers to harvesting in the context of heterotrophic, unicellular, stably suspended algae culture, it basically means a set of processes and strategies deployed to separate water from algae, with the ultimate objective of concentrating the algal biomass by progressively increasing the algae mass fraction. Because heterotrophic unicellular algae grow in very dilute stable suspension [open raceway: 0.25 to 1 g/L (average concentration of 0.3 to 0.4 g/L) and PBR: 2 to 9 g/L (Schlagermann et al. 2012)], the process of harvesting is costly; challenging; and chemical, energy and process intensive, thus making it unsustainable as it stands today. Basically, from a sustainability standpoint a lot more energy goes into concentrating algae than the final amount that comes out of it per unit energy expenditure, thus raising the sustainability question. During the course of the discussion in this section, it will become amply clear as to why currently there are issues with sustainability in the industrial-scale harvesting of microalgae. Having said that, it must also be stressed that significant global effort is currently underway by the scientific community to make the processes more and more sustainable. It is not an easy and straightforward path, but with the current pace of progress (as will be discussed later), it is not at all impossible.

9.5.2 INTRODUCTION

As explained in the previous section, harvesting in the context of microalgae is basically the removal of water from an algae suspension to concentrate the algae fraction. Not only are there several processes, but there are multiple unit operation options within each such process to conduct harvesting. Thus, several permutations and combinations are possible to bring about the concentration of algae cells.

Conventionally, harvesting has been divided into two stages (not types—there are various types within each stage): primary harvesting and secondary harvesting (Show and Lee 2014; Fasaei et al. 2018; Mariam Al, Abdel, and Amal 2015). This segregation is primarily based on concentration, but in practice it has more significance. For example, typically, certain unit operations are more suitable in certain concentration ranges or are economical only when operated to affect a certain fold change in concentration, and this is because cell concentration plays an important role in the performance of a unit operation. There is no rule that exists or a formal convention that applies to this segregation. Conventionally, when virgin algae culture is concentrated to 1% to 6% algae solids, this is called primary harvesting (Bilad et al. 2012; Tran et al. 2013; Shelef, Sukenik, and Green 1984; Eckelberry 2013; Xu et al. 2011; Sims et al. 2014; Hu et al. 2013). The inlet to primary harvesting is the virgin PBR or raceway culture in the concentration range of 0.02% to 1.0% algae solids—at this algae concentration, the algae culture basically behaves like Newtonian fluid. The product coming out of a primary harvesting unit operation is somewhat Newtonian in the lower concentration range and becomes non-Newtonian towards the middle or latter concentration range. At this stage, the algae are not in suspension form anymore, and this is typically referred to as slurry. The slurry coming out from the primary harvesting (1% to 6% algae solids) acts as feed for secondary harvesting. Secondary harvesting is very commonly referred to as dewatering, although some authors make a distinction. In this chapter, secondary harvesting and dewatering will be used interchangeably. The output from secondary harvesting can range anywhere from 10% to 30% algae solids (Gerardo, Oatley-Radcliffe, and Lovitt 2014; Show, Lee, and Chang 2013; Sharma et al. 2013). Towards the lower end of the concentration spectrum, the concentrated algae behaves like a non-Newtonian slurry, whereas towards the middle of the concentration range, the consistency is like paste, and towards the end of the concertation range (sometime even in the middle—this depends on the pre-treatment before dewatering), the consistency is like cake.

9.5.3   Primary Harvesting

Primary harvesting can be classified into three main categories: chemical, physical and electrochemical. Now the mode of harvesting one selects can be purely chemical, purely physical, purely electrochemical or any combination hereof. Different primary harvesting techniques are depicted in Figure 9.3. For the purpose of this

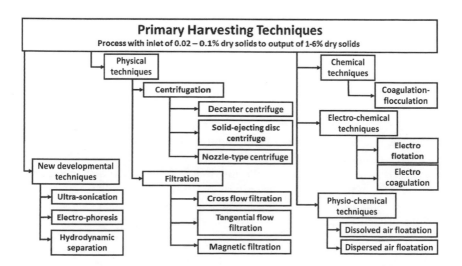

FIGURE 9.3   Different primary harvesting techniques used in the literature.

Source: Shelef, Sukenik, and Green (1984); Sims et al. (2014); Hu et al. (2013); Mariam Al, Abdel, and Amal (2015); Show, Lee, and Mujumdar (2015); Show and Lee (2014)

section, chemical and electrochemical primary harvesting will be grouped together and physical harvesting will be discussed separately.

9.5.3.1   Primary Harvesting—Chemical

Algae in water maintains a negative surface charge, and the repulsion between algae cells stabilizes the algae suspension (Ives 1959; Liu, Zhu, et al. 2013; Gonçalves et al. 2015). The main challenge for chemical-based harvesting is to break the stability of the suspension at the minimal chemical dosage (Das et al. 2016; Garzon-Sanabria, Davis, and Nikolov 2012; Zhang et al. 2016). This is typically done by adding short-to-medium-molecular-weight cationic coagulants or metal salt–based coagulants (for example, alum) in the algae suspension under constant agitation. The coagulant's positively charged polymer chains/metal atoms neutralize the negative surface charge of the algae and impart a very slight residual positive surface charge (Kim et al. 2013). Neutralization of the algae's surface charge drastically reduces the mutual repulsion between algae, thus allowing algae to come to a close enough distance to each other that short-range attractive forces (van der Waals forces) kick in and cause the algae cells to adhere to each other (Kim et al. 2013; Vandamme, Foubert, and Muylaert 2013). This causes destabilization of the algae suspension (breakage of cell suspension stability), which results in the formation of small agglomerates of algae cells, referred to as pin flocs or coagulas—the process up to this stage is called coagulation (Branyikova et al. 2018). At this point very high-molecular-weight polymers called flocculants are added to the de-stabilized algae suspension. The strong negative charge centres on polymers, attracting the slight positive charge of the de-stabilized algae cells, which are engulfed into the polymer chains, resulting in the formation of large flocs, sometime up to few millimetres in size—the process up to this stage post-coagulation is called flocculation. In the literature, on some occasions, the terms coagulation and flocculation are used interchangeably; however, it is important to understand that they are distinct in function and end result. The function of coagulant is limited to breaking the suspension stability, resulting in the formation of coagulas or pin flocs, whereas the function of the flocculation step is to accelerate the rate of settling of algae by increasing the size and density of the small aggregates formed post-coagulation into larger flocs by polymer binding.

Once coagulation and flocculation are completed, there are two main options: the flocs either can be allowed to settle under gravity or can be floated. Typically for settling, settling tanks (clarifiers) are used, and for the flotation technique DAF is used. Within each of these techniques, several options are available, but in general they operate on the same principle as will be explained next.

Clarifiers are basically settling tanks where flocs settle under the force of gravity and accumulate at the bottom in the conical section of the tank. The initial settling is pretty fast, and once the flocs form a settled layer, they are compressed under the hydraulic load of the water column above it. Once the desired biomass concentration is achieved, concentrated slurry is drained from the bottom of the tank. This bleeding of slurry can be continuous (if there is a high degree of automation) or intermittent (if the operation is more or less manual). Various designs are available for the clarifier tank, depending on the nature of the influent, desired product concentration and throughput.

DAF is a very common unit operation in water and wastewater treatment. Particles with low density, such as algae, pose a significant challenge (although some algae species settle out very easily post-flocculation, so it is species specific) to separation by traditional sedimentation methods, even after flocculation (because of the fractal nature of flocs). In this scenario DAF has been reported to be especially useful for algae and other bioparticle removal (Wiley, Brenneman, and Jacobson 2009; Haarhoff and Edzwald 2004; Han 2002; Pieterse and Cloot 1997; Henderson, Parsons, and Jefferson 2009; Hanotu, Bandulasena, and Zimmerman 2012). DAF exploits algae's natural tendency to float and the very low-density flocs that form on coagulation/flocculation (Haarhoff and Edzwald 2004). In DAF, microscopic bubbles are used to bring about the flotation of algae particles. Bubbles in DAF are produced by saturating a recycled clarified water stream with air at high pressure and eventually depressurizing the stream by releasing it at atmospheric pressure. The resulting bubbles adhere to the algae particles' surface, raising them to the surface. The efficiency of the flotation process depends on the effective collision and attachment of bubbles and flocs, which is achieved by coagulating-flocculating influent particles/cells to increase their size and decrease their negative charge, respectively (Haarhoff and Edzwald 2004; Han 2002). Bubbles are strongly negatively charged, and thus coagulation minimizes the repulsive effects between the particles/cells and bubbles, thus facilitating bubble attachment.

Many other possibilities can be applied pre-/post-coagulation-flocculation to further enhance the cell recovery at even lower chemical addition. In any chemical harvesting process, the initial steps are always the same: coagulation followed by flocculation. In the case of some algae species that are prone to forming larger flocs, sometimes only coagulation may be sufficient.

9.5.3.2 Primary Harvesting—Electro-chemical

Electrolysis is a process in which oxidation and reduction reactions take place when an electric current is applied to an electrolytic solution. Electrocoagulation exploits the process of electrolysis and is based on dissolution of the electrode material used as an anode. When the potential difference is applied across the electrodes, the anode (commonly referred to as the sacrificial anode) produces cations, which act as coagulating agents. A typical electrocoagulation setup consists of an anode and a cathode made of metal plates, which are both submerged in the electrolytic solution being treated. All the reactions that take place during a typical electrocoagulation process have not really been fully understood because they are strongly dependent on the nature of the solution in which the electrocoagulation will take place. Electrocoagulation can be envisioned as an accelerated corrosion process, and most illustrations in the literature with regard to the reactions taking place during the electrocoagulation process are an approximate extension of the chemical reactions that take place during chemical corrosion (Garzon-Sanabria, Davis, and Nikolov 2012; Kim et al. 2012; Vandamme et al. 2011; Chen 2004).

Electrocoagulation is an electrochemical method which involves the interaction of suspended particles in water to electric fields and an electrical-induced oxidation and reduction reaction where sacrificial anodes corrode to release reactive coagulant precursors (such as iron or aluminium or any other metal ion, depending on the type of electrode used), which are basically metal cations, into the solution. At the cathode, gas evolves (usually as hydrogen bubbles) accompanying electrolytic reactions. The metal cations released at the anode mask the negative surface charge on the suspended particles and restrict their ability to repel each other. Once these suspended particles cannot repel, they tend to adhere to each other, and as a result the suspension stability is broken. The destabilization of the stable suspension leads to the aggregation of smaller particles into larger particles, which can no longer remain suspended and settle under the influence of gravity.

Electrocoagulation and chemical coagulation may appear similar in terms of what they set out to achieve or even how the chemical reaction between a metal ion and algae takes place; however, the critical difference lies in how the metal ion is introduced into the coagulation system (i.e., the culture). Whereas in chemical coagulation, the metal is introduced into the solution phase by dissociation of the metal salt into its constituent ions, in electrocoagulation, metal is introduced by electrolytic corrosion of the metal rod. Thus, unlike chemical coagulation, in electrocoagulation the accompanying anion is not introduced into the system, which is advantageous because it reduces inorganic addition, which in turn is helpful for improving product and co-product specifications. Even if the CBO is the objective, less inorganic material is very helpful during oil upgradation or catalytic HTL. Additionally, one does not have to pay for the anion of the salt; one just pays for the metal cation in electrocoagulation; however, the cost of electricity needs to be added to the overall electrocoagulation cost.

After electrocoagulation, the coagulation step is finished and the suspension stability has been broken. Now, depending on the type of algae, flocculant may be required in the same quantity as is required after chemical coagulation or in lower amounts. This is because, during electrocoagulation gas is liberated (as explained earlier) and thus sometimes it acts in a way to levitate the small algae coagulas post-coagulation (a process commonly referred to as electroflotation), although at other times it doesn't. For micro-algae, under most circumstances flocculant will be required at least in small dosages.

9.5.3.3 Semantics

Different research groups report coagulant dosage or loading in different units, for different biomass concentration and variable biomass recoveries. It is very important to understand how this affects the interpretation of results and what units will be used in this discussion. Because the amount of coagulant required depends on the type of algae, algae surface charge density, cell concentration, total suspended solids [TSS] and matrix conditions (pH, salinity alkalinity, total organic carbon [TOC] (Hulatt and Thomas 2010; Nguyen et al. 2005), natural organic matter [NOM] (Henderson et al. 2008; Henderson, Parsons, and Jefferson 2010; Pivokonsky, Kloucek, and Pivokonska 2006; Her et al. 2004), algogenic organic matter [AOM] (Henderson et al. 2008; Her et al. 2004; Pivokonska and Pivokonsky 2007; Kim and Yu 2005; Chow et al. 1999; Henderson, Parsons, and Jefferson 2008b) and total organic matter [TOM]), it is more sensible to report dosages as coagulant dosage per unit harvested biomass along with cell recoveries—this will be called C_D/X_{HB}. This is calculated by dividing the coagulant dosed per unit culture volume by the product of biomass concentration and harvesting efficiency (biomass recovery). What this means is that C_D/X_{HB} is the coagulant dosage normalized with the actual amount of biomass harvested. Although C_D/X_{HB} will take care of the variability in biomass concentration reported by a particular research group, to make the comparison truly apple to apple, the biomass recovery must also be taken into consideration. This is because the correlation between C_D/X_{HB} and biomass recovery is not linear: what this means is that the C_D/X_{HB} at lower cell recoveries is not the same at higher cell recoveries.

Consider an example to calculate C_D/X_{HB}: Let's assume that there is a culture of algae having a cell concentration of 1000 mg/L (on AFDW [Ash Free Dry Weight] basis). Consider that it takes a dosage of 100 mg/L of any coagulant X to coagulate algae in the culture and that at the end of coagulation process, 90 % of the initial algae is separated from water (i.e., biomass recovery or harvesting efficiency). For this case, C_D/X_{HB} will be calculated as 100 (mg/L) / [1000 (mg/L) * (90/100)] = 0.1111 (mass/mass [m/m]) = 11.11 % (m/m). If biomass recovery or harvesting efficiency of 90% would not be taken into consideration, coagulant loading will be calculated as 100 (mg/L) / 1000 (mg/L) = 0.10 (mass/mass [m/m]) = 10.0 % (m/m). Thus, clearly C_D/X_{HB} eliminates under-reporting of coagulant loading by taking into consideration the biomass recovery or harvesting efficiency.

Consider another example which demonstrates how coagulant loading reported as coagulant dosage per unit harvested biomass [Harvested Biomass = biomass concentration * harvesting efficiency] i.e., (C_D/X_{HB}) takes care of the variable biomass concentration and harvesting efficiencies reported by different authors/research groups. Consider that three algae research groups have culture of algae having a cell concentration of 1000 mg/L, 500 mg/L and 100 mg/L, respectively (on AFDW [Ash Free Dry Weight] basis). Further assume that it takes a dosage of 100 mg/L of any coagulant X to coagulate algae in the algae culture for each group and that at the end of coagulation process, 90 % of the initial algae is separated from water (i.e., biomass recovery or harvesting efficiency). Now if each of these research groups report only the absolute coagulant dosage (100 mg/L), it would appear that the coagulant loading is the same. If these dosage values are normalized with the product of biomass concentration and harvesting efficiency, and reported it as coagulant loading (C_D/X_{HB}) rather that absolute coagulant dosage, the dosages would be very different for each of the three cases (i.e., [100/(1000*0.9) = 0.1111], [100/(500*0.9) = 0.2222] and [100/(100*0.9) = 1.1111]). Consider another variation of the same example in which the biomass concentration of the algae culture of each research group is identical (1000 mg/L) and so is the coagulant dosage (100 mg/L). However, assume that the harvesting efficiency of each group is different and stands at 90 %, 80 % and 70 %, respectively. Now if coagulant dosage is normalized only with the biomass concentration, then the coagulant loading [= 100 / 1000 = 0.1] will appear identical in case. However, if the coagulant dosage is normalized with the product of biomass concentration and harvesting efficiency, the result of coagulant loading (C_D/X_{HB}) would appear vastly different for the three groups (i.e., [100/(1000*0.9) = 0.1111], [100/(1000*0.8) = 0.125] and [100/(1000*0.7) = 0.1429]).

The examples cited above clearly illustrates why it is essential to report coagulant dosage as coagulant loading (C_D/X_{HB}) and how such reporting method would bring about uniformity in the way different research groups report coagulant dosing values and will allow meaningful and logical comparison between published data.

9.5.3.4   Sustainability Issues with Chemical Coagulation-Flocculation

Core Issue

There are couple of issues with chemical-based harvesting which raise the sustainability question. These issues include the high dosages of chemicals required for coagulation and flocculation; the raw materials for most of these chemicals come from the mining and petrochemical industry, which are non-renewable in nature; the high cost of harvesting chemicals; and the extremely high variability of dosages as a function of the type of algae and the cultivation medium matrix (i.e., the matrix effect). These issues get even more amplified when handling marine algae with a sea water matrix. High coagulant dosages observed in a sea water matrix poses new challenges related to product specifications—the high metal content coming from metal salt–based coagulants is generally not acceptable in either the bio-fuels process chain or other products or co-products meant for animal or human application. With the availability of fresh water reaching critically low levels for even drinking and agriculture, it would be not at all prudent to develop an algae industry that depends on fresh water. Many industries have understood this and have shifted their focus to sea water-based production of algae—at least for open raceway-based cultivation. In view of the same, most of the discussion that will follow will revolve around marine algae.

Literature

Marine algae coagulant loading expressed as % C_D/X_{HB} has been reported in wide ranges, from low loads, such as 1% (Bayat Tork, Khalilzadeh, and Kouchakzadeh 2017), 3% to 4% (Eldridge, Hill, and Gladman 2012), 0.2% to 5% (Roselet et al. 2015) and 5% (Garzon-Sanabria, Davis, and Nikolov 2012), to medium to high loadings, such as 22% to 71% (Eldridge, Hill, and Gladman 2012), 35% (Şirin et al. 2012), 22.97% (Baharuddin et al. 2016), 58% to 127% (Sukenik, Bilanovic, and Shelef 1988) and 9% to 34% (Zhang et al. 2016). Fasaei et al. has suggested that for dilute cultures (~500 mg/L) cultivated in open raceway systems, the operational cost for harvesting and dewatering should be in the range of 0.6 $/kg to 2.3 $/kg of harvested biomass (Fasaei et al. 2018). Slade at al. suggests that present best-case algae biomass production costs for open raceway–based cultivation is in the range of 1.8 $/kg to 2.05 $/kg and makes best-case biomass production projections for the future at 0.34 $/kg to 0.46 $/kg (Slade and Bauen 2013). Another idealized projected cost suggested for biomass production was in the range of 0.132 $/kg to 0.137 $/kg (Mohammed 2012). Norsker et al. suggest that as per the current best-case scenario, the production cost for algae biomass for the open raceways, horizontal tubular PBR and flat-panel PBR are 5.63, 4.72 and 6.78 $/kg, respectively (Norsker et al. 2011). Thus, based on the data presented, it seems that for coagulation-flocculation to fit in the overall cost outlay for harvesting and from an algae biofuel perspective, coagulant loading of ≤5% (w/w) is desirable for marine algae cultivated in natural sea water. At the same time, from a cost perspective, coagulant cost <0.25$/kg would be necessary to keep the economics under control.

If one were to analyse the coagulant loading results presented earlier in view of the desired coagulant loading levels, many interesting facts come to light. Coagulant loading levels demonstrated in the literature, such as 22% to 71% (Eldridge, Hill, and Gladman 2012), 35% (Şirin et al. 2012), 22.97% (Baharuddin et al. 2016), 58% to 127% (Sukenik, Bilanovic, and Shelef 1988) and 9% to 34% (Zhang et al. 2016), are just not viable. Even when low coagulant loading has been demonstrated, such as 1% (Bayat Tork, Khalilzadeh, and Kouchakzadeh 2017), 3% to 4% (Eldridge, Hill, and Gladman 2012), 0.2% to 5% (Roselet et al. 2015) and 5% (Garzon-Sanabria, Davis, and Nikolov 2012), it has not been demonstrated under realistic matrix conditions. A realistic matrix condition would be natural sea water, with salinity in the range of 4% to 6%, total dissolved solids (TDS) in the range of 40,000 to 65,000 mg/L and the presence of metabolites released in the medium during the growth cycle of the organism. However, almost all of the studies that have shown reasonably low coagulant loading of <5% (% C_D/X_{HB}) have not used a realistic matrix condition, as described earlier. For example, (Bayat Tork, Khalilzadeh, and Kouchakzadeh 2017) have shown with a C_D/X_{HB} of 1%, although a marine water strain was used, the algae was cultivated in fresh water using Bold's basal medium—and in fresh water, coagulant dosages are typically far lower than observed for a sea water matrix. Further, they used cationic starch nanoparticles, which just cannot meet the desired harvesting cost of 0.25$/kg because of the high synthesis cost. (Eldridge, Hill, and Gladman 2012) have demonstrated C_D/X_{HB} in the range of 3% to 4%; however, algae was cultivated in artificial sea water formulated with just 30 g/L salt, which is bound to suppress the coagulant requirement. The authors used iron sulphate as a coagulant, which is a reasonably low-value chemical, and if dosages can be shown to be <5% (C_D/X_{HB}) in a real sea water matrix, iron sulphate could provide a feasible option. Roselet et al. have shown C_D/X_{HB} in the range of 0.2% to 5% for *C. vulgaris* and *N. oculata* (Roselet et al. 2015). These dosages are rather low for a sea water matrix; however, this experiment cannot be considered a valid experiment to demonstrate coagulant dosage in a natural growth matrix because the cells were centrifuged and resuspended in fresh medium. When the actual matrix is removed, all the extracellular matter and dissolved organic matter secreted by the cells and natural organic matter and salt buildup in the medium during a growth cycle are removed from the system. Because these components consume a part of the coagulant, if they are completely stripped off, then coagulant dosages are bound to reduce. Garzon et al. have shown harvesting of unwashed *N. oculate* using $AlCl_3$ (Garzon-Sanabria, Davis, and Nikolov 2012). They reported that 100 mg/L $AlCl_3$ was dosed at pH 5.3 in a 2000 mg/L culture with 95% biomass recovery. Initial calculations show a C_D/X_{HB} of only 5.3%. However, the dosage was low because the matrix pH was significantly reduced to 5.3—at lower pH metal salts tend to require lower dosages. However, the HCl loading necessary for lowering the pH is 17 g (of 37% strength) / L of the culture—if the acid loading is considered and the corrected C_D/X_{HB} is calculated, it amounts to a whopping 900% value. Thus, there is cost of the acid required for pH modulation that must be taken into consideration while calculating the total OPEX for chemical harvesting. The authors have also stated that salinity has no effect on the coagulant demand; however, this is not possible because when salt is added, it ionizes to produce negative ions, and then when metal salts are added into such a solution, a part of the metal cation is used to neutralize these negatively charged ions. A metal coagulant cannot discriminate between a negatively charged algae surface and a negative ion of the salt. From the literature stated and the discussion, it is very clear that under the tough matrix conditions of natural sea water containing metabolic products and additional components, it has been a real challenge to bring down the coagulant dosages below 5% (C_D/X_{HB}).

Zhang et al. claim that compared to Fe^{3+}, Al^{+3} or chitosan harvesting of *Chlorella zofingiensis* was both most effective and the dosage required was the least when Mg^{2+} was used (Zhang et al. 2016). Chemical coagulation followed by DAF was used for harvesting. Despite Mg^{2+} being the best cation, the C_D/X_{HB} for Mg^{2+} was 22.6%, which is higher than the sustainable value suggested earlier. The authors have also shown that if the growth medium already contains Mg^{2+}, then basification of the culture in the range of 10.8 to 11.8 would precipitate Mg^{2+} and result in coagulation—the suggestion made here is that coagulant addition is completely eliminated. That may be so; however, base is still being added, and it has a cost. Although the authors did not mention the dosages for basification, it could be in the range of 40% to 70% (mass of base to mass of biomass), assuming that the cultivation pH is in the range of 6.5 to 7.5. This high base loading of NaOH itself makes the process economically challenging.

Key Takeaways

Current coagulant dosages under realistic matrix conditions are simply unsustainable from both the chemical input and cost perspective. It must be stressed that it is not sufficient to only bring down the cost—it is essential that even the dosages of these inorganic metal-based coagulants are brought down to meet end-product specifications, as well as to have a lower life cycle impact and reduction of resource use, and ultimately to boost sustainability. Researchers have tried an alternative approach to eliminate inorganic coagulants altogether to enhance product specifications by using several different types of natural or herbal components or their derivatives as coagulating and flocculating agents (Şirin et al. 2012; Farid et al. 2013; Baharuddin et al. 2016; Roselet et al. 2015; Fuad et al. 2018; Bayat Tork, Khalilzadeh, and Kouchakzadeh 2017; Roselet, Burkert, and Abreu 2016; Zheng et al. 2012). In spite of the fact that in most studies the dosages were high or the cost of the alternative herbal or natural chemistry was significant, one thing was clear: options do exist and performance criterion were met in some cases. Sustained and concentrated effort in this direction may eventually result in a novel chemistry that is both economical and sustainable.

9.5.3.5 Sustainability Issues with Electro-chemical Coagulation

Recently electro-chemical coagulation has emerged as one of the most promising technologies for primary algae harvesting (Gao et al. 2010; Kabdaşlı et al. 2012; Mouedhen et al. 2008; Khemis et al. 2006; Zhou et al. 2016; Marrone et al. 2018; Lee, Lewis, and Ashman 2013; Fayad et al. 2017; Zhang et al. 2015; Zenouzi et al.

2013; Bayat Tork, Khalilzadeh, and Kouchakzadeh 2017; Harif, Khai, and Adin 2012; Souza et al. 2016; Bleeke et al. 2015; Raut, Panwar, and Vaidya 2015). One must be careful not to confuse electro-chemical coagulation with either electro-flotation or electro-flocculation. Whereas electro-chemical coagulation used two reactive metal electrodes to facilitate release of metal ions into the solution phase, electro-flotation consists of a reactive metal anode and an inactive metal cathode, resulting in both, creation of bubbles by electrolysis of water and release of metal ion from cathode (Marrone et al. 2018). On the other hand, electro-flocculation consists of both an inert cathode and anode, and when the potential difference is applied, negatively charged algae is drawn towards the positively charged anode and eventually loses its negative charge, which causes them to form flocs without any metal species liberation into the solution phase (Marrone et al. 2018).

Advantages of Electro-chemical Coagulation

Electro-chemical coagulation has been reported to be very effective in the acidic to neutral pH range from pH 5.0 to pH 7.5. This is a very advantageous situation because most commercial marine microalgae cultivation happens in the pH range of 6.0 to 8.0, and thus this will eliminate the need for pH adjustment and hence result in cost savings (Fayad et al. 2017; Gao et al. 2010; Harif, Khai, and Adin 2012; Souza et al. 2016; Zhang et al. 2015). Another advantage of performing electro-chemical coagulation in the acidic to near-neutral pH range rather than basic range is that in the former pH range, the mode of action of floc formation seems to be charge neutralization, whereas in the latter it is sweep flocculation (Vandamme et al. 2011; Fayad et al. 2017; Gao et al. 2010). Thus, in the former case, less inorganic salts are precipitated with algae, resulting in less sludge volume and lower ash content in the algae slurry.

Several metals (iron, aluminium, stainless steel (SS), magnesium, copper, etc.) were also explored for EC, and it turns out that aluminium and iron seem to be the two best metals for the job, with aluminium being the best (Gao et al. 2010; Bleeke et al. 2015; Marrone et al. 2018). Considering the appropriate weightage for cost vs. efficiency, iron would be a more appropriate choice for the electrode, specifically because of its comparatively much lower cost than the aluminium electrode. Some research groups have suggested that when the same metal used as a metal salt in chemical coagulation is used as an electrode in electro-chemical coagulation, the metal consumption for the same biomass recovery decreases noticeably. This is a desirable characteristic, as it will reduce the metal content or ash content in the final biomass and thus improve product specifications. Vandamme et al. observed that when aluminium was used as an electrode during electro-chemical coagulation, the Al consumption was lower, compared to the scenario when Al-based alum was used during chemical coagulation (Vandamme et al. 2011). Similarly, Souza et al. noted that high biomass recovery efficiencies (in excess of 98%) could be attained at very low metal dosage using electro-chemical coagulation and that the pH of the post-coagulated culture was only very slightly altered (another advantage) (Souza et al. 2016). An important fact to consider is that one of the few advantages the sea water matrix offers is with electro-chemical coagulation. When a potential difference is applied across a pair of electrodes, the amount of current that would flow through the suspension will also strongly depend on the conductivity of the suspension. Conductivity is determined by the concentration of ions coming from the ionized salts in the suspension. Because sea water has a very high concentration of salts (TDS is always >35,000 ppm), it is highly conducting and thus offers much lower resistance to the passage of electricity when compared to fresh water or distilled water. Thus, for the same applied potential difference, the flow of electricity is higher in sea water than fresh water and the power requirement is reduced.

Several authors also investigated the correlation between applied potential difference, current density, time for coagulation, degree of biomass recovery and power consumption. The most common trend that most researchers have observed is that at lower current densities (i.e., at lower applied potential difference), although the rate of algae coagulation is slow, it is more energy efficient compared to a scenario when the current density is kept higher to achieve a higher rate of suspension destabilization (Vandamme et al. 2011; Fayad et al. 2017; Gao et al. 2010; Souza et al. 2016; Bleeke et al. 2015). This may be due to the fact that at lower metal discharge rates, the algae–metal interaction gets sufficient time and thus this helps in reducing the wastage of liberated metal. Another factor that can help to reduce the cost of electrochemical coagulation is to keep the inter-electrode distance as low as possible (Fayad et al. 2017). Reduction in the inter-electrode distance reduces the resistance to flow of current between the electrodes and hence reduces the power drawn per unit metal dissociation or liberation, and hence reduces the cost of operation.

Different research groups have arrived at different energy numbers and hence cost numbers for operating an electro-chemical coagulation unit. Vandamme et al. propose that under ideal conditions, power consumption for electro-chemical coagulation should be 2 KWh/kg of biomass for fresh water algae and about 0.3 KWh/kg for marine algae (Vandamme et al. 2011). Clearly, the salinity of the sea water matrix has a significant impact on power consumption, as indicated earlier. This power consumption is far lower than what is conventionally assumed for centrifugation: around 16 KWh/kg (Vandamme et al. 2011; Danquah et al. 2009). Rodriguez et al. estimate the energy demand per m^3 of suspension processed to be twice that for chemical coagulation; however, the lower metal demand of electro-chemical coagulation puts the total coagulation cost for electro-chemical coagulation at about half that of chemical coagulation (Rodriguez et al. 2007). Similarly Lee et al. estimate the total cost of electro-chemical coagulation (including capital depreciation, electrode cost, power consumption) to 0.19 $/kg of the harvested biomass on an ash free dry weight (AFDW) basis (Lee, Lewis, and Ashman 2013), which is much lower than centrifugation [2.58 $/kg (F. Mohn 1988)], coagulation–flocculation–settling [0.71 to 1.83 $/kg (F. Mohn 1988; Benemann and Oswald 1996)] or even coagulation–flocculation–flotation [0.91 to 1.39 $/kg (Benemann and Oswald 1996)]. Fayad et al. showed that the power consumption for electro-chemical coagulation is about 5.3 KWh/kg, but there was possibility of bringing this down to 1 KWh/kg (Fayad et al. 2017). Assuming electricity cost of 0.1 $/KWh, this would work out to be 0.1 $/kg. Gao et al. have shown energy consumption for complete algae removal by electro-chemical coagulation at around 0.29 KWh/m^3 (Gao et al. 2010). Assuming a biomass concentration of 0.5 kg/m^3 and electricity cost of 0.1 $/KWh, the cost for algae removal by electro-chemical coagulation would work out to 0.058 $/kg. Souza et al. arrived at a figure of 0.10 KWh/m^3

for complete algae removal by electro-chemical coagulation (Souza et al. 2016). Assuming biomass concentration of 0.5 kg/m³ and electricity cost of 0.1 $/KWh, the cost for algae removal by electro-chemical coagulation would work out to 0.02 $/kg. Marrone et al. reports power requirement for electro-chemical coagulation as 3 KWh/m³ of culture, compared to about 16 KWh/m³ for centrifuge (Marrone et al. 2018). Zhang et al. also suggested that electro-chemical coagulation energy consumption is just 0.61 KWh/kg, which translates to 0.061 $/kg (assuming electricity cost of 0.1 $/KWh) (Zhang et al. 2015).

Major Challenges of Electro-chemical Coagulation

Clearly it seems from the data presented by different authors that electro-chemical coagulation is energetically much more economical than chemical coagulation. However, one question that needs to be answered is that if everything about electro-chemical coagulation is so energetically favourable, then why has it not been taken up in a big way? Electro-chemical coagulation, though energetically attractive, has its own set of issues which must be overcome before it can become a commercial success.

One of the critical steps in the economic evaluation of electro-chemical coagulation is electrical energy consumption. The amount of electric energy consumed for a given volume of algae culture in KWh/m³ can be calculated as follows (Vandamme et al. 2011):

$$EEC = \left(\frac{V \times I \times t_h}{U \times 1000} \right) \left[\frac{(V).(A).(h)}{m^3} \right] \qquad \text{Equation 9.1}$$

Where,

V = Voltage applied in V
I = Applied current in A
t_h = Time of application of current or the electrolysis time (h)
U = Volume of water treated in m³
EEC = Amount of electrical energy consumed in KWh/m³ of solution

From the equation it is clear that the current measurement is critical to calculating the power cost. However, none of the authors reviewed earlier in the discussion actually described how the current was measured. The current drawn through the system is correlated to applied voltage and system resistance. System resistance is directly proportional to the inter-electrode distance (m) and inversely proportional to the submerged surface area of the anode or the electrode area in contact with electrolyte (m²) and the system specific conductivity (S/m). The ratio of current to the submerged surface area of the anode is referred to as the current density. The current drawn through electrode setups can be controlled by the applied voltage, inter-electrode distance, electrolyte conductivity and submerged surface area of the anode. As the electrodes electrolytically dissociate to liberate metal ions during electro-chemical coagulation, the ion concentration in the solution changes and hence alters the conductivity. Also, over a period of time as the electrode dissociates, which results in a change in electrode area, the current density would change, and all this would result in a change in the resistance and hence the current drawn through the system. Thus, it seems that during the electro-chemical coagulation process, current drawn through the setup will continuously change, and so will the power being consumed. It doesn't seem like the different authors have taken this into consideration. If not, then the power estimations may have to be reworked.

Another issue with electro-chemical coagulation is the sheer complexity of the process with several interdependent variables. Thus, extensive optimization will be required to put the electro-chemical coagulation process into operation. Also, most of the experiments conducted by various authors to demonstrate the effectiveness of electro-chemical coagulation over chemical coagulation were done at either beaker scale or, at most, a couple of hundred litres. However, once electro-chemical coagulation is implemented at a large scale, several issues will come into play, such as how close the electrodes need to be placed, how far apart each pair will be and whether they will have to be attached in parallel or in series. There are very important questions that are still unanswered.

Thus, clearly the electro-chemical coagulation system is not well optimized and deployable yet; however, a sustained and systematic effort to optimize several operating variables like current, voltage, resistance, conductivity, electrode surface area, current density, electrode submerged surface area, inter-electrode distance, series Vs parallel operation, electrode placement and changing electrode area could result in viable and sustainable electro-chemical coagulation technology in the near future.

9.5.3.6 Physical Harvesting

Figure 9.3 specifies the various types of physical harvesting techniques that have been reported in the literature for the primary harvesting of algae. All the available physical harvesting techniques can be grouped into three main categories: filtration, centrifugation and developmental techniques such as ultrasonication and hydrodynamic separation (HDS). Both ultrasonication and HDS are in their nascent stages of development, and there is no commercial system in operation at any scale to date. There are only a couple of published reports on each technology with relation to application to algae harvesting (Prochazkova, Safarik, and Branyik 2013; Bosma et al. 2003; Hawkes and Coakley 1996; Hill and Harris 2008; Marrone et al. 2018; Zhang, Zhang, and Fan 2009; Bermúdez Menéndez et al. 2014; Sivaramakrishnan and Incharoensakdi 2018; Coons et al. 2014). Thus, there will be only a very brief discussion of these techniques—most of the discussion will be focused on filtration and centrifugation.

Ultrasonication/Ultrasound

This technique is based on the principle of creating a standing wave by superimposing forward and reverse propagating pressure waves in a suspension of algae and water. When algae cells interact with standing waves, primary and secondary radiation forces are created that push algae cells closer to each other, resulting in coagulation of algae (Coons et al. 2014). Marrone et al., over a course of several experiments, showed algae concertation of just 18-fold by ultrasonication (Marrone et al. 2018). Bosama et al. have similarly reported a biomass concentration not exceeding 20-fold (Bosma et al. 2003), and a concentration factor of 50 to 80 times is common with techniques like centrifugation or filtration. Electrical energy cost estimations provided by the author (Marrone et al. 2018) were not representative of the current technology or state of ultrasonication. Although some estimates made by Coons et al. show lower energy consumption than some of the other conventional technologies, such as centrifugation and filtration, there are no published data to support or validate these claims. An assessment made of the current state of the art of ultrasonication shows that currently the power consumption for operating a ultrasonicator for algae harvesting is extremely high when compared to conventional techniques (Bosma et al. 2003). Additionally, the forces in play in ultrasonication are short-range forces, and thus the sonification source has an acting range of only a few centimetres, at most; this will result in a significant spike in the hardware cost, resulting in high CAPEX.

Thus, although ultrasonification is, in principle, capable of providing a chemical-free and fast-acting harvesting solution that preserves the integrity of cells, the current state of art of the technology makes it prohibitively expensive for any kind of practical commercial exploitation for some time to come.

Hydrodynamic Separation

A detailed technical description of all forces governing HDS is quite involved and beyond the scope of this chapter; thus, in this section a rather simplistic picture will be presented with a focus on practical applicability. Very simplistically speaking, HDS is a technique that utilizes customized fluid flow patterns, along with the channel geometry, which create inertial lift forces that focus suspended particles into a concentrated substream of particles in the bulk flow, thus achieving separation (Segré and Silberberg 1961). The concentrated particle stream can then be separated using a splitter at the end of the channel. Particle focusing takes place as a result of the net action of several hydrodynamic forces such as shear, drag and virtual mass, which enables size-dependent rather than density-dependent separation. This endows HDS with the ability to separate particles that are neutrally buoyant, that is, having density equal or very similar to the dispersant (for example, water). Thus, this technique appears to have potential application for the separation of neutrally buoyant algae from fresh water or sea water. The operating cost of HDS compared to filtration or centrifuge seems lower; however, it is more than offset by the cost of the enormous amount of material required to construct these channels and the associated steep cost of fabrication. This is because, in HDS, the channel dimensions must be of the same order of magnitude as the particle size for high-efficiency separation (Di Carlo et al. 2007). So, for cases such as unicellular microalgae, which are small in size (1 to 3 µm), the channel dimension is extremely small. Thus, a large number of small-diameter channels (high material and fabrication cost) will have to be used to achieve HDS, which will increase the cost of algae harvesting exponentially, rendering it currently non-viable for the primary harvesting of algae.

Filtration

Filtration is one of the most widely used techniques for the primary harvesting of algae. However, due to the small size of algae, only microfiltration (MF) or UF can be used for the separation of microalgae from the dispersant (i.e., water). MF for the harvesting of algae offers some excellent benefits: primary harvesting by MF is generally a single-step process, with no multi-chemical addition steps, like chemical harvesting; almost 100% biomass recovery is possible; no chemical is introduced into the system, thus eliminating any chance of residual toxicity in the biomass, which makes it possible to meet the highest and most stringent product specifications; and the harvesting process also cleans the process stream of other microbial loads and thus recyclability of the post-harvested medium/water is significant enhanced, as it is free of any added chemicals or microbial load.

With all the advantages MF offers, there are several disadvantages, as well as complexities associated with the filtration process.

1. *Cost:* One of the major limitations of filtration as a means for harvesting algae for a low-value commodity item, such as algae bio-crude, is cost, which becomes prohibitively expensive, especially if one is talking about low-concentration feed from open raceway cultivation. High-concentration feed from PBRs can offset to a certain extent, but then the CAPEX for PBRs is very high, which has already been discussed in previous sections.

2. *MOC selection:* The second major issue is selecting the appropriate MOC for membrane fabrication from among the several types of materials available. A few major examples include cellulose acetate (CA), polysulphones (PS) [examples include polyether sulphone—PES], polyvinylidenefluoride (PVDF), polyacrylonitrile (PAN), polytetrafluoroethylene (PTFE), polypropylene (PP), polyethylene (PE), polyvinyl chloride (PVC), aliphatic and aromatic polyamides, composite membranes and several other types and subtypes. Several factors determine the applicability of a particular membrane for a given application, and appropriate MOC selection is one of the most important factors. Before discussing MOC selection, it is important to know what characteristics make a good membrane. Some of the desirable characteristics are high porosity, tight pore distribution or sharp molecular weight cut-off (MWCO), good polymer flexibility, wide range of pH stability, good chlorine tolerance, high polymer strength (elongation, high burst and collapse pressure), permanent hydrophilic character, low susceptibility to fouling, easy amenability to cleaning, low transmembrane pressure (TMP) requirement and low cost. No one MOC will offer all or even majority of these desirable membrane characteristics. Thus, the final decision on material selection will depend on the inherent properties that a particular MOC offers (for example, CA has a low price and is less prone to fouling; PS offers exceptional temperature and pH resistance; PVDF offers high resistance to hydrocarbons and oxidizing environments; PTFE offers extraordinary chemical resistance across a wide temperature range), application, matrix properties (like pH, salinity, TOC, NOM, AOM), flux or permeability requirements, susceptibility to fouling, ease of flux recovery, abrasion resistance, longevity, integrity and overall ease of use. Let's consider an example of a study conducted to determine the influence of membrane MOC on performance of microalgae filtration by UF membranes (Sun et al. 2014). Sun et al. considered three kinds

of MOC: PS, PVDF and regenerated CA (RCA). The effect of MOC on filtration was assessed in terms of the following operational parameters: TMP, cross-flow velocity (CFV) and permeate flux (PF). Although the PF increased with increasing TMP and CFV for all membrane types, the PVDF membrane showed the highest increase in PF for a given increase in CFV, probably indicating that the bonding of fouling constituents was weakest with PVDF. However, both PS and PVDF membranes showed the fastest initial drop in PF, with RCA showing the slowest PF decay. Also, the nature of this fouling was such that whereas only flushing with water was sufficient for de-fouling the RCA membrane, chemical cleaning (cleaning in place [CIP]) was essential for the de-fouling of PS and PVDF membranes. Thus, clearly membrane selection is very critical; however, its suitability cannot be assessed in isolation. Rather, a much more comprehensive analysis is required in view of several other considerations and operational parameters. However, this type of analysis seems to be missing in most studies; instead, most studies are either very narrow in their analysis or are simply pushing their own agendas by oversimplifying the observations or making it so narrow in scope that it has almost no practical relevance.

3. *Membrane fouling:* This is another big area which requires considerable research and development efforts for commercially important algae and membrane types under practically feasible operating conditions. The basic commonality between all filtration membrane types is that they have a semi-permeable membrane that isolates particles from the liquid. Membrane fouling is said to occur when undesirable components (contaminants) deposit on the membrane surface, restricting the flow of liquid through the membrane pores such that membrane performance is degraded, resulting in higher operating energy and pressure, frequent cleaning, loss in productivity, higher operational and capital costs and shortened membrane life span. A standard definition for fouling provided by the International Union of Pure and Applied Chemistry is "The process that results in a decrease in performance of a membrane, caused by the deposition of suspended or dissolved solids on the external membrane surface, on the membrane pores, or within the membrane pores".

   Membrane fouling can occur through several mechanisms, namely, inorganic precipitation, pore blocking, concentration polarization, cake formation, organic adsorption and biological fouling. When pure water without any suspended or dissolved solids is filtered through a membrane, the only resistance offered to the liquid flow is the membrane material itself. The flux measured at this point is called the clean water flux. When an actual process fluid containing dissolved and suspended particles is passed through the membrane, cake is formed on the membrane surface due to deposition of particles. When these suspended particles specifically block the pores, it is called pore plugging. The resistance that develops to flow of fluid as a direct consequence of adsorption on or in the membrane is called biofouling. Thus, the total resistance to fluid flow is a combination of these four resistances: membrane resistance + cake layer + pore plugging and scaling + adsorption and biofouling.

When filtering algae culture, the probability of fouling of the membrane is significantly increased because of settling and concentration of algae cells, AOM and other dissolved and particulate organic matter and carbon (TOC, DOC) (Habarou et al. 2005; Liang et al. 2008) on the membrane surface and pores. Wang et al. reported a concentration of *Scenedesmus acuminatus* from 0.5 g/L to 136 g/L by MF, but observed a decrease in average flux as a result of microbial deposition (Wang et al. 2019). The majority of AOM consists of cellular metabolites, protein and polysaccharide (EPS), such as molecules released during active metabolism and other intracellular compounds released during cell death and lysis (Pivokonsky, Kloucek, and Pivokonska 2006; Nguyen et al. 2005). Many of these cellular metabolites, proteins and polysaccharides are hydrophilic in nature and thus easily foul membranes (Elcik, Cakmakci, and Ozkaya 2016). So, fouling is not limited to algae cells, but is a combined effect of the matrix properties, as mentioned earlier (Babel, Takizawa, and Ozaki 2002; Villacorte et al. 2009; Chiou, Hsieh, and Yeh 2010; Ladner, Vardon, and Clark 2010). Membrane fouling comprises two types: reversible and irreversible. The classification is based on the strength of the bond between the foulant and the membrane functional group. It will be explained in the next section how the membrane MOC, operational parameters and nature of fouling interact with each other to determine whether fouling is reversible or irreversible.

4. *Operational parameters:* For highest filtration performance, low operational cost and reduced membrane replacement, an optimal balance between various operational parameters, such as TMP, CFV, PF, direction of liquid flow, duration and frequency of backwash, permeability and flow reversal, is essential (Liu, Wang, et al. 2013; Reyhani et al. 2015). The ideal scenario would be to have the maximum PF for minimum applied TMP. The lower the TMP for a given flux, the lower the pumping cost and hence there is a positive impact on the OPEX. Gerardo et al. has shown that TMP has a far greater influence on PF when the biomass concentration is low; at higher biomass concentrations, TMP did not exert much influence on PF because of less dependence on pressure at this feed concentration (Gerardo, Zanain, and Lovitt 2015). Generally, it is advisable to keep CFV on the higher side to prevent accumulation of particles and thus prevent deposition and fouling. Another operation philosophy to delay the deposition of cells is flow reversal, where in the direction of liquid flow into the filtration module is reversed. The velocity at the entrance of the module is higher and then declines as a function of pressure drop; thus, the fluid velocity is lower at the exit, and this increases the probability of settling the particles towards the end of the module. This probability is enhanced when the module length is higher. Thus, by reversing the direction of fluid flow from time to time, the settled particles are dislodged, which prevents them from forming a permanent layer or cake.

Once filtration is started, the PF will keep decaying with time as the membrane gets fouled and if nothing is done. First there is a fast initial decay in the PF compared to the initial clean water flux and then the flux decays at a comparatively slower rate. The initial rapid decay in flux is understandable because the membrane encounters the process liquid containing suspended particles and matrix containing different components, and the initial rapid decay is basically the new steady-state PF under the actual process conditions and not in the presence of pure water (Elcik, Cakmakci, and Ozkaya 2016). The second phase of PF decay from this new steady-state condition is due to membrane fouling. To delay fouling, several techniques have evolved, such as backwashing the membrane with permeate or bubbling air (Wicaksana, Fane, and Chen 2006) through the feed fluid. In backwashing, a small volume of the permeate is pumped in the opposite direction of filtration to dislodge particles attached on the membrane in the flow direction. The backwash volume, backwash period and backwash frequency are other optimization parameters (Ríos et al. 2012; Wicaksana et al. 2012). The higher the backwash volume, period and frequency, the lower the flux decay; however, the productivity loss will be higher. If the productivity loss is not acceptable in commercial systems, then filtration modules are operated in parallel, such that when one module is undergoing backwash, the fluid flow switches to the other module, thus achieving backwash without a loss in productivity, however this redundancy would increase CAPEX. Another point to consider is that although higher backwash will recover higher PF, it would also dilute the process fluid. Another strategy to reverse fouling is air scouring. This method basically consists of passing air bubbles with the process fluid, which results in shearing of the filtration surface, resulting in dislodging of algae particles and other deposits. The volumetric ratio of air to process fluid, continuous vs. intermittent bubbling and other operation parameters are variables that will need to be optimized based on the fouling nature of the process fluid and the membrane's susceptibility to fouling and the MOC (Hong et al. 2002; Yusuf, Abdul Wahab, and Sahlan 2016).

Broadly speaking there are two types of fouling: reversible (including backwashable and non-backwashable) and irreversible (Guo, Ngo, and Li 2012). Formation of a cake layer or concentration polarization of material on the process fluid side of the membrane causes reversible fouling. Irreversible fouling is caused by pore plugging or chemical adsorption at the membrane surface. Strategies such as backwashing across the membrane surface or hydrodynamic surface scouring, such as air scouring, are effective tools for eliminating backwashable-reversible washing. Non-backwashable-reversible fouling can be removed by mild chemical treatment. However, irreversible fouling may be sometimes partially or fully removed only by full and intense chemical cleaning, but eventually it builds up, with the only solution being membrane replacement. It must be understood that both reversible and irreversible fouling processes occur simultaneously, albeit at different rates, if the intensity and frequency of backwash or hydrodynamic surface souring is insufficient (say, to reduce productivity loss), this approach provides more time for the chemical reaction that causes irreversible fouling to complete, and it encourages a higher degree of irreversible fouling.

5. ***Membrane modification and configuration-based approach to reduce membrane fouling:*** Membrane morphology (surface configuration, pore size and surface finish) and the chemical nature (type of functional group, degree of hydrophobicity or hydrophilicity and surface charge) are the two most important membrane properties that affect the susceptibility of membranes to fouling. Lower fouling probability is encouraged by high hydrophilicity, negative surface charge and high surface evenness. To achieve these properties, several strategies, such as advance surface coatings (Louie et al. 2006), hydrophilization (Ge et al. 2014), surface grafting (Gryta 2012) and zwitterion modification (Guillen-Burrieza et al. 2014), have been attempted. Marbelia et al. tested two membrane surface modification approaches (i.e., Polyvinylpyrrolidone [PVP] modification and sulfonation) on polyvinylidene fluoride (PVDF) membrane to increase their fouling resistance for the filtration of microalgae (Marbelia et al. 2018). PVP modification consists of introducing hydrophilic groups, whereas sulfonation introduces negative charges (sulphonic acid groups) on the membrane surface. The PVP-modified PVDF membrane displayed higher resistance to fouling in comparison to sulphonation, specifically at the start of the filtration process, possibly because the effect of the introduction of the functional group helps in fouling reduction at the initial fouling stage. Nurra et al. demonstrated that by vibrating the membrane assembly (dynamic filtration), performance of conventional tangential cross-flow membrane filtration could be significantly improved with finite reduction of membrane fouling and thus overall cost savings (Nurra et al. 2014). Even if these membrane modification strategies are successful in decreasing the rate of fouling, they often achieve this at the cost of a decrease in flux. In addition, most of these techniques are costly, complicated and at very nascent stages of development, and more often than not, these surface modifications are helpful in delaying early fouling, but in the long term have limited success. Thus, clearly there is a long felt and immediate need for membrane surface modification technologies that are cheap, effective and practical in the long run and offer the optimal balance between anti-fouling efficiency and high permeate flux.

6. ***Filtration economics:*** The cost of filtration would include OPEX and CAPEX components. OPEX would include algae culture pumping cost; cost of membrane replacement; and chemical cost for CIP, backwash and backflush, whereas CAPEX would include the cost of fabrication of the filtration assembly, skid, control system, peripherals and setting up/installation of the module. Some published literature has reported cost in financial terms, whereas others have reported it in terms of energy consumption. Additionally, there is significant variation in the reported cost estimates. This is further complicated by the fact that different research groups have concentrated algae to different extents, and this has a significant bearing on the operating cost of filtration. From an economic sustainability point

of view, the cost of primary harvesting should be in the range of 0.02 $/kg to a maximum of 0.06 $/kg (OPEX). Additionally, a harvesting CAPEX of 0.02 to 0.04 $/kg can be considered, which makes the total cost of primary harvesting between 0.04 $/kg and 0.1 $/Kg (OPEX + CAPEX). Wang et al. reported concentrations of *Scenedesmus acuminatus* from 0.5 g/L to 136 g/L by MF, at a cost of 0.30 $/kg biomass (CAPEX + OPEX) (Wang et al. 2019), which is at least three- to four-fold higher than the desired cost. Gerardo et al. studied the harvesting of *Chlorella minutissima* by concentrating the virgin culture from two initial concentration values of 1 and 2 g/L to 150 g/L using MF [a spiral-wound membrane (Koch part number 3838-K618-HYT)] (Gerardo, Zanain, and Lovitt 2015). The total energy consumption for the initial biomass concentration of 1 and 2 g/L was 2.86 KWh/kg and 0.46 KWh/Kg, assuming an energy cost of 0.1 $/KWh; this works out to be 0.286 $/kg to 0.046 $/kg. Clearly, feed biomass concentration plays a critical role in reducing the cost of harvesting. The cost of 0.046 $/kg is within the target cost set earlier for economic viability. However, one must understand that a feed concentration of 2 g/L is not possible if algae is grown in open raceway ponds—it must come from PBR. When PBR comes into question, then there are other associated costs; however, one can argue that the higher cultivation cost in PBR can be offset by the lower harvesting cost by filtration. Gerardo et al. uses spiral-wound MF to concentrate *Scenedesmus* species from 1.1 g/L to 150 g/L and estimated the overall OPEX to be 0.282 $/kg. This value is again about four to five times higher than desired (Gerardo, Oatley-Radcliffe, and Lovitt 2014). Bilad et al. concentrated PBR culture of *Chlorella vulgaris* and *Phaeodactylum tricornutum* from 0.41 and 0.23 g/L, respectively, to 220 g/L using modified submerged MF. The authors claim that this approach improved flux and decreased fouling, resulting in a lower energy requirement of 0.64 and 0.98 kW h/kg for *C. vulgaris* and *P. tricornutum*, respectively (Bilad et al. 2012). Considering the energy cost of 0.1 $/KWh, this works out to be 0.064 and 0.098 $/kg, and this is just the OPEX. If one adds the CAPEX and additional cost burden for membrane life depreciation because of submersion in sea water, this would be no less than 0.11 to 0.15 $/Kg, which is still higher than the highest acceptable economic numbers.

Although MF and UF technology have come a long way in the last 15 to 20 years, the current stock of these membranes is still not suitable for microalgae harvesting application from a biofuels perspective at the price point that is offered by several commercial manufacturers today. The industry will have to heavily invest in R&D in areas such as novel material synthesis, which offer exceptionally high flux in the region of 100 to 150 LMH Litres per square meter per hour = L/m²/h] at very low TMP, are highly fouling resistant, show very minimal flux decay over an extended period of operation and are easy to restore with mild cleaning agents. On top of this, all of these properties need to come at a very low cost for membranes with exceptional longevity and abrasion resistance. As things stand today, the membrane technology does not seem to be a sustainable option for large industrial-scale harvesting of microalgae, especially marine algae. However, filtration has the potential to become one of the most preferred options for harvesting if the recommended performance criteria is met, as it offers several other additional advantages outlined earlier. Indeed, with some of the groundbreaking efforts outlined earlier (although not fully optimized yet) and rapid advancements in material science and engineering in the past decade, a breakthrough membrane technology may be just a few years away from realization.

Centrifugation

Centrifugation, like filtration, is a physical harvesting technique that offers many of the same advantages as filtration, except that centrifuge may not start with 100% biomass capture efficiency as filtration but under operation can easily reach >95% biomass capture efficiency. Even higher efficiencies are definitely possible, but at a reduced throughput to increase the residence time in the centrifuge so that the biomass is effectively captured. Centrifuges operate on the principle of applying a high gravitational pull by angular acceleration of the centrifuge bowl. Based on the relative density of the suspended particles in reference to the dispersant, the denser particles are pushed towards the wall of the bowl. There are five basic classifications of centrifuges: hydro-cyclones, disc stack centrifuges, imperforated basket centrifuges, decanters and perforated basket centrifuges (Pahl et al. 2013). The application of each is dependent on several operational parameters, such as relative density difference between the suspended and dispersed phases, throughput required, final biomass concentration desired, nature of algae and affordable energy expenditure, among others.

Because centrifugation relies on a relative density difference for the separation of the suspended and dispersed phases, separation of the algae–water suspension becomes difficult by centrifugation because algae and water have similar densities. In algae suspension the density of the suspended and dispersed phases are very close; thus, a higher g-force is required for particle separation and hence a higher energy expenditure per unit biomass separated is required. Thus conventionally, centrifugation has been one of the most expensive commercial algae harvesting methods and is typically reserved for high-value applications (Pienkos and Darzins 2009).

Sim et al. reported an average energy expenditure of 1.3 KWh/m³ for concentrating an open raceway culture from 400 mg/L (0.04%) to 40 g/L (4%). Considering an energy cost of 0.1 $/KWh, this works out to be 0.325 $/kg of biomass and that for a slurry of just 4% concentration. At 4% output concentration itself, the cost is four times higher than acceptable. To make the drying process even borderline economically viable, at least 20% of the biomass must be fed to the dryer (Putt 2007). Mohn et al. reported that get to 22% algae on a dry weight basis, energy consumption could be as high as 8 KWh/m³ (Mohn 1980). Considering an energy cost of 0.1 $/KWh, this works out to be 2 $/kg of biomass, which is 20-fold higher than what would be economically feasible for biofuels applications. Fasaei et al. made an assessment that to concentrate five different marine algae species (*Danuliella parva*, *Picochlorum* sp., *Synechococcus* sp., *Phaeodactylum tricornutum* and *Diacronema lutheri*) and one fresh water species (*Scenedesmus obliquus*) from 0.05% to 15% dry weight, the total cost for centrifuge as an option would be in the range of 0.5 to 2 pounds/kg (0.66 to 2.64 $/kg). Dassey et al. estimated the centrifugation energy consumption

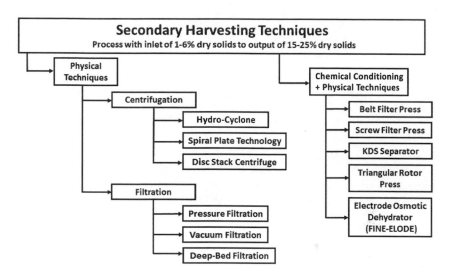

FIGURE 9.4 Secondary harvesting options.

Source: (Chen, Chang, and Lee 2015; Show, Lee, and Mujumdar 2015; Benali and Kudra 2010; Show, Lee, and Chang 2013; Danquah et al. 2009; Rees, Leenheer, and Ranville 1991; Pahl et al. 2013)

for concentrating *Nannochloris* sp. from 0.1 g/L (feed concentration) at about 20 KWh/m^3. Again, given very high energy consumption and considering an energy cost of 0.1 $/KWh, this works out to be 20 $/kg, which is 20- to 30-fold higher than what is economically viable. Without a doubt, primary harvesting of dilute or even concentrated PBR culture (unless one can get 3% to 6% output from PBR) by centrifuge is an economic impossibility, and if used at all, it would only suit secondary harvesting (dewatering) applications (Figure 9.4).

9.6 CONCLUSION

In line with the title of this chapter, there are currently several issues with a sustainable approach in the industrial-scale operations of microalgae. The current practices and state of the art in terms of strain collection, isolation and selection, commercial cultivation of algae and finally harvesting to make algae available for further downstream processing are economically and energetically unsustainable. Although microalgae-based biofuels are considered third-generation biofuels, the pursuit of it has a history of no more than a decade and half. Major commercial pursuits have an even shorter history, and most commercial algae cultivators or technology providers have only emerged in the last decade or so. Among these pursuits, exceptionally few players have invested in an integrated end-to-end research initiative. As a result, there is lack of an adequate fundamental knowledge base, which is hindering the rapid development of algae biofuels even when there is a global need to develop an alternative renewable energy source. Governments, academic institution, corporations and industry–academia partnerships must invest heavily in fundamental research as well as technology to develop a cutting-edge knowledge base in the isolation, selection and development of commercially relevant stains with desirable traits for the long haul; in understanding the key operating variables that govern cultivation, such as the effect of light on growth, light utilization, vessel design, vessel configuration, CO_2 management, nutrient optimization and management, contamination and grazer control; and in understanding and developing ultra-low-dose, metal-free, non-toxic coagulants and flocculants, optimization and deployment of electrochemical systems and system configuration design. As no one harvesting technique can be used in isolation to get the desired algae concentration, several technologies will have to be used in tandem in an innovative combination, such as the development of ultra-low-fouling, high-flux and abrasion-resistant membranes. In the last couple of years, there has been unprecedented growth in the number of start-ups pitching revolutionary technology development in different niche areas. These ideas have received enthusiastic financial support from venture capitalists, angel investors and numerous government renewable energy funding agencies. There has indeed also been some noteworthy developments and breakthroughs in the algal biofuel technology landscape, which have improved the feasibility quotient. The time is absolutely ripe to go all out with a concerted and consistent effort to develop these knowledge bases with in-the-field demonstration of the claimed advantages each technology offers. There has been no better time to be in the algae biofuels business, and if the concerted effort and financial impetus continues at the current rate, a functional and economically viable algal biofuels technology seems to be in sight.

ACKNOWLEDGEMENT

We sincerely thank Reliance Industries Limited, India, which has given us the opportunity to be part of the world's largest algae biofuels initiative, with the objective to make India energy independent and provide a global platform for grassroots-level, algae-specific technology development. This opportunity has enabled us to develop a long-term vision for algae biofuels and allied industries.

REFERENCES

Acién, F. G., J. M. Fernández, J. J. Magán, and E. Molina. 2012. "Production cost of a real microalgae production plant and strategies to reduce it." *Biotechnology Advances* 30 (6):1344–1353. doi: 10.1016/j.biotechadv.2012.02.005.

Adesanya, V. O., E. Cadena, S. A. Scott, and A. G. Smith. 2014. "Life cycle assessment on microalgal biodiesel production using a hybrid cultivation system." *Bioresource Technology* 163:343–355. doi: 10.1016/j.biortech.2014.04.051.

Alabi, A. O., M. Tampier, and E. Bibeau. 2009. Microalgae technologies and processes for bioenergy production in british columbia: Current technology, suitability & barriers to implementation. In *Final Report submitted to the British Columbia Innovation Council*.

Alam, F., A. Date, R. Rasjidin, S. Mobin, H. Moria, and A. Baqui. 2012. "Biofuel from Algae- Is It a Viable Alternative?" *Procedia Engineering* 49:221–227. doi: 10.1016/j.proeng.2012.10.131.

Al-Hasan, R. H., A. M. Ali, H. H. Ka'wash, and S. S. Radwan. 1990. "Effect of salinity on the lipid and fatty acid composition of the halophyteNavicula sp.: Potential in mariculture." *Journal of Applied Phycology* 2 (3):215–222. doi: 10.1007/bf02179778.

Andersen, R. A. 2005. "Algal culturing techniques." In edited by Phycological Society of America Robert A. Andersen. illustrated ed. Burlington, MA: Elsevier/Academic Press.

Arogo, J., P. W. Westerman, and Z. S. Liang. 2003. "Comparing ammonium ion dissociation constant in swine anaerobic lagoon liquid and deionized water." *Transactions of the ASAE* 46 (5):1415. doi: 10.13031/2013.15441.

Azov, Y. 1982. "Effect of pH on inorganic carbon uptake in algal cultures." *Applied and Environmental Microbiology* 43 (6):1300–1306.

Babel, S., S. Takizawa, and H. Ozaki. 2002. "Factors affecting seasonal variation of membrane filtration resistance caused by Chlorella algae." *Water Research* 36 (5):1193–1202. doi: 10.1016/S0043-1354(01)00333-5.

Baharuddin, N., N. S. Aziz, H. N. Sohif, and M. N. Basiran. 2016. "Marine microalgae flocculation using plant: The case of Nannochloropsis oculata and Moringa oleifera." *Pakistan Journal of Botany* 48 (2):831–840.

Bayat T., Rasoul Khalilzadeh, M., and H. Kouchakzadeh. 2017. "Efficient harvesting of marine Chlorella vulgaris microalgae utilizing cationic starch nanoparticles by response surface methodology." *Bioresource Technology* 243:583–588. doi: 10.1016/j.biortech.2017.06.181.

Behera, S., R. Singh, R. Arora, N. K. Sharma, M. Shukla, and S. Kumar. 2015. "Scope of algae as third generation biofuels." *Frontiers in Bioengineering and Biotechnology* 2 (90). doi: 10.3389/fbioe.2014.00090.

Benali, M., and T. Kudra. 2010. "Process intensification for drying and dewatering." *Drying Technology* 28 (10):1127–1135. doi: 10.1080/07373937.2010.502604.

Benemann, J. R., and W. J. Oswald. 1996. Systems and economic analysis of microalgae ponds for conversion of CO_2 to biomass. Final report.; California Univ., Berkeley, CA (United States). Dept. of Civil Engineering.

Bermúdez Menéndez, J. M., A. Arenillas, J. Á. Menéndez Díaz, L. Boffa, S. Mantegna, A. Binello, and G. Cravotto. 2014. "Optimization of microalgae oil extraction under ultrasound and microwave irradiation." *Journal of Chemical Technology and Biotechnology* 89 (11):1779–1784. doi: 10.1002/jctb.4272.

Bilad, M. R., D. Vandamme, I. Foubert, K. Muylaert, and I. F. J. Vankelecom. 2012. "Harvesting microalgal biomass using submerged microfiltration membranes." *Bioresource Technology* 111 (0):343–352. doi: 10.1016/j.biortech.2012.02.009.

Biller, P. 2018. "Hydrothermal liquefaction of aquatic feedstocks." In *Direct Thermochemical Liquefaction for Energy Applications*, edited by Lasse Rosendahl, 101–125. Cambridge: Woodhead Publishing Limited.

Bleeke, F., G. Quante, D. Winckelmann, and G. Klöck. 2015. "Effect of voltage and electrode material on electroflocculation of Scenedesmus acuminatus." *Bioresources and Bioprocessing* 2 (1):36. doi: 10.1186/s40643-015-0064-6.

Bondelind, M., S. Sasic, and L. Bergdahl. 2013. "A model to estimate the size of aggregates formed in a Dissolved Air Flotation unit." *Applied Mathematical Modelling* 37 (5):3036–3047. doi: 10.1016/j.apm.2012.07.004.

Borowitzka, M. A. 1999. "Commercial production of microalgae: Ponds, tanks, and fermenters." In *Progress in Industrial Microbiology*, edited by R. Osinga, J. Tramper, J. G. Burgess and R. H. Wijffels, 313–321. Burlington, MA: Elsevier.

Borowitzka, M. A. 2013. "High-value products from microalgae—their development and commercialisation." *Journal of Applied Phycology* 25 (3):743–756. doi: 10.1007/s10811-013-9983-9.

Bosma, R., W. van Spronsen, J. Tramper, and R. Wijffels. 2003. "Ultrasound, a new separation technique to harvest microalgae." *Journal of Applied Phycology* 15 (2–3):143–153. doi: 10.1023/a:1023807011027.

Branyikova, I., G. Prochazkova, T. Potocar, Z. Jezkova, and T. Branyik. 2018. "Harvesting of microalgae by flocculation." *Fermentation* 4 (4):93.

Brennan, L., and P. Owende. 2010. "Biofuels from microalgae—A review of technologies for production, processing, and extractions of biofuels and co-products." *Renewable and Sustainable Energy Reviews* 14 (2):557–577. doi: 10.1016/j.rser.2009.10.009.

Caiola, M. G., D. Billi, and E. I. Friedmann. 1996. "Effect of desiccation on envelopes of the cyanobacterium Chroococcidiopsis sp. (Chroococcales)." *European Journal of Phycology* 31 (1):97–105. doi: 10.1080/09670269600651251a.

Chen, C.-L., J.-S. Chang, and D.-J. Lee. 2015. "Dewatering and Drying Methods for Microalgae." *Drying Technology* 33 (4):443–454. doi: 10.1080/07373937.2014.997881.

Chen, C.-Y., K.-L. Yeh, R. Aisyah, D.-J. Lee, and J.-S. Chang. 2011. "Cultivation, photobiore-actor design and harvesting of microalgae for biodiesel production: A critical review." *Bioresource Technology* 102 (1):71–81. doi: 10.1016/j.biortech.2010.06.159.

Chen, G. 2004. "Electrochemical technologies in wastewater treatment." *Separation and Purification Technology* 38 (1):11–41. doi: 10.1016/j.seppur.2003.10.006.

Chetsumon, A., F. Umeda, I. Maeda, K. Yagi, T. Mizoguchi, and Y. Miura. 1998. "Broad spectrum and mode of action of an antibiotic produced by scytonema sp. TISTR 8208 in a seaweed-type bioreactor." In *Biotechnology for Fuels and Chemicals: Proceedings of the Nineteenth Symposium on Biotechnology for Fuels and Chemicals Held May 4–8. 1997, at Colorado Springs, Colorado*, edited by Mark Finkelstein and Brian H. Davison, 249–256. Totowa, NJ: Humana Press.

Chiou, Y.-T., M.-L. Hsieh, and H.-H. Yeh. 2010. "Effect of algal extracellular polymer substances on UF membrane fouling." *Desalination* 250 (2):648–652. doi: 10.1016/j.desal.2008.02.043.

Chisti, Y. 2007. "Biodiesel from microalgae." *Biotechnology Advances* 25 (3):294–306. doi: 10.1016/j.biotechadv.2007.02.001.

Chisti, Y. 2008. "Response to Reijnders: Do biofuels from microalgae beat biofuels from terrestrial plants?" *Trends in Biotechnology* 26 (7):351–352. doi: 10.1016/j.tibtech.2008.04.002.

Chisti, Y. 2013. "Raceways-based production of algal crude oil." In *Green—A Systemic Approach to Energy*, edited by Robert Schlögl, 195. Germany: Walter de Gruyter.

Chow, C. W. K., J. A. van Leeuwen, M. Drikas, R. Fabris, K. M. Spark, and D. W. Page. 1999. "The impact of the character of natural organic matter in conventional treatment with alum." *Water Science and Technology* 40 (9):97–104. doi: 10.1016/S0273-1223(99)00645-9.

Cobos, M., J. Paredes, J. Maddox, G. Vargas-Arana, L. Flores, C. Aguilar, J. Marapara, and J. Castro. 2017. "Isolation and characterization of native microalgae from the peruvian amazon with potential for biodiesel production." *Energies* 10 (2):224. doi: 10.3390/en10020224.

Contreras, A., F. García, E. Molina, and J. C. Merchuk. 1998. "Interaction between CO_2-mass transfer, light availability, and hydrodynamic stress in the growth of Phaeodactylum tricornutum in a concentric tube airlift photobioreactor." *Biotechnology and Bioengineering* 60 (3):317–325. doi: 10.1002/(SICI)1097-0290(19981105)60:3<317::AID-BIT7>3.0.CO;2-K.

Coons, J. E., D. M. Kalb, T. Dale, and B. L. Marrone. 2014. "Getting to low-cost algal biofuels: A monograph on conventional and cutting-edge harvesting and extraction technologies." *Algal Research* 6:250–270. doi: 10.1016/j.algal.2014.08.005.

Cooper, G. M., and R. E. Hausman. 2007. *The Cell—A Molecular Approach*. 4th ed. Washington, DC: ASM Press.

Cosgrove, W. J., and D. P. Loucks. 2015. "Water management: Current and future challenges and research directions." *Water Resources Research* 51 (6):4823–4839. doi: 10.1002/2014WR016869.

Cristóbal, G. C. J., and G. L. López Lizárraga. 2016. "Considerations for photobioreactor design and operation for mass cultivation of microalgae." In *Algae: Organisms for Imminent Biotechnology*, edited by Nooruddin Thajuddin and Dharumadurai Dhanasekaran. Croatia: IntechOpen.

Csavina, J. L., B. J. Stuart, R. Guy Riefler, and M. L. Vis. 2011. "Growth optimization of algae for biodiesel production." *Journal of Applied Microbiology* 111 (2):312–318. doi: 10.1111/j.1365–2672.2011.05064.x.

Danquah, M. K., L. Ang, N. Uduman, N. Moheimani, and G. M. Forde. 2009. "Dewatering of microalgal culture for biodiesel production: Exploring polymer flocculation and tangential flow filtration." *Journal of Chemical Technology and Biotechnology* 84 (7):1078–1083. doi: 10.1002/jctb.2137.

Das, P., M. I. Thaher, M. A. Q. M. A. Hakim, H. M. S. J. Al-Jabri, and G. S. H. S. Alghasal. 2016. "Microalgae harvesting by pH adjusted coagulation-flocculation, recycling of the coagulant and the growth media." *Bioresource Technology* 216:824–829. doi: 10.1016/j.biortech.2016.06.014.

Davey, H. M., and D. B. Kell. 1996. "Flow cytometry and cell sorting of heterogeneous microbial populations: The importance of single-cell analyses." *Microbiological Reviews* 60 (4):641–696.

Davis, R., J. Markham, C. Kinchin, N. Grundl, E. C. D. Tan, and D. Humbird. 2016. Process design and economics for the production of algal biomass: Algal biomass production in open pond systems and processing through dewatering for downstream conversion. National Renewable Energy Lab. (NREL), Golden, CO (United States).

de Cruz, C. R., A. Lubrano, and D. M. Gatlin. 2018. "Evaluation of microalgae concentrates as partial fishmeal replacements for hybrid striped bass Morone sp." *Aquaculture* 493:130–136. doi: 10.1016/j.aquaculture.2018.04.060.

de Godos, I., J. L. Mendoza, F. G. Acién, E. Molina, C. J. Banks, S. Heaven, and F. Rogalla. 2014. "Evaluation of carbon dioxide mass transfer in raceway reactors for microalgae culture using flue gases." *Bioresource Technology* 153:307–314. doi: 10.1016/j.biortech.2013.11.087.

De Philippis, R., and M. Vincenzini. 1998. "Exocellular polysaccharides from cyanobacteria and their possible applications." *FEMS Microbiology Reviews* 22 (3):151–175. doi: 10.1111/j.1574–6976.1998.tb00365.x.

Degen, J., A. Uebele, A. Retze, U. Schmid-Staiger, and W. Trösch. 2001. "A novel airlift photobioreactor with baffles for improved light utilization through the flashing light effect." *Journal of Biotechnology* 92 (2):89–94. doi: 10.1016/S0168-1656(01)00350-9.

Di Carlo, D., D. Irimia, R. G. Tompkins, and M. Toner. 2007. "Continuous inertial focusing, ordering, and separation of particles in microchannels." *Proceedings of the National Academy of Sciences of the United States of America* 104 (48):18892–18897. doi: 10.1073/pnas.0704958104.

Duong, V. T., Y. Li, E. Nowak, and P. M. Schenk. 2012. "Microalgae isolation and selection for prospective biodiesel production." *Energies* 5 (6):1835–1849.

Eckelberry, N. 2013. *Systems and Methods for Harvesting and Dewatering Algae*. edited by WIPO: Originoil, Inc.

Elcik, H., M. Cakmakci, and B. Ozkaya. 2016. "The fouling effects of microalgal cells on crossflow membrane filtration." *Journal of Membrane Science* 499:116–125. doi: 10.1016/j.memsci.2015.10.043.

Eldridge, R. J., D. R. A. Hill, and B. R. Gladman. 2012. "A comparative study of the coagulation behaviour of marine microalgae." *Journal of Applied Phycology* 24 (6):1667–1679. doi: 10.1007/s10811-012-9830-4.

Farajzadeh, R., P. L. J. Zitha, and J. Bruining. 2009. "Enhanced mass transfer of CO_2 into water: Experiment and modeling." *Industrial & Engineering Chemistry Research* 48 (13):6423–6431. doi: 10.1021/ie801521u.

Farid, M. S., A. Shariati, A. Badakhshan, and B. Anvaripour. 2013. "Using nano-chitosan for harvesting microalga Nannochloropsis sp." *Bioresource Technology* 131:555–559. doi: 10.1016/j.biortech.2013.01.058.

Farooq, W., M. Moon, B.-G. Ryu, W. I. Suh, A. Shrivastav, M. S. Park, S. K. Mishra, and J.-W. Yang. 2015. "Effect of harvesting methods on the reusability of water for cultivation of Chlorella vulgaris, its lipid productivity and biodiesel quality." *Algal Research* 8 (0):1–7. doi: 10.1016/j.algal.2014.12.007.

Fasaei, F., J. H. Bitter, P. M. Slegers, and A. J. B. van Boxtel. 2018. "Techno-economic evaluation of microalgae harvesting and dewatering systems." *Algal Research* 31:347–362. doi: 10.1016/j.algal.2017.11.038.

Fayad, N., T. Yehya, F. Audonnet, and C. Vial. 2017. "Harvesting of microalgae Chlorella vulgaris using electro-coagulation-flocculation in the batch mode." *Algal Research* 25:1–11. doi: 10.1016/j.algal.2017.03.015.

Fink, F. W. 1960. "Corrosion of Metals in Sea Water." In *SALINE WATER CONVERSION*, 27–39. American Chemical Society.

Fon Sing, S., A. Isdepsky, M. A. Borowitzka, and N. Moheimani. 2013. "Production of biofuels from microalgae." *Mitigation and Adaptation Strategies for Global Change* 18 (1):47–72. doi: 10.1007/s11027-011-9294-x.

Fuad, N., R. Omar, S. Kamarudin, R. Harun, A. Idris, and W. A. K. G. Wan Azlina. 2018. "Effective use of tannin based natural biopolymer, AFlok-BP1 to harvest marine microalgae Nannochloropsis sp." *Journal of Environmental Chemical Engineering* 6 (4):4318–4328. doi: 10.1016/j.jece.2018.06.041.

Gao, S., J. Yang, J. Tian, F. Ma, G. Tu, and M. Du. 2010. "Electro-coagulation—flotation process for algae removal." *Journal of Hazardous Materials* 177 (1–3):336–343. doi: 10.1016/j.jhazmat.2009.12.037.

Garzon-Sanabria, A. J., R. T. Davis, and Z. L. Nikolov. 2012. "Harvesting *Nannochloris oculata* by inorganic electrolyte flocculation: Effect of initial cell density, ionic strength, coagulant dosage, and media pH." *Bioresource Technology* 118 (0):418–424. doi: 10.1016/j.biortech.2012.04.057. http://ac.els-cdn.com/S0960852412006621/1-s2.0-S0960852412006621-main.pdf?_tid=82e58170-3db6-11e2-a776-00000aacb360&acdnat=1354586666_eba2062241661084e8 b87b987100bc58.

Ge, Ju, Y. Peng, Z. Li, P. Chen, and S. Wang. 2014. "Membrane fouling and wetting in a DCMD process for RO brine concentration." *Desalination* 344:97–107. doi: 10.1016/j. desal.2014.03.017.

Gerardo, M. L., D. L. Oatley-Radcliffe, and R. W. Lovitt. 2014. "Minimizing the energy requirement of dewatering Scenedesmus sp. by microfiltration: Performance, costs, and feasibility." *Environmental Science & Technology* 48 (1):845–853. doi: 10.1021/es4051567.

Gerardo, M. L., M. A. Zanain, and R. W. Lovitt. 2015. "Pilot-scale cross-flow microfiltration of Chlorella minutissima: A theoretical assessment of the operational parameters on energy consumption." *Chemical Engineering Journal* 280:505–513. doi: 10.1016/j. cej.2015.06.026.

Ghernaout, B., D. Ghernaout, and A. Saiba. 2010. "Algae and cyanotoxins removal by coagulation/flocculation: A review." *Desalination and Water Treatment* 20 (1–3):133–143. doi: 10.5004/dwt.2010.1202.

Ghernaout, D., A. I. Al-Ghonamy, W. Naceur, A. Boucherit, N. Aït-Messaoudène, M. Aichouni, A. Mahjoubi, and N. Elboughdiri. 2015. "Controlling Coagulation Process: From Zeta Potential to Streaming Potential." *American Journal of Environmental Protection* 4 (5–1):16–27. doi: 10.13140/rg.2.1.4425.6169 10.11648/j.ajeps.s.2015040501.12.

Ghernaout, D., and B. Ghernaout. 2012. "Sweep flocculation as a second form of charge neutralisation—a review." *Desalination and Water Treatment* 44 (1–3):15–28. doi: 10.1080/19443994.2012.691699.

Gonçalves, A. L., C. Ferreira, J. A. Loureiro, J. C. M. Pires, and M. Simões. 2015. "Surface physicochemical properties of selected single and mixed cultures of microalgae and cyanobacteria and their relationship with sedimentation kinetics." *Bioresources and Bioprocessing* 2 (1):21. doi: 10.1186/s40643-015-0051-y.

Griffin, G., D. Batten, T. Beer, and P. Campbell. 2013. "The costs of producing biodiesel from microalgae in the Asia-Pacific region." 2.

Gryta, M. 2012. "Polyphosphates used for membrane scaling inhibition during water desalination by membrane distillation." *Desalination* 285:170–176. doi: 10.1016/j.desal.2011.09.051.

Gudin, C., and D. Chaumont. 1991. "Cell fragility—The key problem of microalgae mass production in closed photobioreactors." *Bioresource Technology* 38 (2):145–151. doi: 10.1016/0960-8524(91)90146-B.

Guillen-Burrieza, E., A. Ruiz-Aguirre, G. Zaragoza, and H. A. Arafat. 2014. "Membrane fouling and cleaning in long term plant-scale membrane distillation operations." *Journal of Membrane Science* 468:360–372. doi: 10.1016/j.memsci.2014.05.064.

Guo, W., H.-H. Ngo, and J. Li. 2012. "A mini-review on membrane fouling." *Bioresource Technology* 122:27–34. doi: 10.1016/j.biortech.2012.04.089.

Gupta, P. L., S.-M. Lee, and H.-J. Choi. 2015. "A mini review: Photobioreactors for large scale algal cultivation." *World Journal of Microbiology and Biotechnology* 31 (9):1409–1417. doi: 10.1007/s11274-015-1892-4.

Haarhoff, J., and J. K. Edzwald. 2004. "Dissolved air flotation modelling: Insights and shortcomings." *Journal of Water Supply: Research and Technology—AQUA* 53 (3):127–150.

Habarou, H., J.-P. Croué, G. Amy, and H. Suty. 2005. Using HPSEC and Pyrolysis GC/MS to characterize organic foulants from MF and UF membranes during algal bloom. Paper read at AWWA—Membrane Technology Conference proceedings.

Hall, D. O., F. G. Acién Fernández, E. Cañizares Guerrero, K. K. Rao, and E. M. Grima. 2003. "Outdoor helical tubular photobioreactors for microalgal production: Modeling of fluid-dynamics and mass transfer and assessment of biomass productivity." *Biotechnology and Bioengineering* 82 (1):62–73. doi: 10.1002/bit.10543.

Han, M. Y. 2002. "Modeling of DAF: The effect of particle and bubble characteristics." *Journal of Water Supply: Research and Technology—AQUA* 51 (1):27–34.

Han, M., Y. Park, J. Lee, and J. Shim. 2002. "Effect of pressure on bubble size in dissolved air flotation." *Water Science and Technology: Water Supply* 2 (5–6):41–46. doi: 10.2166/ws.2002.0148.

Hanotu, J., H. C. H. Bandulasena, and W. B. Zimmerman. 2012. "Microflotation performance for algal separation." *Biotechnology and Bioengineering* 109 (7):1663–1673. doi: 10.1002/bit.24449.

Hanwright, J., J. Zhou, G. M. Evans, and K. P. Galvin. 2005. "Influence of surfactant on gas bubble stability." *Langmuir* 21 (11):4912–4920. doi: 10.1021/la0502894.

Harif, T., M. Khai, and A. Adin. 2012. "Electrocoagulation versus chemical coagulation: Coagulation/flocculation mechanisms and resulting floc characteristics." *Water Research* 46 (10):3177–3188. doi: 10.1016/j.watres.2012.03.034.

Harker, M., A. J. Tsavalos, and A. J. Young. 1996. "Autotrophic growth and carotenoid production of Haematococcus pluvialis in a 30 liter air-lift photobioreactor." *Journal of Fermentation and Bioengineering* 82 (2):113–118. doi: 10.1016/0922-338X(96)85031-8.

Hawkes, J. J., and W. T. Coakley. 1996. "A continuous flow ultrasonic cell-filtering method." *Enzyme and Microbial Technology* 19 (1):57–62. doi: 10.1016/0141-0229(95)00172-7.

Hellebust, J. A. 1985. "Mechanisms of response to salinity in halotolerant microalgae." *Plant and Soil* 89 (1):69–81. doi: 10.1007/bf02182234.

Henderson, R. K., A. Baker, S. A. Parsons, and B. Jefferson. 2008. "Characterisation of algogenic organic matter extracted from cyanobacteria, green algae and diatoms." *Water Research* 42 (13):3435–3445. doi: 10.1016/j.watres.2007.10.032.

Henderson, R. K., S. A. Parsons, and B. Jefferson. 2008a. "Successful removal of algae through the control of zeta potential." *Separation Science and Technology* 43 (7):1653–1666. doi: 10.1080/01496390801973771.

Henderson, R., S. A. Parsons, and B. Jefferson. 2008b. "The impact of algal properties and pre-oxidation on solid–liquid separation of algae." *Water Research* 42 (8–9):1827–1845. doi: 10.1016/j.watres.2007.11.039.

Henderson, R. K., S. A. Parsons, and B. Jefferson. 2009. "The potential for using bubble modification chemicals in dissolved air flotation for algae removal." *Separation Science and Technology* 44 (9):1923–1940. doi: 10.1080/01496390902955628.

Henderson, R. K., S. A. Parsons, and B. Jefferson. 2010. "The impact of differing cell and algogenic organic matter (AOM) characteristics on the coagulation and flotation of algae." *Water Research* 44 (12):3617–3624. doi: 10.1016/j.watres.2010.04.016.

Her, N., G. Amy, H.-R. Park, and M. Song. 2004. "Characterizing algogenic organic matter (AOM) and evaluating associated NF membrane fouling." *Water Research* 38 (6):1427–1438. doi: 10.1016/j.watres.2003.12.008.

Hill, M., and N. R. Harris. 2008. "Ultrasonic microsystems for bacterial cell manipulation." In *Principles of Bacterial Detection: Biosensors, Recognition Receptors and Microsystems*, edited by M. Zourob, S. Elwary and A. Turner, 909–928. New York: Springer.

Holzammer, C., A. Finckenstein, S. Will, and A. S. Braeuer. 2016. "How sodium chloride salt inhibits the formation of CO_2 gas hydrates." *The Journal of Physical Chemistry B* 120 (9):2452–2459. doi: 10.1021/acs.jpcb.5b12487.

Holzinger, A., and U. Karsten. 2013. "Desiccation stress and tolerance in green algae: Consequences for ultrastructure, physiological and molecular mechanisms." *Frontiers in Plant Science* 4:327–327. doi: 10.3389/fpls.2013.00327.

Hong, S. P., T. H. Bae, T. M. Tak, S. Hong, and A. Randall. 2002. "Fouling control in activated sludge submerged hollow fiber membrane bioreactors." *Desalination* 143 (3):219–228. doi: 10.1016/S0011-9164(02)00260-6.

Hu, Q., H. Guterman, and A. Richmond. 1996. "A flat inclined modular photobioreactor for outdoor mass cultivation of photoautotrophs." *Biotechnology and Bioengineering* 51 (1):51–60.

Hu, Y.-R., F. Wang, S.-K. Wang, C.-Z. Liu, and C. Guo. 2013. "Efficient harvesting of marine microalgae *Nannochloropsis maritima* using magnetic nanoparticles." *Bioresource Technology* 138 (0):387–390. doi: 10.1016/j.biortech.2013.04.016.

Huang, Q., F. Jiang, L. Wang, and C. Yang. 2017. "Design of Photobioreactors for Mass Cultivation of Photosynthetic Organisms." *Engineering* 3 (3):318–329. doi: 10.1016/J.ENG.2017.03.020.

Hulatt, C. J., and D. N. Thomas. 2010. "Dissolved organic matter (DOM) in microalgal photobioreactors: A potential loss in solar energy conversion?" *Bioresource Technology* 101 (22):8690–8697. doi: 10.1016/j.biortech.2010.06.086.

Ives, K. J. 1959. "The significance of surface electric charge on algae in water purification." *Journal of Biochemical and Microbiological Technology and Engineering* 1 (1):37–47. doi: 10.1002/jbmte.390010105.

Jia, Z., Y. Liu, M. Daroch, S. Geng, and J. J. Cheng. 2014. "Screening, Growth Medium Optimisation and Heterotrophic Cultivation of Microalgae for Biodiesel Production." *Applied Biochemistry and Biotechnology* 173 (7):1667–1679. doi: 10.1007/s12010-014-0954-7.

Jiménez, C., Cossı, x, Belén R. o, and F. Xavier Niell. 2003. "Relationship between physicochemical variables and productivity in open ponds for the production of Spirulina: A predictive model of algal yield." *Aquaculture* 221 (1):331–345. doi: 10.1016/S0044-8486(03)00123-6.

Jorquera, O., A. Kiperstok, E. A. Sales, M. Embiruçu, and M. L. Ghirardi. 2010. "Comparative energy life-cycle analyses of microalgal biomass production in open ponds and photobioreactors." *Bioresource Technology* 101 (4):1406–1413. doi: 10.1016/j.biortech.2009.09.038.

Kabdaşlı, I., I. Arslan-Alaton, T. Ölmez-Hancı, and O. Tünay. 2012. "Electrocoagulation applications for industrial wastewaters: A critical review." *Environmental Technology Reviews* 1 (1):2–45. doi: 10.1080/21622515.2012.715390.

Kaewpintong, K., A. Shotipruk, S. Powtongsook, and P. Pavasant. 2007. "Photoautotrophic high-density cultivation of vegetative cells of Haematococcus pluvialis in airlift bioreactor." *Bioresource Technology* 98 (2):288–295. doi: 10.1016/j.biortech.2006.01.011.

Karuppasamy, S., A. S. Musale, B. Soni, B. Bhadra, N. Gujarathi, M. Sundaram, A. Sapre, S. Dasgupta, and C. Kumar. 2018. "Integrated grazer management mediated by chemicals for sustainable cultivation of algae in open ponds." *Algal Research* 35:439–448. doi: 10.1016/j.algal.2018.09.017.

Katırcıoğlu, H., B. S. Akın, and T. Atıcı. 2004. "Microalgal toxin(s): Characteristics and importance." *African Journal of Biotechnology* 3 (12):667–674.

Ketheesan, B., and N. Nirmalakhandan. 2012. "Feasibility of microalgal cultivation in a pilot-scale airlift-driven raceway reactor." *Bioresource Technology* 108:196–202. doi: 10.1016/j.biortech.2011.12.146.

Khan, M. I., J. H. Shin, and J. D. Kim. 2018. "The promising future of microalgae: Current status, challenges, and optimization of a sustainable and renewable industry for biofuels, feed, and other products." *Microbial Cell Factories* 17 (1):36. doi: 10.1186/s12934-018-0879-x.

Khemis, M., J.-P. Leclerc, G. Tanguy, G. Valentin, and F. Lapicque. 2006. "Treatment of industrial liquid wastes by electrocoagulation: Experimental investigations and an overall interpretation model." *Chemical Engineering Science* 61 (11):3602–3609. doi: 10.1016/j.ces.2005.12.034.

Khoo, C. G., M. K. Lam, and K. T. Lee. 2016. "Pilot-scale semi-continuous cultivation of microalgae Chlorella vulgaris in bubble column photobioreactor (BC-PBR): Hydrodynamics and gas–liquid mass transfer study." *Algal Research* 15:65–76. doi: 10.1016/j.algal.2016.02.001.

Kim, H.-C., and M.-J. Yu. 2005. "Characterization of natural organic matter in conventional water treatment processes for selection of treatment processes focused on DBPs control." *Water Research* 39 (19):4779–4789. doi: 10.1016/j.watres.2005.09.021.

Kim, J., B.-G. Ryu, B.-K. Kim, J.-I. Han, and J.-W. Yang. 2012. "Continuous microalgae recovery using electrolysis with polarity exchange." *Bioresource Technology* 111 (0):268–275. doi: 10.1016/j.biortech.2012.01.104.

Kim, J., G. Yoo, H. Lee, J. Lim, K. Kim, C. W. Kim, M. S. Park, and J.-W. Yang. 2013. "Methods of downstream processing for the production of biodiesel from microalgae." *Biotechnology Advances* 31 (6):862–876. doi: 10.1016/j.biotechadv.2013.04.006.

Klein-Marcuschamer, D., Y. Chisti, J. R. Benemann, and D. Lewis. 2013. "A matter of detail: Assessing the true potential of microalgal biofuels." *Biotechnology and Bioengineering* 110 (9):2317–2322. doi: 10.1002/bit.24967.

Kumar, D., B. Singh, and Y. Chandra Sharma. 2017. "Challenges and opportunities in commercialization of algal biofuels." In *Algal Biofuels: Recent Advances and Future Prospects*, edited by S. K. Gupta, A. Malik and F. Bux, 421–450. Cham: Springer International Publishing.

Kunjapur, A. M., and R. B. Eldridge. 2010. "Photobioreactor design for commercial biofuel production from microalgae." *Industrial & Engineering Chemistry Research* 49 (8):3516–3526. doi: 10.1021/ie901459u.

Kurokawa, M., P. M. King, X. Wu, E. M. Joyce, T. J. Mason, and K. Yamamoto. 2016. "Effect of sonication frequency on the disruption of algae." *Ultrasonics Sonochemistry* 31:157–162. doi: 10.1016/j.ultsonch.2015.12.011.

Ladner, D. A., D. R. Vardon, and M. M. Clark. 2010. "Effects of shear on microfiltration and ultrafiltration fouling by marine bloom-forming algae." *Journal of Membrane Science* 356 (1):33–43. doi: 10.1016/j.memsci.2010.03.024.

Lee, A. K., D. M. Lewis, and P. J. Ashman. 2013. "Harvesting of marine microalgae by electroflocculation: The energetics, plant design, and economics." *Applied Energy* 108:45–53. doi: 10.1016/j.apenergy.2013.03.003.

Li, Y., and J. G. Qin. 2005. "Comparison of growth and lipid content in three Botryococcus braunii strains." *Journal of Applied Phycology* 17 (6):551–556. doi: 10.1007/s10811-005-9005-7.

Liang, H., W. Gong, J. Chen, and G. Li. 2008. "Cleaning of fouled ultrafiltration (UF) membrane by algae during reservoir water treatment." *Desalination* 220 (1):267–272. doi: 10.1016/j.desal.2007.01.033.

Lim, D. K. Y., S. Garg, M. Timmins, E. S. B. Zhang, S. R. Thomas-Hall, H. Schuhmann, Y. Li, and P. M. Schenk. 2012. "Isolation and evaluation of oil-producing microalgae from subtropical coastal and brackish waters." *PLoS ONE* 7 (7):e40751. doi: 10.1371/journal.pone.0040751.

Liu, J., J. Huang, and F. Chen. 2011. "Microalgae as feedstocks for biodiesel production." In *Biodiesel: Feedstocks and Processing Technologies*, edited by Margarita Stoytcheva and Gisela Montero. Rijeka: Croatia InTech.

Liu, J., Y. Zhu, Y. Tao, Y. Zhang, A. Li, T. Li, M. Sang, and C. Zhang. 2013. "Freshwater microalgae harvested via flocculation induced by pH decrease." *Biotechnology for Biofuels* 6 (1):98.

Liu, Q. F., Q. Wang, S. H. Yan, and J. Zhao. 2013. "The optimization of operation parameters of microfiltration membrane system." *Advanced Materials Research* 726–731:1770–1773. doi: 10.4028/www.scientific.net/AMR.726-731.1770.

Louie, J. S., I. Pinnau, I. Ciobanu, K. Ishida, A. Ng, and M. Reinhard. 2006. "Effects of polyether–polyamide block copolymer coating on performance and fouling of reverse osmosis membranes." *Journal of Membrane Science* 280 (1):762–770. doi: 10.1016/j.memsci.2006.02.041.

Lürling, M., F. Eshetu, E. J. Faassen, S. Kosten, and V. L. M. Huszar. 2013. "Comparison of cyanobacterial and green algal growth rates at different temperatures." *Freshwater Biology* 58 (3):552–559. doi: 10.1111/j.1365-2427.2012.02866.x.

Marbelia, L., M. R. Bilad, S. Maes, H. A. Arafat, and I. F. J. Vankelecom. 2018. "Poly(vinylidene fluoride)-based membranes for microalgae filtration." *Chemical Engineering & Technology* 41 (7):1305–1312. doi: 10.1002/ceat.201700622.

Mariam Al, H., G. Abdel, and H. Amal. 2015. "Microalgae Harvesting Methods for Industrial Production of Biodiesel: Critical Review and Comparative Analysis." *Journal of Fundamentals of Renewable Energy and Applications* 5 (2):1–26. doi: 10.4172/2090-4541.1000154.

Marrone, Babetta L., R. E. Lacey, D. B. Anderson, J. Bonner, J. Coons, T. Dale, C. Meghan Downes, S. Fernando, C. Fuller, B. Goodall, J. E. Holladay, K. Kadam, D. Kalb, W. Liu, J. B. Mott, Z. Nikolov, K. L. Ogden, R. T. Sayre, B. G. Trewyn, and J. A. Olivares. 2018. "Review of the harvesting and extraction program within the National Alliance for Advanced Biofuels and Bioproducts." *Algal Research* 33:470–485. doi: 10.1016/j.algal.2017.07.015.

Masojídek, J., and G. Torzillo. 2008. "Mass cultivation of freshwater microalgae." In *Encyclopedia of Ecology*, edited by Sven Erik Jørgensen and B. D. Fath, 2226–2235. New York: Elsevier Science.

Masojídek, J., G. Torzillo, and M. Koblížek. 2013. "Photosynthesis in microalgae." In *Handbook of Microalgal Culture: Applied Phycology and Biotechnology*, edited by Amos Richmond and Q. Hu. West Sussex, UK: John Wiley & Sons, Ltd.

Massimi, R., and A. E. Kirkwood. 2016. "Screening microalgae isolated from urban storm- and wastewater systems as feedstock for biofuel." *PeerJ* 4:e2396-e2396. doi: 10.7717/peerj.2396.

Melchers, R. E., and R. Jeffrey. 2005. "Early corrosion of mild steel in seawater." *Corrosion Science* 47 (7):1678–1693. doi: 10.1016/j.corsci.2004.08.006.

Mendoza, J. L., M. R. Granados, I. de Godos, F. G. Acién, E. Molina, S. Heaven, and C. J. Banks. 2013. "Oxygen transfer and evolution in microalgal culture in open raceways." *Bioresource Technology* 137:188–195. doi: 10.1016/j.biortech.2013.03.127.

Minillo, A., H. Godoy, and G. Fonseca. 2013. *Growth performance of microalgae exposed to CO$_2$.* Vol. 1.

Mohammed, A. E. 2012. "Modeling Cost Structure for Assessment Production Cost of Algal–Biofuel." *International Journal of Industrial and Manufacturing Engineering* 6 (3):650–657.

Mohn, F. 1988. "Harvesting of microalgal biomass." In *Micro-algal Biotechnology*, edited by Borowitzka M and Borowitzka L, 488. New York: Cambridge University Press.

Mohn, F. H. 1980. "Experiences and strategies in the recovery of biomass from mass cultures of microalgae." In *Algae Biomass*, edited by G. Shelef and C. J. Soeder, 547–571. Amsterdam: Elsevier.

Monahan, E. C. 1969. "Fresh Water Whitecaps." *Journal of the Atmospheric Sciences* 26 (5):1026–1029. doi: 10.1175/1520-0469(1969)026<1026:fww>2.0.co;2.

Monahan, E. C. 1971. "Oceanic Whitecaps." *Journal of Physical Oceanography* 1 (2):139–144. doi: 10.1175/1520-0485(1971)001<0139:ow>2.0.co;2.

Monahan, E. C. 2002. "Oceanic whitecaps: Sea surface features detectable via satellite that are indicators of the magnitude of the air-sea gas transfer coefficient." *Journal of Earth System Science* 111 (3):315–319. doi: 10.1007/bf02701977.

Montes, F., C. A. Rotz, and H. Chaoui. 2009. "Process modeling of ammonia volatilization from ammonium solution and manure surfaces: A review with recommended models." *Transactions of the ASABE* 52 (5):1707–1720. doi: 10.13031/2013.29133.

Moreira, B. J., A. L. M. Terra, Jorge Alberto Costa, and Michele Morais. 2016. "Utilization of CO$_2$ in semi-continuous cultivation of Spirulina sp. and Chlorella fusca and evaluation of biomass composition." 33.

Mouedhen, G., M. Feki, M. D. P. Wery, and H. F. Ayedi. 2008. "Behavior of aluminum electrodes in electrocoagulation process." *Journal of Hazardous Materials* 150 (1):124–135. doi: 10.1016/j.jhazmat.2007.04.090.

Myklestad, S. M. 1995. "Release of extracellular products by phytoplankton with special emphasis on polysaccharides." *Science of the Total Environment* 165 (1–3):155–164. doi: 10.1016/0048-9697(95)04549-G.

Narala, R. R., S. Garg, K. K. Sharma, S. R. Thomas-Hall, M. Deme, Y. Li, and P. M. Schenk. 2016. "Comparison of Microalgae Cultivation in Photobioreactor, Open Raceway Pond, and a Two-Stage Hybrid System." *Frontiers in Energy Research* 4 (29). doi: 10.3389/fenrg.2016.00029.

Ndiaye, M., E. Gadoin, and C. Gentric. 2018. "CO$_2$ gas–liquid mass transfer and kLa estimation: Numerical investigation in the context of airlift photobioreactor scale-up." *Chemical Engineering Research and Design* 133:90–102. doi: 10.1016/j.cherd.2018.03.001.

Nguyen, M.-L., P. Westerhoff, L. Baker, Q. Hu, M. Esparza-Soto, and M. Sommerfeld. 2005. "Characteristics and reactivity of algae-produced dissolved organic carbon." *Journal of Environmental Engineering* 131 (11):1574–1582. doi: 10.1061/(ASCE)0733-9372(2005)131:11(1574).

Nielsen, S. L. 2006. "Size-dependent growth rates in eukaryotic and prokaryotic algae exemplified by green algae and cyanobacteria: Comparisons between unicells and colonial growth forms." *Journal of Plankton Research* 28 (5):489–498. doi: 10.1093/plankt/fbi134.

Norsker, N.-H., M. J. Barbosa, M. H. Vermuë, and R. H. Wijffels. 2011. "Microalgal production—A close look at the economics." *Biotechnology Advances* 29 (1):24–27. doi: 10.1016/j.biotechadv.2010.08.005.

Nurra, C., E. Clavero, J. Salvadó, and C. Torras. 2014. "Vibrating membrane filtration as improved technology for microalgae dewatering." *Bioresource Technology* 157:247–253. doi: 10.1016/j.biortech.2014.01.115.

Ogbonna, J. C., T. Soejima, and H. Tanaka. 1999. "An integrated solar and artificial light system for internal illumination of photobioreactors." *Journal of Biotechnology* 70 (1):289–297. doi: 10.1016/S0168-1656(99)00081-4.

Oncel, S., and F. V. Sukan. 2008. "Comparison of two different pneumatically mixed column photobioreactors for the cultivation of Artrospira platensis (Spirulina platensis)." *Bioresource Technology* 99 (11):4755–4760. doi: 10.1016/j.biortech.2007.09.068.

Oswald, W. J., and C. G. Golueke. 1960. "Biological transformation of solar energy." In *Advances in Applied Microbiology*, edited by Wayne W. Umbreit, 223–262. New York: Academic Press.

Pahl, S. L., A. K. Lee, T. Kalaitzidis, P. J. Ashman, S. Sathe, and D. M. Lewis. 2013. "Harvesting, Thickening and Dewatering Microalgae Biomass." In *Algae for Biofuels and Energy*, edited by M. A. Borowitzka and N. R. Moheimani, 165–185. Dordrecht: Springer Netherlands.

Palaniandy, P., H. Adlan, H. A. Aziz, M. F. Murshed, and Y. Hung. 2017. "Dissolved air flotation (DAF) for wastewater treatment." In *Waste Treatment in the Service and Utility Industries*, edited by Yung-Tse Hung, L. K. Wang, M.-H. S. Wang, N. K. Shammas and J. P. Chen, 145–182. Boca Raton, FL: Taylor & Francis Group.

Paul A., M., J. Patel, and A. Prem Rajan. 2014. "Algae Oil: A Sustainable Renewable Fuel of Future." *Biotechnology Research International* 2014:8. doi: 10.1155/2014/272814.

Peperzak, L., F. Colijn, R. Koeman, W. W. C. Gieskes, and J. C. A. Joordens. 2003. "Phytoplankton sinking rates in the Rhine region of freshwater influence." *Journal of Plankton Research* 25 (4):365–383. doi: 10.1093/plankt/25.4.365.

Pérez-Salado Kamps, Á., M. Jödecke, J. Xia, M. Vogt, and G. Maurer. 2006. "Influence of salts on the solubility of carbon dioxide in (water + methanol). Part 1: sodium chloride." *Industrial & Engineering Chemistry Research* 45 (4):1505–1515. doi: 10.1021/ie050865r.

Pestana, C., M. Holmes, P. Reeves, C. Chow, G. Newcombe, and J. West. 2015. Zeta potential measurement for water treatment coagulation control. In *Conference: Oz Water 2015*.

Pienkos, P. T., and A. Darzins. 2009. "The promise and challenges of microalgal-derived biofuels." *Biofuels, Bioproducts and Biorefining* 3 (4):431–440. doi: 10.1002/bbb.159.

Pieterse, A. J. H., and A. Cloot. 1997. "Algal cells and coagulation, flocculation and sedimentation processes." *Water Science and Technology* 36 (4):111–118. doi: 10.1016/s0273-1223(97)00427-7.

Pivokonska, L., and M. Pivokonsky. 2007. "On the fractionation of natural organic matter during water treatment." *Journal of hydrology and hydromechanics* 55 (4):253–261.

Pivokonsky, M., O. Kloucek, and L. Pivokonska. 2006. "Evaluation of the production, composition and aluminum and iron complexation of algogenic organic matter." *Water Research* 40 (16):3045–3052. doi: 10.1016/j.watres.2006.06.028.

Popper, Zoë A., M.-C. Ralet, and D. S. Domozych. 2014. "Plant and algal cell walls: Diversity and functionality." *Annals of Botany* 114 (6):1043–1048. doi: 10.1093/aob/mcu214.

Posten, C. 2009. "Design principles of photo-bioreactors for cultivation of microalgae." *Engineering in Life Sciences* 9 (3):165–177. doi: 10.1002/elsc.200900003.

Prabakaran, P., and A. D. Ravindran. 2011. "A comparative study on effective cell disruption methods for lipid extraction from microalgae." *Letters in Applied Microbiology* 53 (2):150–154. doi: 10.1111/j.1472-765X.2011.03082.x.

Prochazkova, G., I. Safarik, and T. Branyik. 2013. "Harvesting microalgae with microwave synthesized magnetic microparticles." *Bioresource Technology* 130 (0):472–477. doi: 10.1016/j.biortech.2012.12.060.

Pruvost, J., L. Pottier, and J. Legrand. 2006. "Numerical investigation of hydrodynamic and mixing conditions in a torus photobioreactor." *Chemical Engineering Science* 61 (14):4476–4489. doi: 10.1016/j.ces.2006.02.027.

Pulz, O. 2001. "Photobioreactors: Production systems for phototrophic microorganisms." *Applied Microbiology and Biotechnology* 57 (3):287–293. doi: 10.1007/s002530100702.

Putt, R. 2007. Algae as a biodiesel feedstock: A feasibility assessment In *Algae as a biodiesel feedstock: A feasibility assessment* edited by Ron Putt. Alabama, US: Center for Microfibrous Materials Manufacturing (CM3), Department of Chemical Engineering, Auburn University.

Rajasekhar, P., L. Fan, T. Nguyen, and F. A. Roddick. 2012. "A review of the use of sonication to control cyanobacterial blooms." *Water Research* 46 (14):4319–4329.

Ranjbar, R., R. Inoue, T. Katsuda, H. Yamaji, and S. Katoh. 2008. "High efficiency production of astaxanthin in an airlift photobioreactor." *Journal of Bioscience and Bioengineering* 106 (2):204–207. doi: 10.1263/jbb.106.204.

Raut, N., S. Panwar, and R. Vaidya. 2015. *Electrofloculation for Harvesting Microalgae.* Edited by Scampbell, *Harnessing clean and green energy vi aintegrated treatment of industrial and domestic waste water.* Germany: Lambert Academic Publishing.

Reckermann, M. 2000. "Flow sorting in aquatic ecology." *2000* 64 (2):12. doi: 10.3989/scimar.2000.64n2235.

Rees, T. F., J. A. Leenheer, and J. F. Ranville. 1991. "Use of a single-bowl continuous-flow centrifuge for dewatering suspended sediments: Effect on sediment physical and chemical characteristics." *Hydrological Processes* 5 (2):201–214. doi: 10.1002/hyp.3360050207.

Reyhani, A., K. Sepehrinia, S. M. S. Shahabadi, F. Rekabdar, and A. Gheshlaghi. 2015. "Optimization of operating conditions in ultrafiltration process for produced water treatment via Taguchi methodology." *Desalination and Water Treatment* 54 (10):2669–2680. doi: 10.1080/19443994.2014.904821.

Richardson, J. W., M. D. Johnson, and J. L. Outlaw. 2012. "Economic comparison of open pond raceways to photo bio-reactors for profitable production of algae for transportation fuels in the Southwest." *Algal Research* 1 (1):93–100. doi: 10.1016/j.algal.2012.04.001.

Richardson, J. W., M. D. Johnson, X. Z., P. Zemke, W. Chen, and Q. Hu. 2014. "A financial assessment of two alternative cultivation systems and their contributions to algae biofuel economic viability." *Algal Research* 4:96–104. doi: 10.1016/j.algal.2013.12.003.

Richmond, A., S. Boussiba, A. Vonshak, and R. Kopel. 1993. "A new tubular reactor for mass production of microalgae outdoors." *Journal of Applied Phycology* 5 (3):327–332. doi: 10.1007/bf02186235.

Ríos, S. D., J. Salvadó, X. Farriol, and C. Torras. 2012. "Antifouling microfiltration strategies to harvest microalgae for biofuel." *Bioresource Technology* 119:406–418. doi: 10.1016/j.biortech.2012.05.044.

Rodolfi, L., G. C. Zittelli, N. Bassi, G. Padovani, N. Biondi, G. Bonini, and M. R. Tredici. 2009. "Microalgae for oil: Strain selection, induction of lipid synthesis and outdoor mass cultivation in a low-cost photobioreactor." *Biotechnology and Bioengineering* 102 (1):100–112. doi: 10.1002/bit.22033.

Rodriguez, J., S. Stopić, G. Krause, and B. Friedrich. 2007. "Feasibility assessment of electrocoagulation towards a new sustainable wastewater treatment." *Environmental Science and Pollution Research—International* 14 (7):477–482. doi: 10.1065/espr2007.05.424.

Roselet, F., D. Vandamme, M. Roselet, K. Muylaert, and P. C. Abreu. 2015. "Screening of commercial natural and synthetic cationic polymers for flocculation of freshwater and marine microalgae and effects of molecular weight and charge density." *Algal Research* 10:183–188. doi: 10.1016/j.algal.2015.05.008.

Roselet, F., J. Burkert, and P. Cesar Abreu. 2016. "Flocculation of Nannochloropsis oculata using a tannin-based polymer: Bench scale optimization and pilot scale reproducibility." *Biomass and Bioenergy* 87:55–60. doi: 10.1016/j.biombioe.2016.02.015.

Roselet, F., P. Maicá, T. Martins, and P. C. Abreu. 2013. "Comparison of open-air and semi-enclosed cultivation system for massive microalgae production in sub-tropical and temperate latitudes." *Biomass and Bioenergy* 59:418–424. doi: 10.1016/j.biombioe.2013.09.014.

Rubio, F. Camacho, F. G. Acién Fernández, J. A. Sánchez Pérez, F. García Camacho, and E. M. Grima. 1999. "Prediction of dissolved oxygen and carbon dioxide concentration profiles in tubular photobioreactors for microalgal culture." *Biotechnology and Bioengineering* 62 (1):71–86. doi: 10.1002/(SICI)1097-0290(19990105)62:1 < 71::AID-BIT9 > 3.0.CO; 2-T.

Ruen-ngam, D., P. Wongsuchoto, A. Limpanuphap, T. Charinpanitkul, and P. Pavasant. 2008. "Influence of salinity on bubble size distribution and gas–liquid mass transfer in airlift contactors." *Chemical Engineering Journal* 141 (1):222–232. doi: 10.1016/j. cej.2007.12.024.

Sánchez Roque, Y., Y. D. C. Pérez-Luna, J. M. Acosta, N. F. Vázquez, R. B. Hernández, S. S. Trinidad, and J. S. Pathiyamattom. 2018. "Evaluation of the population dynamics of microalgae isolated from the state of Chiapas, Mexico with respect to the nutritional quality of water." *Biodiversity data journal* (6):e28496-e28496. doi: 10.3897/BDJ.6.e28496.

Sardeing, R., P. Painmanakul, and G. Hébrard. 2006. "Effect of surfactants on liquid-side mass transfer coefficients in gas–liquid systems: A first step to modeling." *Chemical Engineering Science* 61 (19):6249–6260. doi: 10.1016/j.ces.2006.05.051.

Schlagermann, P., G. Göttlicher, R. Dillschneider, R. Rosello-Sastre, and C. Posten. 2012. "Composition of Algal Oil and Its Potential as Biofuel." *Journal of Combustion* 2012:14. doi: 10.1155/2012/285185.

Segré, G., and A. Silberberg. 1961. "Radial particle displacements in poiseuille flow of suspensions." *Nature* 189 (4760):209–210. doi: 10.1038/189209a0.

Sforza, E., D. Simionato, G. M. Giacometti, A. Bertucco, and T. Morosinotto. 2012. "Adjusted light and dark cycles can optimize photosynthetic efficiency in algae growing in photobioreactors." *PLoS One* 7 (6):e38975. doi: 10.1371/journal.pone.0038975.

Sharma, K. K., S. Garg, Y. Li, A. Malekizadeh, and P. M. Schenk. 2013. "Critical analysis of current Microalgae dewatering techniques." *Biofuels* 4 (4):397–407. doi: 10.4155/bfs.13.25.

Shelef, G., A. Sukenik, and M. Green. 1984. Microalgae harvesting and processing: A literature review. United States: Technion Research and Development Foundation Ltd., Haifa (Israel).

Show, K.-Y., and D.-J. Lee. 2014. "Algal biomass harvesting." In *Biofuels from Algae*, edited by Ashok Pandey, Duu-Jong Lee, Yusuf Chisti and Carlos R. Soccol, 85–110. Amsterdam: Elsevier.

Show, K.-Y., D.-J. Lee, and A. S. Mujumdar. 2015. "Advances and Challenges on Algae Harvesting and Drying." *Drying Technology* 33 (4):386–394. doi: 10.1080/07373937.2014.948554.

Show, K.-Y., D.-J. Lee, and J.-S. Chang. 2013. "Algal biomass dehydration." *Bioresource Technology* 135:720–729. doi: 10.1016/j.biortech.2012.08.021.

Silva, T. L. da. 2015. "Scale-up Problems for the Large Scale Production of Algae." In *Algal Biorefinery: An Integrated Approach*, edited by Debabrata Das, 125–149. Cham, Switzerland: Springer International Publishing.

Sims, R., C. Miller, J. T. Ellis, A. Sathish, R. Anthony, and A. Rahman. 2014. Methods for harvesting and processing biomass. Google Patents.

Singh, S. P., and P. Singh. 2014. "Effect of CO_2 concentration on algal growth: A review." *Renewable and Sustainable Energy Reviews* 38:172–179. doi: 10.1016/j.rser.2014.05.043.

Şirin, S., R. Trobajo, C. Ibanez, and J. Salvadó. 2012. "Harvesting the microalgae *Phaeodactylum tricornutum* with polyaluminum chloride, aluminium sulphate, chitosan and alkalinity-induced flocculation." *Journal of Applied Phycology* 24 (5):1067–1080. doi: 10.1007/s10811-011-9736-6.

Sivaramakrishnan, R., and A. Incharoensakdi. 2018. "Microalgae as feedstock for biodiesel production under ultrasound treatment—A review." *Bioresource Technology* 250:877–887. doi: 10.1016/j.biortech.2017.11.095.

Slade, R., and A. Bauen. 2013. "Micro-algae cultivation for biofuels: Cost, energy balance, environmental impacts and future prospects." *Biomass and Bioenergy* 53 (0):29–38. doi: 10.1016/j.biombioe.2012.12.019.

Souza, F. L., S. Cotillas, C. Saéz, P. Cañizares, M. R. V. Lanza, A. Seco, and M. A. Rodrigo. 2016. "Removal of algae from biological cultures: A challenge for electrocoagulation?" *Journal of Chemical Technology and Biotechnology* 91 (1):82–87. doi: 10.1002/jctb.4580.

Spolaore, P., C. Joannis-Cassan, E. Duran, and A. Isambert. 2006. "Commercial applications of microalgae." *Journal of Bioscience and Bioengineering* 101 (2):87–96. doi: 10.1263/jbb.101.87.

Stephenson, A. L., Elena Kazamia, J. S. Dennis, C. J. Howe, S. A. Scott, and A. G. Smith. 2010. "Life-Cycle Assessment of Potential Algal Biodiesel Production in the United Kingdom: A Comparison of Raceways and Air-Lift Tubular Bioreactors." *Energy & Fuels* 24 (7):4062–4077. doi: 10.1021/ef1003123.

Sukenik, A., D. Bilanovic, and G. Shelef. 1988. "Flocculation of microalgae in brackish and sea waters." *Biomass* 15 (3):187–199. doi: 10.1016/0144-4565(88)90084-4.

Sun, A., R. Davis, M. Starbuck, A. Ben-Amotz, R. Pate, and P. T. Pienkos. 2011. "Comparative cost analysis of algal oil production for biofuels." *Energy* 36 (8):5169–5179. doi: 10.1016/j.energy.2011.06.020.

Sun, X., C. Wang, Y. Tong, W. Wang, and J. Wei. 2014. "Microalgae filtration by UF membranes: Influence of three membrane materials." *Desalination and Water Treatment* 52 (28–30):5229–5236. doi: 10.1080/19443994.2013.813103.

Tadros, M. G. 1985. *Screening and characterizing oleaginous microalgal species from the southeastern United States: A final subcontract report.* edited by Robins Mcintosh. Colorado: Solar Energy Research Institute—A Division of Midwest Research Institute.

Tamaru, Y., Y. Takani, T. Yoshida, and T. Sakamoto. 2005. "Crucial role of extracellular polysaccharides in desiccation and freezing tolerance in the terrestrial cyanobacterium Nostoc commune." *Applied and Environmental Microbiology* 71 (11):7327–7333. doi: 10.1128/aem.71.11.7327-7333.2005.

Terry, K. L., and L. P. Raymond. 1985. "System design for the autotrophic production of microalgae." *Enzyme and Microbial Technology* 7 (10):474–487. doi: 10.1016/0141-0229(85)90148-6.

Thangavel, K., P. R. Krishnan, S. Nagaiah, S. Kuppusamy, Senthil Chinnasamy, Jude Sudhagar Rajadorai, Gopal Nellaiappan Olaganathan, and Balachandar Dananjeyan. 2018. "Growth and metabolic characteristics of oleaginous microalgal isolates from Nilgiri biosphere Reserve of India." *BMC Microbiology* 18 (1):1. doi: 10.1186/s12866-017-1144-x.

Thomas, W. H., T. G. Tornabene, and J. Weissman. 1984. Screening for lipid yielding microalgae: Activities for 1983. Final subcontract report. United States.

Ting, H., L. Haifeng, M. Shanshan, Y. Zhang, L. Zhidan, and N. Duan. 2017. "Progress in microalgae cultivation photobioreactors and applications in wastewater treatment: A review." 10.

Tirichine, L., and C. Bowler. 2011. "Decoding algal genomes: Tracing back the history of photosynthetic life on Earth." *The Plant Journal* 66 (1):45–57. doi: 10.1111/j.1365-313X.2011.04540.x.

Tran, D.-T., B.-H. Le, D.-J. Lee, C.-L. Chen, H.-Y. Wang, and J.-S. Chang. 2013. "Microalgae harvesting and subsequent biodiesel conversion." *Bioresource Technology* 140 (0):179–186. doi: 10.1016/j.biortech.2013.04.084.

Uduman, N., Y. Qi, M. K. Danquah, G. M. Forde, and A. Hoadley. 2010. "Dewatering of microalgal cultures: A major bottleneck to algae-based fuels." *Journal of Renewable and Sustainable Energy* 2 (1). doi: 10.1063/1.3294480.

Ugwu, C., J. Ogbonna, and H. Tanaka. 2002. "Improvement of mass transfer characteristics and productivities of inclined tubular photobioreactors by installation of internal static mixers." *Applied Microbiology and Biotechnology* 58 (5):600–607. doi: 10.1007/s00253-002-0940-9.

Ullah, K., Sofia, M. A., V. K. Sharma, P. Lu, A. Harvey, M. Zafar, S. Sultana, and C. N. Anyanwu. 2014. "Algal biomass as a global source of transport fuels: Overview and development perspectives." *Progress in Natural Science: Materials International* 24 (4):329–339. doi: 10.1016/j.pnsc.2014.06.008.

Usher, P. K., A. B. Ross, M. A. Camargo-Valero, A. S. Tomlin, and W. F. Gale. 2014. "An overview of the potential environmental impacts of large-scale microalgae cultivation." *Biofuels* 5 (3):331–349. doi: 10.1080/17597269.2014.913925.

Valério, E., S. Chaves, and R. Tenreiro. 2010. "Diversity and impact of prokaryotic toxins on aquatic environments: A review." *Toxins* 2 (10):2359–2410. doi: 10.3390/toxins2102359.

Valiorgue, P., H. B. Hadid, M. E. hajem, L. Rimbaud, A. Muller-Feuga, and J.-Y. Champagne. 2014. "CO_2 mass transfer and conversion to biomass in a horizontal gas-liquid photobioreactor." *Chemical Engineering Research and Design* 92 (10):1891–1897. doi: 10.1016/j.cherd.2014.02.021.

Vandamme, D., I. Foubert, and K. Muylaert. 2013. "Flocculation as a low-cost method for harvesting microalgae for bulk biomass production." *Trends in Biotechnology* 31 (4):233–239. doi: 10.1016/j.tibtech.2012.12.005.

Vandamme, D., S. C. V. Pontes, K. Goiris, I. Foubert, L. J. J. Pinoy, and K. Muylaert. 2011. "Evaluation of electro-coagulation–flocculation for harvesting marine and freshwater microalgae." *Biotechnology and Bioengineering* 108 (10):2320–2329. doi: 10.1002/bit.23199.

Venteris, E. R., M. S. Wigmosta, A. M. Coleman, and R. L. Skaggs. 2014. "Strain Selection, Biomass to Biofuel Conversion, and Resource Colocation have Strong Impacts on the Economic Performance of Algae Cultivation Sites." *Frontiers in Energy Research* 2 (37):1–10. doi: 10.3389/fenrg.2014.00037.

Venteris, E. R., R. L. Skaggs, A. M. Coleman, and M. S. Wigmosta. 2013. "A GIS Cost Model to Assess the Availability of Freshwater, Seawater, and Saline Groundwater for Algal Biofuel Production in the United States." *Environmental Science & Technology* 47 (9):4840–4849. doi: 10.1021/es304135b.

Villacorte, L. O., M. D. Kennedy, G. L. Amy, and J. C. Schippers. 2009. "The fate of Transparent Exopolymer Particles (TEP) in integrated membrane systems: Removal through pre-treatment processes and deposition on reverse osmosis membranes." *Water Research* 43 (20):5039–5052. doi: 10.1016/j.watres.2009.08.030.

Wang, H., W. Zhang, L. Chen, J. Wang, and T. Liu. 2013. "The contamination and control of biological pollutants in mass cultivation of microalgae." *Bioresource Technology* 128 (0):745–750. doi: 10.1016/j.biortech.2012.10.158.

Wang, L., B. Pan, Y. Gao, C. Li, J. Ye, L. Yang, Y. Chen, Q. Hu, and X. Zhang. 2019. "Efficient membrane microalgal harvesting: Pilot-scale performance and techno-economic analysis." *Journal of Cleaner Production* 218:83–95. doi: 10.1016/j.jclepro.2019.01.321.

Wang, T., H. Yabar, and Y. Higano. 2013. "Perspective assessment of algae-based biofuel production using recycled nutrient sources: The case of Japan." *Bioresource Technology* 128 (0):688–696. doi: 10.1016/j.biortech.2012.10.102.

Watanabe, Y., and H. Saiki. 1997. "Development of a photobioreactor incorporating *Chlorella* sp. for removal of CO_2 in stack gas." *Energy Conversion and Management* 38:S499-S503. doi: 10.1016/S0196-8904(96)00317-2.

Wen, X., K. Du, Z. Wang, X. Peng, L. Luo, H. Tao, Y. Xu, D. Zhang, Y. Geng, and Y. Li. 2016. "Effective cultivation of microalgae for biofuel production: A pilot-scale evaluation of a novel oleaginous microalga Graesiella sp. WBG-1." *Biotechnology for Biofuels* 9 (1):123. doi: 10.1186/s13068-016-0541-y.

Wicaksana, F., A. G. Fane, and V. Chen. 2006. "Fibre movement induced by bubbling using submerged hollow fibre membranes." *Journal of Membrane Science* 271 (1–2):186–195. doi: 10.1016/j.memsci.2005.07.024.

Wicaksana, F., A. G. Fane, P. Pongpairoj, and R. Field. 2012. "Microfiltration of algae (Chlorella sorokiniana): Critical flux, fouling and transmission." *Journal of Membrane Science* 387–388:83–92. doi: 10.1016/j.memsci.2011.10.013.

Wiley, P. F., K. J. Brenneman, and A. E. Jacobson. 2009. "Improved algal harvesting using suspended air flotation." *Water Environ Res* 81 (7):702–708.

Xiangyu, H., G. Lili, C. Zhendong, and Y. Jianhua. 2018. "Corrosion and protection of metal in the seawater desalination." *IOP Conference Series: Earth and Environmental Science* 108 (2):022037.

Xu, L., C. Guo, F. Wang, S. Zheng, and C.-Z. Liu. 2011. "A simple and rapid harvesting method for microalgae by in situ magnetic separation." *Bioresource Technology* 102 (21):10047–10051. doi: 10.1016/j.biortech.2011.08.021.

Xu, L., P. J. Weathers, X.-R. Xiong, and C.-Z. Liu. 2009. "Microalgal bioreactors: Challenges and opportunities." *Engineering in Life Sciences* 9 (3):178–189. doi: 10.1002/elsc.200800111.

Yang, Z., and F. Kong. 2012. "Formation of large colonies: A defense mechanism of *Microcystis aeruginosa* under continuous grazing pressure by *flagellate Ochromonas sp.*" *Journal of Limnology* 71 (1). doi: 10.4081/jlimnol.2012.e5.

Yusuf, Z., N. A. Wahab, and S. Sahlan. 2016. "Fouling control strategy for submerged membrane bioreactor filtration processes using aeration airflow, backwash, and relaxation: A review." *Desalination and Water Treatment* 57 (38):17683–17695. doi: 10.1080/19443994.2015.1086893.

Zenouzi, A., B. Ghobadian, M. A. Hejazi, and P. Rahnemoon. 2013. "Harvesting of microalgae Dunaliella salina using electroflocculation." *Journal of Agricultural Science and Technology* 15 (5):879–887.

Zhang, B., and S. Chen. 2015. "Optimization of culture conditions for Chlorella sorokiniana using swine manure wastewater." *Journal of Renewable and Sustainable Energy* 7 (3):033129. doi: 10.1063/1.4923326.

Zhang, D, Y. Yu, C. Li, C. Chai, L. Liu, J. Liu, and Y. Feng. 2015. "Factors affecting microalgae harvesting efficiencies using electrocoagulation-flotation for lipid extraction." *RSC Advances* 5 (8):5795–5800. doi: 10.1039/c4ra09983d.

Zhang, G., P. Zhang, and M. Fan. 2009. "Ultrasound-enhanced coagulation for Microcystis aeruginosa removal." *Ultrasonics Sonochemistry* 16 (3):334–338. doi: 10.1016/j.ultsonch.2008.10.014.

Zhang, T. 2013. "Dynamics of fluid and light intensity in mechanically stirred photobioreactor." *Journal of Biotechnology* 168 (1):107–116. doi: 10.1016/j.jbiotec.2013.07.007.

Zhang, X., J. C. Hewson, P. Amendola, M. Reynoso, M. Sommerfeld, Y. Chen, and Q. Hu. 2014. "Critical evaluation and modeling of algal harvesting using dissolved air flotation." *Biotechnology and Bioengineering* 111 (12):2477–2485. doi: 10.1002/bit.25300.

Zhang, X., L. Wang, M. Sommerfeld, and Q. Hu. 2016. "Harvesting microalgal biomass using magnesium coagulation-dissolved air flotation." *Biomass and Bioenergy* 93:43–49. doi: 10.1016/j.biombioe.2016.06.024.

Zheng, H., Z. Gao, J. Yin, X. Tang, X. Ji, and H. Huang. 2012. "Harvesting of microalgae by flocculation with poly (γ-glutamic acid)." *Bioresource Technology* 112:212–220. doi: 10.1016/j.biortech.2012.02.086.

Zhou, W., L. Gao, W. Cheng, L. Chen, J. Wang, H. Wang, W. Zhang, and T. Liu. 2016. "Electroflotation of *Chlorella* sp. assisted with flocculation by chitosan." *Algal Research* 18:7–14. doi: 10.1016/j.algal.2016.05.029.

Zittelli, G. C., L. Rodolfi, and M. R. Tredici. 2003. "Mass cultivation of Nannochloropsis sp. in annular reactors." *Journal of Applied Phycology* 15 (2):107–114. doi: 10.1023/a:1023830707022.

10 Trends and Bottlenecks in the Sustainable Mitigation of Environmental Issues Using Microalgae

Debasish Das, Tridib Kumar Bhowmick,
Muthusivaramapandian Muthuraj

CONTENTS

10.1 INTRODUCTION

The major issue that alarmed life on earth over the past decades of the 21st century is the environmental instability caused by water pollution and greenhouse gas emissions (Salama et al. 2017). The freshwater has been incessantly contaminated with pollutants, especially hazardous xenobiotic compounds, attributed to the increased industrialization and population expansion that have challenged environmental sustainability (Debnath 2019). Further, extensive utilization of conventional fossil fuels and urbanization have resulted in the release of greenhouse gases into the atmosphere, thereby leading to global warming (Depra et al. 2018). Anthropogenic CO_2, especially from the burning of conventional fuels, has contributed to more than 68% of the total greenhouse gases emitted (Pacheco et al. 2015). Further, a minimum reduction of 50% to 80% of CO_2 released into the atmosphere will eventually become mandatory to achieve a sustainable environment (Salama et al. 2017). These serious concerns must be addressed with a feasible and sustainable technology that can reduce contaminants in wastewater and the atmosphere, allowing its utilization and recycling for future generations.

Conventional strategies of wastewater treatment have less efficiency in terms of recovery at lower concentrations, require sophisticated plant installation, operate with high-energy inputs, and require regular maintenance, which incurs additional costs (Ummalyma et al. 2018). Thus, these treatment strategies and recycling of wastewater are still capital intensive and remain inviable in terms of techno-economic feasibility and sustainability (Dixit and Singh 2015). Among them, microalgal strains have an inherent capability to utilize atmospheric CO_2 as the major carbon source, with sunlight as the energy source and wastewater as the source of other nutrients and trace elements (Nayak et al. 2016; Bhowmick et al. 2019). These organisms also accumulate numerous value-added chemicals such as biofuels in their biomass, which can be obtained through an integrated biorefinery process, thereby leading to income generation. Thus, the unique features of microalgal cell factories have led to sustainable mitigation of the environmental issues, especially wastewater treatment and carbon sequestration (Ummalyma et al. 2018). However, complete realization of the potentials of microalgae-based wastewater treatment systems remains inaccessible due to several bottlenecks that prevail in both the upstream and downstream processes. To that end, the present study will review the various bottlenecks and the path ahead to achieving sustainability in resolving environmental issues by utilizing microalgal strains.

10.2 UNIQUENESS OF MICROALGAL STRAINS AND PAYBACKS IN BIOREMEDIATION

Microalgae are primitive organisms that existed in the prehistoric earth, which was filled with toxic gases, and remediated the earth to support life. This diversified group of organisms has thrived and tolerated the various environmental conditions that prevailed during the evolution of earth, which is attributed to their inherent ability to tolerate and adapt to the modulating environmental cues (Khan et al. 2009). This evolutionary history and strong adaptability have paved the way for microalgal strains to sustain and remediate hazardous compounds in various wastewaters. The other inherent capabilities of these photosynthetic oxygen producers is to (i) remove nutrients such as carbon, nitrogen, and phosphates from wastewater and effluents rich in organic matter; (ii) sequester CO_2 from the atmosphere and flue gases; (iii) remove heavy metals; (iv) remediate xenobiotic compounds by sorption, accumulation, and conversion technologies; (v) remediate sites with multiple pollutants; (vi) detect toxicity of wastewater with algae-based biosensors; (vii) synthesize a variety of biofuels, such as biodiesel, due to their high neutral lipid contents as compared to plants (Debnath 2019; Muthuraj et al. 2019); (viii) can grow in consortium with other microalgal strains and bacteria; and (ix) are a source of a multiproduct paradigm (Figure 10.1). Over 15,000 novel chemical compounds have been harnessed from algal strains with various degrees of commercial importance (Tabatabaei et al. 2011).

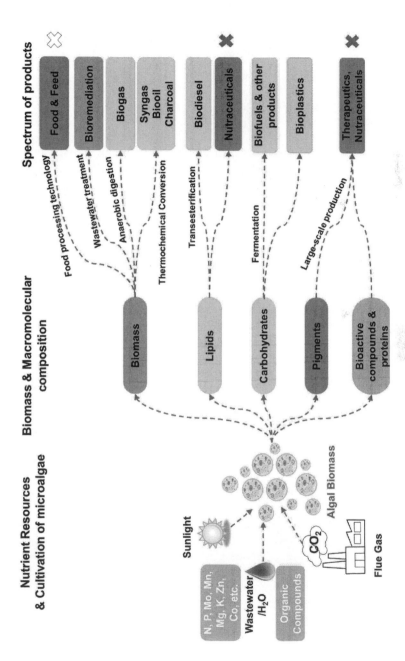

FIGURE 10.1 Potential multiproduct paradigm generated from a microalgal biomass and the macromolecular compositions. The filled cross-marks represent the products that cannot be generated when wastewater is used as the nutrient source, due to the regulatory issues, and the unfilled cross-marks represent the utilization of biomass generated from wastewaters as animal feeds and not as food for humans.

Source: The representation has been adopted and modified from Luque (2010)

With more than 1,54,000 strains identified so far (Guiry and Guiry 2019) and many more yet to be identified, the algal assemblage spreads among the prokaryotic members in two divisions: *Cyanophyta* and *Glaucocystophyta* (Croft et al. 2006), and among eukaryotic members they are grouped into nine divisions: *Glaucophyta, Rhodophyta, Heterokontophyta, Haptophyta, Cryptophyta, Dinophyta, Euglenophyta, Chlorarachinophyta*, and *Chlorophyta*.

Microalgae utilize natural resources, such as freely available sunlight, as the major energy source and adequately available atmospheric CO_2 as the carbon source for growth under photoautotrophic nutrition conditions (Muthuraj et al. 2014). These organisms possess photosynthetic activity that accounts for more than 50% of global photosynthesis, which effectively converts the energy of photosynthetically active radiation (PAR) at a wavelength band of 400 to 700 into a biomass via oxidation and reduction reactions. This accompanies the sequestration of atmospheric CO_2 in the form of a carbon source with a higher efficiency—up 10% to 50% compared to higher plants (Cheng et al. 2013). They also have the ability to grow under heterotrophic conditions by utilizing organic carbon compounds as the carbon and energy source. Photoheterotrophic cultivation requires light as the chief energy source and organic carbon compounds as the source of carbon (i.e., it requires both carbohydrate sugars and light at the same time for maximal productivity) (Chen et al. 2011). On the contrary, few microalgal strains follow mixotrophic growth conditions in which they perform both heterotrophic and light-dependent functions of growth in a simultaneous and independent manner (Muthuraj et al. 2014). With these wide nutritional types and capability to utilize different trophic modes, the organism has evolved well to organize metabolic pathways and transport machinery for efficient functioning under different cultivation conditions and in wastewaters.

In addition, these microbes have simple growth media requirements in the form of macronutrients (nitrogen, phosphorus, sulfur, potassium, magnesium, etc.) and trace elements such as metal salts (Cu, Fe, Zn, Co, Mn, etc.), along with water, which provides the physical environment for the growth of algae (Ummalyma et al. 2018). In general, chemical fertilizers are provided as the source of necessary macronutrients and trace elements to support the microalgal growth. It has been established that municipal wastewater, anaerobic digestion wastewater, piggery wastewater, and other industrial wastewaters have significantly similar macronutrient and trace element compositions that are essential for these photosynthetic microbes to grow (Salama et al. 2017). Thus, the presence of essential nutrients in wastewater supports the commercial-scale use of wastewater as a growth media for microalgae (Muylaert et al. 2015) instead of costlier chemical fertilizers.

It is also important to note that a significantly large amount of water is required to supply the demand of microalgal systems in terms of photosynthesis. All living organisms are made up of roughly 80% to 90% of water, and microalgae are not much different from other organisms. In contrast, these organisms require 1.0 mole of water to fix 1.0 mole of CO_2, and over 1.0 kg of water is required for the production of 1.0 g dry algal biomass (Shen 2014). Water also provides the physical environment for the organism to survive while facilitating the availability of nutrients and CO_2, maintenance of temperature, pH, and waste removal. It has also been estimated that over 6000 L of water will be required to produce 1.0 L of algal oil from the biomass

(Ozkan et al. 2012). Thus, the huge requirement of water for algal growth can be met through the utilization of wastewater and by reusing the utilized water in algal ponds after downstream processes. Therefore, integration of the microalgae-based bioremediation process with simultaneous production of value-added chemicals can enhance process sustainability and economic feasibility.

10.2.1 EFFICIENT NUTRIENT REMOVAL OF MICROALGAL STRAINS

Industrial wastewaters and agricultural waters are usually rich in nitrogen (N) and phosphorus (P) concentrations, with levels three times higher than the levels in freshwater bodies (Osorio et al. 2018), which on release may lead to eutrophication. It is also important to note that the global phosphorus reserve in the form of phosphate rocks are depleting at higher rates, and therefore there is an increased interest in the recovery and reutilization of phosphate sources to produce useful value-added compounds (Rittmann et al. 2011).

Among several strains available for bioremediation, the most common freshwater strains that predominate the wastewater treatment ponds are the *Scenedesmus* sp. and *Chlorella* sp. attributed to their robustness in terms of handling the varying nutritional limitation/repletion conditions and other environmental stresses (Gupta et al. 2017). Evaluation of four strains of *Chlorophyceae*, which includes axenic cultures of *Chlorella vulgaris, Chlorella minutissima, Scenedesmus* sp. and *Chlorococcum* sp., in terms of nutrient removal and growth in domestic wastewater showed higher efficiencies for *Scenedesmus* sp. compared to other strains (Nayak et al. 2016). The effect of the N/P ratio on the nitrate and phosphate removal efficiency of *Chlorella* sp. AUF_802 isolated from the nitrate-rich environment was evaluated in synthetic industrial wastewater by Osorio et al. (2018). The complete phosphorus removal was observed when nitrogen concentrations were five times higher than the phosphate concentration. This also confers the significance of maintaining a strain-dependent optimal N/P ratio for maximal removal efficiency (Osorio et al. 2018; Li et al. 2010). Table 10.1 represents the various studies that evaluated the performance of microalgal strains in the removal of nutrients such as nitrogen, phosphorus, etc., for their growth.

TABLE 10.1

Nutrient Removal Efficiencies of Various Microalgal Strains From Different Wastewater Types

Species	Wastewater type	Removal efficiency			Reference
		Nitrogen	Total Phosphorus	COD	
Chlorella pyrenoidosa	Settled sewage	81% NH_4^+ > 90% NO_3^-	NA	NA	Tam and Wong, (1990)
Botryococcus braunii	Secondary treated sewage	100% NO_{3-} 100% NO_2^-	NA	NA	Sawayama et al. 1992

(Continued)

TABLE 10.1 (*Continued*)

Species	Wastewater type	Removal efficiency			Reference
		Nitrogen	Total Phosphorus	COD	
Chlamydomonas reinhardtii	Municipal waste	43% TN	14%	NA	Kong et al. 2010
Chlorella vulgaris	Dairy manures	99.7% NH_4^+ 89.5% TN	92%	75.5% COD	Wang et al. (2010)
Chlorella vulgaris	Textile wastewater	44% NH_4^+	33% PO_4	38% COD	Lim et al. (2010)
Spirulina platensis	Swine wastewater	NA	41.6%	84.3% COD	Mezzomo et al. (2010)
Chlorella pyrenoidosa	Soybean processing wastewater	88% TN	70.30%	77.8% COD	Hongyang et al. (2011)
Chlorella sp.	Animal wastewater	58%–98% TN	51.9–57.9%	NA	Chen et al. (2012)
Scenedesmus obliquus	Brewery effluent	20.8% TN	NA	57.5% COD	Mata et al. (2012)
Spirulina platensis	Olive-oil mill wastewater	>99% TN	>99%	73% COD	Markou et al. (2012)
Chlorella PY-ZU1	Food waste effluents	99%	99%	99%	Cheng et al. (2013)
Desmodesmus communis	Biological wastewater	NA	100%	NA	Samorì et al. (2013)
Scenedesmus acutus	Municipal wastewater	42%–71% NO_3^- 93% NH_4^+	64% PO_4^{3-}	48–77%	Sacristán et al. (2013)
Scenedesmus obliquus	Municipal wastewater	89.68% TN	NA	NA	Arbib et al. (2013)
Chlorella zofingiensis	Piggery wastewater	82.7% TN	98.17%	79.84% COD	Zhu et al. (2013)
Chlorella vulgaris	Dairy wastewater	91–96% NO_3^-	NA	85–95% COD	Sreekanth et al. (2014)
Chlorella sp.	Mixed domestic-industrial wastewater	98.9% NH_4^+ 87.6% NO_3^-	90%	80.1% COD	Hammouda et al. (2015)
Spirulina platensis	Poultry litter wastewater	NA	>99% PO_4^{-3}	NA	Markou et al. (2016)
Neochloris aquatica CL-M1	swine wastewater	96.2% NH_4^+	NA	81.7% COD	Wang et al. (2017)

Source: Modified from Cuellar-Bermudez et al. (2017)
Note: NA, Not Available; TN, Total Nitrogen; COD, Chemical Oxygen Demand; PO_{4-}, phosphate

Several other operational factors and environmental parameters affect nutrient removal efficiency significantly. For instance, change in the wastewater pH affects the CO_2 uptake by the microalgal strains, and increased pH levels result in precipitation of phosphates, assisting in their recovery from wastewaters. Similarly, changes in temperature also modulate the solubility of nutrients and thereby reduce nutrient access for the organisms to flourish (Li et al. 2010). Being oxygenic microbes, microalgal strains also fix the biological oxygen demands and chemical oxygen demands in the wastewater with high efficiency (Table 10.1).

10.2.2 REMOVAL OF HEAVY METALS, RADIONUCLIDES, AND OTHER XENOBIOTICS BY MICROALGAE

Bioremediation of heavy metals from wastewater using a dead or living microalgal biomass can be economically feasible and environmentally sustainable with high removal efficiencies (Zeraatkar et al. 2016). Heavy metals such as aluminum, arsenic, gold, cadmium, chromium (di-, tri- and hexa-valent compounds), copper, mercury, nickel, lead, selenium, uranium, zinc, ferrous, etc., were reported to be remediated by microalgal strains. The strains utilized for the bioremoval of heavy metal ions include *Chlorella* sp., *Chlamydomonas reinhardtii*, *Chlorococcum* sp., *Cladophora fascicularis*, *Cyclotella cryptica*, *Lyngbyataylorii*, *Phaeodactylum tricornutum*, *Porphyridium purpureum*, *Scenedesmus* sp., *Spirogyra* sp., and *Spirulina platensis*, which comprise both eukaryotic and prokaryotic microalgal systems (Ummalyma et al. 2018; Zeraatkar et al. 2016). Among them, several strains of *Chlorella, Cladophora, Scenedesmsus, Laminaria,* and *Fucus spiralis* were found to uptake higher concentration of heavy metals in per g of biomass with the maximum up to 846 mg g^{-1} of biomass (Zeraatkar et al. 2016). The heavy metal uptake mechanisms vary mainly based on the microalgal strains; nature of the heavy metal ions; medium properties in which biosorption is conducted; operating conditions such as temperature, pressure, agitation, etc.; and the metabolic status of cells— either living or dead (Salama et al. 2017). Bioremediation of cadmium-contaminated wastewater by dried cells of *Hydrodictyon reticulatum* was evaluated by Ammari et al. (2017) and showed higher removal efficiencies, up to 92.8% at pH 4. It was also showed that the alkaline-treated algal biomass resulted in 21% higher efficiency compared to the intact dried algal cells. The non-living microalgal strains uptake the heavy metals through various mechanisms, such as surface adsorption, precipitation, ion exchange, complexation, and chelation or through passive diffusion. The living cells translocate the heavy metals through specific metal ion transporters, such as natural resistance-associated macrophage protein (NRAMP), copper transporter (CTR), zinc transport proteins (ZIP), ferrous transporting protein (FTR), etc., and detoxifies with phytochelatin complexes or metallothionein complexes in vacuoles and cytoplasm (Perales-Vela et al. 2006). However, recovery and reuse of heavy metals from the wastewater will be possible only if dead cells are used to adsorb the metal ions. Post-adsorption treatment such as washing with desorption agents, deionized water, etc., with additional physical and chemical treatments will assist in the recovery of heavy metals without losses. The degree of adsorption and the removal efficiency were also found to be higher in the dead cells compared to the living cells (Arica et al. 2005). Thus, it is evident that microalgae are a potent bioremediation tool for effective elimination of heavy metals from water sources via biosorption.

Over 450 civil nuclear power reactors are in operation and more 55 nuclear power reactors are under construction worldwide, with India alone having 22 functional reactors and 7 nuclear reactors under construction (World Nuclear Association 2019); thus, the potential for environmental contamination with radionuclides has increased significantly. Current remediation and removal methods rely on the resins and filters, which can work better to lower the concentrations in wastewater. However, the methods fail when the concentration of radioactive materials is too low for separation. In such cases, microalgae and cyanobacteria, both living and dead forms, are utilized to reduce and recover the radionuclides from wastewater. For instance, the eukaryotic green alga *Coccomyxa actinabiotis* has shown incredible efficiency in remediating— up to 100% ^{137}Cs in a mixture containing ^{238}uranium, tritium, and ^{14}C (Rivasseau et al. 2013) when a concentration of 67 Bq L^{-1} was used. Similarly, *Vacuoliviride crystalliferum* nak-9 showed a removal rate of 0.047 mg g^{-1} dry-biomass per day of ^{137}cesium and reached over 90% removal (Nakayama et al. 2015). Over 90% reduction in the radioactivity of strontium was observed when *Chlorella vulgaris* was utilized, which facilitated the conversion of strontium to SrCO$_3$. On the other hand, *Hematococcus pluvialis* was found to colonize the radioactive pond containing high amounts of ^{137}Cs and ^{90}Sr. Interestingly, the metabolic fingerprint of the organism was intact even after exposure to such high radiations of about 80-Gy dose (MeGraw et al. 2018) attributed to their high levels of antioxidants, especially astaxanthin, which assisted in evading the consequences of exposure to radioactive materials. The remediation potentials of photosynthetic organisms on the removal of radioactive materials such as ^{134}cesium, ^{139}cesium, ^{90}strontium, ^{131}iodine, ^{239}plutonium, ^{95}zirconium, and many more were conducted by Vanhoudt et al. (2018). The study ranked the dead biomass of *Nostoc carneum, Nostoc insulare, Oscillatoria geminate,* and *Spirulina laxissima* as potential organisms for the remediation of radionuclides and heavy metals with high probability, while living strains of *Microcystis* sp. were identified as organisms with high potential to remove chromium and ferrous from wastewater (Vanhoudt et al. 2018). Living strains of *Ananbena ambigua* were also ranked to be potential candidates that have immense efficiency in terms of the removal of heavy metals such as chromium, ferrous, and manganese (Vanhoudt et al. 2018). Thus, microalgal strains might have a profound capability to evade and reduce radioactive materials (MeGraw et al. 2018).

Oil sands process affected water commonly abbreviated as OSPW released during surface mining of oil sands are strenuous for remediation attributed to their high concentration of xenobitiotic compounds such as complex organic acids and aromatics. Screening of over 21 indigenous algal strains from the mining location at Alberta, identified *Stichococcus* sp. as the topmost reducer of acid-extractable organics at test concentrations of 10 to 100 mg L^{-1} (Ruffel et al. 2016). Hultberg et al. (2016) evaluated the potential of *Chlorella vulgaris* strain 211/11B of CCAP-SAMS to remove 38 different pesticides. The study utilized the living organism adopted with long term exposure to respective pesticides which resulted in the removal efficiency up to 10%

for 10 different pesticides. These studies suggest that microalgae could be a potential candidate for removing organic xenobiotics in wastewater.

10.3 MAJOR CHALLENGES AND TECHNO-ECONOMIC HURDLES IN THE PROCESS ROUTES

Development of a commercial-scale sustainable technology that couples microalgal biomass production with wastewater treatment is still unattainable due to the numerous challenges that prevail in a sequence from the upstream to the downstream process. For instance, the prime challenge is to select a microalgal cell factory with the capability to remove hazardous compounds from wastewater while generating the large amount of biomass required for further production of value-added chemicals. Once a microalgal strain has been selected, the next hurdle lies in the selection of a cultivation system, which will be employed for the large-scale treatment process and for the accumulation of microalgal biomass without contamination by predators. Harvesting of the microalgal biomass from the large-scale open pond or the reactor for the necessary pretreatment and biorefinery process will be the next bottleneck that requires attention. Finally, technologies to convert the algal biomass into value-added products and biorefining must be developed. Thus, sequential challenges in the overall process must be addressed for the complete realization of the technology on a commercial scale.

10.3.1 SELECTION OF WASTEWATER AND APPROPRIATE MICROALGAL STRAIN FOR REMEDIATION

As per the requirements of an industry, any wastewater to be treated can be taken up for phycoremediation after primary and secondary treatment processes, which involve the removal of solid particles, suspended particles, dechlorination, disinfection, and chemical precipitation methods. Microalgal-based bioremediation best suits the place of tertiary and quaternary treatment steps involved in the removal of nutrients and toxic compounds (Salama et al. 2017). Under photoautotrophic growth conditions of microalgae, the cells require sunlight as the major energy source, and therefore the presence of suspended particles and solid particle, might obstruct light penetration in the wastewater ponds. Therefore, prior treatment of wastewater to remove such hindrances will be a better option for maximizing algae-based bioremediation efficiency.

Before moving to the algae-based tertiary or quaternary treatment process, the wastewater composition should be matched with the minimal nutrient requirements for microalgal growth. For instance, the N/P ratio in the wastewater is one of the key factors that modulates the growth rate of microalgal strains. In general, the optimal N/P ratio for the growth of *Chlorella* sp. ranges near 7, whereas many wastewater possesses an N/P ratio less than 4 (Li et al. 2010; Cai et al. 2013). In such cases, the optimal N/P ratio must be maintained with supplementation of additional nutrients in the wastewater. On the other hand, if additional nutrients in the form of trace elements are unavailable, such as cobalt, molybdenum, magnesium, manganese, etc., these must be replenished to enhance the growth rates of microalgal strains.

Selecting the appropriate strain for wastewater treatment is one of the critical steps, as it determines the overall efficiency and process sustainability (Gupta et al. 2017). Finding a unique and ideal cellular factory is a difficult task, as strain-level variations in the remediating efficiencies have been reported. The desirable features required for a strain to be a potential commercial-scale producer are listed by Salama et al. (2017). The most important is the photosynthetic yield and efficiency. Photosynthetic yield is defined as the ratio of energy yield per photon consumed, whereas the photosynthetic efficiency is defined as the efficiency in converting every photon absorbed to energy and carbon. The maximum theoretical efficiency of converting sunlight to biomass is estimated to be 13%, in which the conventional growth conditions are efficient to harness less than half of the theoretical efficiency. Thus, the strains with enhanced photosynthetic yield and efficiency will have the potential to convert maximum photons absorbed into biomass with higher growth rates. The second important criterion is the carbon sequestration capability and tolerance to high CO_2 contents of flue gas compositions released from the boilers of industries. All industrial-scale phycoremediation processes will aim to utilize flue gas as the major source for inorganic carbon other than feeding CO_2. It is also important to note that the CO_2 injection system in large-scale open ponds may hold 27% of total algal production costs (Caia et al. 2018). In general, the flue gas comprises NO_x, SO_x, and heavy metals such as mercury in trace levels, which may eventually get concentrated in the pond. Therefore, the strain should be able to tolerate and should be capable of remediating these toxic compositions. In addition, the temperature of the flue gas is a parameter to be considered, and either the strain selected should have a wide range of temperature resistivity or the flue gas should be passed through cooling systems or heat exchangers to reduce the high temperatures (Caia et al. 2018). The third important criterion for strain selection is the robustness of the strain under outdoor conditions in terms of higher growth rates in continuously fluctuating environmental conditions and the shear that is caused by continuous mixing. The strains must be resistant to pathogens such as bacteria, fungi, and protozoans that feed on algae, which can destabilize even a high-cell-density pond within a day. The fourth criterion should be the ease of harvesting and dewatering. The strain to be used should have self-flocculation capabilities and should be easily separable (Salama et al. 2017; Gupta et al. 2018). The fifth criterion should be the options for biorefining or the multiproduct paradigm of the strains. Further, by modulating the process parameters or by adding elicitors, etc., the augmentation of quality and productivity may be achieved in the strain with high photosynthetic yields. Even though numerous strains have been identified for the removal of various degrees of toxic compounds, it is not certain that the algal species will perform to the same productivities under different environmental conditions. So, spatiotemporal isolation and subsequent mass screening of strains based on multicriteria selection design must be carried out to identify indigenous strains that are resistant to prevailing environmental conditions and pathogens in that respective area, along with substantial biomass productivity and high efficiency in the bioremediation of toxic wastewaters.

The best option for integrating microalgae cultivation and advanced wastewater treatment would be to identify the consortium of microalgae or a microalgae–bacteria consortium rather than utilizing axenic cultures of microalgae. The prime reason behind this is attributed to the complexity in maintaining a unialgal culture or a monoculture in open raceway ponds or in high-rate algal ponds. Several reports signify the necessity for a consortium of microalgal strains over unialgal cultures (Gonclaves et al. 2016). The major advantages include (i) the augmented nutrient and toxic compound removal efficiencies attributed to the probable symbiotic interactions within the consortium; (ii) wide range of substrate specificity and nutrient removal capability, paving way for multiple waste removal possibilities; (iii) offers resistance to invasion by other pathogenic microbes and protozoans; (iv) will be able to replace the secondary treatment process, thereby reducing the cost of chemical precipitations and aerations; and (v) cultivation of microalgae and bacteria together forms flocs, which may reduce the cost and energy needed for harvesting in commercial-scale ponds (Gonclaves et al. 2016). In comparison with microalgal consortiums, a bacteria–microalgal consortium may exhibit outstanding capabilities due to their possibly strong symbiotic and competitive interactions. For instance, the use of an activated microalgae–bacteria consortium will intensify the treatment process by combining the secondary and tertiary treatment processes altogether (Gonclaves et al. 2016). However, when it comes to revenue generation from the biomass, the consortium may reduce the efficiency in terms of product removal from specific systems and incur additional costs and energy requirements. In addition, maintaining an active consortium for a longer time over the seasonal changes remains difficult to achieve (Gonclaves et al. 2016). Thus, instead of monoculture, isolation of indigenous bacteria and a microalgal consortium that prevails in the specific wastewater under the native conditions may be screened for the effective bioremediation with product generation. Still, choosing a top-producing strain with all these criteria is a major challenge.

10.3.2 CULTIVATION SYSTEMS FOR COMMERCIAL-SCALE PRODUCTION OF MICROALGAE UTILIZING WASTEWATER

Commercial-scale growth of microalgal strains depends on open raceway ponds, photobioreactors (PBRs), turf scrubbers, and hybrid systems that combine the three other systems (Gupta et al. 2018). PBRs use a transparent closed vessel in which the natural energy source (sunlight) is replaced with an electrical light to run the system under phototrophic conditions. Stirred tank, vertical column, horizontal tubular, and flat-panel are the major types of PBRs that are used commonly for algal cultivation (Salama et al. 2017). The major advantages of PBRs over raceway ponds is that the environmental parameters and process variables can be continuously monitored and controlled efficiently while maintaining higher growth rates (Kumar et al. 2015). Being a closed environment, the chance of contamination is completely reduced in the case of PBR. Moreover, the photoconversion efficiency has been reported to be three-fold higher than that of the open pond systems. However, the energy consumption to run a PBR may go up to 15 W m^{-3} with the energy demand up to approximately 207 MJ m^{-2} $year^{-1}$, which is approximately 3.6-fold higher than that required for the open pond systems (Schlagermann et al. 2012). The algal turf scrubber (ATS) was designed by Dr. Walter Adey in early 1970, which is an engineering system that allows the flow of wastewater over the sloping surface in a pulsed manner, thereby allowing the growth of algal strains on the surface. The microalgal strains growing on the surface of slopes remove the nutrients from the wastewater flowing over them in pulses, thereby remediating the river streams, lakes or ponds, farm wastes, etc. (Adey et al. 2011; Salama et al. 2017). This treatment strategy handles over 4000 to 8000 m³ of water per day and generates an algal biomass at rates 5 to 10 times higher than that of an open raceway pond system. The ATS system has a low operating cost, with low monitoring and maintenance requirements. The system also possesses less contamination risk and higher productivity as compared to open pond systems. The major drawbacks of such a system are the requirements of space and infrastructure for construction (Salama et al. 2017).

Irrespective of different cultivation systems and hybrid systems, open raceway ponds are the all-time favorite due to their simple design, easy scalability, large production volumes, less operating costs, and easier maintenance (Salama et al. 2017), except for rainy seasons and poor climatic conditions. From the early 1950s, high-rate algal ponds (HRAPs) are used extensively for wastewater treatment using microalgal systems (Oswald and Golueke 1960), which comprise raceway ponds, circular ponds, tanks, and shallow big ponds. Located in Australia, the largest commercial open pond system spreads for 200 ha each and grows *Dunaliella salina* for β-carotene synthesis. The major challenges in utilizing raceway ponds for microalgal growth are the high risk of contamination, reduced photoconversion efficiencies up to 1.5% with a theoretical estimation of 12.8% to 14.4% conversion of solar energy into biomass, reduced light penetration, shadow effects, continuous mixing, seasonal and weather changes, high-temperature fluctuation during the diurnal cycles, low growth rates, large land requirements, and high harvesting costs (Kumar et al. 2015). The other parameter to be monitored is the cost for CO_2 injection into open raceway pond system, which holds over 27% of the total cost of microalgal production if separate arrangements are required, contrary to primitive surface aeration (Caia et al. 2018). Analysis of the techno-economic feasibility for the production of a microalgal biomass utilizing ATS and open raceway ponds resulted in a biomass production cost of US$510 t⁻¹ and US$673 t⁻¹, respectively (Hoffman et al. 2017). A study on the energy balance of microalgal biomass production using the activated sludge processed wastewater in four raceway ponds with a volume of ~9500 L each showed that the coupling of wastewater treatment with microalgal cultivation can be energetically favorable (Sturm et al. 2011). Similarly, it was reported that when HRAPs are utilized for the growth of biomass in wastewater, the overall production cost for 1 m³ algal oil was found to be 14.6-fold less than the cost acquired when HRAP was used for algal growth without wastewater (Lundquist et al. 2010). Thus, open raceway pond systems, ATS, or hybrid systems, when using wastewater as the nutrient source for microalgal growth, could be economically feasible and sustainable. However, purging of CO_2 and other critical parameters, such as light, temperature, nutrient concentration, cultivation mode, hydraulic retention time, agitation for homogenous mixing, algal recycling, microalgal growth, biomass productivity, contamination by protozoans and other grazers, are to be optimized and managed for a maximal reduction in cost.

10.3.3 CHALLENGES IN THE SELECTION OF APPROPRIATE DOWNSTREAM PROCESSES

Sustainability in microalgae-based processes is limited by many factors, with the major challenge being the downstream process. Harvesting and dewatering of algal cells, pretreatment of algal biomass, and further utilization in the biorefinery to convert the biomass components into value-added products are the major cost-consuming steps (Williams and Laurens 2010; Depra et al. 2018) in which harvesting alone consumes 20% to 30% of the total project cost (Rawat et al. 2011). The efficiency and methodology used in these three steps can have a major impact on the economics of the commercial-scale process, and still there exists huge uncertainty in process modeling and cost estimation. The high cost of harvesting is attributed to the dilute nature of the broth, with less biomass fraction (10^3 to 10^8 cells mL^{-1}), smaller cell size, and the negative charge of cells, with excessive extracellular algal materials that keep them stable in a dispersed state (Danquah et al. 2009). The major techniques available for harvesting and to concentrate the algal cells include gravity settling, centrifugation, flocculation, filtration, flotation, electrocoagulation, electrolysis, and electrophoresis, which can be segregated broadly as mechanical, chemical, biological, and electrical methods (Pragya et al. 2013). The harvesting method selected should be applicable to a wide range of algal species and should utilize less energy and fewer chemicals.

Table 10.2 compares various harvesting techniques available based on their energy utilization and recovery efficiencies. Gravity sedimentation is the most energy-efficient harvesting method that settles out the suspended cells using gravitational force, forming a concentrated slurry at the bottom and clear liquid at the top (Uduman et al. 2010). The efficiency of gravity sedimentation is enhanced by using lamellar separators and sedimentation tanks for oleaginous algal systems. Addition of flocculants, on the other hand, increases the rate of sedimentation to a greater extent than is being commercially used (Pragya et al. 2013). Microalgal cells carry an electronegative charge on their surface, which may vary between 2.5 and 11.5, and therefore, the addition of flocculants neutralizes their surface charge and enables particle bridging that results in cell aggregation (Pragya et al. 2013). Different flocculating agents are available that have a varying influence on the flocculation process, which includes inorganic (alum, ferric sulfate, and lime) and polymeric forms such as Purifloc, Zetag 51, Dow 21M, Dow C-31, Chitosan, etc. (Uduman et al. 2010). The polymeric form of flocculants include both ionic and non-ionic molecules, which work by forming electrostatic or chemical bonding forces, and the efficiency depends upon their charge density and polymer chain length. Addition of iron or aluminium-based coagulants causes the charge neutralization in the algal cells to be based upon their charge density. Another interesting method is autoflocculation, in which the algal cells spontaneously arrange in sediments to form flocs, and is often associated with elevated pH or excretion of polymeric macromolecules (Park et al. 2011).

Changing the environmental pH or low-temperature condition alters the cell wall composition of the algae, thereby inducing aggregation of the cells. Addition of NaOH increases the pH to the alkali side, which induces many algal cells to

TABLE 10.2

Comparison of Various Harvesting Techniques Based on Their Efficiencies in Concentrating Microalgae and Energy Requirements

Methods	Microalgal biomass recovery (%)	Energy requirement (kW h m⁻³)	References
Centrifugation	94	0.8	Dassey and Theegala (2013)
Filtration (polyvinylidene difluoride membrane)	98–99	0.91	Bilad et al. (2012)
Filtration (polypropylene nonwoven fabric membrane)	94	NA	Sahoo et al. (2017)
Filtration (polyethersulfone)	99.9	0.9	Gerardo et al. (2014)
Ultrafiltration and centrifugation	80	0.169	Baerdemaeker et al. (2013)
Membrane filtration and centrifugation	80	0.284	Baerdemaeker et al. (2013)
Tangential flow filtration	89	2.06	Danquah et al. (2009)
Pressure filtration	99	0.18	Grima et al. (2003)
Vacuum filtration	99	1.23	Grima et al. (2003)
UV-C induced primary settling	NA	0.576	Sharma et al. (2014)
Flocculation	95	NA	Uduman et al. (2010)
Flocculation-flotation	90	10–20	Uduman et al. (2010)
Flotation	49.5	0.16–0.44	Barrut et al. (2013)
Electrocoagulation	90	1.08	Kim et al. (2012)
Electrocoagulation-flocculation	90	2	Vandamme et al. (2012)
Electrochemical (carbon electrode)	94.52	1.6	Misra et al. (2014)

Source: Table obtained and modified from Gupta et al. (2018)

aggregate themselves (Chen et al. 2011; Vandamme et al. 2012). Studies on biofloc-culation induced by the co-culturing of *Nannochloropsis* cells with bacterium show it to be an energy-efficient strategy for algal harvesting (Wang et al. 2012). Similarly, algal-fungal, algal-algal, and alga-bacterial flocculation strategies are being developed which are showing promising economic effects (Gupta et al. 2018). In sustainable mitigation of environmental issues such as large-scale wastewater treatment, the algal-bacterial consortium utilized in the wastewater treatment are selected as activated algae. Tiron et al. (2017) utilized activated granules of algae instead of algal flocs to achieve high efficiency in separation and cost reduction. Addition of the filamentous microalga *Phormidium* sp. to the open pond with *Chlorella* sp. resulted in the formation of activated granules with *Chlorella* cells trapped in the filamentous microalgae. The activated granules measured 0.6 to 2 mm, which settled at rates 21.6 m h^{-1} with a complete microalgal removal efficiency of 99% (Tiron et al. 2017). Centrifugation is an alternative method that utilizes centripetal acceleration to separate the algal cells with removal efficiencies up to 99% in a time frame of 2 to 5 minutes, with a high energy consumption up to 3000 kW h t^{-1} (Tiron et al. 2017). However, the process is highly energy intensive and not suitable for large commercial-scale biodiesel production (Uduman et al. 2010). Filtration is another process which retains the algal cells and allows the water to pass through the filters. Depending on the type of filters used and the flow pattern, several filtration forms can be used, which include microfiltration, ultrafiltration, dead-end filtration, vacuum filtration, tangential flow filtration, and pressure filtration. It is economical to filter large and filamentous algal cells through simple vacuum filtration, whereas the filtration of small cells is too expensive, as it requires complex filters with very small pore size and frequent backwashing and cleaning (Wyatt et al. 2012). However, the operation of membrane filtration and ultrafiltration requires less energy input, in the range 0.1 to 0.3 kW h m^{-3}, as shown in Table 10.2 (Baerdemaeker et al. 2013). Tangential and pressure filtration methods are considered energy-efficient methods that have the ability to concentrate up to 90% of microalgal cells with minimal membrane fouling (Uduman et al. 2010). Flotation is a phenomenon in which the air bubbles generated carry the solid particles, such as algal cells, to the upper surface of a suspension, which can be skimmed off (Uduman et al. 2010). This method has been found to more efficient than sedimentation for many microalgal systems (Pragya et al. 2013). Depending upon the bubble size, the flotation can be divided into dissolved air flotation, dispersed flotation, and electrolytic flotation (Chen et al. 2011). Dissolved air flotation is commonly used along with chemical flocculation, as the effectiveness of this method depends on the particulate size (i.e., with larger particle size, higher efficiency is achieved). Dispersed air flotation forms the bubbles ranging from 700 to 1500 μm, which interact with the negatively charged algal cells. By increasing the cationic charge on the bubbles, the interaction and effectiveness of algal cell removal is increased (Rawat et al. 2011). Ozone is used in the dispersed air flotation, and harvesting *Chlorella vulgaris* using the ozone flotation method resulted in an increase in lipid content also. However, the use of ozone for flotation is not economically feasible for biodiesel production (Rawat et al. 2011). The electrolytic flotation method involves the use of a cathode from which H$_2$ ions are released that attract the negatively charged algal cells and move them to the surface.

Electrocoagulation and flocculation involve the dissolution of the anode to form metal cations, which interact with the negatively charged algal mass and enable aggregation. The method is suitable for high cell densities and marine species, as saltwater lowers the power input required (Vandamme et al. 2011). Fouling of the cathode or anode is the major problem in using such electrode-based methods for harvesting. Singh and Patidar (2018) developed a strategy to identify the best type of harvesting method that may be economically feasible for wastewater remediation and for biofuel production by considering six different criteria, which include biomass quantity, biomass quality, cost, processing time, species, and toxicity. The ranking strategy showed coagulation and flocculation as the best strategy, with ranking of 1 for the wastewater treatment, followed by filtration, flotation, centrifugation, and electrolytic processes with, rankings of 2, 3, 4, 5, and 6, respectively. Whereas for biofuel production from microalgae, operations with better economic feasibility were identified as coagulation and flocculation, with ranking of 1 followed by filtration, centrifugation, flotation, and electrolytic processes, with rankings of 2, 3, 4, 5, and 6, respectively (Singh and Patidar 2018). However, if we target a high-value product, the centrifugation is the best strategy to employ for harvesting the microalgal biomass (Singh and Patidar 2018). Thus, selection of an economically feasible process with high removal efficiency is a must to attain sustainability in the whole microalgae-based bioremediation process.

10.3.4 CHALLENGES OF THE ALGAL BIOMASS GENERATED FOR THE PRODUCTION OF VALUE-ADDED PRODUCTS

The harvested wet algal biomass should be further fractionated to obtain the carbohydrates, protein, lipids, and pigments required for the biorefinery process, or the whole biomass can be used as the fertilizer or feed for animals, or can be converted into biofuels (Figure 10.1). For instance, the conversion of biomass to biofuel relies either on the biochemical route or the thermochemical conversion route. In thermochemical conversions, the biomass will be processed in the presence or absence of a catalyst at higher temperatures in a pressurized vessel devoid of oxygen, which includes gasification, pyrolysis, and hydrothermal conversions (Barreiro et al. 2013). Hydrothermal liquefaction is one of the recent advancements that has increased interest in the industry for the production of bio-oil. This technology processes wet algal biomass into bio-oil under supercritical or subcritical temperatures ranging from 280 to 380°C and pressure ranges from 70 to 300 bar (Barreiro et al. 2013). From bio-oil the biodiesel can be refined in the form of fatty esters, from the mixture containing different products such as polymers, aromatics, lubricants, etc. (Kumar et al. 2018). However, for the recovery of biodiesel from biocrude or bio-oil, the biochemical composition of the microalgal strains obtained should contain high levels of neutral lipids. With low levels of neutral lipid content, the quality of biodiesel generated will not be as per the ASTM standards.

On the other hand, the biochemical route involves using the pretreated microalgal biomass as a feedstock or as the nutrient medium for the growth of beneficial microbes. The pretreatment process usually involves cell lysis followed by breaking down of the complex metabolites into simpler molecules, which can be further consumed by the beneficial microbes for their growth. Such a process has been demonstrated for ethanol production from *Zymomonas mobilis* utilizing the microalgal biomass as the nutrient medium (Ho et al. 2013). Thus, the biochemical route leads to consolidated bioprocessing, which utilizes multiple bacterial or fungal strains for the synthesis of various products while using pretreated microalgal strains as nutrient feedstock.

In addition, biorefinery represents the process through which the biomass is separated into several components and further used for the production of different products. As the microalgal strains obtained from wastewater cannot be used for high-value-added products such as pharmaceuticals or food products, they are utilized for the production of bioplastics, biofuels, etc. Several studies investigated the feasibility of microalgal biorefinery and the efficiency in the extraction of different metabolites from the biomass (Ansari et al. 2017; Depra et al. 2018). Still, numerous challenges need to be overcome in the overall conversion of biomass into several fractions through sequential extraction procedures, which remains one of the cost-consuming steps in addition to the harvesting of microalgal systems. It is important to note whether the sequential extraction method uses the wet algal biomass or the dry algal biomass. If the method works better in the dry algal biomass, an additional step will be required in the downstream process for drying the wet algal biomass harvested, which is once again a high-energy-consuming step. Thus, the valorization of the whole biomass in the biorefinery is one of the difficult bottlenecks to be addressed (Gifuni et al. 2019). Cascade extraction is the approach normally used, which involves the sequential recovery of the different macromolecular compositions using two-phase extraction, filtration, chromatography, and final processing—either transesterification of the lipid fraction for biodiesel production or fermentation of the other residues for the production of biofuels (Figure 10.1). In addition, recovery of pigments such as astaxanthin, lutein, and β-carotene have gained increased interest due to their high prices (Gifuni et al. 2019). In the present scenario, even though several bottlenecks need to be addressed for the commercial-scale extenuation of environmental issues using microalgal systems, extensive research, and an understanding of microalgal strains have opened new avenues for sustainability, which needs to be realized.

10.4 CONCLUSIONS

Integration of the microalgal cultivation process with bioremediation of wastewater and carbon sequestration possesses great potential for the sustainable mitigation of environmental issues. Compared to the conventional treatment process, the microalgal-based bioremediation has shown high potential in remediating different hazardous compounds, heavy metals, radioactive substances, etc., while generating wastewater safe for release into the environment and producing value-added compounds. However, the industrial-scale processes are not still economically feasible due to the immature technologies in biorefinery and in algal biomass generation/harvesting procedures. Thus, augmented research activities are required, with the goal to develop a feasible and sustainable technology for the future. To sum up, the microalgal biomass is the best source with intrinsic potential for the sustainable mitigation of environmental issues.

REFERENCES

Adey, W. H., P. C. Kangas, and W. Mulbry. 2011. "Algal turf scrubbing: Cleaning surface waters with solar energy while producing a biofuel." *Bioscience* 61:434–441.

Ammari, T. G., M. Al-Atiyat, E. S. Abu-Namesh, A. Ghrair, D. Jaradat, and S. Abu-Romman. 2017. "Bioremediation of cadmium-contaminated water systems using intact and alkaline-treated alga (*Hydrodictyon reticulatum*) naturally grown in an ecosystem." *International Journal of Phytoremediation* 19 (5):453–462.

Ansari, F. A., A. Shriwastav, S. K. Gupta, I. Rawat, and F. Bux. 2017. "Exploration of microalgae biorefinery by optimizing sequential extraction of major metabolites from *Scenedesmus obliquus.*" *Industrial & Engineering Chemistry Research* 56 (12):3407–3412.

Arbib, Z., J. Ruiz, P. Álvarez-Díaz, C. Garrido-Pérez, J. Barragan, and J. A. Perales. 2013. "Long term outdoor operation of a tubular airlift pilot photobioreactor and a high rate algal pond as tertiary treatment of urban wastewater." *Ecological Engineering* 52:143–153.

Arıca, M. Y., I. Tüzün, E. Yalçın, O. Ince, and G. Bayramoglu. 2005. "Utilisation of native, heat and acid-treated microalgae *Chlamydomonas reinhardtii* preparations for biosorption of Cr(VI) ions." *Process Biochemistry* 40 (7):2351–2358.

Baerdemaeker, T. D., B. Lemmens, C. Dotremont, J. Fret, L. Roef, K. Goiris, and L. Diels. 2013. "Benchmark study on algae harvesting with backwashable submerged flat panel membranes." *Bioresource Technology* 129:582–591.

Barreiro, D. L., W. Prins, F. Ronsse, and W. Brilman. 2013. "Hydrothermal liquefaction (HTL) of microalgae for biofuel production: State of the art review and future prospects." *Biomass and Bioenergy* 53:113–127.

Barrut, B., J. P. Blancheton, M. Callier, J. Y. Champagne, and A. Grasmick. 2013. "Foam fractionation efficiency of a vacuum airlift-Application to particulate matter removal in recirculating systems." *Aquaculture Engineering* 54:16–21.

Bhowmick, G. D., A. K. Samrah, and R. Sen. 2019. "Zero-waste algal biorefinery for bioenergy and biochar: A green lap towards achieving energy and environmental sustainability." *Science of the Total Environment* 650:2467–2482.

Bilad, M. R., G. Mezohegyi, P. Declerck, and I. F. Vankelecom. 2012. "Novel magnetically induced membrane vibration (MMV) for fouling control in membrane bioreactors." *Water Research* 46 (1):63–72.

Cai, T., S. Y. Park, and Y. Li. 2013. "Nutrient recovery from wastewater streams by microalgae: Status and prospects." *Renewable and Sustainable Energy Reviews* 19:360–369.

Caia, M., O. Bernard, and Q. Bechet. 2018. "Optimizing CO_2 transfer in algal ponds." *Algal Research* 35:530–538.

Chen, C.-Y., K.-L. Yeh, R. Aisyah, D.-J. Lee, and J.-S. Chang. 2011. "Cultivation, photobioreactor design and harvesting of microalgae for biodiesel production: A critical review." *Bioresource Technology* 102:1649–1655.

Chen, R., R. Li, L. Deitz, Y. Liu, Y. Stevenson, and W. Liao. 2012. "Freshwater algal cultivation with animal waste for nutrient removal and biomass production." *Biomass Bioenergy* 39:128–138.

Cheng, J., Y. Huang, J. Feng, J. Sun, J. Zhou, and K. Cen. 2013. "Improving CO_2 fixation efficiency by optimizing *Chlorella* PY-ZU1 culture conditions in sequential bioreactors." *Bioresource Technology* 144:321–327.

Croft, M. T., M. J. Warren, and A. G. Smith. 2006. "Algae need their vitamins." *Eukaryotic Cell* 5:1175–1183.

Cuellar-Bermudez, S. P., G. S. Aleman-Nava, R. Chandra, J. S. Garcia-Perez, J. R. Contreras-Angulo, G. Markou, K. Muylaert, B. E. Rittmann, and R. Parra-Saldivar. 2017. "Nutrients utilization and contaminants removal. A review of two approaches of algae and cyanobacteria in wastewater." *Algal Research* 24:438–449.

Danquah, M. K., B. Gladman, N. Moheimani, and G. M. Forde. 2009. "Microalgal growth characteristics and subsequent influence on dewatering efficiency." *Chemical Engineering Journal* 151:73–78.

Dassey, A. J., and C. S. Theegala. 2013. "Harvesting economics and strategies using centrifugation for cost effective separation of microalgae cells for biodiesel applications." *Bioresource Technology* 128:241–245.

Debnath, S. 2019. "Characterization of extracellular proteins to explore their role in bio-flocculation for harvesting algal biomass for wastewater treatment." In *The Role of Microalgae in Wastewater Treatment*, edited by L. B. Sukla, et al. 229–266. Singapore: Springer Nature Pte. Ltd.

Deprá, M. C., A. M. dos Santos, I. A. Severo, A. B. Santos, L. Q. Zepka, and E. Jacob-Lopes. 2018. "Microalgal biorefineries for bioenergy production: Can we move from concept to industrial reality?" *Bioenergy Research* 11:727–747.

Dixit, S., and D. P. Singh. 2015. "Phycoremediation: Future perspective of green technology." In *Algae and Environmental Sustainability, Developments in Applied Phycology 7*, edited by B. Singh, K. Bauddh, and F. Bux, 9–22. India: Springer India Publishing.

Gerardo, M. L., D. L. Oatley-Radcliffe, and R. W. Lovitt. 2014. "Minimizing the energy requirement of dewatering *scenedesmus* sp. by microfiltration: Performance, costs, and feasibility." *Environmental Science & Technology* 48:845–853.

Gifuni, I., A. Pollio, C. Safi, A. Marzocchella, and G. Olivieri. 2019. "Current bottlenecks and challenges of the microalgal biorefinery." *Trends in Biotechnology* 37 (3):242–252.

Goncalves, A. L., C. M. Rodrigues, J. C. M. Pires, and M. Simoes. 2016. "The effect of increasing CO_2 concentrations on its capture, biomass production and wastewater bio-remediation by microalgae and cyanobacteria." *Algal Research* 14:127–136.

Grima, E. M., E. H. Belarbi, F. A. Fernández, A. R. Medina, and Y. Chisti. 2003. "Recovery of microalgal biomass and metabolites: Process options and economics." *Biotechnology Advances* 20 (7):491–515.

Gupta, S. K., K. Dhandayuthapani, and F. A. Ansari. 2018. "Techno-economic perspectives of bioremediation of wastewater, dewatering, and biofuel production from microal-gae: An overview." In *Phytomanagement of polluted sites, Market Opportunities in Sustainable Phytoremediation*, edited by V. C. Pandey, and K. Bauddh, 471–499. New York: Elsevier Inc.

Gupta, S. K., A. Sriwastav, F. A. Ansari, M. Nasr, A. K. Nema. 2017. "Phycoremediation: An eco-friendly algal technology for bioremediation and bioenergy production." In *Phytoremediation Potential of Bioenergy Plants*, edited by K. Bauddh, et al. 431–456. Singapore: Springer Nature Singapore Pvt Ltd.

Hammouda, O., N. Abdel-Raouf, M. Shaaban, M. Kamal, and, B. S. W. T. Plant. 2015. "Treatment of mixed domestic-industrial wastewater using microalgae *Chlorella* sp." *Journal of American Science* 11 (12):303–315.

Ho, S.-H., S.-W. Huang, C.-Y. Chen, T. Hasunuma, and A. Kondo. 2013. "Bioethanol pro-duction using carbohydrate-rich microalgae biomass as feedstock." *Bioresource Technology* 135:191–198.

Hoffman, J., R. C. Pate, T. Drennen, and J. C. Quinn. 2017. "Techno-economic assessment of open microalgae production systems." *Algal Research* 23:51–57.

Hongyang, S., Z. Yalei, Z. Chunmin, Z. Xuefei, and L. Jinpeng. 2011. "Cultivation of *Chlorella pyrenoidosa* in soybean processing wastewater." *Bioresource Technology* 102 (21):9884–9890.

Hultberg, M., H. Bodin, E. Ardal, and H. Asp. 2016. "Effect of microalgal treatments on pesticides in water." *Environmental Technology* 37 (7):893–898.

Khan, S. A., M. Z. Hussain, S. Prasad, and U. Banerjee. 2009. "Prospects of biodiesel pro-duction from microalgae in India." *Renewable and Sustainable Energy Reviews* 13 (9):2361–2372.

Kim, J., B. G. Ryu, B. K. Kim, J. I. Han, and J. W. Yang. 2012. "Continuous microalgae recov-ery using electrolysis with polarity exchange." *Bioresource Technology* 111:268–275.

Kong, Q., L. Li, B. Martinez, P. Chen, and R. Ruan. 2010. "Culture of microalgae *Chlamydomonas reinhardtii* in wastewater for biomass feedstock production." *Applied Biochemistry and Biotechnology* 160:9–18.

Kumar, K., S. K. Mishra, A. Shrivastav, M. S. Park, and J.-W. Yang. 2015. "Recent trends in the mass cultivation of algae in raceway ponds." *Renewable and Sustainable Energy Reviews* 51:875–885.

Kumar, P. K., S. V. Krishna, K. Verma, K. Pooja, D. Bhagwan, K. Srilatha, and V. Himabindu. 2018. "Bio oil production from microalgae via hydrothermal liquefaction technology under subcritical water conditions." *Journal of Microbiological Methods* 153:108–117.

Li, X., H. Y. Hu, K. Gan, and Y. X. Sun. 2010. "Effects of different nitrogen and phosphorus concentrations on the growth, nutrient uptake, and lipid accumulation of a freshwater microalga *Scenedesmus* sp." *Bioresource Technology* 101:5494–5500.

Lim, S.-L., W.-L. Chu, and S.-M. Phang. 2010. "Use of *Chlorella vulgaris* for bioremediation of textile wastewater." *Bioresource Technology* 101:7314–7322.

Lundquist, T. J., I. C. Woertz, N. W. T. Quinn, and J. R. Benemann. 2010. "A realistic technology and engineering assessment of algae biofuel production." *Energy Bioscience Institute*. 1.

Luque, R. 2010. "Algal biofuels: The eternal promise." *Energy and Environmental Science* 3:254–257.

Markou, G., I. Chatzipavlidis, and D. Georgakakis. 2012. "Cultivation of *Arthrospira* (*Spirulina*) *platensis* in olive-oil mill wastewater treated with sodium hypochlorite." *Bioresource Technology* 112:234–241.

Markou, G., D. Iconomou, and K. Muylaert. 2016. "Applying raw poultry litter leachate for the cultivation of *Arthrospira platensis* and *Chlorella vulgaris*." *Algal Research* 13:79–84.

Mata, T. M., A. C. Melo, M. Simões, and N. S. Caetano. 2012. "Parametric study of a brewery effluent treatment by microalgae *Scenedesmus obliquus*." *Bioresource Technology* 107:151–158.

MeGraw, V. E., A. R. Brown, C. Boothman, R. Goodacre, K. Morris, D. Sigee, L. Anderson, and J. R. Lloyd. 2018. "A novel adaptation mechanism underpinning algal colonization of a nuclear fuel storage pond." *MBio* 9 (3):e02395–17.

Mezzomo, N., A. G. Saggiorato, R. Siebert, P. Oliveira, M. C. Lago, M. Hemkemeier, et al. 2010. "Cultivation of microalgae *Spirulina platensis* (*Arthrospira platensis*) from biological treatment of swine wastewater." Ciência *E Tecnologia de* Alimentos. 30:173–178.

Misra, R., A. Guldhe, P. Singh, I. Rawat, and F. Bux. 2014. "Electrochemical harvesting process for microalgae by using nonsacrificial carbon electrode: A sustainable approach for biodiesel production." *Chemical Engineering Journal* 255:327–333.

Muthuraj, M., V. Kumar, B. Palabhanvi, and D. Das. 2014. "Evaluation of indigenous microalgal isolate *Chlorella* sp. FC2 IITG as a cell factory for biodiesel production and scale up in outdoor conditions." *Journal of Industrial Microbiology and Biotechnology* 41 (3):499–511.

Muthuraj, M., B. Selvaraj, B. Palabhanvi, V. Kumar, and D. Das. 2019. "Enhanced lipid content in *Chlorella* sp. FC2 IITG via high energy irradiation mutagenesis." *Korean Journal of Chemical Engineering* 36 (1):63–70.

Muylaert, K., A. Beuckels, O. Depraetere, I. Foubert, G. Markou, and D. Vandamme. 2015. "Wastewater as a source of nutrients for microalgae biomass production." In *Biomass and Biofuels from Microalgae, Biofuel and Biorefinery Technologies 2*, edited by N. R. Moheimani, et al. Switzerland: Springer International Publishing Switzerland.

Nakayama, T., A. Nakamura, A. Yokoyama, T. Shiratori, I. Inouye, and K. Ishida. 2015. "Taxonomic study 1093 of a new eustigmatophycean alga, *Vacuoliviride crystalliferum* gen. et sp. nov." *Journal of Plant Research* 128:249–257.

Nayak, M., A. Karemore, and R. Sen. 2016. "Performance evaluation of microalgae for concomitant wastewater bioremediation, CO_2 biofixation and lipid biosynthesis for biodiesel application." *Algal Research* 16:216–233.

Guiry, M. D., and G. M. Guiry. 2019. AlgaeBase. World-wide electronic publication, National University of Ireland. Web page: www.algaebase.org. (Accessed March 12, 2019).

Osorio, J. H. M., V. Luongo, A. D. Monod, G. Pinto, A. Pollio, L. Frunzo, P. N. L. Lens, and G. Esposito. 2018. "Nutrient removal from high strength nitrate containing industrial wastewater using *Chlorella* sp. strain ACUF_802." *Annals of Microbiology* 68 (12):899–913.

Oswald, W. J., and C. G. Golueke. 1960. "Biological transformation of solar energy." *Advanced Applied Microbiology* 11:223–242.

Ozkan, A., K. Kinney, L. Katz, and H. Berberoglu. 2012. "Reduction of water and energy requirement of algae cultivation using an algae biofilm photobioreactor." *Bioresource Technology* 114:542–548.

Pacheco, M. M., M. Hoeltz, M. S. A. Moraes, and R. C. S. Schneider. 2015. "Microalgae: Cultivation techniques and wastewater phycoremediation." *Journal of Environmental Science and Health, Part A: Toxic/Hazardous Substances and Environmental Engineering* 50 (6):585–601.

Park, J. B. K., R. J. Craggs, and A. N. Shilton. 2011. "Recycling algae to improve species control and harvest efficiency from a high rate algal pond." *Water Research* 45:6637–6649.

Perales-Vela, H. V., J. M. Peña-Castro, and R. O. Cañizares-Villanueva. 2006. "Heavy metal detoxification in eukaryotic microalgae." *Chemosphere* 64:1–10.

Pragya, N., K. K. Pandey, and P. K. Sahoo. 2013. "A review on harvesting, oil extraction and biofuels production technologies from microalgae." *Renewable and Sustainable Energy Reviews* 24:159–171.

Rawat, I., R. R. Kumar, T. Mutanda, and F. Bux. 2011. "Dual role of microalgae: Phycoremediation of domestic wastewater and biomass production for sustainable biofuels production." *Applied Energy* 88:3411–3424.

Rittmann, B. E., B. Mayer, P. Westerhoff, and M. Edwards. 2011. "Capturing the lost phosphorus." *Chemosphere* 84:846–853.

Rivasseau, C., E. Farhi, A. Atteia, A. Coute, M. Gromova, D. D. Saint Cyr, A. M. Boisson, A. S. Feret, E. Compagnon, and R. Bligny. 2013. "An extremely radioresistant green eukaryote for radionuclide bio-decontamination in the nuclear industry." *Energy and Environmental Science* 6:1230–1239.

Ruffell, S. E., R. A. Frank, A. P. Woodworth, L. M. Bragg, A. E. Bauer, L. E. Deeth, K. M. Muller, A. J. Farwell, D. G. Dixon, M. R. Servos, and B. J. McConkey. 2016. "Assessing the bioremediation potential of algal species indigenous to oil sands process-affected waters on mixtures of oil sands acid extractable organics." *Ecotoxicology and Environmental Safety* 133:373–380.

Sacristán D. A. M., V. M. Luna-Pabello, E. Cadena, and E. Ortíz. 2013. "Green microalga *Scenedesmus acutus* grown on municipal wastewater to couple nutrient removal with lipid accumulation for biodiesel production." *Bioresource Technology* 146:744–748.

Sahoo, N. K., S. K. Gupta, I. Rawat, F. A. Ansari, P. Singh, S. N. Naik, et al. 2017. "Sustainable dewatering and drying of self-flocculating microalgae and study of cake properties." *Journal of Cleaner Production* 159:248–256.

Salama, E., B. M. Kurade, R. A. I. Abou-shanab, and M. M. Eldalatony. 2017. "Recent progress in microalgal biomass production coupled with wastewater treatment for biofuel generation." *Renewable and Sustainable Energy Reviews* 79:1189–1211.

Samorì, G., C. Samorì, F. Guerrini, and R. Pistachios. 2013. "Growth and nitrogen removal capacity of *Desmodesmus communis* and of a natural microalgae consortium in a batch culture system in view of urban wastewater treatment: Part I." *Water Research* 47:791–801.

Sawayama, S., T. Minowa, Y. Dote, and S. Yokoyama. 1992. "Growth of the hydrocarbon-rich microalga *Botryococcus braunii* in secondarily treated sewage." *Applied Microbiology and Biotechnology* 38 (1):135–138.

Schlagermann, P., G. Gottlicher, R. Dillschneider, R. Rosello-Sastre, and C. Posten. 2012. "Composition of algal oil and its potential as biofuel." *Journal of Combustion*. 285185. doi: 10.1155/2012/285185.

Sharma, K., Y. Li and P. M. Schenk. 2014. "UV-C-mediated lipid induction and settling, a step change towards economical microalgal biodiesel production." *Green Chemistry* 16:3539–3548.

Shen, Y. 2014. "Carbon dioxide bio-fixation and wastewater treatment via algae photochemical synthesis for biofuels production." *RSC Advances* 4:49672–49722.

Singh, G., and S. K. Patidar. 2018. "Microalgae harvesting techniques: A review." *Journal of Environmental Management* 217:499–508.

Sreekanth, D., D. Sreekanth, K. Pooja, Y. Seeta, V. Himabindu, and P. M. Reddy. 2014. "Bioremediation of dairy wastewater using microalgae for the production of biodiesel." *International Journal of Science Engineering and Advance Technology* 2:783–791.

Sturm, B. S. M., and S. L. Lamer. 2011. "An energy evaluation of coupling nutrient removal from wastewater with algal biomass production." *Applied Energy* 88:3499–3506.

Tabatabaei, M., M. Tohidfar, G. S. Jouzani, M. Safarnejad, and M. Pazouki. 2011. "Biodiesel production from genetically engineered microalgae: Future of bioenergy in Iran." *Renewable and Sustainable Energy Reviews* 15:1918–1927.

Tam, N. F. Y., and Y. S. Wong. 1990. "The comparison of growth and nutrient removal efficiency of *Chlorella pyrenoidosa* in settled and activated sewages." *Environmental Pollution* 65:93–108.

Tiron, O., C. Bumbac, E. Manea, M. Stefanescu, and M. N. Lazar. 2017. "Overcoming microalgae harvesting barrier by activated algae granules." *Scientific Reports* 7 (4646):1–11.

Uduman, N., Y. Qi, M. K. Danquah, G. M. Forde, and A. Hoadley. 2010. "Dewatering of microalgal cultures: A major bottleneck to algae-based biofuels." *Journal of Renewable and Sustainable Energy* 2:1–15.

Ummalyma, S. B., A. Pandey, R. K. Sukumaran, and D. Sahoo. 2018. "Bioremediation by microalgae: Current and emerging trends for effluents treatments for value addition of waste streams." In *Biosynthetic Technology and Environmental Challenges, Energy, Environment, and Sustainability*, edited by S. J. Varjani, et al. 355–375. Singapore: Springer Nature Pte. Ltd.

Vandamme, D., I. Foubert, I. Fraeye, B. Meeschaert, and K. Muylaert. 2012. "Flocculation of *Chlorella vulgaris* induced by high pH: Role of magnesium and calcium and practical implications." *Bioresource Technology* 105:114–119.

Vandamme, D., S. C. V. Pontes, K. Goiris, I. Foubert, L. J. J. Pinoy, and K. Muylaert. 2011. "Evaluation of electrocoagulation-flocculation for harvesting marine and freshwater microalgae." *Biotechnology and Bioengineering* 108:2320–2330.

Vanhoudt, N., H. Vandenhove, N. Leys, and P. Janssen. 2018. "Potential of higher plants, algae, and cyanobacteria for remediation of radioactively contaminated waters." *Chemosphere* 207:239–254.

Wang, H., H. D. Laughinghouse, M. A. Anderson, F. Chen, E. Williams, A. R. Place, O. Zmora, Y. Zohar, T. Zheng, and R. Hill. 2012. "Novel bacterial isolate from Permian groundwater, capable of aggregating potential biofuel-producing microalga *Nannochloropsis oceanica* IMET1." *Applied and Environmental Microbiology* 78:1445–1453.

Wang, L., Y. Wang, P. Chen, and R. Ruan. 2010. "Semi-continuous cultivation of *Chlorella vulgaris* for treating undigested and digested dairy manures." *Applied Biochemistry and Biotechnology* 162:2324–2332.

Wang, Y., S. H. Ho, C. L. Cheng, D. Nagarajan, W. Q. Guo, C. Lin, et al. 2017. "Nutrients and COD removal of swine wastewater with an isolated microalgal strain." *Bioresource Technology* 242:7–14.

Williams, P. J. B., and L. M. L. Laurens. 2010. "Microalgae as biodiesel and biomass feedstocks: Review and analysis of the biochemistry, energetics an economics." *Energy and Environmental Science* 3:554–590.

World Nuclear Association. 2019. "Reactor Database." Web Page: www.world-nucelar.org/ (Accessed March 12, 2019).

Wyatt, N. B., L. M. Gloe, P. V. Brady, J. C. Hewson, A. M. Grillet, M. G. Hankins, and P. I. Pohl. 2012. "Critical conditions for ferric chloride-induced flocculation of freshwater algae." *Biotechnology and Bioengineering* 109:493–501.

Zeraatkar, A. K., H. Ahmadzadeh, A. F. Talebi, N. R. Moheimani, and M. P. McHenry. 2016. "Potential use of algae for heavy metal bioremediation, a critical review." *Journal of Environmental Management* 181:817–831.

Zhu, L., Z. Wang, Q. Shu, J. Takala, E. Hiltunen, P. Feng, and Z. Yuan. 2013. "Nutrient removal and biodiesel production by integration of freshwater algae cultivation with piggery wastewater treatment." *Water Research* 47 (13):4294–4302.

11 Optimization of the Integrated Downstream Processing of Microalgae for Biomolecule Production

Soumyajit Sen Gupta, Yogendra Shastri, Sharad Bhartiya

CONTENTS

11.1 INTRODUCTION

Fossil fuels such as coal and petroleum products have been extensively used over the past centuries as sources of energy. Although these traditional resources are mostly energy-dense products, their widespread applications have led to climate change and subsequent implications on the future survival and well-being of human civilization. The mean global concentration of carbon dioxide in the atmosphere is estimated to be 410 ppm in 2018, compared with around 310 ppm in the 1950s (CO_2 Earth, 2018). Moreover, the rate of replenishment of fossil fuels is too slow, and thus, they cannot be considered sustainable resources. These facts culminate into a definite and critical need to develop renewable sources of energy that are affordable and abundant in nature to arrest the incremental damage on the climate and with the aim of having a sustained availability over a long enough time in the future. Energy harnessed from the wind, tide, and biomass are some examples of non-conventional resources with limited or in some cases even a net positive impact on the environment. The biomass as an energy source has a specific advantage over the other non-conventional sources, as it is mostly independent of the intensity of solar irradiation and other climatic conditions of a location, which ensures its wide applicability. It can be transported across regions relatively easily, thereby facilitating its usage across the globe. Moreover, biomass is a source of transportation fuels such as ethanol and biodiesel.

There are multiple potential sources of biomass-based energy. Carbohydrate fraction of food crops such as soybean, palm, and sugarcane can be utilized to produce ethanol as an energy product. However, this avenue engenders a food versus fuel debate and thus is not suitable from the viewpoint of a comprehensive economy. On the other hand, sources of non-food biomass include lignocellulosic biomass, agricultural residue, and dedicated energy-based crops such as switchgrass and *Jatropha curcas*. However, most of these species require extensive resource inputs in terms of land, nutrients, and water during their cultivation, thereby rendering these options non-viable for a large-scale application. Accordingly, the most desired characteristics of the potentially most suited source of biomass would be marginal requirements in terms of land, water, and nutrients, while exhibiting a large enough productivity. Microalgae, with all these salient features, hold great promise as a biomass source.

Microalgae are unicellular or simple multicellular species, with their sizes ranging from a few microns (µm) to a few millimetres (mm). *Chlorella, Dunaliella, Spirulina*, and *Nannochloropsis* are a few common microalgae species. The metabolism in the microalgae cells may be of phototrophic, heterotrophic, mixotrophic, or photoheterotrophic type. The growth rate of most of the species is quite high, with the doubling time of the biomass being as low as 24 hours (Chisti 2007). The cells consist of carbohydrate, lipid, and protein, and each of these is a precursor to different value-added chemicals. In terms of energy products, the lipid, specifically its neutral fraction consisting of the saturated and unsaturated fatty acid esters, can be processed through a transesterification reaction to yield biodiesel, a potential replacement for diesel. The theoretical yield of oil from microalgae is also quite high, 13.69 L/m^2, compared to a maximum yield of 0.6 L/m^2 for the case of food crops such as soybean, coconut, and palm oil (Chisti 2007). However, implementation of microalgae-derived fuel has not yet been realized on a large scale (Hannon et al. 2010; Amaro et al. 2011; Christenson and Sims 2011; Yang et al. 2011; Lam and Lee 2012), due to the challenges in terms of high expenditures and various technological bottlenecks at the processing steps. Reduction in cost by a factor of ten, coupled with a concomitant improvement in productivity by a scale of three orders, has been estimated to be an essential requirement for microalgae biofuel to be economically competitive with conventional energy sources as a transportation fuel (Wijffels and Barbosa 2010).

Since the utilization of microalgae to produce only fuel is not viable, their other potential applications deserve special attention. Carbohydrates in microalgae cells can be fermented or gasified to ethanol, and the protein segment can be consumed as nutraceuticals. *Spirulina*, *Chlorella*, and *Dunaliella* have been popular as food supplements (Brennan and Owende 2010). Moreover, polysaccharides, carotenoids, β-carotene, and other such bioactive compounds are also quite abundant in different microalgae species, and most of these have pharmaceutical applications (Harun et al. 2010; Mata et al. 2010). Thus, microalgae can be conceived as a resource with far-reaching potential in terms of providing a suite of end products with varied utilities. Even on the upstream, cultivation of microalgae sequesters carbon dioxide, amounting to 1.8 times the amount of the biomass (Wijffels and Barbosa 2010), thereby assisting in moving towards a carbon-neutral economy. Further, they can scavenge wastewater streams for salts (nitrates and phosphates), the requisite nutrients for growth (Mata et al. 2010), thus treating the wastewater, while at the same time, reducing the requirement for freshwater and fertilizers. This aids in curbing global warming because the use of chemical fertilizers in microalgae cultivation has been found to account for 50% of the associated greenhouse gases production and energy use (Wiley et al. 2011). On the other hand, use of wastewater can potentially reduce freshwater usage by 3726 kg per kg of biodiesel produced, considering no recycling, to the extent of 90% (Yang et al. 2011).

Thus, a biorefinery can be conceptualized to ensure proper realization of the diverse rich potential of microalgae. The schematic for a possible biorefinery is presented in Figure 11.1. After being cultivated microalgae are harvested and water removed from the medium. The harvested and dried biomass paste is then subjected to different downstream processing steps as per the choice of the end product. For biodiesel production, lipid is extracted and transesterified with alcohol, whereas other products

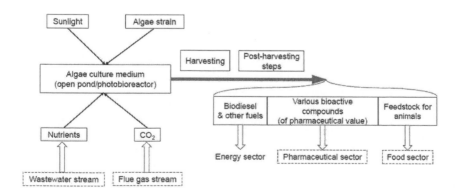

FIGURE 11.1   Resource inputs to and end products from an integrated microalgae biorefinery.

Source: Sen Gupta et al. (2014); by kind permission of CAB Reviews

have their customized processing steps. The review articles by Chen et al. (2009), Brennan and Owende (2010), Greenwell et al. (2010), Harun et al. (2010), Mata et al. (2010), Chen et al. (2011), Wiley et al. (2011), Halim et al. (2012), Lam and Lee (2012), and Sen Gupta et al. (2014) present detailed and critical reviews of these process steps.

The co-production of multiple products from microalgae has been estimated to be quite profitable, and this compensates for the expensive extraction protocol required for specific bioactive compounds. For instance, the microalgae biomass market, excluding processed products, has been found to generate an annual turnover of US\$1.25 × 10^9 (Pulz and Gross 2004). The market value of β-carotene varies from €215 to 2150/kg (Brennan and Owende 2010), and that of natural astaxanthin is €7,150/kg (Brennan and Owende 2010). One thousand kilograms of algal biomass, on being refined, is estimated to give rise to a suite of products, equivalent to a value of €1,646, of which only €150 is from biofuels (Wijffels et al. 2010).

11.1.1   MODEL-BASED STUDIES FOR A MICROALGAE BIOREFINERY

The systematic and accurate design of an integrated biorefinery is not direct and obvious as per the literature (Subhadra 2010). The principal challenge emanates from the availability of multiple alternatives, each with its own benefits and pitfalls at each step of the process. These alternatives have an impact on system performance and economics. Additionally, the dynamic market conditions and diverse environmental constraints need to be considered. The mutual trade-offs among economics, environmental impact, and operability pose a great hurdle to the design of a biorefinery. Moreover, different processing steps of the biorefinery are at different maturity levels. With the techno-economic challenges impending, a scientific approach to its design needs to involve modeling and analysis from a systems engineering perspective (Taylor 2008). To this end, superstructure optimization is a proven technique that aids us in identifying the most optimized configuration of a process among all the various alternatives. Solution to a superstructure optimization model provides us with optimal design and operation of the process, quantification of sensitivities to system parameters, and identification of trade-offs among different factors, thereby presenting itself as a comprehensive decision support tool. Application of superstructure optimization for analyzing the system has been on the rise across the different engineering disciplines. For energy sectors, Baliban et al. (2012) has studied the process of thermochemical conversion of hybrid coal, biomass, and natural gas to liquid fuels. Ethanol production through gasification of switchgrass has been studied by Martin and Grossmann (2011). This strategy has also been applied for studying microalgae processes.

Rizwan et al. (2013) has formulated a mixed integer non-linear programming (MINLP) model to attain the optimal biodiesel production protocol from microalgae for different cases. The scenarios in their study include maximization of gross operating margin, simultaneous maximization of gross operation margin, and minimization of waste. Mass balance equations for different streams across the stages have been modeled as constraints. Rizwan et al. (2015) has expanded the model by including the cultivation of microalgae in the modeling framework and considering usages of lipid-depleted biomass as energy products. Gebreslassie et al. (2013) has modeled an algae-based hydrocarbon biorefinery for the potential biomitigation of carbon dioxide obtained from the power plant, with alternatives in the downstream processing of algal oil. Gong and You (2014) have formulated a multiobjective life cycle optimization problem for simultaneous analysis of the mitigation of carbon dioxide and the production of biofuel. The model incorporates the life cycle environmental impact as constraints in addition to the mass and energy balance and process economics. The economic viability and environmental sustainability of the microalgae-based bioproducts, such as hydrogen, propylene glycol, glycerol-tert-butyl ether, and poly-3-hydroxybutyrate, compared to their petrochemical alternatives have been investigated by Gong and You (2015). A cradle-to-gate life cycle analysis is integrated with a multiobjective optimization framework. Cheali et al. (2015) has optimized the network for simultaneous potential production of protein, ethanol, and biodiesel while considering uncertainty in the oil content in microalgae and cost, as well as selling prices of biodiesel and ethanol. Torres et al. (2013) have studied multicriteria analysis with respect to economics and environmental factors to trace the optimal topology for microalgae-based biodiesel production. Akin to many other studies, the authors have predicted the necessity to improve the process economics of microalgae-based biodiesel production.

In addition to the superstructure optimization studies, there have been other model-based studies focused on microalgae processes. Economic potentials of microalgae-derived products and the impacts that model uncertainties have on them have been reported by Davis et al. (2011). Life cycle assessment studies on microalgae systems (Campbell et al. 2011; Yang et al. 2011) have also been gathering considerable attention. Crucial region-specific variations on microalgae productivity based on solar irradiation as well as that on the market demand for different products have been studied by Ruiz et al. (2016).

Although these prior studies indeed encompass a comprehensive viewpoint on the microalgae biorefinery from a systems engineering standpoint, we present a superstructure-based optimization model in this chapter to arrive at the optimized design and operational protocol. In particular, our model focuses on the following crucial factors and incorporates them.

1. State of microalgae cells: The growth parameters such as number of days of growth, cultivation medium, and growth equipment have a significant impact on the growth rate of microalgae and their cell composition. Hence, the decisions in terms of growth directly affect the state of all the downstream steps and the final product yield.
2. Performance of operations: The efficiency and yield in the biorefinery steps, such as harvesting, lipid (and other metabolite) extraction, and reaction, are functions of decisions at these steps as well as upstream steps. Hence, the integrated model needs to consider performance indicators as functions of the appropriate decisions.
3. Scheduling: With most of the processing steps for attaining value-added products from microalgae being batch steps, the necessary operational details need to be attained through a systematic scheduling analysis.
4. Sizing of equipment: Factors such as number and size of equipment, as well as their operating strategy, are also critical details featured in our study.

11.2   GENESIS OF THE OPTIMIZATION MODEL

A superstructure-based optimization model is formulated for investigating the prospects of a microalgae biorefinery. In this section, the structure of the model is briefly described. The details on the Mixed Integer Linear Programming (MILP) model and other understandings can be found in Sen Gupta et al. (2016, 2017a).

Different alternatives in terms of design and the operating parameters of each of the steps are considered in the superstructure. The model expresses the process variables, such as biomass, at output of a step, or volume of water in the cultivation medium, or amount of lipid extraction solvent as functions of those model decisions that affect them. The constraint equations of the model can be categorized as follows:

- Volume balance constraints: modeling the water balance across the steps
- Process synthesis constraints: modeling the selection of an option for each step of the process
- Flowsheet connectivity constraints: maintaining a logical sequence of decisions
- Equipment selection constraints: determining the amount of equipment for each of the processing steps
- Material balance constraints: computing the amount of necessary chemicals, such as flocculant, extraction solvent, reactant alcohol, and catalyst
- Biodiesel and/or other products' demand constraints: relating the concerned process variables with the demand value
- Cost constraints: determining the fixed and operating costs associated with each step

The variables are integer, continuous, or binary in nature. Instances of integer variables are numbers of equipment. Volume of streams, that is, the cultivated solution or the solvent or reactant added at the relevant steps are modeled as continuous variables. Binary variables are used to model options among all the available alternatives and helps to maintain linearity of the model. All the decision variables for the entire system are optimized simultaneously through the integrated model.

The research presented in this chapter is focused on three broad scenarios, as discussed in the subsequent sections. Although the structure of the model remains largely similar for all these cases, there are some critical differences too. The salient features of these individual models are discussed next with a reference to the specific scenarios on which they are applied.

11.2.1   BASE MODEL: BIODIESEL PRODUCTION FROM MICROALGAE

The modeling exercise is initiated with its application to the case of biodiesel as the only product with a specified target demand. Microalgae is cultivated until a desired duration (which is being optimized) is achieved, following which the medium is harvested. The lipid fraction of the harvested biomass is extracted, followed by transesterification of the neutral lipid with methanol to yield biodiesel. The superstructure

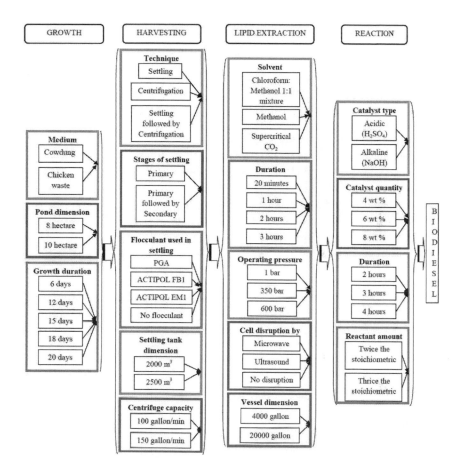

FIGURE 11.2   Superstructure for the study on biodiesel as the only product.

Source: Sen Gupta et al. (2016)

for this model is depicted in Figure 11.2. For the growth step, two different growth media, two different dimensions of ponds, and five possible durations of growth are considered. The same understanding applies to the other three steps of the biorefinery as well.

The batch time is governed by the slowest of all the post-cultivation steps and is computed for all the available combinations of time. As per the model, microalgae is cultivated as per the demand on the downstream, and no storage is assumed between the steps. Thus, the superstructure decisions on the time index for each of the post-cultivation steps appear in expressions for almost all the entities, thereby infusing the model with a large-scale characteristic. The model, while being optimized, selects the best combination of options for which the annualized life cycle cost (ALCC) is minimized. The detailed formulation can be found in Sen Gupta et al. (2016). The key findings from this model are discussed in Section 11.3.

11.2.2 ADVANCED MODEL: INTEGRATED BIOREFINERY TO CO-PRODUCE MULTIPLE PRODUCTS

The scope of the model is expanded to include the possibility of a suite of products from lipids, proteins, and carbohydrates. Figure 11.3 presents a schematic of the flowsheet, and Figures 11.4 and 11.5 present the superstructure applicable for this modeling horizon for biodiesel production and co-product generation steps, respectively.

As can be noticed in Figure 11.3, apart from the target product biodiesel, the proteins and carbohydrates from the lipid-extracted biomass are processed to their respective end products—protein and reduced sugar—and the polar fraction of lipid is also realized as an end product. The intent is to study the trade-off between the added expenses with their extraction and their selling potential. Their purification steps are not considered in the model due to lack of detailed data on those steps. Thus, the decisions on recycle and bypass streams and the split fractions are also included in the model. The recycling of the lipid-extracted biomass for a potential second round of extraction accounts for the extraction of polar lipid in the second round, in case the first round of extraction is only with non-polar solvents. After extraction of all the necessary metabolites, the biomass is considered as residual

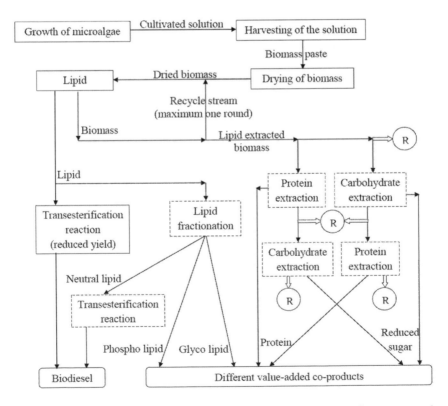

FIGURE 11.3 Schematic of the process for integrated biorefinery operations to co-produce multiple end products.

Source: Sen Gupta et al. (2017a)

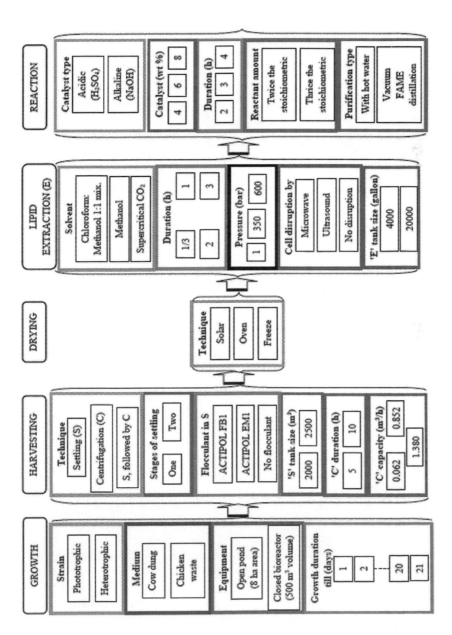

FIGURE 11.4 Superstructure for the mandatory steps for biodiesel production in the study on co-generation of products.

Source: Sen Gupta et al. (2017a)

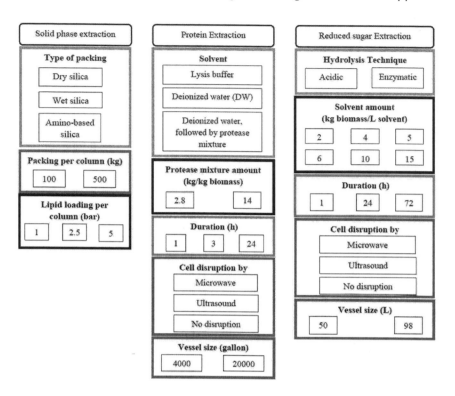

FIGURE 11.5   Superstructure for the co-product generation steps in the study.

Source: Sen Gupta et al. (2017a)

stream, marked as R in Figure 11.3. The superstructure for the steps is also more detailed than the base model. For example, in case of growth (Figure 11.4), the option for heterotrophic strain, as well as enclosed bioreactor (photobioreactor, for the case of phototrophic strains) is incorporated. The ALCC of biodiesel, the target end product, remains the objective function of the model, as it was in the base model. Apart from modeling the boundary and detailing in the superstructure, the advanced model is different from the base model in terms of the modeling pedagogy as well. The batch time is defined as a variable based on the post-cultivation steps for the production of biodiesel, the principal target, and, thus, the signature of time index of all the steps does not remain explicit anymore; accordingly, the number of variables and constraints is effectively reduced. But, on the other hand, due to this modification in the model, we encounter bilinear terms, arising out of the multiplication of two variables. Because bilinear terms can be potentially non-convex, the standard techniques available in the literature (Harjunkoski et al. 1999) are followed to linearize the model by means of additional variables and constraints. The detailed formulation can be found in Sen Gupta et al. (2017a). The key findings from this model for different scenarios of product distribution, process specifications, and storage protocols are discussed in Section 11.4.

11.2.3   ADVANCED MODEL: INTEGRATED BIOREFINERY FOR CARBON CAPTURE

The advanced model is next applied for studying the prospects of microalgae as a potential medium for carbon capture. Figure 11.6 depicts the schematic of the design of this biorefinery. Although the model remains mostly similar, the key difference with the previous approach is that the model is no longer product demand driven; rather, it is influenced by carbon dioxide provided from a power plant. Accordingly, as can be noticed from Figure 11.6, the lipid being no longer a mandatory product, the cultivated biomass can be processed for the extraction of its intracellular metabolites in any sequence.

The batch time is defined as a variable based on each of the post-cultivation steps. The rate of actual production of any of the end products is a function of the

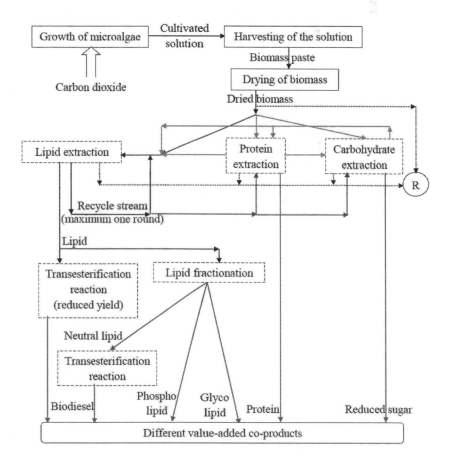

FIGURE 11.6   Schematic of the process for integrated biorefinery operations aimed at carbon capture.

Source: Sen Gupta et al. (2017b)

processing cost for its production, its demand and selling price, and availability of carbon dioxide. In case microalgae is cultivated, thereby sequestering carbon dioxide, the model enforces the selection of harvesting and drying steps, notwithstanding the absence of any further process steps. This is indicated by blocks with a solid boundary in Figure 11.6, signifying mandatory steps. This is to ensure availability of the dried biomass (having a larger shelf-life) for any downstream operations in case they are feasible. Thus, any novel downstream processing strategy, aside from the ones incorporated in the modeling framework, can be seamlessly integrated with the model.

The objective function for this problem is expressed as net annualized expenses associated with the system, with a benefit from the sale of products and additional benefit from the sequestration of carbon dioxide, which, if emitted to the atmosphere, is assumed to involve a monetary penalty. Sen Gupta et al. (2017b) presents more details on this scenario, and the key results are presented in Section 11.5.

The models are provided with process and economic parameter values, which are available in published literature; the crucial ones are enlisted in Table 11.1. The models are solved in General Algebraic Modeling System (GAMS) (GAMS, 2014) with the CPLEX optimization solver for MILP, and the results are discussed in the subsequent sections.

TABLE 11.1
Source of Crucial Process Parameters Used as Model Inputs

Process step	Key parameters	Reference sources
Growth	Concentration of cell biomass and its individual constituents	Agwa et al. (2012, 2013), Liu et al. (2011), Weissman and Goebel (1987), Tapie and Bernard (1988)
Settling	Harvesting efficiency, concentration factor	Granados et al. (2012), Tapie and Bernard (1988)
Centrifugation	Harvesting efficiency, concentration factor	Dassey and Theegala (2013)
Drying	Mass fraction of water in the outlet stream, drying time	Ansari et al. (2015)
Lipid extraction	Solvent charge, extraction efficiency	D'Oca et al. (2011), Safi et al. (2014)
Solid phase extraction	Solvent charge, recovery of each fraction of lipid	Callahan et al. (2015), Pernet et al. (2006)
Transesterification reaction	Reactant charge, yield	Prommuak et al. (2012)
Protein extraction	Solvent charge, extraction efficiency	Ansari et al. (2015), Sari et al. (2013)
Carbohydrate extraction	Solvent charge, extraction efficiency	Ansari et al. (2015), Ho et al. (2013)

Source: Sen Gupta et al. (2017a)

11.3   BIODIESEL PRODUCTION FROM MICROALGAE

This section focuses on the production of biodiesel as the only target product from microalgae, as was introduced in Section 11.2.1. The results are presented for the base case, along with a few crucial scenarios, as well as sensitivity studies with respect to model parameters.

11.3.1   BASE CASE

The model was solved for a case of daily production of 30 Mg (metric ton) of biodiesel from phototrophic strains of *Chlorella*. The optimized model decisions, available from the model solution, are presented in Figure 11.7.

The ALCC for this base case was US\$13.286/L of biodiesel produced. The distribution of this cost among the various components across the different process steps is detailed in Table 11.2.

As per the values reported in Table 11.2, the growth step was responsible for almost 88% of the total production cost. The main contributing factors were the dilute nature of the microalgae solution and low lipid content in the cells. Thus, a substantial quantity of solution had to be cultivated to supply the fixed target demand of biodiesel. Moreover, because the batch time for the downstream processes was 4 hours, with the reaction step being the slowest step, 3,024 ponds were operated in a staggered fashion to ensure availability of 5.45×10^8 L of solution, as was required to be cultivated per each batch. In terms of the decisions related to growth, the options of 18 days as the growth period and chicken waste as the medium can be attributed to higher biomass and lipid availability.

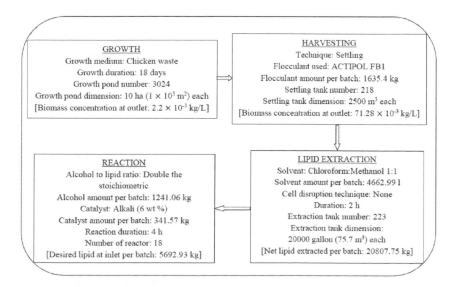

FIGURE 11.7   Optimized process decisions for the base case on biodiesel as the only product.

Source: Sen Gupta et al. (2016)

TABLE 11.2

Break-Down of Net Expenditures Among Different Contributors for the Base Case Study on Biodiesel as the Only Product

Step	Cost component	Corresponding annualized cost in US$/L of biodiesel produced	Relative amount (in %), as compared to the net cost in the same category
Growth	Growth pond (capital) Electricity for paddle wheel	9.855 1.807	84.50 15.50
Harvesting	Settling tank (capital) Settling tank (O&M) Flocculant	0.102 0.027 0.700	12.30 3.26 84.44
Lipid extraction	Extraction tank (capital) Extraction solvent	0.082 0.048	63.08 36.92
Reaction	Reactant Catalyst Reactor	0.620 0.004 0.041	93.23 0.61 6.16

Source: Sen Gupta et al. (2016)

The choice of settling as the only mode of harvesting was determined to be optimal, and the associated cost was only US$0.83/L. The model had no imposed limit on the concentration of microalgae paste into the lipid extraction step. ACTIPOL FB1 was chosen as the optimal choice of flocculant due to its higher value of harvesting efficiency. However, the concentration factor achieved with ACTIPOL FB1 was lower than that with the other flocculants. Thus, it indicated the significant amount of biomass available compared to the processing volume on the process economics of the downstream steps, at least for the numerical values for our model parameters. The necessary charge of flocculant for ACTIPOL FB1 was also less than that for the others. On the other hand, the option of flocculant-unaided settling would have required cultivation of a higher volume of solution per batch due to higher settling time, and thus higher batch time, while also requiring a fewer number of batches per day. Accounting for trade-offs such as these exemplifies the critical usefulness of superstructure optimization studies.

For the case of lipid extraction, the optimum decision was extraction with conventional solvent chloroform:methanol (1:1) despite its lower yield. this was attributed to the presence of the expensive step of compression in the other option of supercritical carbon dioxide. Extraction was for a duration of 2 hours to ensure higher yield, because it would not affect the batch time, which was 4 hours as per the slowest step of the reaction, and 4663 L of lipid extraction solvent was used per batch. The decisions for the transesterification reaction step, that is, amount of alcohol and catalyst, as well as the reaction duration, corroborate the highest possible yield for their combination compared to the other ones. Lower yields at the reaction step for other alternative decisions might have lowered the reaction cost. However, this would have enforced an enhanced requirement of lipid and biomass, and thus volume of solution being cultivated, thereby increasing the cost at all upstream steps. As can be noticed in Figure 11.7, only 27.36% of the extracted lipid was considered an input to the reaction step, as the polar lipid is not a precursor to biodiesel.

The computed ALCC of US$13.286/L would increase on incorporation of the intermediate steps, such as separation of extracted lipid into different streams prior to the reaction and biodiesel purification. Moreover, costs associated with storage, land procurement, labor, and other contingencies also would add to the ALCC. Accordingly, this also highlights the distinct necessity of technological improvement required at each of these steps for efficient processing, thereby aiding in process economics.

Meanwhile, in the current literature, one notices a wide variability in reported estimates of biofuel prices based on the specific scenarios considered. A value of US$ 5.954/L was estimated by Hannon et al. (2010) on considering 0.21 g/L/day of biomass productivity and 21% lipid content in the cells. As per estimates of algae biofuel startup Solix, the production cost of biodiesel from microalgae was US$32.81/gallon, that is, US$8.667/L (Kanellos 2009).

11.3.2 Scenario Analysis

As evident from the results presented in the previous section, the optimal solution did not predict use of carbon dioxide as a lipid extraction solvent or use of microwave or ultrasound to aid in lipid extraction, even though these alternatives were available for selection in the superstructure (Figure 11.2). Further, the option of heterotrophic strains was not considered in the base case due to its high glucose demand (0.01 kg/L of cultivated solution) during cultivation (Agwa et al. 2013). Hence, we studied three separate scenarios where only one decision at a time was enforced upon the model in the form of an additional constraint. This exercise was done to ascertain the trade-offs among different alternatives and their potential implications on the optimized decision making and process economics. Figure 11.8 presents the comparative assessment of the process economics. The batch time for all these cases remained at 4 hours. The optimized decisions and the values of the variables for these cases can be traced in Sen Gupta et al. (2016).

Case A: For case A, with a heterotrophic strain, the optimized decisions remained the same as the base case. The net ALCC (US$3.331/L) was lesser than the base case value due to the requirement of a lesser volume of solution to be cultivated, due to the greater availability of neutral lipid in the microalgae cells. The number of ponds for this case was 648, compared to 3,024 for the base case, whereas the biomass concentration at the pond outlet was 2.6 gm/L, compared to 2.2 gm/L in the base case. This was facilitated by an increased biomass and lipid productivity value for the strain. However, the high requirement for glucose would enhance the cost, while, at the same time, triggering the food versus fuel debate which otherwise generally does not eclipse the prospect of microalgae. As a result of differences in the

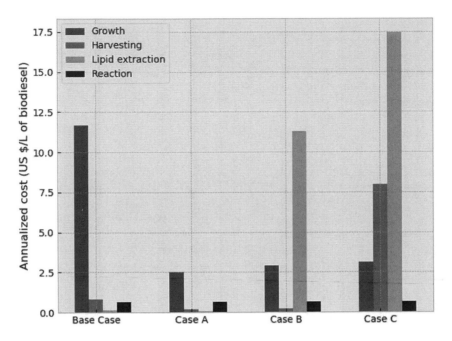

FIGURE 11.8 Distribution of cost across the various process steps for different scenarios with biodiesel as the only product.

relative content of various lipid fractions in phototrophic and heterotrophic strains, there was a slight difference in the necessary quantity of alcohol and catalyst in the transesterification reaction.

Case B: For case B, with supercritical carbon dioxide as the lipid extraction solvent, the decisions for all the other steps, except lipid extraction, remained same, while the cost values for those steps were reduced. This was due to a higher lipid extraction yield, thereby necessitating a lesser amounts of biomass as input. The number of growth ponds reduced from 3,024 in the base case to 756 for this case, and 9.6×10^5 kg of CO_2 was used per batch and the extraction operation was for 3 hours every batch. The net ALCC was US\$15.032/L and the lipid extraction step was the costliest among all the steps for this scenario (Figure 11.8) due to the expensive step of compression of the solvent. However, with carbon dioxide being non-polar solvent, the lipid extracted for this case is pure neutral lipid, whereas for the base case, with a mixture of polar and non-polar solvent, the extracted lipid should ideally be fractionated prior to reaction. Thus, inclusion of this step in the modeling framework may have an impact on the optimal decisions.

Case C: For the case C, with mandatory cell disruption during lipid extraction, we notice a major change in the harvesting step, with the selection of settling, followed by centrifugation. Moreover, the lipid extraction solvent for this case was carbon dioxide compared to the chloroform–methanol mixture for the base case. With the expensive step of cell disruption being made mandatory, the model selected those decisions for which the volume of solution at the time of cell disruption was the least.

Microwave was the optimally chosen mode for cell disruption. The net ALCC for this case was US$29.173/L. The cost for the growth step was lower than the base case, as the number of ponds was 960.

These case studies provide us with more insight about all the decisions under consideration and their mutual trade-offs. Proper values of the process parameters for the lipid extraction step is the most critical alongside the decision making with this step. This also highlights the salient benefits of the superstructure optimization model. Sen Gupta et al. (2016) presents a detailed discussion of the results for each scenario.

11.3.3 PARAMETRIC SENSITIVITY ANALYSIS OF THE BASE CASE

The model was also subjected to parametric sensitivity studies to ascertain the relative prominence of parameters on the decision making and process economics. Accordingly, different parameters related to the steps of growth, harvesting, lipid extraction, and reaction were subjected to a change of ±10%, one at a time, with respect to the base case value, and the optimization model was solved for the new parameter value. The detailed result of this study can be found in Sen Gupta et al. (2016).

The model was most sensitive to those parameters that had a direct impact on the lipid content in the input stream to the reactor. Accordingly, the concentration of biomass in microalgae cells, lipid content in cells, harvesting efficiency, lipid extraction efficiency, and reaction yield had the most pronounced impact on the net ALCC. For a change of 10% in these parameter values, ALCC changed in a reverse direction by 10.05% to 10.25%. The slight variations were attributed to the finite capacity of the equipment. The reverse trend was obvious because the cost would reduce with higher lipid content in the reactant stream, as a lesser volume of solution would be cultivated. The sensitivity on the reaction yield was non-uniform for the two levels of changes (+10% and −10%) with respect to the base case. This was due to assumption of a maximum yield of 95%. The growth rate also had a similar impact (change in ALCC of 8.8%). The sensitivity of ALCC with respect to reaction rate was different from the other parameters. Net ALCC dropped by 0.88% and 2.07%, respectively, for increase and decrease of reaction rate by 10%. This was a consequence of a change in batch time, thereby changing the optimized values of upstream variables, specifically those in the growth step, substantially. The impact of other parameters such as dose of flocculant, concentration factor, lipid extraction solvent charge, alcohol and catalyst charge was largely insignificant.

The study on production of biodiesel as the sole product having provided us with insights about the trade-offs in decision making, we investigated the modeling aspect in order to ensure a realistic representation of the system in terms of parameters and variables to the maximum possible extent. Thus, a few of the entities which were considered parameters in the base model were redesigned as variables. This led to terms involving the multiplication of variables, thereby introducing non-convexity. Accordingly, different standard techniques available in the literature (Harjunkoski et al. 1999) were applied for a test case of our model. The most efficient one was followed in the advanced model formulation, as has been discussed in Section 11.2.2.

11.4 CO-GENERATION OF MULTIPLE END PRODUCTS

The prospect of utilizing microalgae as the source of different value-added products was studied with the advanced model discussed in Section 11.2.2, and the salient features of the optimized design and operation of the biorefinery are discussed in this section. The detailed discussions can be traced in Sen Gupta et al. (2017a, 2017b). Polar lipid, protein, and carbohydrate are the co-products alongside the principal product, biodiesel (Figure 11.3). These being high-value, low-volume products compared to biodiesel which is low-value/high-volume, the demand for these co-products was considered as the respective upper bound of production throughput. Considering a minimum daily demand of 30 Mg for biodiesel, the maximum annual demand of protein was 11,650 Mg; for reduced sugar, it was 6,657 Mg; and the same for polar lipids was 832 kg, on the basis of the ratio of global demand of diesel and these co-products. The selling price values, obtained from the literature (Ruiz et al. 2016), were adjusted to account for possible downstream purification that may be required to attain the final products but was not considered in the model. The corresponding values for our model were US$610.5/Mg for protein, US$416.25/Mg for reduced sugar, and US$3,330/Mg for polar lipids.

11.4.1 BASE CASE

The optimized decision making for the co-production of multiple products was investigated for the base case involving the absence of any storage facilities between the stages. Figure 11.9 presents the optimal configuration of the biorefinery alongside

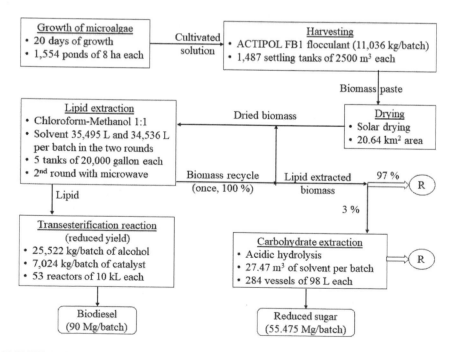

FIGURE 11.9 Optimized configuration of the biorefinery for the base case on co-generation of multiple products.

information on the major design and operation decisions. Drying was the slowest step, and thus, the batch time of the biorefinery was 72 hours, the same as the duration of solar drying, selected as the optimal option among available alternatives. Reduced sugar was co-produced alongside biodiesel, for which 3% of the lipid-extracted biomass was sent for the carbohydrate extraction process. This indicates the significant potential of producing more reduced sugar beyond the specified demand. Ninety-two percent of the biomass cultivated (70.99×10^4 Mg per batch) was treated as residual. No other co-product was produced from either polar lipid or protein, due to more added expenditure in their processing in comparison to their value additions.

The net ALCC of biodiesel production for this scenario was US 8.53/L. The ALCC would have been US$8.74/L without any co-production of reduced sugar, which, as a co-product, facilitated the economics by 2.4%. The ALCC drops from the previous case (see Section 11.3) largely due to two rounds of lipid extraction, which made the reactant inlet stream a lipid-rich one.

The selection of optimal process decisions for this case is discussed in Sen Gupta et al. (2017a). Here we highlight the key findings. Except for two cycles of lipid extraction, the design decisions for all the steps remained same as the base case for the study with biodiesel as the only target (see Section 11.3), with the growth step turning out to be the most expensive step along similar lines, accounting for 74.14% of the total expenditure. The number of ponds for cultivating microalgae solution was 1,554 for this case compared to 3,024 for biodiesel as the only product (see Section 11.3), due to more lipid being extracted in two rounds. As a result, the capital expenditure for the downstream steps also reduced. The additional step for co-production of reduced sugar involved a cost of US$0.014/L, and the revenue from the co-product was US$0.223/L of biodiesel produced. Figure 11.10 depicts the step-wise distribution of cost across the steps for this base case and other scenarios analyzed in Section 11.4.2.

11.4.2   Scenario Studies on Process Decision Alternatives

Separate scenarios were also studied in order to understand the selection of optimized decisions in the base case. The detailed results are available in Sen Gupta et al. (2017a).

Case D—Cultivation of the heterotrophic strain: The heterotrophic strain, despite its higher biomass productivity and higher neutral lipid content, was not selected for cultivation in the base case. For this scenario, the net ALCC was US$18.85/L which was 2.21 times the corresponding base case value. The high cost was due to the high amount of glucose (4.84×10^5 Mg/y) in the cultivation medium. The volume being cultivated was lesser than the base case due to higher lipid content per unit volume, and thus, the downstream processes were also scaled down. In terms of process decisions, the application of supercritical carbon dioxide as the lipid extraction solvent was the major difference with the base case. The more efficient protocol of lipid extraction was selected, albeit at a higher cost, to ensure the requirement of lesser cultivation volume, thereby reducing the amount of glucose as a substrate in the growth step to the largest extent possible.

Similar to the base case, reduced sugar was produced as a co-product. However, because the biomass was more enriched in lipid, 28% of the lipid-extracted biomass

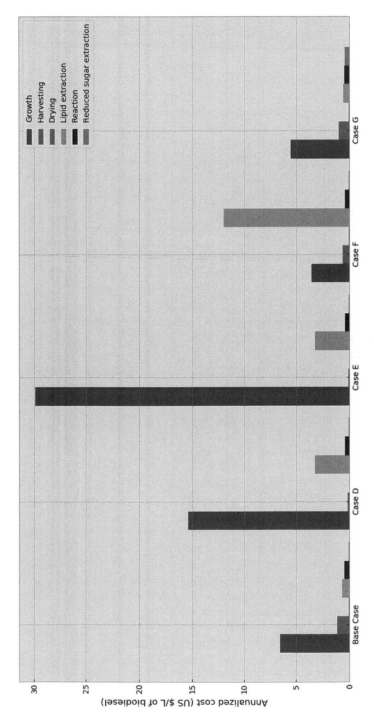

FIGURE 11.10   Distribution of cost across the various process steps for different scenarios involving the co-generation of multiple products.

was processed to meet the demand of reduced sugar, compared to only 3% having been used for the base case. Moreover, the potential of integrating the output sugar beyond the demand level with the input requirement was also investigated; the net ALCC dropped to US$17.62/L for such a scenario. The substrate requirement at the growth stage reduced by 8.88% to a value of 4.41×10^5 Mg/y.

Case E—Cultivation in a closed bioreactor: The case of cultivating microalgae in closed bioreactor, ensuring higher productivity at higher expenses, was also analysed. For this case, the model selected cultivation of a heterotrophic strain and lipid extraction with supercritical carbon dioxide in spite of them being more expensive. Through these selections, the lipid content in the cultivated biomass and lipid extraction efficiency were greatly improved, thereby requiring lesser throughput of microalgae solution at the growth step (around 10% of the base case value), thus reducing the expenses for closed bioreactors. The net ALCC for biodiesel production was $33.37/L, which was around 190% higher than the base case value. Reduced sugar was a co-product, similar to the base case. Further, it was also noticed that a reduction in the cost of the closed bioreactor by even ten times would not result in its selection in the base case.

Case F—Supercritical extraction of lipid: For the scenario of supercritical extraction of lipid, the net ALCC was US$16.32/L, 91% higher than the base case value. The process decisions were similar to the base case ones, as also has been the case for the study with biodiesel as the only product. The model in the base case did not favor the selection of supercritical extraction of lipid. However, the situation might change based on the impact of the polar lipid content in the input stream to the transesterification reactor on the reaction yield (the model assumed a reduction of yield by 10%) because carbon dioxide is a non-polar solvent, extracting only neutral lipid, whereas the chloroform–methanol mixture was the solvent for the base case extracts of both polar and non-polar lipids. Polar lipids, though available as potential co-products, were not produced in the optimized design.

Case G—Co-cultivation of phototrophic and heterotrophic strains: The base case co-produced reduced sugar as the added expenditure in its extraction was less than the value addition achieved. Thus, the scenario involving co-cultivation of a phototrophic and heterotrophic strain was studied to potentially leverage the higher carbohydrate content in the former to help grow the latter with a higher lipid content, thereby facilitating biodiesel production. The hypothesis of potential mutual benefit between these two strains was supported by a 10.2% reduction in net ALCC to a value of US$7.66/L. The cultivation of heterotrophic strain involved only 49 Mg/y of substrate—the same would have been as high as 1.86×10^5 Mg/y in the absence of any recycling from the product stream of the phototrophic strain. We noticed a scale-up by a factor of 28.31 in the infrastructure in reduced sugar extraction process, as the entire lipid-extracted biomass was subjected to that step to maximize production.

11.4.3 Scenario Studies on an Alternative Demand Profile of Co-products

Maximum potential of reduced sugar production: Reduced sugar was a viable co-product; yet in the base case, only 3% of the lipid-extracted biomass was used to meet the demand of reduced sugar; thus, we investigated the impact of the maximum potential of reduced sugar co-production. The net ALCC for biodiesel production for such a scenario was US$ – 27.23/L, thereby indicating a net profit. The total annualized expenditure was US$35.35/L, and a revenue of US$62.58/L was generated. The total production of reduced sugar was 280 times the base case. More crucially, many of the design decisions of the biorefinery were significantly different from the base case. For instance, microalgae cells were cultivated only for ten days to have the cultivated biomass be relatively rich in carbohydrate. The lipid extraction step did not involve any cell disruption due to the same purpose. The batch time (72 hours) of the biorefinery remained the same as the base case one. On the other hand, considering a target biodiesel selling price of US$0.48/L (Bart et al. 2010), we noticed that the net profit of US$27.71/L from the biodiesel was equivalent to a reduction in the selling price of reduced sugar by US$51.1/kg. This presents a strong case for potential applications of the microalgae biorefinery in reduced sugar industries. These insights emphasize the necessity of exploring the promising potential of co-production from a comprehensive perspective.

Mandatory co-production of protein and polar lipids: We separately investigated the exclusion of protein and polar lipid as co-products in the optimized base case by considering the demand as a lower bound on production quantity, compared to the upper bound in the base case. For the scenario with mandatory protein production, the net ALCC increased by 77% to a value of US$15.11/L. The decisions for cell growth, harvesting, drying, and lipid extraction remained the same as the base case. The distribution of a lipid-extracted biomass for downstream operations was different from the base case. Fourteen percent of the lipid-extracted biomass was used as an input to the carbohydrate extraction step, and 64% of the residual biomass from the outlet stream of carbohydrate extraction was processed for the extraction of protein. The added expenditure for the protein extraction step was US$7.107/L, and the product generated a revenue of US$0.575/L. Thus, a hike in the selling price of protein by 12.5% or an equivalent improvement in process efficiency was necessary for it to be economically viable. On the other hand, for polar lipid as a co-product, the net ALCC increased by 55% from the base case to a value of US$13.20/L of the biodiesel produced. The additional expenses arising out of the step of solid-phase extraction to separate the lipid fractions were US$4.67/L, and the revenue was only US$0.0002/L of the biodiesel produced, thereby highlighting the major improvement necessary for polar lipid co-production to be a viable option.

Residual biomass as fertilizer: Almost 92% of the cultivated biomass remained unused in the base case, thereby inspiring us to look into its potential value-added end usages. We considered its application as fertilizer. Because protein is the source of nitrogen in the biomass, only the available protein in the residual biomass was assumed to have economic potential. Based on the nitrogen content in the fertilizer, its selling price was US$880/Mg, and the adjusted annual demand, analogous to other co-products, was 4626.72 Mg. For this scenario, the ALCC reduced from US$8.53/L to US$8.2/L. The demand of fertilizer was fulfilled, while all the process decisions in the upstream were the same as their base case levels. For unlimited production of fertilizer, the biorefinery would generate a profit of US$3.83/L. For this case, the design of the biorefinery was very different from the base case design, as it was for the scenario of unlimited production of reduced sugar. Cultivation for seven days and the absence of cell disruption in the lipid extraction step ensured a relatively higher content of protein in the residual biomass.

The model was also capable of studying the temporal variation of demand and selling price values for each product with respect to specified mean and variance levels.

11.4.4 SENSITIVITY ANALYSIS

The model was also subjected to sensitivity analysis with respect to model parameters to ascertain their relative prominence on the objective function and thus the design and operation of the biorefinery. A marginal change of ±10% in parameter values (with each parameter at a time) with respect to the base case values was enforced, and the results were analysed. Sen Gupta et al. (2017a) presents a detailed discussion on this.

The sensitivity studies for the case with biodiesel as the only product provided us with an understanding of the non-trivial parameters to this end, and the relevant parameters were chosen for the present study accordingly. The process economics were more sensitive to parameters such as cell productivity, growth rate, lipid extraction efficiency, and reaction yield, as these had a direct impact on the production route of biodiesel, the principal target product. For a 10% increase in values for these parameters, the ALCC dropped by 7.65%, 7.62%, 7.93%, and 8.85%, respectively. On the other hand, for a 10% decrease in those parameter values, the ALCC increased by 10.02%, 7.57%, 9.65%, and 10.95%, respectively. The asymmetric nature of impacts for a positive and negative change in parameter values was due to the discrete nature of variables, such as split fraction of streams, as well as maximum reaction yield of 90%. Drying being the slowest among all the steps, a reduction in its duration would reduce batch time; thus, the idle time of equipment for other processes would also drop, aiding in the economics. A drying time value of half the base case value would result in reduction of ALCC by 10%. Although reduced sugar was a co-product, the parameters for the carbohydrate extraction step, such as solvent-to-biomass ratio or extraction efficiency, did not have any significant impact on the process economics. This was obvious, given the relatively lower contribution of the carbohydrate extraction steps in the base case. The design and operation decisions of the biorefinery remained same for all these instances.

Moreover, the model was also modified to incorporate quarterly variation of demand and selling price of the products. For each of these entities, a normal distribution was considered with mean as the base case value and standard deviation of 10% of the mean. These values were known to the optimization problem as a set of parameters, and the problem was not posed as a stochastic optimization problem due to its high computational load. The design decisions remained time invariant, and the operational ones were functions of time. For variation in either demand or selling price of products, the average change noticed in ALCC was around a 7% rise compared to the base case, due to a drop in reduced sugar demand and selling price values. Parameter values for polar lipid and protein did not have any role in the economics, as they were not co-produced. In terms of configuration, flocculant-unaided settling was chosen for those quarters with comparatively lesser demand of biodiesel. Two-stage settling without any flocculant lowered the expenses, whereas the reduction in harvesting efficiency did not have an impact due to the decreased demand of biomass on the downstream. The results for the scenario, involving a variation in market conditions, would be case specific and not generic. Thus, the main contribution of the model rests in its robust characteristic, necessary to forecast the optimized decision making under such conditions.

The sub-optimal solutions were also investigated to ascertain their level of sub-optimality and the concerned decision variable and any change in the biorefinery. In terms of equipment specification decisions, the most sensitive was vessel size for the lipid extraction step, though it imparted an insignificant $7.85 \times 10^{-4}\%$ increment on the ALCC. Among the process decisions, the amount of solvent used for carbohydrate extraction was the most sensitive, and it facilitated a 0.08% increase in ALCC. The first sub-optimal option in terms of synthesis decisions was the biorefinery without any co-product, and ALCC increased by 2.34%. These sub-optimal solutions help us in identifying the decisions whose alterations would have an immediate impact on the economics and quantifying the same.

11.4.5 Batch Scheduling Strategies

The base case, as well as all the scenarios studied up to now, considered no storage facility in between the stages. Thus, we studied additional scenarios with differences in the scheduling strategy—in the presence of infinite intermediate storage (Case H) or through debottlenecking of the slowest step (Case I) and a combination of both these strategies (Case J), thereby ensuring potentially efficient usages of equipment. The distribution of cost across the process steps is represented in Figure 11.11. More details on the findings for these cases can be found in Sen Gupta et al. (2017a).

The provision of intermediate storage (Case H) allows for potentially staggered operations of the faster steps, thereby lowering their idle time and reducing the necessary amount of equipment for these faster steps. For our case, the model was modified by including additional variables and constraints to make the effective duration of operation for all the post-cultivation steps the same as the batch time. With the demand of products remaining the same as the base case, the throughput of microalgae did not change, and thus, the operational variables also remained at the same levels as the base case. The design for the biorefinery, with biodiesel and reduced sugar as co-products, remained identical as well. Because the throughput remained the same, there was no impact on any of the decision variables or the cost in the growth step. At the same time, on account of efficient use of equipment for the faster steps in the downstream, their requisite numbers were lesser compared to the base case. This resulted in a reduction in the net ALCC by 11.25% from the base case to a value of US$7.57/L. Figure 11.11 reflects the reduced cost in almost all the post-cultivation downstream steps.

On the other hand, debottlenecking the slowest and successively slower steps through more equipment (or area), as studied in Case I, helps in increasing the number of batches over time. If the rate of demand for products remains constant, as in our case, the increased number of batches implies a reduced throughput of production per batch. Hence, the net capital cost is reduced at all the steps, including

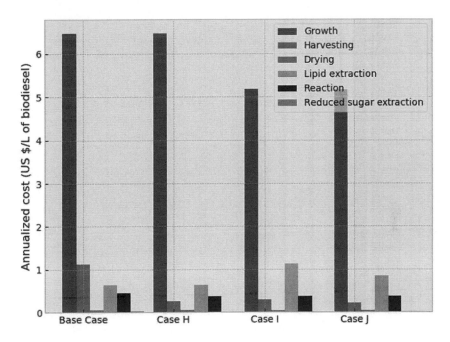

FIGURE 11.11 Distribution of cost across the various process steps for different batch scheduling strategies.

growth. However, batch time reduction beyond a threshold may not yield any further benefit due to the fixed capacity of the equipment. Further, some of the operations may have higher electricity requirements. Accordingly, the model was modified to investigate this trade-off and estimate the optimal level of debottlenecking. The net ALCC for this study was US$6.78/L, which was 20.5% more economical than the base case. This saving was largely facilitated by the requirement of a smaller number of ponds (1,242) compared to the base case (1,554). A reduction in expenditure with respect to the base case was observed for most of the steps because the demand rate of the products remained the same (Figure 11.11). The only difference was noticed for the lipid extraction step. Optimized decisions remained the same as the base case, with the only exception being the application of microwave for cell disruption in the first round of lipid extraction, compared to the second round in the base case. Application of cell disruption in the first round vis-à-vis in the second round enhances the amount of lipid extract because the amount of biomass being treated is higher in the first round and, for the same reason, the expenses with the process also rise. Accordingly, cell disruption in the first round would be viable only below a threshold value of biomass being treated. For this very reason, for the study presented in Section 11.4.2, the first round of lipid extraction was with microwave assistance for the scenarios of heterotrophic strain and closed bioreactor. For this present case (Case I), the selection of cell disruption mode in first round of lipid extraction ensured better extraction, thereby reducing the volume requirement and growth cost even further. Thus, a minimal increase in lipid extraction helped in reducing the growth cost substantially. The biorefinery operations were debottlenecked to the extent of having an effective batch time of 10.43 hours, which implied that the 69 batches were operated in a month, whereas the base case operation involved only 10 batches per month.

Simultaneous utilization of both the strategies of infinite storage and debottlenecking (Case J) facilitated the process economics even further, as expected, and the net ALCC reduced by 25% from the base case to a value of US$6.4/L. One hundred and twenty-four batches were operated every month. Although infinite storage is practically infeasible, we attain a theoretical limit on the reduction in ALCC that can be realized on adding provisions for intermediate storage.

11.5 MICROALGAE AS A MEDIUM OF CARBON SEQUESTRATION

The modified advanced model, introduced in Section 11.2.3, was simulated with the aim of minimizing the net annualized expenditure for sequestering 3.5×10^6 Mg of carbon dioxide on an annual basis. The flue gas containing carbon dioxide was assumed to be available from a 500-MW coal-fired power plant. The price of flue gas emission to the atmosphere was considered as US$5.53/Mg of carbon dioxide (Carbon Commodities, 2017). Although microalgae require carbon dioxide for growth, microalgae's productivity has been found to drop at high values of carbon dioxide concentration; the uptake rate of carbon dioxide by microalgae was assumed to affect the volume of cultivation, and uptake efficiency values of 5% and 25% were assumed for cultivation in an open pond and closed bioreactor, respectively.

Two separate scenarios were studied with differences in the selling price of biodiesel; the selling price of other products remained the same. Biodiesel, being a high-volume product, was not assigned with any demand level, whereas for other products, demands were adjusted to a value of 3.5 times the base case values on the basis of the ratio of carbon dioxide available and demand of these products for a certain region. The detailed understanding, as well as the results, are discussed in Sen Gupta et al. (2017b).

For the scenario on the low selling price of biodiesel (US$0.48/L) (Bart et al. 2010), the schematic of the optimized biorefinery is depicted in Figure 11.12 which also presents information on the critical design and operational decisions. Reduced sugar was the only product. Table 11.3 presents the income and expenditure for each of the steps. Net ALCC of the biorefinery was US$14.13 million, indicating a net loss. The annual deficit stood at US$9.71 million, inclusive of the operating expenses over the year. On the other hand, net capital investment was US$36.62 million.

On the upstream, only 2.77% of the available carbon dioxide (9722.22 Mg/d) was captured because the reduced sugar market was saturated with the produced amount. The extra expenditure in sequestering additional carbon dioxide was not economical for the present level of carbon penalty. The cultivation continued for seven days to make use of the carbohydrate-rich biomass. The other decisions for growth, as well as those for the downstream steps, were optimized in a way so as to maximize the yield of reduced sugar. However, microalgae settling did not involve any flocculant, as the flocculation efficiency was no longer a critical factor due to the abundance of carbon dioxide in comparison to the amount necessary to meet the demand of

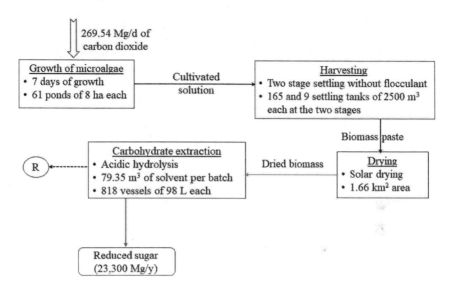

FIGURE 11.12 Configuration of the biorefinery for carbon capture for the low price of biodiesel.

TABLE 11.3

Step-wise Economics for Carbon Capture (for the Low Price of Biodiesel)

Steps	Contribution to expenditure in terms of ALCC ($\times 10^6$ US $)	Revenue generated/ Penalty accrued per year ($\times 10^6$ US $)
Growth	• Capital cost: 2.87 • Operating (electricity) cost: 0.28	• Carbon penalty: −18.82 • Emission into atmosphere 9452.68 Mg/d
Harvesting	• Capital cost: 1.03 • Operating (equipment) cost: 0.27	
Drying	• Capital cost: 0.05	
Carbohydrate extraction for production of reduced sugar	• Capital cost: 0.49 • Solvent cost: 0.012	• Revenue from reduced sugar (23,300 Mg/y): 9.7
	Gross expenditure: 5.01	Gross revenue: −9.12
	Net ALCC = 5.01 − (−9.12) = 14.13	
	Net revenue = Gross revenue − Operating expenses = −9.71	

---

reduced sugar, which turned out to be the only product. Batch time was 72 hours. The sensitivity of the design was studied with respect to the carbon penalty value. It was observed that 10% change in the carbon penalty did not have any impact on the model decisions, whereas the net annualized expenditure changed by 13.3% in an inversely proportional trend, as expected.

Next, the selling price of biodiesel was increased to a value of US$8.74/L, the value computed as ALCC in the base case (Section 11.4.1) without the cost reduction resulting from the reduced sugar co-production. The schematic of the biorefinery with the major decisions is depicted in Figure 11.13. Table 11.4 presents the results on process economics.

Biodiesel (35,580 Mg/y) and reduced sugar (23,300 Mg/y) were the co-products of the biorefinery, with a net ALCC of US$−5.77 million, indicating a net profit. The net capital expenses were US$2465.23 million, and with the net annual revenue being US$303.27 million, including operating expenses, the biorefinery would have a payback period of 8.13 years. At the growth stage, 99.89% of the available carbon dioxide was sequestered in the biorefinery. Sequestration beyond this point would result in an increased capital expenditure (due to the fixed capacity of the equipment) beyond the possible trade-off potential of the sale of products. The design decisions of the biorefinery were the same as the previous study on the biodiesel demand-driven design of the biorefinery because for this case also, production of biodiesel turned out to be predominating factor. A change in carbon penalty by 10% did not bear any impact on the configuration of the biorefinery and had a negligible impact (0.03%) on the annualized profit, because almost the entire amount of the available carbon dioxide was sequestered at the base value of the carbon penalty.

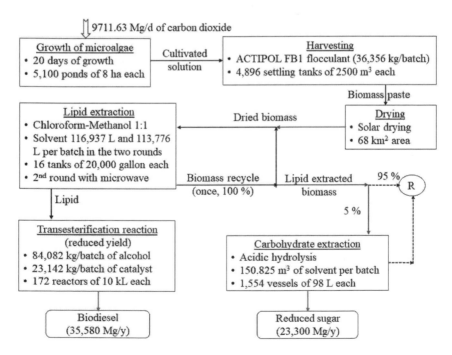

FIGURE 11.13   Configuration of the biorefinery for carbon capture for the high price of biodiesel.

TABLE 11.4
Step-wise Economics for Carbon Capture (for the High Price of Biodiesel)

Steps	Contribution to expenditure in terms of ALCC (×10⁶ US $)	Revenue generated/ Penalty accrued per year (×10⁶ US $)
Growth	• Capital cost: 240.14 • Operating (electricity) cost: 23.9	• Carbon penalty: −0.021 • Emission into atmosphere 10.59 Mg/d
Harvesting	• Capital cost: 28.87 • Operating (equipment) cost: 7.53 • Flocculant cost: 9.69	
Drying	• Capital cost: 2.04	
Lipid extraction	• Capital cost: 0.07 • Solvent cost: 1.48 • Cell disruption mechanism cost: 24.63	
Trans-esterification reaction	• Capital cost: 3.19 • Alcohol cost: 14.6 • Catalyst cost: 0.17	• Revenue from biodiesel (35,580 Mg/y): 353.37
Carbohydrate extraction for production of reduced sugar	• Capital cost: 0.935 • Solvent cost: 0.024	• Revenue from reduced sugar (23,300 Mg/y): 9.7
	Gross expenditure: 357.28	Gross revenue: 363.05
	Net ALCC= 357.28 − 363.05 = −5.77	
	Net revenue= Gross revenue − Operating expenses = 303.27	

11.6 CONCLUSIONS

The bright prospect of microalgae as a renewable source of energy and as a precursor to different value-added products has already been established in theory. However, its bright potential has not yet been realized to the fullest. Model-based studies such as the one presented in this chapter present a systematic approach to analyse the mutual trade-off among different combinations of process options and identify techno-economic challenges. The superstructure optimization study through the formulation of an MILP problem has facilitated in attaining the optimal design and operational protocols of the integrated microalgae biorefinery from various alternatives at each process step. Different scenarios as functions of the main objective of the biorefinery system, as well as the demand profile of the products, have also been studied to develop more insight into the biorefinery system.

The net ALCC for the production of biodiesel as the sole product is US$13.286/L, with growth and lipid extraction steps being the most critical in terms of decision making. The growth step accounted for 88% of the total expenditure. The prominence of the parameters having a direct impact on the lipid content in the biomass at the reactant input on the process economics is established from the sensitivity analysis. Thus, these parameters are most crucial from the perspective of the biorefinery operations. The base model is modified for a better representation of the system and the superstructure expanded to incorporate more process alternatives. The scope of the biorefinery is also broadened to include co-generation of multiple end products. The ALCC for biodiesel production for this case is US$8.53/L, with reduced sugar, the co-product, facilitating a reduction in cost by 2.4%. Co-cultivation of phototrophic and heterotrophic strains results in a 10.2% reduction in this value, and a scheduling strategy with infinite intermediate storage, coupled with debottlenecking of slower steps, leads to a 25% improvement. In terms of sensitivity of demand profile of products, reduced sugar, as the only viable co-product, has the potential to generate revenue equivalent to ALCC of US$27.23/L of biodiesel. Thus, it would be worth investigating the possible avenues for such a high amount of reduced sugar (280 times the base case value) and exploring a new horizon for the sugar industry, as well as promoting the cause of the biorefinery. On the other hand, temporal variations in market conditions of the products are analysed with an upgraded version of the model. The prospect of using microalgae for atmospheric carbon dioxide sequestration is also analyzed; the design of the biorefinery in such cases is a strong function of the selling price of biodiesel.

The potential challenge for such a modeling analysis rises from the necessity of keeping consistency of data across the process steps. For example, the transesterification reaction yield values for the extracted lipid should be based on the lipid profile of the specific cultivated strain whose growth data are considered in the model. In the present study, due to the unavailability of the relevant set of data for all the process steps for a single strain, consistency of data has been assumed for different *Chlorella* strains. For the model to be more accurate, all the different alternative process options for each of the steps of the biorefinery should be performed as a complete package with a single strain. Additionally, the computational complexity can pose a challenge, thereby requiring one to look into efficient ways of problem formulation and solution. Decomposition techniques in which the model variables are segregated into two divisions aid in achieving faster convergence to the solution, and may be one potential approach.

The study can be potentially extended to include other schemes of microalgae biomass processing and valorization to different products. Thermochemical avenues of processing, such as hydrothermal liquefaction into liquid fuel, need to be compared with the biochemical routes studied in this chapter prior to deciding on the scope and configuration of the biorefinery. Further, the residual biomass (92% for the base case) needs to be properly characterized in order to ascertain its potential end usages. For instance, apart from fertilizer, it can be converted to biogas through anaerobic digestion (Jarvis et al. 2013; Ward et al. 2014). On the other hand, for a comprehensive understanding, the environmental impact, caused by the generation of products from the microalgae biorefinery, should be evenly compared with the impact in obtaining these products from other precursors. Moreover, to this end, a multiobjective optimization involving economic and environmental consideration also deserves special attention.

Our systems engineering-based process optimization study appropriately complements the contemporary research on a biotechnological level, aimed at improving the microalgae strain profile, and on a process level, aimed at enhancing the efficiency of the steps of microalgae production and processing. As an extension of the integrated biorefinery, the idea of an integrated renewable energy park comprising technologies to generate various forms of renewable energy such as solar, wind, bio, and geothermal in a sustainable way has also been conceived (Subhadra 2010; Subhadra et al. 2010). An extended version of our present model can serve as a potential decision-making tool in the investigation of such an integrated system.

REFERENCES

Agwa, O. K., S. N. Ibe, and G. O. Abu. 2012. "Economically effective potential of *Chlorella sp.* for biomass and lipid production." *Journal of Microbiology and Biotechnology Research* 2 (1):35–45.

Agwa, O. K., S. N. Ibe, and G. O. Abu. 2013. "Heterotrophic cultivation of *Chlorella sp.* using different waste extracts." *International Journal of Biochemistry and Biotechnology* 2 (3):289–297.

Amaro, H. M., A. C. Guedes, and F. X. Malcata. 2011. "Advances and perspectives in using microalgae to produce biodiesel." *Applied Energy* 88 (10):3402–3410.

Ansari, F. A., A. Shriwastav, S. K. Gupta, I. Rawat, A. Guldhe, and F. Bux. 2015. "Lipid extracted algae as a source for protein and reduced sugar: A step closer to the biorefinery." *Bioresource Technology* 179:559–564.

Baliban, R. C., J. A. Elia, R. Misener, and C. A. Floudas. 2012. "Global optimization of a MINLP process synthesis model for thermochemical based conversion of hybrid coal, biomass, and natural gas to liquid fuels." *Computers & Chemical Engineering* 42:64–86.

Bart, J. C. J., N. Palmeri, and S. Cavallaro. 2010. *Biodiesel Science and Technology, 1st edition, From Soil to Oil.* Cambridge: Woodhead Publishing Limited.

Brennan, L., and P. Owende. 2010. "Biofuels from microalgae—A review of technologies for production, processing, and extractions of biofuels and co-products." *Renewable & Sustainable Energy Reviews* 14 (2):557–577.

Callahan, D. L., G. J. O. Martin, D. R. A. Hill, I. L. D. Olmstead, and D. A. Dias. 2015. "Analytical approaches for the detailed characterization of microalgal lipid extracts for the production of biodiesel." In *Marine Algae extracts Processes, products and applications Volume 1*, edited by S. E. Kim and K. Chojnacka, 331–346. Weinheim: Wiley VCH.

Campbell, P. K., T. Beer, and D. Batten. 2011. "Life cycle assessment of biodiesel production from microalgae in ponds." *Bioresource Technology* 2 (1):50–56.

Carbon Commodities. www.investing.com/commodities/carbon-emissions-historical-data (Accessed March 29, 2017).

Cheali, P., A. Vivion, K. V. Gernaey, and G. Sin. 2015. "Optimal design of algae biorefinery processing networks for the production of protein, ethanol and biodiesel." *Computer aided Chemical Engineering* 37:1151–1156.

Chen, P., M. Min, Y. Chen, et al. 2009. "Review of the biological and engineering aspects of algae to fuels approach." *International Journal of Agricultural and Biological Engineering* 2 (4):1–30.

Chen, C.-Y., K.-L. Yeh, R. Aisyah, D.-J. Lee, and J.-S. Chang. 2011. "Cultivation, photobioreactor design and harvesting of microalgae for biodiesel production: A critical review." *Bioresource Technology* 102 (1):71–81.

Chisti, Y. 2007. "Biodiesel from microalgae." *Biotechnology Advances* 25 (3):294–306.

Christenson, L., and R. Sims. 2011. "Production and harvesting of microalgae for wastewater treatment, biofuels, and bioproducts." *Biotechnology Advances* 29 (6):686–702.

CO_2 Earth. www.co2.earth (Accessed December 7, 2018).

Dassey, A. J., and C. S. Theegala. 2013. "Harvesting economics and strategies using centrifugation for cost effective separation of microalgae cells for biodiesel applications." *Bioresource Technology* 128:241–245.

Davis, R., A. Aden, and P. T. Pienkos. 2011. "Techno-economic analysis of autotrophic microalgae for fuel production." *Applied Energy* 88 (10):3524–3531.

D'Oca, M. G. M., C. V. Viêgas, J. S. Lemões, et al. 2011. "Production of FAMEs from several microalgal lipidic extracts and direct transesterification of the *Chlorella pyrenoidosa*." *Biomass & Bioenergy* 35 (4):1533–1538.

General Algebraic Modeling System (GAMS). www.gams.com/dd/docs/solvers/cplex.pdf (Accessed on April 6, 2014).

Gebreslassie, B. H., R. Waymire, and F. You. 2013. "Sustainable design and synthesis of algae-based biorefinery for simultaneous hydrocarbon biofuel production and carbon sequestration." *AIChE Journal* 59 (5):1599–1621.

Gong, J., and F. You. 2014. "Global Optimization for sustainable design and synthesis of algae processing network for CO_2 mitigation and biofuel production using life cycle optimization." *AIChE Journal* 60 (9):3195–3321.

Gong, J., and F. You. 2015. "Value-added chemicals from microalgae: Greener, more economical, or both?" *ACS Sustainable Chemistry & Engineering* 3 (1):82–96.

Granados, M. R., F. G. Acién, C. Gómez, J. M. Fernández-Sevilla, and E. M. Grima. 2012. "Evaluation of flocculants for the recovery of freshwater microalgae." *Bioresource Technology* 118:102–110.

Greenwell, H. C., L. M. L. Laurens, R. J. Shields, R. W. Lovitt, and K. J. Flynn. 2010. "Placing microalgae on the biofuels priority list: A review of the technological challenges." *Journal of the Royal Society Interface* 7 (46):703–726.

Halim, R., M. K. Danquah, and P. A. Webley. 2012. "Extraction of oil from microalgae for biodiesel production: A review." *Biotechnology Advances* 30 (3):709–732.

Hannon, M., J. Gimpel, M. Tran, B. Rasala, and S. Mayfield. 2010. "Biofuels from algae: Challenges and potential." *Biofuels* 1 (5):763–784.

Harjunkoski, I., T. Westerlund, and R. Pörn. 1999. "Numerical and environmental considerations on a complex industrial mixed integer non-linear programming (MINLP) problem." *Computers & Chemical Engineering* 23:1545–1561.

Harun, R., M. Singh, G. M. Forde, and M. K. Danquah. 2010. "Bioprocess engineering of microalgae to produce a variety of consumer products." *Renewable & Sustainable Energy Reviews* 14 (3):1037–1047.

Ho, S. H., S. W. Huang, C. Y. Chen, T. Hasunuma, A. Kondo, and J. S. Chang. 2013. "Bioethanol production using carbohydrate-rich microalgae biomass as feedstock." *Bioresource Technology* 135:191–198.

Jarvis, E., R. Davis, and C. Frear. 2013. Efficient use of algal biomass residues for biopower production with nutrient recycle. NREL Report 2013. WBS # 9.2.2.3. http://csanr.wsu.edu/wp-content/uploads/2013/10/TOTAL-ALGAE-PROJECT-FINAL- REPORT.pdf

Kanellos, M. 2009. Algae Biodiesel: It's $ 33 a gallon. Article, Greentech Media. www.greentechmedia.com/articles/read/algae-biodiesel-its-33-a-gallon-5652   (Accessed October 31, 2013).

Lam, M. K., and K. T. Lee. 2012. "Microalgae biofuels: A critical review of issues, problems and the way forward." *Biotechnology Advances* 30 (3):673–690.

Liu, J., J. Huang, Z. Sun, Y. Zhong, Y. Jiang, and F. Chen. 2011. "Differential lipid and fatty acid profiles of photoautotrophic and heterotrophic *Chlorella zofingiensis*: Assessment of algal oils for biodiesel production." *Bioresource Technology* 102 (1):106–110.

Martin, M., and I. E. Grossmann. 2011. "Energy optimization of bioethanol production via gasification of switchgrass." *AIChE Journal* 57 (12):3408–3428.

Mata, T. M., A. A. Martins, and N. S. Caetano. 2010. "Microalgae for biodiesel production and other applications: A review." *Renewable & Sustainable Energy Reviews* 14 (1):217–232.

Pernet, F., C. J. Pelletier, and J. Milley. 2006. "Comparison of three solid-phase extraction methods for fatty acid analysis of lipid fractions in tissues of marine bivalves." *Journal of Chromatography A* 1137:127–137.

Prommuak, C., P. Pavasant, A. T. Quitain, M. Goto, and A. Shotipruk. 2012. "Microalgal lipid extraction and evaluation of single-step biodiesel production." *Engineering Journal* 16 (5):157–166.

Pulz, O., and W. Gross. 2004. "Valuable products from biotechnology of microalgae." *Applied Microbiology and Biotechnology* 65 (6):635–648.

Rizwan, M., J. H. Lee, and R. Gani. 2013. "Optimal processing pathway for the production of biodiesel from microalgal biomass: A superstructure based approach." *Computers & Chemical Engineering* 58:305–314.

Rizwan, M., J. H. Lee, and R. Gani. 2015. "Optimal design of microalgae-based biorefinery: Economics, opportunities and challenges." *Applied Energy* 150:69–79.

Ruiz, J., G. Olivieri, J. de Vree, et al. 2016. "Towards industrial products from microalgae." *Energy and Environmental Science* 9:3036–3043.

Safi, C., S. Camy, C. Frances, et al. 2014. "Extraction of lipids and pigments of *Chlorella vulgaris* by supercritical carbon dioxide: Influence of bead milling on extraction performance." *Journal of Applied Phycology* 26 (4):1711–1718.

Sari, Y. W., M. E. Bruins, and J. P. M. Sanders. 2013. "Enzyme assisted protein extraction from rapeseed, soybean, and microalgae meals." *Industrial Crops and Products* 43:78–83.

Sen Gupta, S., S. Bhartiya, and Y. Shastri. 2014. "The practical implementation of microalgal biodiesel: Challenges and potential solutions." *CAB Reviews* 9,020:1–12.

Sen Gupta, S., Y. Shastri, and S. Bhartiya. 2016. "Model-based optimisation of biodiesel production from microalgae." *Computers & Chemical Engineering* 89:222–249.

Sen Gupta, S., Y. Shastri, and S. Bhartiya. 2017a. "Optimization of integrated microalgal biorefinery producing fuel and value-added products." *Biofuels, Bioproducts & Biorefining* 11 (6):1030–1050.

Sen Gupta, S., Y. Shastri, and S. Bhartiya. 2017b. "Integrated microalgae biorefinery: Impact of product demand profile and prospect of carbon capture." *Biofuels, Bioproducts & Biorefining* 11 (6):1065–1076.

Subhadra, B. G. 2010. "Sustainability of algal biofuel production using integrated renewable energy park (IREP) and algal biorefinery approach." *Energy Policy* 38 (10):5892–5901.

Subhadra, B., and M. Edwards. 2010. "An integrated renewable energy park approach for algae biofuel production in United States." *Energy Policy* 38 (9):4897–4902.

Tapie, P., and A. Bernard. 1988. "Microalgae production: Technical and economic evaluations." *Biotechnology and Bioengineering* 32:873–885.

Taylor, G. 2008. "Biofuels and the biorefinery concept." *Energy Policy* 36 (12):4406–4409.

Torres, C. M., S. D. Ríos, C. Torras, J. Salvadó, J. M. Mateo-Sanz, and L. Jiménez. 2013. "Microalgae- based biodiesel: A multicriteria analysis of the production process using realistic scenarios." *Bioresource Technology* 147:7–16.

Ward, A. J., D. M. Lewis, and F. B. Green. 2014. "Anaerobic digestion of algae biomass: A review." *Algal Research* 5:204–214.

Weissman, J. C., and R. P. Goebel. 1987. Design and analysis of microalgal open pond systems for the purpose of producing fuels. A Subcontract Report. Dept. of Energy (US), Solar Energy Research Institute. www.nrel.gov/docs/legosti/old/2840.pdf

Wijffels, R. H., and M. J. Barbosa. 2010. "An outlook on microalgal biofuels." *Science* 329 (5993):796–799.

Wijffels, R. H., M. J. Barbosa, and M. H. M. Eppink. 2010. "Microalgae for the production of bulk chemicals and biofuels." *Biofuels, Bioproducts and Biorefining* 4:287–295.

Wiley, P. E., J. E. Campbell, and B. McKuin. 2011. "Production of biodiesel and biogas from algae: A review of process train options." *Water Environment Research* 83 (4):326–338.

Yang, J., M. Xu, X. Zhang, Q. Hu, M. Sommerfeld, and Y. Chen. 2011. "Life-cycle analysis on biodiesel production from microalgae: Water footprint and nutrients balance." *Bioresource Technology* 102 (1):159–165.

Index

351

Printed and bound by CPI Group (UK) Ltd, Croydon, CR0 4YY

17/10/2024

01775681-0012